Grundlehren der mathematischen Wissenschaften

A Series of Comprehensive Studies in Mathematics

Volume 354

Grundlehren der mathematischen Wissenschaften (subtitled Comprehensive Studies in Mathematics), Springer's first series in higher mathematics, was founded by Richard Courant in 1920. It was conceived as a series of modern textbooks. A number of significant changes appear after World War II. Outwardly, the change was in language: whereas most of the first 100 volumes were published in German, the following volumes are almost all in English. A more important change concerns the contents of the books. The original objective of the Grundlehren had been to lead readers to the principal results and to recent research questions in a single relatively elementary and accessible book. Good examples are van der Waerden's 2-volume Introduction to Algebra or the two famous volumes of Courant and Hilbert on Methods of Mathematical Physics.

Today, it is seldom possible to start at the basics and, in one volume or even two, reach the frontiers of current research. Thus many later volumes are both more specialized and more advanced. Nevertheless, most books in the series are meant to be textbooks of a kind, with occasional reference works or pure research monographs. Each book should lead up to current research, without over-emphasizing the author's own interests. There should be proofs of the major statements enunciated, however, the presentation should remain expository. Examples of books that fit this description are Maclane's Homology, Siegel & Moser on Celestial Mechanics, Gilbarg & Trudinger on Elliptic PDE of Second Order, Dafermos's Hyperbolic Conservation Laws in Continuum Physics ... Longevity is an important criterion: a GL volume should continue to have an impact over many years. Topics should be of current mathematical relevance, and not too narrow.

The tastes of the editors play a pivotal role in the selection of topics.

Authors are encouraged to follow their individual style, but keep the interests of the reader in mind when presenting their subject. The inclusion of exercises and historical background is encouraged.

The GL series does not strive for systematic coverage of all of mathematics. There are both overlaps between books and gaps. However, a systematic effort is made to cover important areas of current interest in a GL volume when they become ripe for GL-type treatment.

As far as the development of mathematics permits, the direction of GL remains true to the original spirit of Courant. Many of the oldest volumes are popular to this day and some have not been superseded. One should perhaps never advertise a contemporary book as a classic but many recent volumes and many forthcoming volumes will surely earn this attribute through their use by generations of mathematicians.

More information about this series at http://www.springer.com/series/138

Barry Simon

Loewner's Theorem on Monotone Matrix Functions

Barry Simon
Division of Physics, Math, and Astronomy
Caltech, Pasadena, CA, USA

ISSN 0072-7830 ISSN 2196-9701 (electronic)
Grundlehren der mathematischen Wissenschaften
ISBN 978-3-030-22424-0 ISBN 978-3-030-22422-6 (eBook)
https://doi.org/10.1007/978-3-030-22422-6

Mathematics Subject Classification: 26A48, 26A51, 47A56, 47A63

This Springer imprint is published by the registered company Springer Nature Switzerland AG.
The registered company address is: Gewerbestrasse 11, 6330 Cham, Switzerland

Preface

This book is a love poem to Loewner's theorem. There are other mathematical love poems, although not many. One sign, not always present and not foolproof, is that like this book, the author has included pictures of some of the main figures in the development of the subject under discussion. The telltale sign is that the reader's initial reaction is "how can there a whole book on that subject" (although not all narrow books are love poems).

Loewner's theorem concerns the theory of monotone matrix functions, i.e. functions, f so that $A \leq B \Rightarrow f(A) \leq f(B)$ for pairs of selfadjoint matrices. That this is a subtle notion is seen by the fact (see Corollary 14.3) that if $f : \mathbb{R} \to \mathbb{R}$ is monotone on all pairs of 2×2 matrices, then f is affine! So Loewner realized one needed to fix a proper interval $(a, b) \subset \mathbb{R}$, demand that $f : (a, b) \to \mathbb{R}$, and only demand the monotonicity result for pairs A and B all of whose eigenvalues are in (a, b). In 1934, Charles Loewner proved the remarkable result that f on (a, b) is matrix monotone on all such $n \times n$ pairs (for all n) if and only if f is real analytic on (a, b) and has an analytic continuation to the upper half plane with a positive imaginary part there. That functions with this property are matrix monotone follows in a few lines from the Herglotz representation theorem, so I call that half the easy half. The other direction is the hard half. Many applications of Loewner's notion of matrix monotone functions involve explicit examples and so only the easy half.

Matrix monotonicity is an algebraic statement, but Loewner's theorem says it is equivalent to an analytic fact. One fascination of the subject is the mix of the algebraic and the analytic in its study.

Interestingly enough, this is not the first love poem to Loewner's theorem. Forty-five years ago, in 1974, Springer published W. F. Donoghue's *Monotone Matrix Functions and Analytic Continuation* [81] in the same series where I am publishing this book. Donoghue explained that, at the time of writing, there were three existing proofs of (the hard half of) Loewner's theorem, all very different: Loewner's original proof, the proof of Bendat–Sherman, and the proof of Korányi. In fact, as we'll discuss several times in this book, there was a fourth proof which workers in the field didn't seem to realize was there—it was due to Wigner–von Neumann and appeared in the *Annals of Mathematics*! The main goal of Donoghue's book was to expose those three proofs with their underlying mathematical background. He also had several new results including converses of two theorems of Loewner's student Otto Dobsch. Donoghue also discussed applications to the theory of schlicht and Herglotz functions.

Similarly, the main goal of the current book is to expose many of the proofs of the hard half that now exist. Indeed, we will give 11 proofs in all. As I'll explain later, there is a 12th proof which I don't expose. Of course, it is not always clear when two proofs are really different and when one proof should be regarded as a variant of another. I am prepared to defend the notion that these proofs are different (although there are relations among them) and that a proof like Hansen's in [133] should be viewed as a variant (albeit an interesting variant) of Hansen's earlier proof with Pedersen. But I'd agree some might differ.

In any event, it is striking that there are no really short proofs and that the underlying structure of the proofs is so different. I like to joke that it is almost as if every important result in analysis is the basis of some proof of Loewner's theorem. To mention a few of the underlying machines that lead to proofs: the moment problem, Pick's theorem, commutant lifting theorems, the Krein–Milman theorem, and Mellin transforms.

Shortly after Loewner's 1934 paper, his student Fritz Kraus defined and began the study of matrix convex functions. Donoghue includes Kraus' 1936 paper in his bibliography but never refers to it in the text and doesn't discuss the subject of matrix convex or concave functions. In the years since his book, it has become clear that the two subjects are intertwined in deep ways. So my book also has a lot of material on matrix convex functions.

We let $\mathcal{M}_n(a, b)$ be the functions monotone on $n \times n$ matrices. We'll see that $\mathcal{M}_{n+1}$ is a strictly proper subset of $\mathcal{M}_n$. Loewner's theorem gives an effective description of $\mathcal{M}_\infty \equiv \cap \mathcal{M}_n$, and we'll see in Chapter 14 that there is an effective description of $\mathcal{M}_2$ but there is no especially direct description of $\mathcal{M}_n$. One other subject we explore involves properties of $\mathcal{M}_n$ and its convex analog for fixed finite n.

I should say something about the history of this book. I first learned of Loewner's theorem in graduate school 50 years ago, probably from Ed Nelson, one of my mentors. I learned the proof of the easy half and a variant of it appears in Reed–Simon. The easy half seemed more useful since it told one certain function was matrix monotone and that was what could be applied. About 2000, I became curious how one proved the hard half and looked at the expositions in Donoghue and Bhatia's books. I was intrigued that both quoted a paper of Wigner–von Neumann claiming it had applications to quantum physics. I looked up the paper and was surprised to discover that there was no application to quantum physics but there was a complete and different proof of Loewner's theorem, which Loewner, Donoghue, and Bhatia didn't seem to realize was there! For several years, I gave a mathematics colloquium called "The Lost Proof of Loewner's Theorem." I found my new proof that appears in Chapter 20 which is, in many ways, the most direct proof. After one of the colloquia, around 2005, I learned of Boutet de Monvel's unpublished proof and started on this book. I finished about half and then put it aside in 2007 to focus on my five-volume Comprehensive Course [325–329]. After completing that, I returned to this book and completed it. The early part of the book was typed in TEX by my wonderful secretary, Cherie Galvez, who we lost to cancer before I returned to the project. I often miss her.

There is some literature on multivariable extensions of Loewner's theorem. The earliest such papers are a series (Korányi [187], Singh–Vasudeva [315], and Hansen [131]) that considers operators on distinct spaces, A_j on $\mathcal{H}_j$, and defines $f(A_1, \ldots, A_n)$ on the tensor product $\mathcal{H}_1 \otimes \cdots \otimes \mathcal{H}_n$. Recently, there have been several papers (Agler et al. [3], Najafi [230], Pascoe and Tully-Doyle [259], and Pálfia [256]) that involve functions of several variables, even non-commuting variables, on a single Hilbert space. These papers are complicated, and the subject is in flux, so it seemed wisest not to discuss the subject in detail here. However, we should mention that Pálfia [256] says that his multivariable proof specialized to one variable provides a new proof of the classical result, the 12th proof referred to above. In correspondence, he told me that the restriction to one variable doesn't especially simplify his proof (his preprint is 40 pages). He didn't believe it could be shortened to less than about 25 pages and that doesn't count any mathematical background on the tools he uses, so I decided not to try to include it.

The 11 proofs appear in Part II of this book which also has background on tools like Pick's theorem and rational approximation. There is also a discussion of the analog of Loewner's theorem when (a, b) is replaced by a more general open set. Part I sets the background for the theorem and also for the theory of matrix convex functions. Part III discusses some applications, many due to Ando. Each part begins with a brief introduction that summarizes the content and context of that part.

Karel Löwner was born near Prague in 1893 and proved his great theorem while a professor at the Charles University in Prague. In 1938, he fled from there to the United States (he was arrested when the Nazis entered Prague probably because of his left-wing political activities rather than the fact that he was Jewish) where he made a decision to change his name to the less German Charles Loewner. I have decided to respect his choice by calling the result Loewner's theorem. The reader should be warned that much of the literature refers to Löwner's theorem. Of course, we use the original spelling in the bibliography when that is what appeared in the author's name or in the published title of an article or book.

We warn the reader that in our complex inner product, we use a universal convention in the physics community, but the minority one among mathematicians namely, our inner products, are linear in the second vector and antilinear in the first. Other symbols are listed in Appendix C just before the bibliography. We also warn the reader that when speaking of matrices, M, we will call them "positive" if and only if $\langle \varphi, M\varphi \rangle \geq 0$ for all φ and use "strictly positive" when that inner product is strictly positive for all non-zero φ.

Two individuals should be especially thanked for their aid in improving this book. Mitsuru Uchiyama sent me extensive corrections to an early draft of this book. Detailed corrections of the kind he sent are like gold to an author. Otte Heinävaara was kind enough to share his many insights into the subject of this book. It is a pleasure to also thank Anne Boutet de Monvel, Larry Brown, Fritz Gesztesy, Frank

Hansen, Ira Herbst, Bill Helton, Elliott Lieb, John McCarthy, James Pascoe, James Rovnyak, and Beth Ruskai for their useful discussions and/or correspondence in connection with this book.

And, as always, thanks to my wife, Martha, for her love and support.

Los Angeles, CA, USA Barry Simon
March 2019

Contents

Part I
Tools

A real-valued continuous function, f, on (a, b) is said to be monotone on $n \times n$ matrices, if and only if

$$A \leq B \Rightarrow f(A) \leq f(B) \tag{p1.1}$$

for all $n \times n$ Hermitian matrices, A and B, with eigenvalues in (a, b). The set of all such functions will be denoted $\mathcal{M}_n(a, b)$ and we define

$$\mathcal{M}_\infty(a, b) = \cap_{n=1}^{\infty} \mathcal{M}_n(a, b) \tag{p1.2}$$

These definitions require one know what the order is on matrices and what $f(A)$ means—we provide precise definitions at the start of Chapter 1.

Loewner's theorem says that a real-valued continuous function, f, on (a, b) lies in $\mathcal{M}_\infty(a, b)$ if and only if f is real analytic on (a, b) and has an analytic continuation to the upper half plane, $\mathbb{C}_+$, with $\text{Im}(z) > 0 \Rightarrow \text{Im} f(z) > 0$. Such analytic functions have an integral representation which we present in Chapter 3 and that representation easily implies that such functions lie in $\mathcal{M}_\infty(a, b)$, a result we call the easy half of Loewner's theorem. In particular, since $f(x) = \sqrt{x}$ defined on $(0, \infty)$ has such an analytic continuation, it is matrix monotone. Despite this immediate consequence of the easy half, various authors have found direct proofs of the monotonicity of $\sqrt{x}$ which we present in Chapter 4. Proofs of the hard half are the subject of Part II.

In Chapter 2, we discuss two separate issues. First we show that one can easily go from Loewner's theorem for one proper (a, b) to any other. All finite intervals are related to each other by affine maps and all semi-infinite intervals by translation or $f(x) \mapsto -f(-x)$. Finally $x \mapsto -x^{-1}$ interchanges $(0, \infty)$ and $(-1, 0)$ and one can prove directly that $f(x) = -x^{-1}$ is a matrix monotone map. All proofs of the hard part use one of $(0, 1)$, $(-1, 1)$, or $(0, \infty)$ for (a, b).

The other issue in Chapter 2 concerns operators on an infinite dimensional Hilbert space. It is a simple fact that matrix monotonicity for finite matrices of arbitrary dimension allows one to prove monotonicity for self-adjoint operators on a Hilbert space and vice versa. While we mainly discuss finite matrix versions here, it will be occasionally useful to deal with infinite dimensional versions instead. This will sometimes (and, in particular, in Chapter 2) require us to use results for operators on Hilbert space. We will give textbook references for these results, most often Simon [329].

Most proofs of the hard direction use a preliminary result of Loewner that completely describes $\mathcal{M}_n(a, b)$ in terms of certain $n \times n$ matrices built out of f and n points $a < x_1 < \ldots < x_n < b$. This equivalence is the subject of Chapter 5 where two proofs are given of the equivalence: one using a formula of Daleckiĭ and Krein and one using rank one perturbation theory. In Chapter 14, this equivalence is used to give an explicit description of $\mathcal{M}_2(a, b)$ and then use that to show that $\mathcal{M}_2(-\infty, \infty)$ is precisely the affine functions (with non-negative linear coefficient).

A theorem of Dobsch–Donoghue gives a local criteria for a function to lie in $\mathcal{M}$. Chapters 6 and 8 provide two recent elegant proofs of this result by the young Finnish mathematician, Otte Heinävaara. Given the local theorem, it is easy to construct functions in $\mathcal{M}_n(a, b)$ that are not in $\mathcal{M}_{n+1}(a, b)$ and we do this in Chapter 7.

For scalar functions on $\mathbb{R}$, there is a simple connection between convex and monotone functions. g is convex on $(a, b) \ni 0$ if and only if there is a monotone functions, f, on (a, b) with

$$g(x) = g(0) + \int_0^x f(y)dy \qquad \text{(p1.3)}$$

equivalently, the distribution derivative of g is a monotone function. There are similar relations between matrix convex and matrix monotone functions which we discuss in Chapter 9. Chapter 11 discusses analogs of Jensen's inequality for matrices which is much richer because Jensen's convex combinations $\theta x + (1 - \theta)y$ can be replaced by $a^*xa + b^*yb$ where a and b are matrices with $a^*a + b^*b = 1$. Chapter 10 deals with a striking new connection: a scalar positive function on $(0, \infty)$ which is concave is monotone as is easy to see but, of course, the converse is false. However, for functions on matrices, a positive function on $(0, \infty)$ is matrix concave for all n if and only if it is matrix monotone for all n! Chapter 12 provides a criteria for matrix convexity in terms of Loewner matrices. Chapter 13 discusses a notion stronger than matrix convexity and includes the result that if $f \in \mathcal{M}_\infty$, so its second finite difference $[x, x_1, x_2; f]$.

The equivalence of matrix monotonicity on $(0, \infty)$ and matrix concavity is one of the three remarkable equivalences of matrix monotonicity with some other property so that Loewner's theorem allows the complete description of some other object. Another such case involves the notion of quadratic interpolation, the theory of Marcinkiewicz real interpolation for inner products norms. The classification of such norms is given by a theorem of Foiaş–Lions which can be related to matrix monotone functions and we discuss this in Chapter 15. The third equivalence concerns the theory of operator means where the equivalence is called the Kubo–Ando theorem. We postpone the discussion of this subject to Chapters 36–37.

Chapter 1
Introduction: The Statement of Loewner's Theorem

Most of this monograph will consider finite matrices, but since we will be interested in positivity defined in terms of a (Euclidean) inner product, we make definitions in a general context. Thus $\mathcal{H}$ will be a complex separable Hilbert space (perhaps finite dimensional) with inner product $\langle \cdot, \cdot \rangle$ which is linear in the second factor and antilinear in the first. Many mathematicians use the opposite convention on which factor is linear but I use the almost universal convention among physicists. For more on the basics of inner products, see [325, Section 3.1].

Definition A bounded operator on $\mathcal{H}$ is called *positive*, written $A \geq 0$, if

$$\langle \varphi, A\varphi \rangle \geq 0 \tag{1.1}$$

for all $\varphi \in \mathcal{H}$.

We will use *strictly positive* if (1.1) has >0 for all $\varphi \neq 0$. Some call this positive definite and (1.1) positive semi-definite. We will drop "definite."

Since the space, $\mathcal{H}$, is complex,

$$\langle \varphi, A\psi \rangle = \frac{1}{4} \sum_{\sigma=\pm 1, \pm i} \bar{\sigma} \langle (\varphi + \sigma\psi), A(\varphi + \sigma\psi) \rangle$$

so any A obeying (1.1) obeys

$$\langle \varphi, A\psi \rangle = \overline{\langle \psi, A\varphi \rangle} \tag{1.2}$$

that is, A is self-adjoint.

We write

$$A \leq B \tag{1.3}$$

B. Simon, *Loewner's Theorem on Monotone Matrix Functions*, Grundlehren der mathematischen Wissenschaften 354, https://doi.org/10.1007/978-3-030-22422-6_1

as shorthand for $B - A \geq 0$. Loewner's theorem involves the following simple question: For which real-valued functions f of a real variable is it true that

$$A \leq B \Rightarrow f(A) \leq f(B) \tag{1.4}$$

Of course, to make sense out of (1.4), one needs to define $f(A)$ given a function, f, of a real variable. It is natural to restrict to self-adjoint A and B, so we need to define $f(A)$ for A self-adjoint. There are two equivalent ways to make this definition for finite matrices:

(i) For a polynomial $P(x) = \sum_{j=0}^{n} \alpha_j x^j$, define $P(A) = \sum_{j=0}^{n} \alpha_j A^j$. If P and Q are two polynomials, it is easy to see that $P(A) = Q(A)$ if and only if $P(\lambda_j) = Q(\lambda_j)$ for each eigenvalue, λ_j, of A. Thus, given f, we find any polynomial, P, so that $f(\lambda_j) = P(\lambda_j)$ and let $f(A) \equiv P(A)$. This is equivalent to

$$f(A) = \sum_{j=1}^{\ell} f(\lambda_j) \Big[\prod_{k \neq j} (A - \lambda_k)(\lambda_j - \lambda_k)^{-1} \Big] \tag{1.5}$$

if $\lambda_1, \ldots, \lambda_\ell$ are the *distinct* eigenvalues of A.

(ii) If $\lambda_1 \leq \lambda_2 \leq \cdots \leq \lambda_n$ are the eigenvalues of A counting multiplicity (subtle shift from the meaning in (i)), there is a U obeying

$$A = U \begin{pmatrix} \lambda_1 & & \\ & \ddots & \\ & & \lambda_n \end{pmatrix} U^{-1} \tag{1.6}$$

One defines

$$f(A) = U \begin{pmatrix} f(\lambda_1) & & \\ & \ddots & \\ & & f(\lambda_n) \end{pmatrix} U^{-1} \tag{1.7}$$

U is not unique nontrivially if some eigenvalue is degenerate, but (1.7) is independent of U—for example, by showing all choices lead to (1.5).

For A, a bounded self-adjoint operator on an infinite dimensional space, $f(A)$ can again be defined analogous to either of these methods: polynomial approximation corresponding to (i) or the spectral theorem analogous to (ii). Indeed, one way of proving the spectral theorem is to first control polynomial approximation and use that to define spectral measures. In any event, since the infinite dimensional case is peripheral to our concerns, we leave the construction of $f(A)$ in that case to other references, for example, Reed–Simon [279, Chapter VII] or Simon [329, Chapter 5].

Now that we have settled what $f(A)$ means, we can return to (1.4) and note that if it is intended in too strong a sense, then very few f's work. In Chapter 14 and also in Chapter 28, we will prove

Theorem 1.1 (= Corollary 14.3 = Corollary 10.7) *Let $f\colon \mathbb{R} \to \mathbb{R}$ be such that for any pair of 2×2 self-adjoint matrices, (1.4) holds. Then for some $b \in \mathbb{R}$ and $a \geq 0$,*

$$f(x) = ax + b \tag{1.8}$$

This result says that if one looks at (1.4) for too general a family of A, B, the set of f's is not very rich. Loewner had the good idea to look at f's defined on an interval $(a, b) \subset \mathbb{R}$. Thus

Definition A function $f\colon (a, b) \to \mathbb{R}$ is said to lie on $\mathcal{M}_n(a, b)$ if and only if (1.4) holds for any pair of self-adjoint $n \times n$ matrices, both of whose eigenvalues lie in (a, b). We will sometimes use the phrase "monotone matrix function."

If A, B are $(n-1) \times (n-1)$ matrices with eigenvalues in (a, b), and if

$$\widetilde{A} = \begin{pmatrix} A & 0 \\ 0 & \frac{a+b}{2} \end{pmatrix} \qquad \widetilde{B} = \begin{pmatrix} B & 0 \\ 0 & \frac{a+b}{2} \end{pmatrix}$$

then $A \leq B \Leftrightarrow \widetilde{A} \leq \widetilde{B}$ and $f(A) \leq f(B) \Leftrightarrow f(\widetilde{A}) \leq f(\widetilde{B})$. This implies that

$$\mathcal{M}_{n-1}(a, b) \supset \mathcal{M}_n(a, b) \tag{1.9}$$

It is natural to define

$$\mathcal{M}_\infty(a, b) = \bigcap_n \mathcal{M}_n(a, b) \tag{1.10}$$

Thus, $f \in \mathcal{M}_\infty(a, b)$ if and only if (1.4) holds for any pair of matrices with eigenvalues in (a, b). In Chapter 2, we will prove that if $f \in \mathcal{M}_\infty(a, b)$, then (1.4) even holds for pairs of self-adjoint operators on an infinite dimensional Hilbert space whose spectrum lies in (a, b).

Example 1.2 Consider the function $f(x) = x^2$ on $[0, \infty)$. Let P and Q be orthogonal projections. Of course, $P + Q \geq P$. One can ask when $(P+Q)^2 \geq P^2$, that is, when

$$C \equiv Q^2 + QP + PQ \geq 0 \tag{1.11}$$

Let ψ, φ obey $Q\psi = 0$, $Q\varphi = \varphi$. Let $\psi_\varepsilon = \psi + \varepsilon\varphi$, so $\langle \psi_{\varepsilon=0}, C\psi_{\varepsilon=0} \rangle = 0$ and

$$\langle \psi_\varepsilon, C\psi_\varepsilon \rangle = 2\,\mathrm{Re}(\varepsilon \langle \psi, PQ\varphi \rangle) + O(\varepsilon^2) \tag{1.12}$$

Since we can pick ε so $\varepsilon\langle\psi, PQ\varphi\rangle = -\varepsilon|\langle\psi, PQ\varphi\rangle|$, (1.11) implies $\langle\psi, PQ\varphi\rangle = 0$, that is,

$$(1 - Q)PQ = 0$$

Thus, $PQ = QPQ$, so taking adjoints, $QP = PQ$. Thus $(P + Q)^2 \geq P^2$ if and only if $[P, Q] \equiv PQ - QP = 0$ (the "if" is easy). Thus $f(x) = x^2$ is not matrix monotone on $[0, \infty]$ even for 2×2 matrices. We will see later (Theorem 4.1) that $f(x) = x^\alpha$ is in $\mathcal{M}_\infty(0, \infty)$ if and only if $0 \leq \alpha \leq 1$.

We will see below (Proposition 2.2) that $A \leq B \Leftrightarrow -A^{-1} \leq -B^{-1}$, so that this and Theorem 4.1 imply that $f(x) = -x^\alpha$ is in $\mathcal{M}_\infty$ if and only if $-1 \leq \alpha \leq 0$.

This example shows that it is false that $A \leq B$ and $C \leq D$ imply that $\frac{1}{2}(AC + CA) \leq \frac{1}{2}(BD + DB)$ even if A commutes with C and B commutes with D. □

Loewner's theorem exactly describes $\mathcal{M}_\infty(a, b)$. For simplicity of exposition, we will state it initially for $(a, b) = (-1, 1)$:

Theorem 1.3 *Let f be a function from $(-1, 1)$ to $\mathbb{R}$. The following are equivalent:*

(a) $f \in \mathcal{M}_\infty(-1, 1)$
(b) *For some positive measure $d\nu$ on $[-1, 1]$,*

$$f(x) = f(0) + \int_{-1}^{1} \frac{x}{1 + \lambda x}\, d\nu(\lambda) \tag{1.13}$$

(c) *f is the restriction to $(-1, 1)$ of a function, f, analytic in $(\mathbb{C}\backslash\mathbb{R}) \cup (-1, 1)$ that obeys*

$$\operatorname{Im} z > 0 \Rightarrow \operatorname{Im} f(z) > 0 \tag{1.14}$$

The big surprise, of course, is that $f \in \mathcal{M}_\infty$ implies analyticity. In some proofs, this just falls out while in others, its cause is clearer. This book is partially devoted to discussing several proofs of this theorem. This theorem has an "easy half" and a "hard half." We note first that (b) $\Rightarrow$ (c) is obvious since (1.13) can be used for $z \in (\mathbb{C}\backslash\mathbb{R}) \cup (-1, 1)$ and

$$\operatorname{Im}\left(\frac{z}{1 + \lambda z}\right) = \operatorname{Im}\left(\frac{z(1 + \lambda \bar{z})}{|1 + \lambda z|^2}\right) = \frac{\operatorname{Im} z}{|1 + \lambda z|^2} \tag{1.15}$$

showing (1.14). That (c) $\Rightarrow$ (b) is not due to Loewner; it is a variant of the Herglotz representation theorem. Note that (b) $\Rightarrow$ (a) is equivalent to

$$-1 < A \leq B < 1 \Rightarrow \frac{A}{1 + \lambda A} \leq \frac{B}{1 + \lambda B} \tag{1.16}$$

which will be an easy calculation. We will make all these aspects explicit in Chapter 3.

We want to note that the measure $d\nu$ in (1.13) is unique. Recall that a set in a normed vector space is called *total* if the linear combinations of finite subsets are dense.

Lemma 1.4 *Let $S \subset (0, 1)$ be a subset with a countable subset $\{x_n\}$ with $x_n \downarrow 0$. Then $\{f_x(\lambda) = \frac{x}{1+\lambda x}\}_{x \in S}$ is total in $C([-1, 1])$.*

Proof Let Q be the closure of the linear span of $\{f_x\}_{x \in S}$ and, for $m = 0, 1, \ldots,$ let $g_m(\lambda) = \lambda^m$. By Weierstrass' theorem [325, Theorem 2.4.2] on the density of polynomials in $C([-1, 1])$, it suffices to prove that each $g_m \in Q$. Note that, for each $x \in S$, uniformly in x bounded away from 1, $f_x = \sum_{m=0}^{\infty} (-1)^m x^{m+1} g_m$. Pick $x_n \downarrow 0$ with $x_n \in S$. By the last formula, $x_n^{-1} f_{x_n} \to g_0$ so $g_0 \in Q$. Similarly, $x_n^{-2}[f_{x_n} - x_n g_0] \to g_1$ so $g_1 \in Q$. By an obvious induction, each $g_m \in Q$. □

Theorem 1.5 *The measure $d\nu$ in (1.13) is unique, that is, there is at most one measure obeying (1.13).*

Proof By (1.13) $\int f(\lambda)\, d\nu(\lambda)$ is determined for $f_x(\lambda) = \frac{x}{1+\lambda x}$. Thus, ν is unique by Lemma 1.4. □

Therefore, the bulk of the book will be to focus on the hard half. One surprise is that while the problem (1.4) is purely algebraic, the proofs are largely analytic, connected of course to the analytic form of (1.13).

For general (a, b), the result of Loewner is

Theorem 1.6 *Let $a < b$ ($a = -\infty$ and/or $b = \infty$ are allowed). Let f be a function from (a, b) to $\mathbb{R}$. The following are equivalent:*

(a) $f \in \mathcal{M}_\infty(a, b)$
(b) *For some finite measure $d\mu$ on $\mathbb{R}\backslash(a, b) \equiv J$ and some constants C, real, and $A \geq 0$, we have*

$$f(x) = C + Ax + \int_J \frac{1 + xy}{y - x}\, d\mu(y) \tag{1.17}$$

(c) *f is the restriction to (a, b) of a function, f, analytic in $(\mathbb{C}\backslash\mathbb{R}) \cup (a, b)$ which obeys (1.14).*

Remarks

1. Equations (1.17) and (1.13) seem distinct if $(a, b) = (-1, 1)$, but we will see in Chapter 3 that they are equivalent.
2. If $a = -\infty$, $b = \infty$, we must interpret J as empty, and so (1.17) becomes $C + Ax$ for $A \geq 0$, and we recover a weaker form of Theorem 1.1.
3. The measure $d\mu$ and constants A, C in (1.17) are unique by a variant of the argument that proved uniqueness in (1.13).

In Chapter 2, we will show (a) $\Leftrightarrow$ (c) for $(-1, 1)$ implies it for general (a, b). In Chapter 3, we will prove (c) $\Leftrightarrow$ (b) in the general (a, b) case.

While we will focus on Theorem 1.3, we will consider convex and concave matrix functions in Chapters 6, 8, 9, 10, 11, 12, and 13 and extensions of Theorem 1.6 when (a, b) is replaced by more general sets in Chapter 35. There are actually two distinct extensions, as we will see. For the simplest case, here are two extensions:

Theorem 1.7 *Let $a < b < c < d$ and let f be a real-valued function on $\Omega = (a, b) \cup (c, d)$. Suppose that for any n,* (1.4) *holds for any pair of $n \times n$ self-adjoint matrices with eigenvalues in Ω. Then, $f \in \mathcal{M}_\infty(a, d)$.*

Theorem 1.8 *Let $a < b < c < d$ and let f be a real-valued function on $\Omega = (a, b) \cup (c, d)$. Then the following are equivalent:*

(a) *For any j, n with $j \leq n$,* (1.4) *holds for any pair of $n \times n$ self-adjoint matrices, each of which has j eigenvalues in (a, b) and $(n - j)$ eigenvalues in (c, d).*
(b) *f is the restriction to Ω of a function, f, analytic in $(\mathbb{C}\backslash\mathbb{R}) \cup \Omega$ which obeys* (1.14).

We will prove these theorems in Chapter 35.

Notes and Historical Remarks Theorem 1.6 is due to Loewner [209] in 1934. Karel Löwner was born in 1893 in a small town near Prague which is now the Czech Republic. After getting his Ph.D. under Pick at the German University of Prague (during this period, Karel became Germanized to Karl), he did a postdoc in Berlin, where his contemporaries included von Neumann and Szegő. During this period, he did his celebrated work on the Bieberbach conjecture for $n = 3$, work in which he introduced certain differential equations on dynamical conformal maps. Many years later, these equations were critical both in the de Branges' solution of the full Bieberbach conjecture and Schramm's work on stochastic conformal maps.

In 1929, Löwner left Berlin to accept a professorship in Prague. In the mid-1930s, he saw the gathering storm and started to learn English. But he was not fast enough and was arrested (probably because of his leftist political connections rather than his Jewish parentage) when the Nazis marched into Prague. His wife was tireless in winning his release, and after paying considerable money, he was able to flee to the USA. In revulsion of all things German, he insisted on changing his name to Charles Loewner. We have acknowledged his choice by referring to "Loewner's theorem" rather than "Löwner's theorem" but we warn the reader that the bibliography lists the author's name on his papers in the way it appeared in the journal when it was written.

von Neumann helped him in his initial job searches, and in the early 1950s, Szegő, by then chair at Stanford, brought him to that university, where he ended his career.

Loewner was extremely careful and set himself such a high standard that he only published five papers prior to his leaving Prague. Two, of course, were the one that is the basis of this book, and another the basis of the Loewner differential equations.

These biographic notes are based on the introduction that Bers, a student of Loewner, wrote to Loewner's complete works [213], which can be consulted for further details. In particular, Bers has the following remarkable description of Loewner: *a man whom everyone liked, perhaps because he was a man at peace with himself. He conducted a lifelong passionate love affair with mathematics, but was neither competitive, nor jealous, nor vain. His kindness and generosity in scientific matters, to students and colleagues alike, were proverbial. He seemed to be incapable of malice. His manners were mild and even diffident, but those hid a will of steel. But first and foremost, he was a mathematician.*

Although Theorem 1.1 follows in a few lines from Theorem 14.1 (due to Loewner) in Chapter 14, it seems to have only been explicitly noted in 2009 by Hansen–Tomiyama [139, Theorem 3.4].

Theorem 1.7 is due to Chandler [58]; see also Donoghue [82]. Chandler used a result of Šmul'jan [331]. Theorem 1.8 is due to Rosenblum–Rovnyak [287, 288]; see also Donoghue [83].

Chapter 2
Some Generalities

In this chapter, we will discuss several issues that will help us prove the easy half of Loewner's theorem in the next chapter and then, we will extend Theorem 1.6 to infinite dimensions:

(i) Reduction of (a) $\Leftrightarrow$ (c) for general (a, b) to $(-1, 1)$
(ii) Some conformal maps
(iii) Going from $\mathcal{M}_\infty$ to Hilbert space operators.
(iv) Canonical decomposition of Hilbert space operators.

Theorem 2.1 *Suppose* (a) $\Leftrightarrow$ (c) *is proven in Theorem 1.3. Then it holds in Theorem 1.6.*

Proof Suppose first that $-\infty < a < b < \infty$. Let

$$T(x) = \frac{b-a}{2}\,x + \frac{a+b}{2} \tag{2.1}$$

so T is a monotone map of $(-1, 1)$ to (a, b). As a linear map with positive linear coefficient, it is obvious that for any two self-adjoint matrices A and B, we have

$$A \le B \Leftrightarrow T(A) \le T(B) \tag{2.2}$$

Thus, $f\colon (a, b) \to \mathbb{R}$ has

$$f \in \mathcal{M}_\infty(a, b) \Leftrightarrow f \circ T \in \mathcal{M}_\infty(-1, 1) \tag{2.3}$$

On the other hand, T extends to a map of $\mathbb{C} \to \mathbb{C}$ with $T\colon \mathbb{C}_+ \to \mathbb{C}_+$ where

$$\mathbb{C}_+ = \{z \mid \operatorname{Im} z > 0\} \tag{2.4}$$

B. Simon, *Loewner's Theorem on Monotone Matrix Functions*, Grundlehren der mathematischen Wissenschaften 354, https://doi.org/10.1007/978-3-030-22422-6_2

It is obvious that f is the restriction to (a, b) of a function analytic on $(\mathbb{C}\backslash\mathbb{R})\cup(a, b)$ obeying (1.14) if and only if $f \circ T$ is the restriction to $(-1, 1)$ of a function analytic on $(\mathbb{C}\backslash\mathbb{R}) \cup (-1, 1)$ obeying (1.14). For $T: \mathbb{C}_+ \to \mathbb{C}_+$, and $T^{-1}: \mathbb{C}_+ \to \mathbb{C}_+$ says (1.14) is preserved if T is used to map to functions on $\mathbb{C}\backslash\mathbb{R}$. This shows that (a) $\Leftrightarrow$ (c) for $(-1, 1)$ implies it for any bounded (a, b).

Consider now $(0, \infty)$. Let $I_m = (0, m)$. If $f \in \mathcal{M}_\infty(0, \infty)$, then $f \in \mathcal{M}_\infty(I_m)$, so for each m, f is the restriction of a function analytic on $(\mathbb{C}\backslash\mathbb{R}) \cup (0, m)$. By uniqueness of analytic continuation, the extension of f is analytic in $\mathbb{C}\backslash(-\infty, 0]$.

Conversely, if f is the restriction of a function analytic on $\mathbb{C}\backslash(-\infty, 0]$, it is of a function analytic in $(\mathbb{C}\backslash\mathbb{R}) \cup I_m$ for each m, so in $\cup_m \mathcal{M}_\infty(I_m)$. If $0 < A \leq B$, for matrices there is an m with $0 < A \leq B < m$. Since $f \in \mathcal{M}_\infty(I_m)$, $f(A) \leq f(B)$. Thus, $f \in \mathcal{M}_\infty(0, \infty)$. Thus, (a) $\Leftrightarrow$ (c) for $(0, \infty)$. A similar argument works for any other unbounded interval. □

There is another, enlightening, way to go from bounded to proper unbounded intervals of $\mathbb{R}$. It is not hard to see that, by using affine maps, it suffices to go from $(-1, 0)$ to $(0, \infty)$.

Proposition 2.2 *Let A, B be two finite matrices with eigenvalues in* $(-1, 0)$. *Then* $-A^{-1}, -B^{-1}$ *have eigenvalues in* $(0, \infty)$ *and*

$$A \leq B \Leftrightarrow -A^{-1} \leq -B^{-1} \tag{2.5}$$

Remarks (Remarks 2 and 3 Are Courtesy O. Heinävaara)

1. This Proposition says that the function $f(x) = -x^{-1}$ lies in $\mathcal{M}_\infty(-\infty, 0)$. By using $(-C)^{-1} = -C^{-1}$, one sees that this f is also in $\mathcal{M}_\infty(0, \infty)$ which is the form that appears more often and will be used below in many places.
2. Equation (2.5) also follows from a special case of Theorem 13.3 (a) which implies that if $C, D \geq 0$

$$0 < D \leq C \iff \begin{pmatrix} C & \mathbf{1} \\ \mathbf{1} & D^{-1} \end{pmatrix} \geq 0 \tag{2.6}$$

For $\begin{pmatrix} C & \mathbf{1} \\ \mathbf{1} & D^{-1} \end{pmatrix} \geq 0 \iff \begin{pmatrix} D^{-1} & \mathbf{1} \\ \mathbf{1} & C \end{pmatrix} \geq 0$ gives (2.5) if we let $C = -A$, $D = -B$.

3. One can also obtain (2.5) from the following identity for all invertible C, D:

$$C^{-1} - D^{-1} = D^{-1}[(D-C)C^{-1}(D-C) + (D-C)]D^{-1} \tag{2.7}$$

so $0 < C \leq D \Rightarrow D^{-1} \leq C^{-1}$. Taking $C = -B$, $D = -A$, we see that $A \leq B < 0 \Rightarrow 0 < C \leq D \Rightarrow -A^{-1} \leq -B^{-1}$ proving (2.5). To prove (2.7), we note that $C^{-1} - D^{-1} = D^{-1}(D-C)C^{-1} = C^{-1}(D-C)D^{-1}$ and thus

$$
\begin{aligned}
C^{-1} - D^{-1} &= D^{-1}(D - C)C^{-1} \\
&= D^{-1}(D - C)D^{-1} + D^{-1}(D - C)(C^{-1} - D^{-1}) \\
&= D^{-1}[(D - C)C^{-1}(D - C) + (D - C)]D^{-1}
\end{aligned}
$$

Proof If $\lambda_1, \dots, \lambda_n$ are the eigenvalues of C, $-\lambda_1^{-1}, \dots, -\lambda_n^{-1}$ are the eigenvalues of $-C^{-1}$, so the first conclusion is immediate. To see (2.5), notice that, by the lemma below, when A and B have their eigenvalues in $(-1, 0)$, we have that

$$
\begin{aligned}
A \le B \Leftrightarrow |B| \le |A| &\Leftrightarrow \||B|^{1/2}|A|^{-1/2}\| \le 1 \\
&\Leftrightarrow \||A|^{-1/2}|B|^{1/2}\| \le 1 \Leftrightarrow |A|^{-1} \le |B|^{-1} \\
&\Leftrightarrow -A^{-1} \le -B^{-1}
\end{aligned}
$$

where the third $\Leftrightarrow$ uses $\|F^*\| = \|F\|$ and the fourth $\Leftrightarrow$ uses the lemma with $D = |B|^{-1}$ and $C = |A|^{-1}$. □

Lemma 2.3 *If C, D are both strictly positive finite matrices, then*

$$
0 < C \le D \Leftrightarrow \|C^{1/2}D^{-1/2}\| < 1 \tag{2.8}
$$

Proof Since $D^{-1/2}$ is a bijection,

$$
\begin{aligned}
\forall\varphi,\ (\varphi, C\varphi) \le (\varphi, D\varphi) &\Leftrightarrow \forall\psi,\ (D^{-1/2}\psi, CD^{-1/2}\psi) \le \|\psi\|^2 \\
&\Leftrightarrow \|D^{-1/2}CD^{-1/2}\| \le 1 \\
&\Leftrightarrow \|C^{1/2}D^{-1/2}\| \le 1
\end{aligned}
$$

where the last step used $\|F^*F\| = \|F\|^2$. □

By Proposition 2.2, the map from f on $(-1, 0)$ to g on $(0, \infty)$ by $g(x) = f(-x^{-1})$ sets up a one-to-one correspondence between $\mathcal{M}_\infty(-1, 0)$ and $\mathcal{M}_\infty(0, \infty)$. Since $z \to -z^{-1}$ maps $\mathbb{C}_+$ to $\mathbb{C}_+$, it also sets a map from a function analytic on $(\mathbb{C}\backslash\mathbb{R}) \cup (-1, 0)$ and $\mathbb{C}\backslash(-\infty, 0]$ obeying (1.14). Thus, we get another proof of (a) $\Leftrightarrow$ (c) for $(-1, 0)$, implying (a) $\Leftrightarrow$ (c) for $(0, \infty)$. Moreover, unlike our proof of Theorem 2.1, this shows that the implication goes the other way. Clearly, the same argument shows that if T is a fractional linear transformation that maps $\mathbb{R}$ to $\mathbb{R}$ and $\mathbb{C}_+$ to $\mathbb{C}_+$, and $T[(a, b)] = (c, d)$, then $f \mapsto f \circ T$ maps $\mathcal{M}_\infty(c, d)$ to $\mathcal{M}_\infty(a, b)$. In fact, by the same argument, it also maps $\mathcal{M}_n(c, d)$ to $\mathcal{M}_n(a, b)$. We summarize with

Theorem 2.4 *If T is a fractional linear transformation mapping $(a, b) \subset \mathbb{R}$ monotonically to $(c, d) \subset \mathbb{R}$, then $f \in \mathcal{M}_n(c, d)$ if and only if $f \circ T \in \mathcal{M}_n(a, b)$.*

As the next topic in this chapter, we want to consider Hilbert space extensions. We will first consider bounded intervals. Since this is an aside, we will use some notions and theorems without proof (giving references, though). If A_n, A_∞ are operators on a separable Hilbert space, $\mathcal{H}$, we say A_n converges to A_∞ strongly if and only if $\|A_n\varphi - A_\infty\varphi\| \to 0$ for each $\varphi \in \mathcal{H}$ ($\sup_n \|A_n\|$ is then bounded by the uniform boundedness principle [325, Theorem 5.4.9]). We write $A_n \xrightarrow{s} A_\infty$. One has that (Theorem VIII.20 of [279]; Theorem 7.2.10 of [329])

Theorem 2.5 *If A_n, A_∞ are self-adjoint operators, $A_n \xrightarrow{s} A_\infty$ and f is continuous on $\cup_n$ spec$(A_n) \cup$ spec(A_∞), then $f(A_n) \xrightarrow{s} f(A_\infty)$.*

Lemma 2.6 *Given a (bounded) self-adjoint operator, A, on a separable Hilbert space, $\mathcal{H}$, and a self-adjoint projection, P, let $A_P \equiv PAP \restriction \text{Ran}\, P$. If spec$(A) \subset (a, b)$, then spec$(A_P) \subset (a, b)$.*

Remarks

1. If $A = \begin{pmatrix} 0 & 1 \\ 1 & 0 \end{pmatrix}$, and P is the projection onto all vectors whose second component is 0, then $\text{spec}(A) = \{-1, 1\}$ while $\text{spec}(A_P) = \{0\}$, so it is not true that $\text{spec}(A_P)$ is a subset of $\text{spec}(A)$, but only in the convex hull of that set.
2. We emphasize that A_P involves the restriction to Ran P. If we don't restrict and $P \neq \mathbf{1}$, then $0 \in \text{spec}(PAP)$ so that without the restriction, we'd need that $a < 0 < b$, which is true for the application that we make.

Proof $\text{spec}(B) \subset [c, d] \iff c\mathbf{1} \le B \le d\mathbf{1}$. Thus, $\text{spec}(A) \subset [c, d] \Rightarrow cP \le PAP \le cP \Rightarrow c\mathbf{1}_{\text{Ran}\, P} \le A_P \le d\mathbf{1}_{\text{Ran}\, P} \Rightarrow \text{spec}(A_P) \subset [c, d]$. Now use the fact that for a general self-adjoint operator, B, one has that $\text{spec}(B) \subset (a, b)$ if and only if, for some $a < c < d < b$, one has that $\text{spec}(B) \subset [c, d]$. □

This immediately leads to

Theorem 2.7 *Let A, B be self-adjoint operators on a separable Hilbert space, $\mathcal{H}$, with A and B having spectrum in a bounded interval (a, b). Let $f \in \mathcal{M}_\infty(a, b)$. Then*

$$A \le B \Rightarrow f(A) \le f(B) \tag{2.9}$$

Proof Without loss, we can suppose $0 \in (a, b)$ by translating the interval. Let $\{\varphi_n\}_{n=1}^\infty$ be an orthonormal basis for $\mathcal{H}$ and let P_N be the projection onto $\{\varphi_n\}_{n=1}^N$. Define

$$A_N = P_N A P_N \qquad B_N = P_N B P_N \tag{2.10}$$

Since $P_N \xrightarrow{s} \mathbf{1}$ (by the convergence of abstract Fourier expansions), $A_N \xrightarrow{s} A$ and $B_N \xrightarrow{s} B$. So, by Theorem 2.5, $f(A_N) \to f(A)$ and $f(B_N) \to f(B)$. This is because f is continuous by Loewner's theorem (see also Theorem 5.2).

Thus, it suffices to prove for each N that

$$f(A_N) \leq f(B_N) \tag{2.11}$$

By the Lemma, $\operatorname{spec}(A_N) \cup \operatorname{spec}(B_N) \subset (a, b)$. Since $f(A_N) = f(A_N \restriction \operatorname{Ran} P_N) \oplus f(0) \restriction \operatorname{Ran}(1 - P_N)$ and similarly for B_N, and since $A_N \leq B_N$, monotonicity of f on finite matrices implies (2.11). □

There are also versions for unbounded intervals. For simplicity, we consider only $(0, \infty)$, that is, $0 \leq A < B$. For unbounded operators, one needs to define what this means.

Definition If A is a positive unbounded self-adjoint operator, its form domain $Q(A)$ is $D(|A|^{1/2})$.

See [279, Chapter 8] or [329, Chapter 7] for a discussion of forms, unbounded self-adjoint operators, etc.

Definition If $0 \leq A, B$ for unbounded self-adjoint operators, we say

$$A \leq B$$

if and only if $Q(B) \subset Q(A)$ and $\langle \varphi, A\varphi \rangle \leq \langle \varphi, B\varphi \rangle$ for all φ in $Q(A)$.

Remarks

1. $\langle \varphi, A\varphi \rangle$ is defined to be $\| |A|^{1/2} \varphi \|^2$.
2. This is $\langle \varphi, A\varphi \rangle \leq \langle \varphi, B\varphi \rangle$ for all φ if we think of $\langle \varphi, A\varphi \rangle = \infty$ if $\varphi \notin Q(A)$.

One has

Theorem 2.8 ([178, Theorem VI.2.21] or [329, Theorem 7.5.14]) *If A and B are positive self-adjoint operators, then $A \leq B$ if and only if $(B + 1)^{-1} \leq (A + 1)^{-1}$.*

Remark $A \leq B$ is intended in the sense above. $(B + 1)^{-1} \leq (A + 1)^{-1}$ is intended in the sense of bounded operators, that is, (1.3).

From this theorem, Theorem 2.7, and Proposition 2.2, as discussed above we have that

Theorem 2.9 *If A and B are positive self-adjoint operators and $A \leq B$ and if $f \in \mathcal{M}_\infty(0, \infty)$, then $f(A) \leq f(B)$.*

For this reason, some authors call operators in $\mathcal{M}_\infty$, operator monotone rather than matrix monotone but we will not.

As a final topic, we want to consider the canonical decomposition of operators on a Hilbert space and of matrices on $\mathbb{C}^n$. For more on this subject, see [320] or [329, Sections 2.4 and 3.5]. Given any bounded operator, A, one defines $|A| = \sqrt{A^*A}$. The polar decomposition then says that there is a partial isometry, U, with $A = U|A|$, called the *polar decomposition* of A. If $\ker(A) = 0$, U is unique. We have that $|A| = U^*A$.

$|A|$ is a positive operator. We can write its eigenvalues counting multiplicity as $\mu_1(A) \geq \mu_2(A) \geq \dots$ (where if the Hilbert space is infinite dimensional, we suppose A is compact to be talking about its eigenvalues). These are called the *singular values* of A. If we pick an orthonormal set of eigenvectors $\{\varphi_j\}_{j=1}^N$ with $|A|\varphi_j = \mu_j(A)\varphi_j$, then $\psi_j = U\varphi_j$ is also an orthonormal family and one has

$$A = \sum_{j=1}^{N} \mu_j(A)\langle\varphi_j, \cdot\rangle\psi_j \tag{2.12}$$

called the *canonical decomposition* of A.

It is not hard to show [329, Theorem 3.5.3] that

$$\mu_j(A^*) = \mu_j(A) \tag{2.13}$$

$$\mu_j(BA) \leq \|B\|\mu_j(A) \tag{2.14}$$

$$\mu_j(AB) \leq \|B\|\mu_j(A) \tag{2.15}$$

Notes and Historical Remarks The first part of this chapter is standard folklore. Theorem 2.7 is due to Bendat–Sherman [22].

I have actually cheated in the proofs of Theorems 2.7–2.9, since Theorem 2.5 is only true for bounded continuous functions and one wants results like Theorem 2.9 for functions like $f(x) = \sqrt{x}$ or $f(x) = \log(x)$. There are several ways to fill in this gap:

(i) Use the integral representation formula (1.13) and cut off the integral and remove the cutoff.
(ii) Cut off A and B. For example, let $P_N \equiv P_{[0,N)}(B)$ be a spectral projection for B, and let A_N, B_N be given by (2.10). In place of Theorem 2.5, one uses monotone convergence theorems for forms (see [318, 319]).

Since this issue is peripheral, we have been less than complete on this point.

It is easy to see that if $f \in \mathcal{M}_\infty(a, b)$, then for self-adjoint operators, A, B in a general C^*-algebra, one has that $\sigma(A), \sigma(B) \subset (a, b)$, $A \leq B \Rightarrow f(A) \leq f(B)$, essentially because C^*-algebras have lots of representations on Hilbert spaces. But for a given C^*-algebra, there might be $f \notin \mathcal{M}_\infty(a, b)$ for which the above implications holds because there might be restrictions on the type of A, B that can occur. For example, it is known for a type II_n von Neumann algebra, it suffices that $f \in \mathcal{M}_n(a, b)$. This problem is studied and solved in Osaka, Silvestov, and Tomiyama [251]. The analogous problem for matrix convex functions is studied in Hoa and Tikhonov [163].

Chapter 3
The Herglotz Representation Theorems and the Easy Direction of Loewner's Theorem

In this chapter, we will show (b)⇔(c) in Theorems 1.3 and 1.6 and also (b)⇒(a) and the equivalent part of Theorem 1.8. We'll have a different proof of (c)⇒(a) in Chapter 16. We want to study functions f on $\mathbb{C}_+$ with $\text{Im } f(x) > 0$, but begin with functions f on $\mathbb{D} = \{z \in \mathbb{C} \mid |z| < 1\}$ with $\text{Re } f > 0$ and use conformal mapping. The basic input is a complex Poisson representation:

Theorem 3.1 (Complex Poisson Representation) *Let f be analytic in a neighborhood of $\overline{\mathbb{D}}$. Then*

$$f(z) = i \text{ Im } f(0) + \int K(e^{i\theta}, z) \text{ Re } f(e^{i\theta}) \frac{d\theta}{2\pi} \tag{3.1}$$

where

$$K(w, z) = \frac{w + z}{w - z} \tag{3.2}$$

Remarks

1. Of course, one is interested in (3.1) in cases where f is only analytic in $\mathbb{D}$ and $f(e^{i\theta})$ is some kind of boundary value. Indeed, we will prove such a result below in case $\text{Re } f > 0$. The key, though, is to prove this result in the regular case first.
2. We will give four proofs that illustrate these different aspects of the kernel K and of analytic functions. This is overkill—even in a pedagogic monograph—but I am fond of this result.

First Proof Write

$$K(e^{i\theta}, z) = \frac{1 + ze^{-i\theta}}{1 - ze^{-i\theta}} = 1 + 2\sum_{n=1}^{\infty} z^n e^{-in\theta} \tag{3.3}$$

B. Simon, *Loewner's Theorem on Monotone Matrix Functions*, Grundlehren der mathematischen Wissenschaften 354, https://doi.org/10.1007/978-3-030-22422-6_3

with the sum converging uniformly in θ for each fixed $z \in \mathbb{D}$. Write

$$f(z) = \sum_{n=0}^{\infty} a_n z^n \tag{3.4}$$

converging uniformly on $\overline{\mathbb{D}}$ since f is analytic in a neighborhood of $\overline{\mathbb{D}}$. Thus

$$\operatorname{Re} f(e^{i\theta}) = \operatorname{Re}(a_0) + \tfrac{1}{2} \sum_{n=1}^{\infty} (a_n e^{in\theta} + \bar{a}_n e^{-in\theta}) \tag{3.5}$$

converging uniformly in θ. Since

$$\int e^{-ik\theta} e^{in\theta} \frac{d\theta}{2\pi} = \delta_{kn} \tag{3.6}$$

using the uniform convergence of (3.3) and (3.5) to interchange the sums and integrals, one finds

$$\begin{aligned}\int K(e^{i\theta}, z) \operatorname{Re} f(e^{i\theta}) \frac{d\theta}{2\pi} &= \operatorname{Re}(a_0) + \sum_{n=1}^{\infty} 2(\tfrac{1}{2} a_n) z^n \\ &= f(z) - i \operatorname{Im} f(0)\end{aligned}$$

□

Second Proof Let

$$g(z) = \int K(e^{i\theta}, z) \operatorname{Re} f(e^{i\theta}) \frac{d\theta}{2\pi} \tag{3.7}$$

g is clearly analytic in $\mathbb{D}$ and we have

$$\operatorname{Re} g(re^{i\varphi}) = \int P_r(\theta, \varphi) \operatorname{Re} f(e^{i\theta}) \frac{d\theta}{2\pi} \tag{3.8}$$

where P_r

$$P_r(\theta, \varphi) = \frac{1 - r^2}{1 + r^2 - 2r\cos(\theta - \varphi)} \tag{3.9}$$

Since $h(re^{i\varphi}) \equiv P_r(\theta, \varphi) = \operatorname{Re} K(e^{i\theta}, re^{i\varphi})$ is harmonic in $re^{i\varphi}$, we have

$$\int P_r(\theta, \varphi) \frac{d\varphi}{2\pi} = h(0) = 1 \tag{3.10}$$

Since $P_r(\theta, \varphi)$ is symmetric under interchange of θ and φ, the same is true of the θ integral. For each $\varepsilon > 0$, $\lim_{r\uparrow 1} \sup_{|\theta-\varphi|>\varepsilon} |P_r(\theta, \varphi)| = 0$, and we see P_r is an approximate delta function. Thus, since $\operatorname{Re} f(re^{i\theta})$ is smooth, $\operatorname{Re} g(e^{i\theta})$ has a continuous extension to $\overline{\mathbb{D}}$ and

$$\operatorname{Re} g(e^{i\theta}) = \operatorname{Re} f(e^{i\theta}) \tag{3.11}$$

Let $w(z) = \operatorname{Re} f(z) - \operatorname{Re} g(z)$ for $z \in \overline{\mathbb{D}}$ and define w for $z \in \mathbb{C}\backslash\mathbb{D}$ by $w(1/z) = -w(z)$. Then w is harmonic on $\mathbb{C}$ and bounded, so constant. Since $w(e^{i\theta}) = 0$, $w \equiv 0$. Thus, on $\mathbb{D}$, $f(z) - g(z)$ is an analytic function with vanishing real part, so by the Cauchy–Riemann equations, $f(z) - g(z)$ is an imaginary constant which is $i \operatorname{Im} f(0)$, since $\operatorname{Im} g(z) = 0$. This proves (3.1). □

Third Proof The Cauchy integral formula says

$$f(z) = \frac{1}{2\pi i} \oint_{|w|=1} \frac{f(w)}{w - z}\, dw \tag{3.12}$$

Using $w = e^{i\theta}$, we see $dw = ie^{i\theta}\, d\theta$, so (3.12) says

$$f(z) = \int \frac{e^{i\theta}}{e^{i\theta} - z} f(e^{i\theta}) \frac{d\theta}{2\pi} \tag{3.13}$$

$$= \int \frac{1}{2}\left[\frac{e^{i\theta} + z}{e^{i\theta} - z} + 1\right] f(e^{i\theta}) \frac{d\theta}{2\pi} \tag{3.14}$$

$$= \frac{1}{2} \int K(e^{i\theta}, z) f(e^{i\theta}) \frac{d\theta}{2\pi} + \frac{1}{2} f(0) \tag{3.15}$$

For now, we note that the step from the RHS of (3.12) to (3.15) only uses $f(e^{i\theta})$ harmonic.

If $w = e^{i\theta}$, then $\overline{dw} = -ie^{-i\theta}\, d\theta = -w^{-2}\, dw$. Thus

$$-\frac{1}{2\pi i} \oint_{|w|=1} \frac{f(w)}{\bar{w} - \bar{z}}\, \overline{dw} = \frac{1}{2\pi i} \oint_{|w|=1} \frac{f(w)}{w(1 - \bar{z}w)}\, dw \tag{3.16}$$

$$= f(0) \tag{3.17}$$

where (3.16) uses $\bar{w}w = 1$ and (3.17) uses that since $|\bar{z}| < 1$, $f(w)/w(1 - \bar{z}w)$ has a pole only at $w = 0$.

Taking complex conjugates,

$$\overline{f(0)} = \frac{1}{2\pi i} \oint_{|w|=1} \frac{\overline{f(w)}}{w - z}\, dw$$

$$= \frac{1}{2} \int K(e^{i\theta}, z)\, \overline{f(e^{i\theta})} \frac{d\theta}{2\pi} + \frac{1}{2} \overline{f(0)} \tag{3.18}$$

where (3.18) follows by using the steps from the right side of (3.12) to (3.15) which, as noted, only need that $\overline{f(z)}$ is harmonic. Thus, adding (3.15) to (3.18), we get

$$f(z) = \int K(e^{i\theta}, z) \operatorname{Re} f(e^{i\theta}) \frac{d\theta}{2\pi} + \frac{1}{2}(f(0) - \overline{f(0)}) \tag{3.19}$$

which is (3.1). □

Fourth Proof Let $u(z) = \operatorname{Re} f(z)$. Since u is harmonic,

$$u(0) = \frac{1}{2\pi} \int_0^{2\pi} u(e^{i\theta}) \frac{d\theta}{2\pi} \tag{3.20}$$

For $z \in \mathbb{D}$, let $T_z : \overline{\mathbb{D}} \to \overline{\mathbb{D}}$ by

$$T_z(w) = \frac{w+z}{1+\bar{z}w} \tag{3.21}$$

and

$$v_z(w) = u(T_z(w)) \tag{3.22}$$

so v_z is also harmonic. Thus, (3.20) becomes

$$\begin{aligned} u(z) = v_z(0) &= \frac{1}{2\pi} \int_0^{2\pi} v_z(e^{i\theta}) \frac{d\theta}{2\pi} \\ &= \frac{1}{2\pi} \int_0^{2\pi} u(e^{i\varphi}) \frac{d\theta}{d\varphi} \frac{d\varphi}{2\pi} \end{aligned} \tag{3.23}$$

where

$$T_z(e^{i\theta}) = e^{i\varphi}$$

so

$$e^{i\theta} = T_{-z}(e^{i\varphi})$$

and

$$\frac{d\theta}{d\varphi} = \frac{1-|z|^2}{|1-ze^{-i\varphi}|^2} \tag{3.24}$$

by a direct calculation. Thus, (3.23) becomes

$$u(z) = \int \operatorname{Re} K(e^{i\theta}, z) \operatorname{Re} f(e^{i\theta}) \frac{d\theta}{2\pi} \tag{3.25}$$

Thus,

$$\operatorname{Re}\left(f(z) - \int K(e^{i\theta}, z) \operatorname{Re} f(e^{i\theta}) \frac{d\theta}{2\pi}\right) = 0 \tag{3.26}$$

which implies the function in (...) is an imaginary constant, thus Im $f(0)$. □

We can now analyze all analytic functions on $\mathbb{D}$ with Re $f > 0$:

Theorem 3.2 (Herglotz Representation for $\mathbb{D}$) *Let f be an analytic function on $\mathbb{D}$ with* Re $f(z) > 0$ *for $z \in \mathbb{D}$. Then there exists a measure $d\mu$ on $\partial\mathbb{D}$ so that*

$$f(z) = i \operatorname{Im} f(0) + \int K(e^{i\theta}, z)\, d\mu(\theta) \tag{3.27}$$

Remarks

1. Our proof shows that μ is unique.
2. Since $\operatorname{Re} K(e^{i\theta}, re^{i\varphi}) = P_r(\theta, \varphi) > 0$, any non-zero measure $d\mu$ leads via (3.27) to an analytic function with Re $f > 0$.
3. Thus, there is a one-to-one correspondence between functions f with Re $f > 0$ and pairs $(a, d\mu)$ with $a = \operatorname{Im} f(0) \in \mathbb{R}$ and $d\mu$ a positive measure.
4. Notice that

$$\operatorname{Re} f(0) = \mu(\partial\mathbb{D}) \tag{3.28}$$

Proof By subtracting $i \operatorname{Im} f(0)$ from f and multiplying by a constant, we can suppose that $f(0) = 1$ (functions with Re $f > 0$ and $f(0) = 1$ are called *Carathéodory functions*).

By Theorem 3.1, for any $z \in \mathbb{D}$ and $r \in (0, 1)$,

$$f(rz) = \int K(e^{i\theta}, z)\, d\mu_r(\theta) \tag{3.29}$$

where

$$d\mu_r(\theta) = \operatorname{Re} f(re^{i\theta}) \frac{d\theta}{2\pi} \tag{3.30}$$

Since $f(0) = 1$, $\{d\mu_r\}_{0<r<1}$ is a family of probability measures. By (3.29), for each $z \in \mathbb{D}$,

$$\lim_{r\uparrow 1} \int K(e^{i\theta}, z)\, d\mu_r(\theta) = f(z) \tag{3.31}$$

exists. By the lemma below, $\{K(e^{i\theta}, z)\}_{z\in\mathbb{D}}$ and its conjugates are total in $\mathbb{C}(\partial\mathbb{D})$. So $d\mu_r$ has a weak limit $d\mu$ and (3.27) holds by (3.31). □

Remark The proof shows that

$$d\mu(\theta) = \text{w-}\lim_{r\uparrow 1} \operatorname{Re} f(re^{i\theta}) \frac{d\theta}{2\pi} \tag{3.32}$$

Lemma 3.3 *The linear combinations of* $\{K(\cdot, z)\}_{z\in\mathbb{D}}$ *and their conjugates are dense in* $\mathbb{C}(\partial\mathbb{D})$.

Proof By (3.3) and writing derivatives as limits, we see that $\{e^{-in\theta}\}_{n=0}^{\infty}$ are in the closure of the span of $\{K(\cdot, z)\}_{z\in\mathbb{D}}$. Its conjugates give $\{e^{in\theta}\}_{n=0}^{\infty}$. By Weierstrass's theorem on the density of trigonometric polynomials [325, Theorem 2.4.2], the lemma is proven. □

By using a conformal map from $\mathbb{C}_+$ to $\mathbb{D}$, we shall prove

Theorem 3.4 (Herglotz Representation for $\mathbb{C}_+$) *Let* $f : \mathbb{C}_+ \to \mathbb{C}_+$ *be analytic. Then there exists a finite (positive) measure,* μ, *on* $\mathbb{R}$ *and constant* $A \geq 0$ *so that*

$$f(z) = \operatorname{Re} f(i) + Az + \int \frac{1+xz}{x-z}\, d\mu(x) \tag{3.33}$$

Conversely, any function obeying (3.33) *for* $A \geq 0$ *obeys* (1.14).

Remarks

1. As $x \to \infty$, $(1 + xz)/(x - z) \to z$, so one can drop the Az term in (3.33) if we take $d\mu$ as a measure on $\mathbb{R} \cup \{\infty\}$ with $\mu(\{\infty\}) = A$ and interpret $\left.\frac{1+xz}{x-z}\right|_{x=\infty}$ as z.
2. A function f on $\mathbb{C}_+$ obeying (1.14) is called a *Herglotz function*, a *Pick function*, or a *Nevanlinna function*.

Proof For the converse, we write

$$\frac{1+xz}{x-z} = -x + \frac{1+x^2}{x-z} \tag{3.34}$$

so that, in particular,

$$\operatorname{Im}\left(\frac{1+xz}{x-z}\right) = \frac{1+x^2}{|x-z|^2} \operatorname{Im} z \tag{3.35}$$

proving that any function of the form (3.33) obeys (1.14).

Consider the fractional linear maps $T : \mathbb{C} \cup \{\infty\} \to \mathbb{C} \cup \{\infty\}$ and T^{-1} given by

$$T(z) = \frac{z-i}{z+i} \qquad T^{-1}(w) = \frac{w+1}{i(w-1)} \tag{3.36}$$

It is easily checked that they are inverse to one another. Moreover,

$$T\left(-\cot\left(\frac{\theta}{2}\right)\right)=e^{i\theta} \tag{3.37}$$

so $T\colon \mathbb{R}\cup\{\infty\}\to\partial\mathbb{D}$. Since $T(i)=0$, T maps $\mathbb{C}_+\to\mathbb{D}$ and, of course, T^{-1} maps $\mathbb{D}\to\mathbb{C}_+$.

Thus, if $f\colon\mathbb{C}_+\to\mathbb{C}_+$, then $-if\circ T^{-1}\colon\mathbb{D}\to\{z\mid \operatorname{Re}z>0\}$ and, therefore, by Theorem 3.2, there is a measure $\tilde{\mu}$ on $\partial\mathbb{D}$ so that

$$-if(T^{-1}(w))=i\operatorname{Im}(-if(i))+\int K(e^{i\theta},w)\,d\tilde{\mu}(\theta) \tag{3.38}$$

In this equation, set $z=T^{-1}(w)$ and use (3.37) which is a bijection of $\mathbb{R}\cup\{\infty\}$ to $\partial\mathbb{D}$ to drag $\tilde{\mu}$ to a measure μ on $\mathbb{R}\cup\{\infty\}$ (i.e., $\mu(A)=\tilde{\mu}(T[A])$ to obtain

$$f(z)=\operatorname{Re}f(i)+\int iK(T(x),T(z))\,d\mu(x) \tag{3.39}$$

For $x\neq\infty$, $iK(T(x),T(z))=(1+xz)/(x-z)$ by a direct computation from (3.3) and (3.36). Since $T(\infty)=1$, one sees $iK(T(\infty),T(z))=z$. Thus, (3.39) is (3.33). □

Using (3.34), one can rewrite (3.33) as

$$f(z)=\operatorname{Re}f(i)+az+\int\left(\frac{1}{x-z}-\frac{x}{1+x^2}\right)d\nu(x) \tag{3.40}$$

where

$$d\nu=(1+x^2)\,d\mu \tag{3.41}$$

is no longer necessarily a finite measure but rather

$$\int\frac{d\nu(x)}{1+x^2}<\infty \tag{3.42}$$

(3.40) is the more common form in the literature, but (3.33) will be more useful for us.

We now turn to seeing about functions on $(\mathbb{C}\backslash\mathbb{R})\cup(a,b)$. It will help to write first the analog of (3.32):

Proposition 3.5 *The measure $d\mu$ of* (3.33) *is given by*

$$d\mu(x)=\operatorname*{w-lim}_{\varepsilon\downarrow 0}\frac{1}{\pi}(1+x^2)^{-1}\operatorname{Im}f(x+i\varepsilon)\,dx \tag{3.43}$$

in the sense that for any $C_0^\infty(\mathbb{R})$ function $g(x)$,

$$\int g(x)\,d\mu(x) = \lim_{\varepsilon\downarrow 0} \int (1+x^2)^{-1} \frac{1}{\pi}\, g(x) \operatorname{Im} f(x+i\varepsilon)\,dx \tag{3.44}$$

Proof We have that

$$\frac{1}{\pi} \operatorname{Im}\left(\frac{1}{x-(y+i\varepsilon)}\right) = \frac{\varepsilon}{\pi[(x-y)^2+\varepsilon^2]}$$

The right side of this function has integral 1 (independent of ε) and goes to zero uniformly in y on each interval $\{y \mid |x-y| > \delta\}$. It follows that it is an approximate delta function. This, plus (3.40) and (3.41), implies (3.44). □

Corollary 3.6 *A function $f\colon \mathbb{C}_+ \to \mathbb{C}_+$ has an analytic continuation through an interval $(a,b) \subset \mathbb{R}$ with $\operatorname{Im} f(x) = 0$ for $x \in (a,b)$ (i.e., is the restriction to $\mathbb{C}_+$ of a function on $(\mathbb{C}\setminus\mathbb{R}) \cup (a,b)$ with f real on (a,b)) if and only if in the representation* (3.33), *we have*

$$\mu(a,b) = 0 \tag{3.45}$$

Proof If $\operatorname{Im} f(x) = 0$ on (a,b), then (3.45) holds, by (3.43). Conversely, if $\mu(a,b) = 0$, then for any $[c,d] \subset (a,b)$, the integrand in (3.33) is uniformly bounded in $\{(z,x) \mid x \in \operatorname{supp}(d\mu),\ \operatorname{Re} z \in [c,d],\ |\operatorname{Im} z| \le 1\}$, so we get convergence of the integral to a function analytic in that region. Clearly, by (3.35), $\operatorname{Im} f(z) = 0$ if $z \in [c,d]$. □

With this, we have (b) $\Leftrightarrow$ (c) in Theorem 1.6.

Theorem 3.7 *A function f is analytic in $(\mathbb{C}\setminus\mathbb{R}) \cup (a,b)$, real on (a,b), with* (1.14) *holding if and only if* (1.17) *holds with $J = \mathbb{R}\setminus(a,b)$.*

Proof Immediate from Theorems 3.4 and Corollary 3.6. □

Depending on the application, there are wide variety of rewritings of (3.40) that are useful. We'll present a number of them below and use them later in the book. We begin with a special form for $(a,b) = (-1,1)$ that leads to (b) $\Leftrightarrow$ (c) in Theorem 1.3:

Theorem 3.8 *A function f is analytic in $(\mathbb{C}\setminus\mathbb{R}) \cup (-1,1)$, real on $(-1,1)$ with* (1.14) *holding if and only if* (1.13) *holds.*

Proof By Corollary 3.6, we need only show that (1.13) is implied by

$$f(z) = \operatorname{Re} f(i) + az + \int_{x\notin(-1,1)} \frac{1+xz}{x-z}\,d\mu(x) \tag{3.46}$$

since (1.13) clearly implies (c) of Theorem 1.3. Since

$$\frac{1+xz}{x-z} = x^{-1} + \frac{xz}{x-z}\,\frac{1+x^2}{x^2}$$

to then rewrite (3.46) as

$$f(z) = c + az + \int_{x\notin(-1,1)} \frac{xz}{x-z}\,\frac{1+x^2}{x^2}\,d\mu(x) \tag{3.47}$$

where c is a suitable constant. Now change variables to $\lambda = -x^{-1}$ and transfer $\frac{1+x^2}{x^2}\,d\mu(x)$ to a measure $d\tilde{\nu}(\lambda)$ on $[-1,1]\backslash\{0\} = \{-\lambda^{-1} \mid \lambda \in \mathbb{R}\backslash(-1,1)\}$:

$$f(z) = c + az + \int_{\lambda\neq 0} \frac{z}{1+\lambda z}\,d\tilde{\nu}(\lambda)$$

Define $\nu = \tilde{\nu} + a\delta_{\lambda=0}$ to obtain (1.13) with c a constant. Setting $z = 0$, we find $c = f(0)$. □

In line with our focus in the last two theorems, we define a *rational Herglotz function* to be a rational function, f, which is Herglotz and which is real on the real axis (away from any pole of f). Note without the reality condition, for every $z_0 \in \mathbb{C}_+$, and every rational Herglotz function, f, $z \mapsto f(z+z_0)$ is rational and Herglotz but not a rational Herglotz function as we just defined it. By the reality condition, $f(\bar{z}) = \overline{f(z)}$ so f has no poles in $\mathbb{C}\setminus\mathbb{R}$. Thus, a rational function which is bounded near ∞ is a rational Herglotz function if and only if all its poles lie on $\mathbb{R}$, are simple, and have negative residues.

An immediate consequence of Theorem 3.8 is

Theorem 3.9 *Let $\mathcal{H}(-1,1)$ denote the set of real-valued functions, f, on $(-1,1)$ which are restrictions of analytic functions on $\mathbb{C}\setminus[(-\infty,-1]\cup[1,\infty)]$ with $\operatorname{Im} f > 0$ on $\mathbb{C}_+$. Then $\{f \in \mathcal{H}(-1,1) \mid f(0) = 0,\ f'(0) = 1\}$ and, for any c_1, c_2, $\{f \in \mathcal{H}(-1,1) \mid |f(0)| \le c_1,\ |f'(0)| \le c_2\}$ are compact in the topology of uniform convergence on compact subsets of $\mathbb{C}\setminus[(-\infty,-1]\cup[1,\infty)]$. Similarly, for any two distinct $\alpha, \beta \in (-1,1)$, $\{f \in \mathcal{H}(-1,1) \mid |f(\alpha)| \le c_1,\ |f(\beta)| \le c_2\}$ is compact.*

Remarks

1. This result is conceptually critical for considering many proofs of the hard part of Loewner's theorem. These proofs approximate f with f_n's in $\mathcal{H}(-1,1)$ obeying $|f_n(0)| \le c_1$, $|f_n'(0)| \le c_2$, so compactness shows these f_n have limit points.
2. By Vitali's theorem (see [326, Section 6.2]), this convergence is equivalent to pointwise convergence on a dense subset of $(-1,1)$.
3. Our proof will rely on the compactness of the set of probability measures. One could instead appeal to Montel's theorem (see [326, Section 6.2])

4. Similarly, for any $z_0 \in \mathbb{C}_+$, $\{f \in \mathcal{H}(-1, 1) \mid |f(z_0)| \leq c_1\}$ is compact, for the bound on $\operatorname{Im}(f(z_0))$ gives a bound on the total mass of the measures and then the bound on $|\operatorname{Re}(f(z_0)|$ gives a bound on $|f(0)|$.

Proof By the last theorem, f has a representation of the form (1.13) with a measure $d\nu_f$ obeying $\int_{-1}^{1} d\nu_f(\lambda) = f'(0)$. Thus, the first result follows from the compactness of the set of probability measures on $[-1, 1]$ [325, Section 5.8] and the second from a more general application of the Banach–Alaoglu theorem.

For the $\alpha > \beta$ result note that

$$\frac{\alpha}{1 + \lambda\alpha} - \frac{\beta}{1 + \lambda\beta} = \frac{\alpha - \beta}{(1 + \lambda\alpha)(1 + \lambda\beta)}$$

so using $(1 + \lambda\alpha)(1 + \lambda\beta) > (1 - |\alpha|)(1 - |\beta|)$, we get a bound on $\nu_f(-1, 1)$ and then on $|f(0)|$. With these, we repeat the above proof. □

Next, we want to rewrite (3.40) when f is real analytic on $(a, b) \subsetneq \mathbb{R}$ so that $d\nu$ is supported on $J_- \cup J_+$ where

$$J_- = (-\infty, b], \qquad J_+ = [a, \infty) \tag{3.48}$$

Pick $x_0 \in (a, b)$ and subtract (3.40) for x_0 from (3.40) for z and then set $\gamma = \operatorname{Re} f(i) + f(x_0)$ to find that

$$\begin{aligned} f(z) &= \alpha z + \gamma + \int \left[\frac{1}{y - z} - \frac{1}{y - x_0} \right] d\nu(y) \\ &= \alpha z + \gamma + \int \frac{z - x_0}{(y - z)(y - x_0)} d\nu(y) \end{aligned} \tag{3.49}$$

Since $\pm(y - x_0) > 0$ for $y \in J_\pm$, we define

$$d\nu_\pm = \pm(y - x_0)\, d\nu(y) \restriction J_\pm \tag{3.50}$$

so $d\nu_\pm$ are positive measures obeying

$$\int \frac{d\nu_\pm(y)}{(1 + y^2)^{3/2}} < \infty \tag{3.51}$$

$$\begin{aligned} f(z) = \alpha z + \gamma &+ \int_b^\infty \frac{z - x_0}{(y - z)(y - x_0)^2} d\nu_+(y) \\ &+ \int_{-\infty}^{a} \frac{z - x_0}{(z - y)(y - x_0)^2} d\nu_-(y) \end{aligned} \tag{3.52}$$

which has positive integrands for $z \in (x_0, b)$ and negative for $z \in (a, x_0)$. It is sometimes useful to write this as

$$f(z) = \alpha z + \gamma + \int_b^\infty \left[\frac{1}{(y-z)(y-x_0)} - \frac{1}{(y-x_0)^2} \right] d\nu_+(y) \\ + \int_{-\infty}^a \left[\frac{1}{(y-z)(x_0-y)} + \frac{1}{(y-x_0)^2} \right] d\nu_-(y) \tag{3.53}$$

There are formulae like (1.13) whenever $0 \in (a, b)$. For example, if $0, c \in (a, b)$, then

$$f(x) = f(c) + \int_{-b^{-1}}^{-a^{-1}} \frac{(x-c)}{1+\lambda x} \, d\nu(\lambda) \tag{3.54}$$

There is also a version when f is analytic on $\mathbb{C}\backslash(-\infty, 0]$:

Theorem 3.10 *A function f is analytic on $\mathbb{C}\backslash(-\infty, 0]$ with $f : \mathbb{C}_+ \to \mathbb{C}_+$ if and only if*

$$f(z) = f(1) + \int_0^1 \frac{(z-1)}{\lambda + (1-\lambda)z} \, d\nu(\lambda) \tag{3.55}$$

for a measure $d\nu$ on $[0, 1]$.

Proof We start with

$$f(z) = \operatorname{Re} f(i) + az + \int_{x \in (-\infty, 0]} \frac{1+xz}{x-z} \, d\mu(x) \tag{3.56}$$

If $Q(x, z) = (1+xz)/(x-z)$, then

$$Q\left(-\frac{\lambda}{1-\lambda}, z\right) - Q\left(-\frac{\lambda}{1-\lambda}, 1\right) = \frac{(z-1)(\lambda^2 + (\lambda-1)^2)}{\lambda + (1-\lambda)z} \tag{3.57}$$

by straightforward algebra. It follows that

$$f(z) = f(1) + a(z-1) + \int \frac{(z-1)}{\lambda + (1-\lambda)z} (\lambda^2 + (\lambda-1)^2) \, d\mu\left(\frac{-\lambda}{1-\lambda}\right)$$

Since $\lambda \to \frac{-\lambda}{1-\lambda}$ maps $[0, 1)$ into $(-\infty, 0]$ by setting

$$d\nu(\lambda) = (\lambda^2(\lambda-1)^2) \, d\mu\left(\frac{-\lambda}{1-\lambda}\right) + a\delta_{\lambda=1} \tag{3.58}$$

we obtain (3.55). □

Theorem 3.11 *A function f is analytic on $\mathbb{C}\backslash(-\infty, 0]$ with* (1.14) *and $f(x) > 0$ on $(0, \infty)$ if and only if there is a measure $d\nu_1$ on $[0, 1]$ with*

$$f(z) = \int \frac{z}{(1-\lambda)z + \lambda} \, d\nu_1(\lambda) \tag{3.59}$$

Remarks

1. When $\lambda = 0$, we have that $\frac{z}{(1-\lambda)z+\lambda} = 1$ at least if $z \neq 0$ so we interpret it as 1 even if $z = 0$. Thus, $f(0) = \nu_1(\{0\})$.
2. Notice that if (3.59) holds, then

$$f(1) = \int d\nu_1(\lambda) \tag{3.60}$$

so $f(1) = 1$ is equivalent to $d\nu_1$ being a probability measure.

Proof $f(x)$ is given by (3.55), so

$$f'(x) = \int_0^1 \frac{1}{[(1-\lambda)x + \lambda]^2} \, d\nu(x) > 0$$

so f is monotone and $f(x) > 0$ implies $\lim_{x \downarrow 0} f(x) \equiv f(0)$ exists. Moreover, taking $x \downarrow 0$ in (3.55) implies

$$f(1) - f(0) = \int_0^1 \frac{d\nu(\lambda)}{\lambda} \tag{3.61}$$

so, in particular, $d\nu(\lambda)/\lambda$ is a finite measure. Subtracting (3.61) from (3.55) shows

$$f(z) = f(0) + \int_0^1 \frac{z}{\lambda + (1-\lambda)z} \frac{d\nu(\lambda)}{\lambda} \tag{3.62}$$

Setting

$$d\nu_1(\lambda) = \frac{d\nu(\lambda)}{\lambda} + f(0)\delta_{\lambda=0} \tag{3.63}$$

we get (3.59). □

We will sometimes need a rephrasing of the Herglotz representation, (3.62), for such f's, viz.

$$f(z) = f(0) + \int_0^1 \frac{z}{\alpha + (1-\alpha)z} \frac{d\nu(\alpha)}{\alpha} \tag{3.64}$$

Change variables to $\lambda = \alpha/(1-\alpha)$ which maps $(0, 1)$ bijectively to $(0, \infty)$. Equivalently, $\alpha = \lambda/(1+\lambda)$, $(1-\alpha) = 1/(1+\lambda)$. Then

$$\alpha + (1-\alpha)z = (1-\alpha)(z+\lambda) = (z+\lambda)/(1+\lambda)$$

Picking $d\rho(\lambda) = \frac{d\nu(\alpha)}{\alpha}$ on $(0, \infty)$, $f(0) = a$ and $\nu(\{1\}) = b$, we get

$$f(z) = a + bz + \int_0^\infty \frac{z(1+\lambda)}{z+\lambda} d\rho(\lambda) \tag{3.65}$$

$$f(1) = a + b + \int_0^\infty d\rho(\lambda) \tag{3.66}$$

As with (3.59), we can absorb the point masses into ρ and define a measure ρ_1 on $[0, \infty]$ and write (3.65) as

$$f(z) = \int_0^\infty \frac{z(1+\lambda)}{z+\lambda} d\rho_1(\lambda) \tag{3.67}$$

where $\frac{z(1+\lambda)}{z+\lambda}$ is interpreted as z when $\lambda = \infty$ and as 1 when $\lambda = 0$ for all z including $z = 0$. Thus, $b = \rho_1(\{\infty\})$ and $a = \rho_1\{0\}$. Another way that some rewrite (3.65) is as

$$f(z) = a + bz + \int_0^\infty \frac{z\lambda}{z+\lambda} d\mu(\lambda) \tag{3.68}$$

where instead of $d\rho$ being a finite measure, we have that

$$\int \frac{\lambda}{1+\lambda} d\mu(\lambda) < \infty \tag{3.69}$$

Remark The last three theorems (i.e., (1.13), (3.55), and (3.59)) can be understood in terms of the Krein–Milman theorem discussed in Chapter 28. Essentially,

(a) $\{f \mid f$ analytic in $(\mathbb{C}\backslash\mathbb{R}) \cup (-1, 1)$ with $\pm\operatorname{Im}(f) > 0$ when $\pm\operatorname{Im}(z) > 0$; $f(0) = 0$, $f'(0) = 1\}$ is a compact convex set (in the topology of locally uniform convergence) with extreme points $z/(1+\lambda z)$.
(b) $\{f \mid f$ analytic in $\mathbb{C}\backslash(-\infty, 0)$ with $\pm\operatorname{Im}(f) > 0$ when $\pm\operatorname{Im}(z) > 0$; $f(1) = 0$, $f'(1) = 1\}$ is a compact convex set with extreme points $(z-1)/(\lambda+(1-\lambda)z)$.
(c) $\{f \mid f$ analytic in $\mathbb{C}\backslash(-\infty, 0)$ with $\pm\operatorname{Im}(f) > 0$ when $\pm\operatorname{Im}(z) > 0$; $f > 0$ on $(0, \infty)$, $f(1) = 1\}$ is a compact convex set with extreme points $z/(\lambda+(1-\lambda)z)$.

Finally, we note that (1.13) implies the easy half of Loewner's theorem:

Theorem 3.12 (b) $\Rightarrow$ (a) *in Theorems 1.3 and 1.6.*

Proof By Theorem 2.1, it suffices to prove the result if $(a, b) = (-1, 1)$. By (1.13), we need only show that

$$\lambda \in [-1, 1], \ -\mathbf{1} < A \leq B < \mathbf{1} \Rightarrow \frac{A}{1+\lambda A} \leq \frac{B}{1+\lambda B} \tag{3.70}$$

Let $\mu = |\lambda|^{-1}$. Multiplying the right side of (3.70) by μ^{-1}, we see that we need

$$|\mu| \geq 1, \ -\mathbf{1} < A \leq B < \mathbf{1} \Rightarrow \frac{A}{|\mu| \pm A} \leq \frac{B}{|\mu| \pm B} \tag{3.71}$$

Since

$$\frac{C}{|\mu| \pm C} = \pm \mathbf{1} \mp \frac{|\mu|}{|\mu| \pm C} \tag{3.72}$$

we need only show that

$$|\mu| \geq 1, \ -\mathbf{1} < A \leq B < \mathbf{1} \Rightarrow \frac{\mp 1}{|\mu| \pm A} \leq \frac{\mp 1}{|\mu| \pm B} \tag{3.73}$$

This follows from (2.5). For the $+$ case, note

$$A \leq B \Rightarrow |\mu| + A \leq |\mu| + B \Rightarrow -(|\mu| + A)^{-1} \leq -(|\mu| + B)^{-1}$$

while for the $-$ case,

$$A \leq B \Rightarrow |\mu| - B \leq |\mu| - A \Rightarrow (|\mu| - A)^{-1} \leq (|\mu| - B)^{-1}$$

□

For the analogous half of Theorem 1.8, we need the following preliminary:

Lemma 3.13 *Let A and B be $n \times n$ matrices with $A \leq B$. Let $(a_1, b_1), \ldots, (a_\ell, b_\ell)$ be distinct open intervals (say, $b_j < a_{j+1} < b_{j+1}$) and $k_1, \ldots, k_\ell$ positive integers with $\sum_{j=1}^{\ell} k_j = n$ and*

$$\#\ \textit{eigenvalues of } A \textit{ in } (a_j, b_j) = k_j \tag{3.74}$$

$$\#\ \textit{eigenvalues of } B \textit{ in } (a_j, b_j) = k_j \tag{3.75}$$

For $\theta \in [0, 1]$, define

$$A(\theta) = (1-\theta)A + \theta B \tag{3.76}$$

Then, for every $\theta \in [0, 1]$,

$$\#\ \textit{eigenvalues of } A(\theta) \textit{ in } (a_j, b_j) = k_j \tag{3.77}$$

Proof Let $e_1(\theta) \le e_2(\theta) \le \cdots \le e_n(\theta)$ be the eigenvalues of $A(\theta)$. They are continuous in θ (by Corollary 5.29 below) and monotone since $A(\theta_1) \ge A(\theta_2)$ if $\theta_1 \ge \theta_2$ (since $(B - A) \ge 0$).

For $i = 1, \dots, k_1$, $e_i(0), e_i(1) \in (a_1, b_1)$ by (3.74)/(3.75), so $e_i(\theta) \in [e_i(0), e_i(1)] \subset (a_1, b_1)$. For $i = k_1 + 1, \dots, k_1 + k_2$, $e_i(0), e_i(1) \in (a_2, b_2)$, etc. □

Theorem 3.14 *Let* A, B *be* $n \times n$ *matrices obeying* (3.74)/(3.75). *Let* f *have the form* (3.33) *where*

$$\mathrm{supp}(d\mu) \subset \mathbb{R} \setminus \bigcup_{j=1}^{k} (a_j, b_j) \tag{3.78}$$

Then

$$A \le B \Rightarrow f(A) \le f(B) \tag{3.79}$$

Proof Suppose first $\int (1 + x^2)\, d\mu < \infty$. From (3.35),

$$\frac{1 + xA(\theta_1)}{x - A(\theta_1)} - \frac{1 + xA(\theta_2)}{x - A(\theta_2)} = (1+x^2)(x - A(\theta_1))^{-1}(\theta_2 - \theta_1)(B - A)(x - A(\theta_2))^{-1}$$

Thus, by the lemma, $f(A(\theta))$ is differentiable and $\frac{d}{d\theta} f(A(\theta)) \ge 0$ since $(x - A(\theta))^{-1}(B - A)(x - A(\theta))^{-1} \ge 0$. For general μ, the result follows by a limiting argument. □

Notes and Historical Remarks The representation (3.8) for harmonic functions is universally known as the Poisson representation. What we have called the complex Poisson representation, (3.1), is sometimes called the Cauchy representation. Poisson did have an analog for harmonic functions on the ball in $\mathbb{R}^3$, but it was Fatou [95] who had (3.11) and (3.8). I would have preferred calling them the Fatou representations but, as noted, it has become standard to use Poisson's name.

The Herglotz representation on $\mathbb{D}$ is due to Herglotz [153] and Riesz [283]. Their work was one of the explosion of papers [54, 55, 96, 98, 283, 284, 308, 309, 345] that followed Carathéodory's initial paper [53] on analytic functions with positive real part on $\mathbb{D}$. This is responsible for the name "Carathéodory function."

One of my bugaboos is that many discussions of the Herglotz representation use compactness (a.k.a. Helly's theorem) to prove the existence of a weak limit (3.32). In general, if a limit point is unique, there is likely to be a direct way to prove convergence, and in this case, Lemma 3.3 provides that. It is inelegant to appeal to compactness when such a simple direct argument is available.

Extensions to analytic functions on $\mathbb{C}_+$ with $\mathrm{Im}\, f > 0$ are due to Nevanlinna [241] who used them to study the moment problem (see Simon [322] or [329, Section 7.7]) and in his work on analytic functions.

Chapter 4
Monotonicity of the Square Root

As explained in the preface, most applications of Loewner's theorem involve the easy half of the theorem. This chapter is an aside involving the two most significant and common matrix monotone functions: fractional powers and the log. We begin with:

Theorem 4.1 *Let* $s \in \mathbb{R}$. *On* $(0, \infty)$, *let*

$$f_s(x) = x^s \tag{4.1}$$

Then $f_s \in \mathcal{M}_\infty(0, \infty)$ *if and only if* $0 \leq s \leq 1$. *The function* ℓ *on* $(0, \infty)$ *given by*

$$\ell(x) = \log(x) \tag{4.2}$$

lies in $\mathcal{M}_\infty(0, \infty)$.

Remarks

1. Since $y \mapsto -y^{-1}$ is matrix monotone (Proposition 2.2), this implies that $x \mapsto -x^s$ is matrix monotone if and only if $-1 \leq \text{s} \leq 0$.
2. Our official proof will rely on the easy half of Loewner's theorem, i.e. a Herglotz representation, but given the historical importance, we give five other proofs!
3. For $s \notin [0, 1]$, not only is $f_s \notin \mathcal{M}_\infty(0, \infty)$, but, either by looking at $\det(B_2)$ and using Theorem 5.18 or by Theorem 14.1, $f_s \notin \mathcal{M}_2(a, b)$ for any $(a, b) \subset (0, \infty)$.

Proof f_s has an analytic continuation to $\mathbb{C}\backslash(-\infty, 0]$,

$$f_s(z) = z^s$$

that is, if $z = re^{i\varphi}$, then

$$f_s(re^{i\varphi}) = r^s e^{i\varphi s}$$

B. Simon, *Loewner's Theorem on Monotone Matrix Functions*, Grundlehren der mathematischen Wissenschaften 354, https://doi.org/10.1007/978-3-030-22422-6_4

Thus,

$$\operatorname{Im} f_s(re^{i\varphi}) = r^s \sin(\varphi s) \tag{4.3}$$

This obeys (1.14) if and only if $0 \le s \le 1$. Thus, Theorem 1.3 implies the first assertion. Similarly,

$$\operatorname{Im} \ell(re^{i\varphi}) = \varphi \tag{4.4}$$

obeys (1.14) and so, by Theorem 1.3, $\ell \in \mathcal{M}_\infty(0, \infty)$. □

Basically, the reason x^s and $\log(x)$ are monotone is because of the integral representations

$$A^s = \frac{\sin \pi s}{\pi} \left[\int_0^\infty w^{s-1} A(A+w)^{-1}\, dw \right] \tag{4.5}$$

$$\log A = \int_0^\infty \frac{xA-1}{A+x} \frac{dx}{1+x^2} \tag{4.6}$$

Example 4.2 The function $f(z) = \pi \tan(\pi x)$ on $(-1/2, 1/2)$ is in $\mathcal{M}_\infty(-\frac{1}{2}, \frac{1}{2})$ for one has the formula (see [326, Problem 9.2.3])

$$f(z) = \sum_{n=0}^\infty \frac{2z}{(n+\frac{1}{2})^2 - z^2} \tag{4.7}$$

Since

$$\begin{aligned} \operatorname{Im}\left(\frac{z}{a^2 - z^2}\right) &= \frac{\operatorname{Im}(a^2 z - |z|^2 \bar{z})}{|a^2 - z^2|^2} \\ &= \frac{a^2 + |z|^2}{|a^2 - z^2|^2} \operatorname{Im}(z) \end{aligned} \tag{4.8}$$

$\tan(\pi z)$ has a positive imaginary part in the upper half plane. In essence, (4.7) is an "integral" representation for $\tan(\pi x)$ where the measure is a positive point measure. □

For historical reasons, we want to describe four other ways of proving matrix monotonicity of x^s: in three cases for $s = 1/2^k$ and in the other case for $0 < s \le 1$. Since

$$\log(x) = \lim_{s \downarrow 0} \frac{x^s - 1}{s} \tag{4.9}$$

these x^s results immediately imply matrix monotonicity of $\log(x)$.

Lemma 4.3 *Let C and D be arbitrary $n \times n$ matrices. Then*

(i) *We have that*

$$spec(CD) = spec(DC) \tag{4.10}$$

(ii) *If CD is self-adjoint, then*

$$\|CD\| \leq \|DC\| \tag{4.11}$$

Proof

(i) Suppose C is invertible. Since

$$CD - z = C[DC - z]C^{-1} \tag{4.12}$$

we have that

$$\det(CD - z) = \det(DC - z) \tag{4.13}$$

Replacing C by $C + \varepsilon \mathbf{1}$ and taking $\varepsilon \to 0$, we see (4.13) holds even if C is not invertible. Since $\text{spec}(E) = \{\text{zeros of } \det(E - z)\}$, (4.13) implies (4.10).

(ii) If E is self-adjoint,

$$\|E\| = \sup\{|\lambda| \mid \lambda \in \text{spec}(E)\} \tag{4.14}$$

while, in general,

$$\sup\{|\lambda| \mid \lambda \in \text{spec}(E)\} \leq \|E\| \tag{4.15}$$

Since (4.10) says

$$\{|\lambda| \mid \lambda \in \text{spec}(CD)\} = \{|\lambda| \mid \lambda \in \text{spec}(DC)\} \tag{4.16}$$

(4.10) implies (4.11).

□

Here is the second proof of matrix monotonicity of $\sqrt{x}$:

Theorem 4.4 $0 < A \leq B \Rightarrow \sqrt{A} \leq \sqrt{B}$

Proof By Lemma 2.3,

$$0 < A \leq B \Rightarrow \|A^{1/2}B^{-1/2}\| \leq 1$$

$$\Rightarrow \|B^{-1/4}A^{1/2}B^{-1/4}\| \leq 1 \tag{4.17}$$

$$\Rightarrow \|A^{1/4}B^{-1/4}\| \leq 1 \tag{4.18}$$

$$\Rightarrow A^{1/2} \leq B^{1/2} \tag{4.19}$$

Here (4.17) is (4.11) with $C = B^{-1/4}$, $D = A^{1/2}B^{-1/4}$ and (4.18) comes from $\|E^*E\| = \|E\|^2$. Finally, (4.19) comes from (2.8). □

By induction, $A^{1/2^k} \leq B^{1/2^k} \Rightarrow A^{1/2^{k+1}} \leq B^{1/2^{k+1}}$, so we get matrix monotonicity of $x^{1/2^k}$. Here is the third proof:

Theorem 4.5 *If $0 < C, D$, then for all $s \in (0, 1)$,*

$$\|CD^{-1}\| \leq 1 \Rightarrow \|C^s D^{-s}\| \leq 1 \tag{4.20}$$

Remark By (2.8) with $C = A^{1/2}$, $D = B^{1/2}$, we see (4.20) shows that for $0 < s \leq 1$, $x \mapsto x^s$, lies in $\mathcal{M}_\infty(0, \infty)$.

Proof For $z \in \mathbb{C}$, define $f(z) \equiv C^z D^{-z}$. Since C^{iy} and D^{-iy} are unitary for all real y,

$$\|f(x+iy)\| = \|f(x)\| \tag{4.21}$$

Thus, $\|f(z)\|$ is bounded in $\{z \mid 0 \leq \operatorname{Re} z \leq 1\}$. By the maximum principle,

$$\sup_{0 \leq \operatorname{Re} z \leq 1} \|f(z)\| = \sup_{\substack{\operatorname{Re} z = 0 \\ \text{or } \operatorname{Re} z = 1}} \|f(z)\| \tag{4.22}$$

By (4.21),

$$\text{LHS of (4.22)} = \max \|f(0)\|, \ \|f(1)\| = 1$$

since $f(0) = \mathbf{1}$ while, by hypothesis, $\|f(1)\| \leq 1$. □

$f(z)$ is bounded in the strip, so $f_\varepsilon(z) \equiv e^{\varepsilon z^2} f(z)$ is bounded in the strip and goes to zero as $|y| \to \infty$. Equation (4.22) follows from applying the usual maximum principle to f_ε and taking $\varepsilon \downarrow 0$.

The fourth proof depends on

Lemma 4.6 *Let C be a strictly positive Hermitian matrix and D a Hermitian matrix. If $T \equiv CD + DC$ is positive, so is D.*

Proof Shift to a basis in which D is diagonal. Then $0 \leq T_{ii} = 2D_{ii}C_{ii}$. Since $C_{ii} > 0$, we see that D is a diagonal matrix with positive elements, so a positive matrix. □

Here is the fourth proof

Fourth Proof of Theorem 4.4 By adding $\epsilon\mathbf{1}$ to A and B, we can suppose both are strictly positive. Let $C = \sqrt{B} + \sqrt{A}$ and $D = \sqrt{B} - \sqrt{A}$. Then $T = CD + DC = 2(B - A) \geq 0$. Since C is strictly positive, the lemma implies that $D \geq 0$. □

Here is the fifth proof (which can be viewed as a variant of the fourth):

Fifth Proof of Theorem 4.4 Suppose $0 < A \leq B$ but that $S \equiv \sqrt{A} \leq T \equiv \sqrt{B}$ is false. Then let P be the projection onto an eigenvector for $S - T$ with eigenvalue $e > 0$ so $P(S - T) = P(S - T) = eP$. Thus, $\text{Tr}(P(S - T)(S + T)) = e\text{Tr}(P(S + T)P) \geq 0$.

On the other hand, by cyclicity of the trace (and $[P, S - T] = 0$)

$$\text{Tr}(P(S - T)(S + T)) = \text{Tr}((S + T)(S - T)P) = \text{Tr}(P(S + T)(S - T))$$

so $2\text{Tr}(P(S - T)(S + T)) = \text{Tr}(P(S^2 - T^2)) = \text{Tr}(P(A - B)P) \leq 0$. It follows that $\text{Tr}(P(S + T)P) = 0$ so $\text{Tr}(PSP) = \text{Tr}(PTP) = 0$ so $\text{Tr}(P(S - T)P) = 0$ so $e = 0$ contrary to hypothesis. This contradiction establishes the result. □

Here is the sixth proof (which is clearly related to the last two proofs)

Sixth Proof of Theorem 4.4 Let $X = B - A$. By taking limits we can suppose that A and X are strictly positive. Define for $t \geq 0$

$$F(t) = \sqrt{A + tX} \tag{4.23}$$

Clearly, it suffices to prove that $F'(t)$ is positive, equivalently, if $F'(t)\varphi = \lambda\varphi$ that $\lambda > 0$. Differentiating $F(t)F(t) = A + tX$, we see that

$$F'(t)F(t) + F(t)F'(t) = X$$

Put this equation in $\langle \varphi, \cdot\, \varphi \rangle$ to see that

$$2\lambda\langle \varphi, F(t)\varphi \rangle = \langle \varphi, X\varphi \rangle$$

Since $F(t)$ and X are strictly positive, so is λ. □

Having seen five proofs of the matrix monotonicity of $\log(x)$, we give an application:

Theorem 4.7 (Segal's Lemma) *Let A and B be finite self-adjoint matrices. Then*

$$\|e^{A+B}\| \leq \|e^A e^B\| \tag{4.24}$$

Proof By adding a constant to B, we can suppose

$$\|e^A e^B\| = 1 \tag{4.25}$$

By (2.8),

$$\begin{aligned} (4.25) &\Rightarrow e^{2A} \leq e^{-2B} \\ &\Rightarrow 2A \leq -2B \end{aligned} \tag{4.26}$$

$$\begin{aligned} &\Rightarrow A + B \leq 0 \\ &\Rightarrow \text{spec}(A + B) \subset (-\infty, 0] \\ &\Rightarrow \text{spec}(e^{A+B}) \subset (0, 1] \\ &\Rightarrow \|e^{A+B}\| \leq 1 \end{aligned} \tag{4.27}$$

where (4.26) uses the matrix monotonicity of $\log(x)$, and (4.27) the spectral mapping theorem. □

Segal's lemma is important in the study of models of quantum field theory; see the Notes.

That f_s of (4.1) is matrix monotone on operators has been called by some the *Loewner–Heinz* or *Heinz inequality* (even though Heinz's rediscovery followed Loewner's work by 17 years)! There are a number of variants that have been given names and there are papers proving the equivalence to each other and to the Loewner–Heinz inequality. One is the *Cordes inequality* that

$$0 \leq A, B; \quad 0 \leq s \leq 1 \Rightarrow \|A^s B^s\| \leq \|AB\|^s \tag{4.28}$$

which is just (4.20) if $A = C$ and $B = D^{-1}$. Another is the *Heinz–Kato* inequality: given A, B positive and T another operator, if for all φ, ψ, we have that $\|T\varphi\| \leq \|A\varphi\| \quad \|T^*\psi\| \leq \|B\psi\|$ then

$$|\langle\psi, T\varphi\rangle| \leq \|A^s\varphi\|\|B^{1-s}\psi\| \tag{4.29}$$

for all $0 \leq s \leq 1$.

To prove (4.29), we use the singular value decomposition (see [329, Section 3.5]) which in the finite dimensional case says that if T is an n-dimension operator, there are orthonormal bases $\{\kappa_j\}_{j=1}^n$ and $\{\eta_j\}_{j=1}^n$ and non-negative reals $\{e_j\}_{j=1}^n$ so that $|T|\kappa_j = e_j\kappa_j$ and $|T^*|\eta_j = e_j\eta_j$ and so that if U is the unitary with $U\kappa_j = \eta_j$, we have the polar decomposition $T = U|T| = |T^*|U$. This implies that $U|T|U^* = |T^*|$ so that for any positive t, $U|T|^tU^* = |T^*|^t$ which implies that

$$|T|^tU^* = U^*|T^*|^t \tag{4.30}$$

It follows that

$$\begin{aligned} \langle\psi, T\varphi\rangle &= \langle\psi, U|T|^{1-s}|T|^s\varphi\rangle = \langle|T|^{1-s}U^*\psi, |T|^s\varphi\rangle \\ &= \langle U^*|T^*|^{1-s}\psi, |T|^s\varphi\rangle \end{aligned} \tag{4.31}$$

using (4.30) with $t = 1 - s$. Thus,

$$|\langle\psi, T\varphi\rangle| \leq \||T^*|^{1-s}\psi\|\||T|^s\varphi\| \tag{4.32}$$

With these preliminaries, we can prove the Heinz–Kato inequality (4.29). For the left side of (4.29) and the Loewner–Heinz inequality imply that for $0 \le s, t \le 1$ and all φ, ψ, we have that $\| |T|^s \varphi \| \le \| A^s \varphi \|$ and $\| |T^*|^t \psi \| \le \| B^t \psi \|$ which, with (4.32), implies the right side of (4.29). This proves the Heinz–Kato inequality.

Another equivalent form is the *Chan–Kwong inequality*: If we have four positive matrices and if A commutes with C and B commutes with D then $A \ge B$ and $C \ge D \Rightarrow A^{1/2}C^{1/2} \ge B^{1/2}D^{1/2}$. To see this we first note we can suppose all the operators are strictly positive since we can add $\epsilon \mathbf{1}$ and take $\epsilon \downarrow 0$. Assuming that, let $X \equiv A^{1/2}C^{1/2}$ and $Y \equiv B^{1/2}D^{1/2}$. Then $XA^{-1}X = C \ge D = YB^{-1}Y \ge YA^{-1}Y$ since $A \ge B \Rightarrow B^{-1} \ge A^{-1}$. Multiplying on both sides by $A^{-1/2}$ we see that $(A^{-1/2}XA^{-1/2})^2 \ge (A^{-1/2}YA^{-1/2})^2$ so, by the monotonicity of square root, $A^{-1/2}XA^{-1/2} \ge A^{-1/2}YA^{-1/2}$. Multiplying on both sides by $A^{1/2}$, we get that $X \ge Y$, the conclusion of the Chan–Kwong inequality.

Taking $C = D = \mathbf{1}$, we recover the monotonicity of the square root. But we get even more: taking $A = X^s$, $C = X^t$, $B = Y^s$, $D = Y^t$, the Chan–Kwong inequality implies that the set of s for which $x \mapsto x^s$ is matrix monotone is closed under averages so it contains all dyadic rationals in $[0, 1]$ and so, by a continuity argument all of $[0, 1]$. Since the above proof of the Chan–Kwong inequality only needed monotonicity of the square root, we see one again a way to go from monotonicity of the square root to all fractional powers.

As a final equivalent inequality we mention (there are still a few others in the Notes), we consider the *Fujii–Furuta inequality*: for any pair of positive operators, C and D, one has that $\|CDC\| \le \|C^2D\|$. If $C = A^{1/2}$ and $D = B$, we note that $\|CDC\| = \|A^{1/2}B^{1/2}\|$ and the inequality is just (4.28) for $s = 1/2$. The point in singling this out is that Fujii–Furuta [107] have a compact discussion of the equivalences centered on this form. Moreover, there is a vast generalization due to McIntosh (see the Notes):

$$A, B \ge 0,\ X \text{ arbitrary } n \times n \text{ matrices} \Rightarrow \|A^s X B^{1-s}\| \le \|AX\|^s \|XB\|^{1-s} \tag{4.33}$$

for $0 \le s \le 1$. The special case $A = B = C^2$, $D = X$, $s = 1/2$ is the Fujii–Furuta inequality.

Notes and Historical Remarks Despite the fact that the easy half of Loewner [209] implies monotonicity of square root, the work was so little known in the early 1950s that Heinz [151] made a splash in 1951 proving this in the context of certain PDE bounds. The proof given after our statement of Theorem 4.4 is essentially due to Kato [176]. Kato noted (and Pedersen [262] rediscovered) that the same idea proves that if $0 \le A^\gamma \le B^\gamma$ for $\gamma = \alpha$ and $\gamma = \beta$, then it is true for $\gamma = \frac{1}{2}(\alpha + \beta)$ which proves that $x \mapsto x^s$ is matrix monotone for $0 \le s \le 1$.

The fourth proof is borrowed from Bhatia [34] who notes that the arguments are variants of those in Lax [198]. The fifth proof is from Ogawasara [247] who also proved that a C^*-algebra in which $0 \le A \le B \Rightarrow A^2 \le B^2$ is abelian (see Example 1.2). The sixth proof appears in Hiai–Petz [157].

$AB + BA$ is sometimes called the *symmetric product* and $\frac{1}{2}(AB + BA)$ is called the *Jordan product* after the physicist, Pascual Jordan, who used it in quantum theory (not the mathematician, Camille Jordan, of the Jordan normal form and Jordan curve theorem). Moslehian–Najafi [226] have proven an interesting connection between these products and matrix monotone functions. If $A, B \geq 0$, then their symmetric product is positive if and only if for every positive matrix monotone function, f on $(0, \infty)$, one has that

$$f(A + B) \leq f(A) + f(B)$$

In particular, this implies for such f, $0 < p \leq 1/2$ and $0 \leq A \leq B$, we have that $f(B^p) \leq f((B^p + A^p)/2) + f((B^p - A^p)/2)$.

Segal's lemma is due to Segal [312]; that it follows from matrix monotonicity of log is a remark of Simon and Høegh-Krohn [330]. It is an abstraction of an argument of Nelson [237]. For analogs concerning traces (Golden–Thompson inequalities), see Simon [320].

Equation (4.10) is an extremely useful result as discussed, for example, by Deift [74]. It extends to the Hilbert space case in a slightly different form

$$\text{spec}(CD)\setminus\{0\} = \text{spec}(DC)\setminus\{0\} \tag{4.34}$$

Removing 0 is necessary, for if $\{\varphi_n\}$ is an orthonormal basis and $C\varphi_n = \varphi_{n+1}$, then $0 \in \text{spec}(CC^*)$, but $0 \notin \text{spec}(C^*C)$. Equation (4.34) follows from

$$(CD - \lambda)^{-1} = -\lambda^{-1} + \lambda^{-1}C(DC - \lambda)^{-1}D$$

The Heinz–Kato inequality is from Kato [176], the Cordes inequality appeared in Cordes [64] (and independently in Furuta [115]), and the Chan–Kwong inequality in Chan–Kwong [57]. The McIntosh inequality was unpublished but appeared without proof in an often quoted technical report that does not seem to be available online. A generalization to general unitarily invariant norms (see Chapter 42) with a proof can be found in Kittaneh [184] (see also [334]). Papers that discuss equivalent forms of the Heinz–Loewner inequality include [104, 105, 107, 115, 118, 184, 366]. Fujii–Furuta [107] discuss some other equivalent forms such as $\|AX + XB\| \geq \|A^s X B^{1-s} + A^{1-s} X B^s\|$ (for $A, B \geq 0$ and $0 \leq s \leq 1$) and $\|A^2 B + BA^2\| \geq \|ABA\|$ (for $A \geq 0$ and $B^* = B$). Another equivalent form is the result of Furuta [115] that for $A, B \geq 0$, one has that $s \mapsto \|A^s B^s\|$ is a convex, increasing function.

Furuta [113] has proven a generalization of the matrix monotonicity of x^s; $0 \leq s \leq 1$—namely if $0 \leq A \leq B$, then $(B^r A^p B^r)^{1/q} \leq (B^r B^p B^r)^{1/q}$ for $r, p, q > 0$ obeying $(1 + 2r)q \geq p + 2r$ (the x^s result is $q = 1, r = 0, p = s$). For further literature on this inequality, see [106, 108–112, 114, 174, 367].

For other direct constructions of matrix monotone functions, (i.e., functions in $\mathcal{M}_\infty(0, \infty)$), see Furuta [117], Uchiyama [348, 349], and Uchiyama–Hasumi [352]. For a general discussion of matrix inequalities, see the book of Furuta [116].

An operator, perhaps unbounded, on a Hilbert space, $\mathcal{H}$, is called maximal accretive if $\mathrm{Re}\langle\varphi, A\varphi\rangle \geq 0$ for all $\varphi \in D(A)$ and $\mathrm{Ran}(A+1) = \mathcal{H}$ (it is known [280, Theorem X.48] this is equivalent to being the generator of a contraction semigroup). Kato [177] has a generalization of the Heinz–Loewner inequality: If A and B are maximal accretive operators on a Hilbert space with $D(A) \subset D(B)$ so that $\|B\varphi\| \leq \|A\varphi\|$ for all $\varphi \in D(A)$, then for $0 \leq s \leq 1$ and all $\varphi \in D(A^s)$, one has that $\|B^s\varphi\| \leq \exp(\pi^2 s(1-s)/2)\|A^s\varphi\|$ (it is known how to define fractional powers of maximal accretive operators in a natural way).

In this chapter, we proved directly that several functions ($x \mapsto x^\alpha$; $0 < \alpha < 1$ and $x \mapsto \log(x)$) are matrix monotone and later we'll discuss others. We should mention there is a literature on still other functions, many of use in statistical physics and/or quantum information theory. Some proofs are direct and others prove the applicability of Loewner's theorem. For example, Szabó [337] proved that the function

$$f(x) = \frac{(x-1)^2}{(x^p-1)(x^{1-p}-1)} \tag{4.35}$$

lies in $\mathcal{M}_\infty(0,\infty)$ if $0 < p < 1$. (It is matrix monotone decreasing if $p \in (-1,0) \cup (1,2)$). For other examples, see the book of Hiai–Petz [157] and the papers of Nagisa [228], Nagisa–Wada [229], Nakamura [231], and Uchiyama [351].

Chapter 5
Loewner Matrices

In this chapter, we will reduce matrix monotonicity to the positivity of certain matrices and determinants. Seven of the eleven proofs we'll give of the hard part of Loewner's theorem rely on this reduction. It is a long chapter in part because we'll discuss two variants of the basic Loewner matrix (multipoint and extended Loewner matrices) and a local variant (Dobsch matrix) as well as analogs that will be needed when we study matrix convex functions. We'll also present tools including eigenvalue perturbation theory, the Daleckiĭ–Krein formula and divided differences that will be useful later. Finally, we'll prove an important smoothness result that any $f \in \mathcal{M}_n(a, b)$ is a C^{2n-3} function on (a, b).

Suppose f is C^1 on (a, b) and $a < x_1 < x_2 < \cdots < x_k < b$. The *Loewner matrix* associated to these points is the $k \times k$ matrix,

$$(L_k(x_1, \ldots, x_k))_{ij} = \begin{cases} \frac{f(x_i)-f(x_j)}{x_i-x_j} & i \neq j \\ f'(x_i) & i = j \end{cases} \tag{5.1}$$

This formula suggests it is useful to define $[x_i, x_j; f]$ by the right side of (5.1). Here is the key reduction:

Theorem 5.1 (Loewner) *Let f be a C^1 function on (a, b). Then $f \in \mathcal{M}_n(a, b)$ if and only if for all $a < x_1 < \cdots < x_n < b$, we have*

$$L_n(x_1, \ldots, x_n) \geq 0 \tag{5.2}$$

We'll give two proofs of this basic result. The first will involve an explicit formula for $\frac{df(A+\lambda B)}{d\lambda}$ and the second, close to Loewner's proof will use the theory of rank one perturbations. It is not really a restriction to suppose f is C^1. For we will also prove in this chapter that:

Theorem 5.2 *Any $f \in \mathcal{M}_n(a, b)$ is C^{2n-3}. In particular, any f in $\mathcal{M}_n(a, b)$ for some $n \geq 2$ is C^1.*

B. Simon, *Loewner's Theorem on Monotone Matrix Functions*, Grundlehren der mathematischen Wissenschaften 354, https://doi.org/10.1007/978-3-030-22422-6_5

We will see in Chapter 7, that there exist f in $\mathcal{M}_n(-1,1)$ where f is not classically C^{2n-2}.

Not only does Theorem 5.2 illuminate Theorem 5.1, it turns out that Theorem 5.1 will be the key to proving Theorem 5.2. Our proof of Theorem 5.2 will involve an explicit formula for $\frac{d}{d\lambda} f(A+\lambda C)\big|_{\lambda=0}$ when A is the diagonal matrix

$$A = \begin{pmatrix} x_1 & & 0 \\ & \ddots & \\ 0 & & x_n \end{pmatrix} \tag{5.3}$$

To state and exploit this theorem, it will be convenient to have

Definition Let X, Y be two n-dimensional matrices. Their *Schur product* $X \odot Y$ is defined by

$$(X \odot Y)_{mk} = X_{mk} Y_{mk} \tag{5.4}$$

Lemma 5.3 *The Schur product of two positive matrices is positive.*

Proof For any vector φ, let $Q^{(\varphi)}$ be the matrix

$$Q^{(\varphi)}_{mk} = \bar{\varphi}_m \varphi_k \tag{5.5}$$

which is clearly positive. Since any positive matrix has an orthonormal basis of eigenvectors with non-negative eigenvalues, any positive X is a sum of $Q^{(\varphi)}$'s. Thus, it suffices to prove the result for $X = Q^{(\varphi)}$, $Y = Q^{(\psi)}$. Since

$$Q^{(\varphi)} \odot Q^{(\psi)} = Q^{(\varphi \odot \psi)}$$

where $(\varphi \odot \psi)_m = \varphi_m \psi_m$, the special case is obvious. □

As a second preliminary, it will be useful to have two approximation theorems: One is a simple extension of the Weierstrass approximation theorem [325, Theorem 2.4.2]. The other is specific to matrix monotone functions.

Proposition 5.4

(a) *Let f be a continuous function on some bounded open interval (a, b). Then there exist polynomials P_m so that for any $[c, d] \subset (a, b)$,*

$$\sup_{c \le x \le d} |f(x) - P_m(x)| \to 0 \tag{5.6}$$

as $m \to \infty$.

Moreover, if f is C^k, then for $\ell = 0, 1, \dots, k$,

$$\sup_{c \le x \le d} \left| \frac{d^\ell}{dx^\ell} f - \frac{d^\ell}{dx^\ell} P_m \right| \to 0 \tag{5.7}$$

as $m \to \infty$.

(b) *Let (a, b) be a finite interval and $f \in \mathcal{M}_n(a, b)$ be continuous. Then there exist C^∞ functions f_m on $(a + \frac{1}{m}, b - \frac{1}{m})$ so that $f_m \in \mathcal{M}_n(a + \frac{1}{m}, b - \frac{1}{m})$ and* (5.6) *holds with P_m replaced by f_m. If f is C^k,* (5.7) *holds with P_m replaced by f_m. If f is assumed matrix monotone, but not a priori continuous, then f_m exists in $\mathcal{M}_n(a + \frac{1}{m}, b - \frac{1}{m})$ so $\sup_{c \le x \le d; m} |f_m(x)| < \infty$ and $f_m(x) \to f(x)$ for almost all x.*

Proof

(a) If we prove the result for closed intervals, $[c, d]$, uniformly on $[c, d]$, it follows for open intervals (a, b) by a two-step approximation. By scaling, we can take $[c, d] = [0, 1]$, so suppose f is continuous on $[0, 1]$.

Define the *Bernstein polynomials* $B_m(x)$ by

$$B_m(x) = \sum_{j=0}^{m} \binom{m}{j} x^j (1 - x)^{m-j} f\left(\frac{j}{m}\right) \tag{5.8}$$

Introduce the shorthand

$$\mathbb{E}_{m,x}(Q(j, x)) = \sum_{j=0}^{m} \binom{m}{j} x^j (1 - x)^{m-j} Q(j, x) \tag{5.9}$$

$\mathbb{E}$ stands for expectation since

$$\mathbb{E}_{m,x}(1) = 1; \qquad Q \ge 0 \Rightarrow \mathbb{E}_{m,x}(Q(j, x)) \ge 0$$

by the binomial theorem (see also the Notes). We have that

$$\frac{d^\ell}{da^\ell} (1 + a)^m = \frac{d^\ell}{da^\ell} \left[\sum_{j=0}^{m} \binom{m}{j} (x + a)^j (1 - x)^{m-j} \right]$$

so evaluating both sides at $a = 0$, we get

$$\mathbb{E}_{m,x}(j(j - 1) \dots (j - \ell + 1)) = m(m - 1) \dots (m - \ell + 1)x^\ell \tag{5.10}$$

In particular,

$$\mathbb{E}_{m,x}(j) = mx \qquad \mathbb{E}_{m,x}(j(j - 1) + j) = m(m - 1)x^2 + mx \tag{5.11}$$

so

$$\mathbb{E}_{m,x}\left(\left(x-\frac{j}{m}\right)^2\right) = x^2 + \left(1-\frac{1}{m}\right)x^2 + \frac{x}{m} - 2x^2$$
$$= \frac{x(1-x)}{m} \tag{5.12}$$

Thus,

$$|B_m(x) - f(x)| = \left|\mathbb{E}_{m,x}\left(f\left(\frac{j}{m}\right) - f(x)\right)\right|$$
$$\leq 2\sup_x |f(x)|\, \mathbb{E}_{m,x}(\chi_{\{j||x-\frac{j}{m}|>\delta\}}) + \sup_{|x-y|\leq\delta} |f(x)-f(y)|$$
$$\leq 2\delta^{-2}\frac{x(1-x)}{m}\sup_x |f(x)| + \sup_{|x-y|\leq\delta}|f(x)-f(y)|$$

It follows that

$$\limsup_{m\to\infty}\sup_x |B_m(x) - f(x)| \leq \sup_{|x-y|\leq\delta} |f(x)-f(y)| \tag{5.13}$$

Taking $\delta \downarrow 0$ and using uniform continuity of f, we see $B_m \to f$ uniformly.

Now suppose f is C^k. Since

$$\frac{d}{dx}\binom{m}{j}x^j(1-x)^{m-j} = m\left[\binom{m-1}{j-1}x^{j-1}(1-x)^{m-j} - \binom{m-1}{j}x^j(1-x)^{m-1-j}\right]$$

we see that

$$\frac{d}{dx}\mathbb{E}_{m,x}(Q(j)) = m\,\mathbb{E}_{m-1,x}(Q(j+1) - Q(j)) \tag{5.14}$$

Letting δ be the operator $(\delta Q)(j) = Q(j+1) - Q(j)$, we see, by induction, that

$$\frac{d^\ell}{dx^\ell}\mathbb{E}_{m,x}(Q(j)) = \frac{m(m-1)\dots(m-\ell+1)}{m^\ell}\mathbb{E}_{m-\ell,x}\left(\frac{(\delta^\ell Q)(j)}{(\frac{1}{m})^\ell}\right) \tag{5.15}$$

Since $Q(j) \equiv f(j/m)$ and $\ell \leq k$ implies

$$\frac{(\delta^\ell Q)(j)}{(\frac{1}{m})^\ell} - \frac{d^\ell f}{dx^\ell}\left(\frac{j}{m}\right) \to 0$$

uniformly in j, m, the same argument that led to (5.14) shows that $\frac{d^\ell}{dx^\ell}B_m(x) \to \frac{d^\ell f}{dx^\ell}(x)$.

(b) Let j be an approximate identity, that is,

$$j_m(x) = \frac{1}{m} j_1\left(\frac{x}{m}\right)$$

where j_1 is supported on $(-1, 1)$, $j_1 \geq 0$, j_1 is C^∞, and $\int_{-1}^{1} j_1(x)\, dx = 1$. Let $f_m(x)$ be defined for $x \in (a + \frac{1}{m}, b - \frac{1}{m})$ by

$$f_m(x) = \int f(x - y) j_m(y)\, dy \tag{5.16}$$

f_m is C^∞ since j is, and in $\mathcal{M}_n(a + \frac{1}{m}, b - \frac{1}{m})$ since for $|y| \leq \frac{1}{m}$, $f(\cdot - y) \in \mathcal{M}_n(a + \frac{1}{m}, b - \frac{1}{m})$. By a standard argument, if f is continuous, $f_m \to f$ uniformly, and if f is C^k, $d^\ell f_m/dx^\ell \to d^\ell f/dx^\ell$. If f is monotone, $f_m(x) \to f(x)$ at points of continuity for f which include a.e. x.

□

Remark One can also prove (a) by using convolution (first with scaled Gaussians and then approximating the Gaussians with polynomials), but the Bernstein polynomials have a more explicit feel.

Lemma 5.5 *Let f_k, f_∞ be C^1 functions on (a, b) so that $f_k \to f_\infty$ and $df_k/dx \to df_\infty/dx$ uniformly on subintervals $[c, d] \subset (a, b)$. Then*

(a) *If A and C are self-adjoint matrices and A has distinct eigenvalues in (a, b), then for λ near 0, $f_k(A + \lambda C)$ is differentiable in λ and*

$$\lim_{k\to\infty} \left.\frac{df_k(A + \lambda C)}{d\lambda}\right|_{\lambda=0} = \left.\frac{df_\infty(A + \lambda C)}{d\lambda}\right|_{\lambda=0} \tag{5.17}$$

(b) *If $a < x_1 < \cdots < x_m < b$, then $L_m(x_1, \ldots, x_m; f_k) \to L_m(x_1, \ldots x_m; f_\infty)$.*

Proof We see below (see Corollary 5.29) that for λ near 0, there are eigenvalues $x_1(\lambda), \ldots, x_m(\lambda)$ and one-dimensional eigenprojections $P_1(\lambda), \ldots, P_m(\lambda)$, so that

$$A + \lambda C = \sum_{j=1}^{m} x_j(\lambda) P_j(\lambda) \tag{5.18}$$

and x_j and P_j are analytic near $\lambda = 0$. Then

$$f(A + \lambda C) = \sum_{j=1}^{m} f(x_j(\lambda)) P_j(\lambda) \tag{5.19}$$

and so $f(A + \lambda C)$ is C^1 near $\lambda = 0$ and

$$\frac{df}{d\lambda}(A+\lambda C)=\sum_{j=1}^{m}\frac{df}{dx}(x_j(\lambda))\frac{dx_j}{d\lambda}P_j(\lambda)+f(x_j(\lambda))\frac{dP_j}{d\lambda} \tag{5.20}$$

This immediately implies (5.17).

(b) is immediate from the definition. □

We can now find an explicit formula for $\frac{df}{d\lambda}(A+\lambda C)$ that makes it clear where Loewner matrices enter:

Theorem 5.6 (Daleckiĭ–Krein Formula) *Let $a < x_1 < \cdots < x_m < b$ and let f be a C^1 function on (a,b). Let A be the diagonal matrix*

$$A=\begin{pmatrix} x_1 & & \\ & \ddots & \\ & & x_m \end{pmatrix} \tag{5.21}$$

and C an arbitrary $m \times m$ self-adjoint matrix. Then

$$\left.\frac{d}{d\lambda}f(A+\lambda C)\right|_{\lambda=0}=L_m(x_1,\dots,x_m;f)\odot C \tag{5.22}$$

Remark We'll provide two more proofs of (5.22) later, one after the proof of Theorem 5.13 and the other implicitly in the proof of Theorem 5.16.

Proof (Loewner!) By eigenvalue perturbation theory, for small real λ, there exists an orthonormal basis, $\{\varphi_k(\lambda)\}_{k=1}^m$, and eigenvalues, $\{x_k(\lambda)\}_{k=1}^m$, so that

$$(A+\lambda C)\varphi_k(\lambda)=x_k(\lambda)\varphi_k(\lambda) \tag{5.23}$$

$$\varphi_k(0)=\delta_k; \qquad x_k(0)=x_k \tag{5.24}$$

Thus, using $\langle\varphi_k(\lambda), f(A+\lambda C)\varphi_\ell(0)\rangle=\langle f(A+\lambda C)\varphi_k(\lambda),\varphi_\ell(0)\rangle$,

$$\langle\varphi_k(\lambda),[f(A+\lambda C)-f(A)]\varphi_\ell(0)\rangle=[f(x_k(\lambda))-f(x_\ell(0))]\langle\varphi_k(\lambda),\varphi_\ell(0)\rangle \tag{5.25}$$

so taking $f(u)=u$, we get

$$\langle\varphi_k(\lambda),\lambda C\varphi_\ell(0)\rangle=(x_k(\lambda)-x_\ell(0))\langle\varphi_k(\lambda),\varphi_\ell(0)\rangle \tag{5.26}$$

Combining these last two equations yields

$$\left\langle\varphi_k(\lambda),\frac{f(A+\lambda C)-f(A)}{\lambda}\varphi_\ell(0)\right\rangle=[x_k(\lambda),x_\ell(0);f]\langle\varphi_k(\lambda),C\varphi_\ell(0)\rangle \tag{5.27}$$

Taking λ to zero, we see that

$$\left[\frac{d}{d\lambda} f(A+\lambda C)\Big|_{\lambda=0}\right]_{k\ell} = [x_k, x_\ell; f] C_{k\ell} \tag{5.28}$$

which is (5.22) □

Remark While $f'(A)$ doesn't appear, it is not hard to show that if f is matrix monotone, then $\|\left[\frac{d}{d\lambda} f(A+\lambda C)\big|_{\lambda=0}\right]\| \leq \|f'(A)\| \|C\|$. More is true. Under the monotonicity assumption, the sup over all C with $\|C\| = 1$ is equal to $\|f'(A)\|$. This is a result of [32].

We have the tools to prove Theorem 5.1, but to state a third equivalence, we need to analyze the relationship between positivity of a matrix and of certain determinants.

Proposition 5.7 *Let A be a $d \times d$ self-adjoint matrix and $A^{(d-1)}$ the matrix obtained by dropping the last row and column. Then their eigenvalues interlace, that is, if $\lambda_1 \leq \lambda_2 \leq \cdots \leq \lambda_d$ are the eigenvalues of A and $\mu_1 \leq \mu_2 \leq \cdots \leq \mu_{d-1}$ of $A^{(d-1)}$, then*

$$\lambda_j \leq \mu_j \leq \lambda_{j+1} \tag{5.29}$$

Proof Since

$$\lambda_1 = \min_{\|\varphi\|=1} \langle \varphi, A\varphi \rangle \qquad \mu_1 = \min_{\substack{\|\varphi\|=1 \\ \varphi_d=0}} \langle \varphi, A\varphi \rangle \tag{5.30}$$

clearly, $\lambda_1 \leq \mu_1$. Let ψ_1 be a unit eigenvector for λ_1 and A. Then

$$\lambda_2 = \min_{\substack{\|\varphi\|=1 \\ \langle \varphi, \psi_1 \rangle = 0}} \langle \varphi, A\varphi \rangle \leq \min_{\substack{\|\varphi\|=1 \\ \varphi_d=0 \\ \langle \varphi, \psi_1 \rangle = 0}} \langle \varphi, A\varphi \rangle \tag{5.31}$$

There is a linear combination, φ, of the eigenvectors associated to μ_1, μ_2 orthogonal to ψ_1, and it has $\varphi_d = 0$, so RHS of (5.31) $\leq \mu_2$.

By similar arguments,

$$\lambda_j \leq \mu_j$$

Replacing *min*'s by max's, we get

$$\lambda_{j+1} \geq \mu_j$$

□

Definition Let A be a $d \times d$ matrix. Let $I \subset \{1, \ldots, d\}$. The *principal determinant*, $d_I(A)$, is the determinant of the $\#I \times \#I$ matrix $\{A_{ij}\}_{i,j \in I}$. The *main principal determinants* are $d_j(A) = d_{\{1,\ldots,j\}}(A)$ for $j = 1, 2, \ldots, d$.

Proposition 5.8 *A self-adjoint matrix A is strictly positive if and only if each main principal determinant is strictly positive. A self-adjoint matrix A is positive if and only if each principal determinant is non-negative.*

Remark $\left(\begin{smallmatrix} 0 & 0 \\ 0 & -1 \end{smallmatrix}\right)$ has all main principal determinants non-negative, but is not positive. Thus, we need all principal and not just main principal in the second sentence.

Proof If A is (strictly) positive, so is each $A^I = \{A_{ij}\}_{i,j\in I}$, so its eigenvalues are (strictly) positive, so its determinant, $d_I(A)$, is (strictly) positive.

We prove the converse in the strictly positive case inductively in $d = \dim(A)$. $d = 1$ is trivial. If the result holds for dimension $d-1$ and $d_j(A) > 0$ for all j, then $d_j(A^{(d-1)}) > 0$ for $j = 1, \dots, d-1$. So, by induction, $A^{(d-1)}$ is strictly positive and its eigenvalues obey $0 < \mu_1 \le \dots \le \mu_{d-1}$. By (5.29), $0 < \mu_1 \le \lambda_2 \le \dots \le \lambda_d$. Since $\lambda_1\lambda_2 \dots \lambda_d > 0$, we conclude that $\lambda_1 > 0$ also, so A is strictly positive.

For the general converse, let $B(\varepsilon) = A + \varepsilon\mathbf{1}$. Then, with $d = \dim(A)$,

$$\det(B(\varepsilon)) = \det(A) + \varepsilon \sum_{\#(I)=d-1} d_I(A) + \varepsilon^2 \sum_{\#(I)=d-2} d_I(A) + \dots + \varepsilon^d \ge 0 \tag{5.32}$$

The eigenvalues of $B(\varepsilon)$ are $\{\lambda_j + \varepsilon\}_{j=1}^d$ are nonzero for all sufficiently small ε, that is, $\det(B(\varepsilon)) \ne 0$, so $\det(B(\varepsilon)) > 0$. Similarly, $d_j(B(\varepsilon)) > 0$ for ε small, so by the strictly positive case, $B(\varepsilon) > 0$, and thus $\lim_{\varepsilon\downarrow 0} B(\varepsilon) = A \ge 0$. □

The following includes Theorem 5.1:

Theorem 5.9 *Let f be a real-valued function on (a, b). Then, the following are equivalent:*

(a) $f \in \mathcal{M}_n(a, b)$
(b) *For each*

$$a < x_1 < \dots < x_n < b \tag{5.33}$$

we have

$$L_n(x_1, \dots, x_n; f) \ge 0 \tag{5.34}$$

(c) *For each $\ell \le n$ and $a < x_1 < \dots < x_\ell < b$,*

$$\det(L_\ell(x_1, \dots, x_\ell; f)) \ge 0 \tag{5.35}$$

Proof (b) $\Leftrightarrow$ (c) by Proposition 5.8 and the fact that any principal determinant is $\det(L_\ell(x_{i_1} \dots x_{i_\ell}; f))$.

(a) $\Rightarrow$ (b) Pick $x_1, \dots, x_n$ obeying (5.33) and A as given by (5.21). Let $C \ge 0$. Then $A + \lambda C \ge A$ for $\lambda \ge 0$, so $f(A + \lambda C) \ge f(A)$, so $\frac{df(A+\lambda C)}{d\lambda} \ge 0$. Thus, by (5.22),

$$C \geq 0 \Rightarrow L \odot C \geq 0 \tag{5.36}$$

If C is the rank one matrix, $C_{ij} = \bar{\varphi}_i \varphi_j$,

$$\mathrm{Tr}(L \odot C) = \langle \varphi, L\varphi \rangle$$

so (5.36) implies L is positive.

(b) $\Rightarrow$ (a) By Lemma 5.3 and (5.22), if $C \geq 0$, (5.33) holds and A has the form (5.21), then

$$\left. \frac{df}{d\lambda}(A + \lambda C) \right|_{\lambda=0} \geq 0 \tag{5.37}$$

But positivity is basis independent, so (5.37) holds for any A with distinct eigenvalues (by shifting to a basis where A is diagonal). From this, we see that if $A \leq B$ and $A + \lambda(B - A)$ have distinct eigenvalues for all $\lambda \in [0, 1]$, then $f(B) \geq f(A)$.

If A has distinct eigenvalues and $B \geq A$ and $C(\lambda) = A + \lambda(B - A)$, then the eigenvalues of $C(\lambda)$ are analytic in λ in a neighborhood of $[0, 1]$ (see Corollary 5.29 and the Notes), so there exist finitely many $0 = \lambda_0 < \lambda_1 < \cdots < \lambda_\ell < \lambda_{\ell+1} = 1$ with $C(\lambda)$ having degenerate eigenvalues only at $\lambda_1, \lambda_2, \ldots, \lambda_\ell$, and perhaps at $\lambda_{\ell+1}$. By the above,

$$f(C(\lambda_j + \varepsilon)) \leq f(C(\lambda_{j+1} - \varepsilon))$$

so taking $\varepsilon \downarrow 0$, we see that

$$f(C(\lambda_j)) \leq f(C(\lambda_{j+1}))$$

and thus, $f(A) \leq f(B)$ so long as A has distinct eigenvalues.

Since any A with eigenvalues in (a, b) is a limit of such matrices, A_m, with distinct eigenvalues and $f(A_m) \leq f(A_m + (B - A))$ for m large, we see $f(A) \leq f(B)$ in general. □

The next key development from Theorem 5.1 is to prove an infinitesimal version of (5.35). In particular, we are heading towards showing that if f is C^{2n-1}, the matrix $B_n(x_0)$ with

$$(B_n(x_0; f))_{k\ell} = \frac{f^{k+\ell-1}(x_0)}{(k + \ell - 1)!} \tag{5.38}$$

is positive. We will call B the *Dobsch matrix*. It is a Hankel matrix (see [266]), i.e. its matrix elements are only a function of $i+j$. Hankel matrices are relevant to the study of the moment problem (see [325, Section 4.17]). In particular, if $c_j = \int x^j \, d\mu(x)$ for a positive measure and $f(x) = c_0 x + c_1 x^2 + \ldots c_{2n-2} x^{2n-1}$, then $B_n(0; f)$ is a positive matrix and, if the support of μ has at least n points, a strictly positive matrix(see [325, Theorem 4.17.3]).

Intuitively the Dobsch matrix will come from taking $x_0 < x_1 < \cdots < x_n$ and letting $x_{j+1} - x_j \rightarrow 0$ suitably. It will help to have the machinery of *divided differences*. We'll discuss the basics here and say a lot more about them in Chapters 22 and 23

Definition Given a continuous function f on (a, b) and $x_1, \dots, x_n$ distinct in (a, b), we define $[x_1, \dots, x_n; f]$ inductively by

$$[x_1; f] = f(x_1) \tag{5.39}$$

$$[x_1, \dots, x_\ell; f] = \frac{[x_1, \dots, x_{\ell-1}; f] - [x_2, \dots, x_\ell; f]}{x_1 - x_\ell} \tag{5.40}$$

The reader will note that the off-diagonal matrix elements of $L(x_1, \dots, x_n; f)$ are precisely $[x_i, x_j; f]$. We want to show that $[x_1, \dots, x_n; f]$ is symmetric in $x_1, \dots, x_n$ and can be extended continuously to coincident points if f is C^{n-1}, and in that case,

$$[x, \dots, x; f]_{k\text{-times}} = \frac{f^{(k-1)}(x)}{(k-1)!} \tag{5.41}$$

This suggests that $[x_1, \dots, x_{n+1}; f]$ called the nth divided difference is a kind of generalized nth derivative with the advantage that for it to be defined (for all x's unequal), we don't need any kind of a priori smoothness or even measurability hypothesis on f. For example, recall that a function $f : (a, b) \rightarrow \mathbb{R}$ is called convex if and only if for all $x, y \in (a, b)$ and $\theta \in [0, 1]$, we have that

$$f(\theta x + (1 - \theta)y) \leq \theta f(x) + (1 - \theta) f(y) \tag{5.42}$$

It is well known that if f is C^2, it is convex if and only if $f^{(2)}(x) \geq 0$ for all $x \in (a, b)$. But what if f is a priori not C^2? Writing $x_1 = x, x_2 = y$, and $x_3 = \theta x + (1 - \theta)y$, so that $\theta = (x_3 - x_2)/(x_1 - x_2)$ a little manipulation shows that

$$(5.42) \iff \forall_{x_1, x_2, x_3 \in (a,b)} [x_1, x_2, x_3; f] \geq 0 \tag{5.43}$$

Once (5.41) and the mean value theorem for divided differences (Theorem 5.15 below) are proven, this immediately provides a proof that when f is C^2, then f is convex $\iff f'' \geq 0$.

The following will be very useful

Theorem 5.10 (Genocchi–Hermite Formula) *Let $f \in C^{n-1}(a, b)$. Then for any distinct $x_1, \dots, x_n \in (a, b)$, one has that*

$$[x_1, \ldots, x_n; f] = \int_0^1 dt_1 \int_0^{t_1} dt_2 \ldots \int_0^{t_{n-2}} dt_{n-1} f^{(n-1)}(x(t_j, x_j)) \tag{5.44}$$

$$x(t_j, x_j) = (1 - t_1)x_1 + (t_1 - t_2)x_2 + \cdots + (t_{n-2} - t_{n-1})x_{n-1} + t_{n-1}x_n$$

Remarks

1. Letting

$$s_j = \begin{cases} 1 - t_1, & \text{if } j = 1 \\ t_{j-1} - t_j, & \text{if } j = 2, \ldots, n-1 \\ t_{n-1}, & \text{if } j = n \end{cases}$$

we have that $s_j \geq 0$, $\sum_{j=1}^n s_j = 1$ which are standard coordinates for the $(n-1)$ dimensional simplex, Δ_{n-1} and

$$y(s_j, x_j) \equiv x(t_j, x_j) = \sum_{j=1}^n s_j x_j \tag{5.45}$$

This is a convex combination of the x_j's so it lies in $[\min_j x_j, \max_j x_j]$ and (5.46) can be written in the symmetric form

$$[x_1, \ldots, x_n; f] = \int \underset{\Delta_{n-1}}{\ldots} \int f^{(n-1)} \left(\sum_{j=1}^n s_j x_j \right) ds_2 \ldots ds_n \tag{5.46}$$

2. Sometimes $x(t_j, x_j)$ is written

$$x(t_j, x_j) = x_1 + (x_2 - x_1)t_1 + \cdots + (x_n - x_{n-1})t_{n-1} \tag{5.47}$$

Proof We use induction on n. For $n = 1$, we have setting $x = x_1 + t(x_2 - x_1)$

$$\int_0^1 f'((1 - t_1)x_1 + t_1 x_2)\, dt_1 = (x_2 - x_1)^{-1} \int_{x_1}^{x_2} f'(x)\, dx = [x_1, x_2; f] \tag{5.48}$$

Letting

$$y_j = x_1 + (x_2 - x_1)t_1 + \cdots + (x_{n-2} - x_{n-3})t_{n-3} + (x_{n-2+j} - x_{n-2})t_{n-2}; \quad j = 1, 2$$

we have that, as in (5.48),

$$\int_0^1 f'(y_1 + (x_n - x_{n-1})t_{n-1})\, dt_{n-1} = \frac{f^{(n-2)}(y_2) - f^{(n-2)}(y_1)}{x_n - x_{n-1}} \tag{5.49}$$

Integrating $dt_1 \ldots dt_{n-2}$ and using the inductive result for $n-1$ yields

$$\text{RHS of (5.44)} = \frac{[x_1, \ldots, x_{n-2}, x_n; f] - [x_1, \ldots, x_{n-2}, x_{n-1}; f]}{x_n - x_{n-1}}$$
$$= [x_1, \ldots, x_n; f] \tag{5.50}$$

□

Corollary 5.11 *Let $f \in C^n(a, b)$. Then $[x_1, \ldots, x_n; f]$ has a continuous extension to all $(x_1, \ldots, x_n) \in \prod_{j=1}^n (a, b)$, i.e. to some coincident points. Moreover,* (5.41) *holds.*

Proof The integral in (5.44) makes sense for all such x's and is continuous. If all $x_j = x_0$, then $x(t_j, x_j) = x_0$ for all t so the integral is $f^{(n-1)}(x_0)$ times the volume of the simplex which, by doing the integral (or using symmetry), is $1/(n-1)!$ □

Corollary 5.12 *$[x_1, \ldots, x_n; f]$ is symmetric in the x_j's.*

Proof The measure $ds_2 \ldots ds_n$ on Δ_{n-1} is equal to $\prod_{j \neq k} ds_j$ for all k showing that (5.46) is invariant under permutations of the x_j's. □

There is another tool, due to Frobenius, that leads to a second proof of symmetry as well as an explicit formula for $[x_1, \ldots, x_n; f]$ in terms of the values of f at x_j, or, if some of the x_j are coincident, the derivatives of f at the coincident points. This key formula

$$[x_1, \ldots, x_n; f] = \sum_{j=1}^n \frac{f(x_j)}{\prod_{k \neq j}(x_j - x_k)} \tag{5.51}$$

can be established inductively and then used to provide another proof of (5.41), but it is easier, following Frobenius, to develop formulae for polynomial f (in fact, f analytic in a neighborhood of (a, b) will do) and then use Proposition 5.4 to prove what we want for general f.

Theorem 5.13

(i) $[x_1, \ldots, x_n; f]$ *obeys* (5.51) *and is a symmetric function of* $(x_1, \ldots, x_n)$.
(ii) *If f is C^{n-1}, then $[x_1, \ldots, x_n; f]$ has a continuous extension to $(a, b)^n$ given by*

$$[x_1, \ldots, x_n; f] = \sum_{j=1}^{\ell} \frac{1}{(m_j - 1)!} \left(\frac{d}{dx}\right)^{m_j - 1} \left[f(x) \prod_{k \neq j} (x - y_k)^{-m_k} \right] \Bigg|_{k=0} \tag{5.52}$$

if $(x_1, \dots, x_n)$ *includes* y_1 m_1*-times,* y_2 m_2*-times,* $\dots$*, where* $y_1, \dots, y_\ell$ *are distinct and* $\sum_{j=1}^{\ell} m_j = n$*. In particular, taking* $\ell = 1$*, (5.41) holds.*

Remark If $X = (x_1, \dots, x_n)$ with y_j, m_j defined as in (ii) and if $m = \max(m_j)$, then one only needs that $f \in C^{m-1}$ for $[x_1, \dots, x_n; f]$ to have a continuous extension from non-coincident points to a neighborhood of $X \in (a, b)^n$.

Proof We will exploit the fact that if

$$h_z(x) = \frac{1}{z - x} \tag{5.53}$$

then, by a direct computation,

$$[x_1, \dots, x_n; h_z] = \prod_{j=1}^{n} (z - x_j)^{-1} \tag{5.54}$$

Thus, if f is analytic in a simply connected neighborhood, N, of (a, b) (and, in particular, if f is a polynomial) and if C is a contour in N that surrounds (a, b) once in a counterclockwise sense, then

$$f(x) = \frac{1}{2\pi i} \oint_C \frac{f(z)}{z - x} \, dz \tag{5.55}$$

so (5.54) implies

$$[x_1, \dots, x_n; f] = \frac{1}{2\pi i} \oint \frac{f(z)}{\prod_{j=1}^{n} (z - x_j)} \, dz \tag{5.56}$$

For such f, (5.51) is immediate by the residue calculus and (5.56). By Proposition 5.4, for any continuous and distinct $x_1, \dots, x_n$, (5.51) holds. By the residue calculus, we also obtain (5.52) and so (5.41) for such f.

From (5.51) or (5.56), it is obvious that $[x_1, \dots, x_n; f]$ is symmetric in the x_j's for analytic f's. Fix $f \in C^\ell(a, b)$. By convoluting with Gaussian that shrink to a point, we can find entire analytic functions $\{f_m\}_{m=1}^{\infty}$ so that for $k = 0, 1, \dots, \ell$, $f_m^{(k)} \to f^{(k)}$ uniformly on each $(a + \epsilon, b - \epsilon)$. If f is merely continuous this implies symmetry in the x_j's at non-coincident x's. Moreover, if $f \in C^{n-1}$, by (5.44), we have that (5.52) for f_m implies it for f. □

Here is an alternate proof of the Daleckiĭ–Krein formula (5.22) that appears in Horn–Johnson [169] and which relies on (5.56)

Second Proof of Theorem 5.6 Suppose first that f is analytic in a simple connected neighborhood of $[a, b]$ so that (5.55) holds. Then

$$f(A+tC) = \frac{1}{2\pi i} \oint \frac{f(z)}{z-(A+tC)}\, dz \tag{5.57}$$

A simple use of the second resolvent formula shows that

$$\frac{d}{dt}[z-(A+tC)]^{-1}\bigg|_{t=0} = (z-A)^{-1}C(z-A)^{-1} \tag{5.58}$$

Thus, for f obeying (5.55)

$$\frac{d}{dt} f(A+tC)_{ij}\bigg|_{t=0} = \frac{1}{2\pi i} \oint \frac{f(z)}{(z-x_i)(z-x_j)} C_{ij}\, dz \tag{5.59}$$

$$= \begin{cases} f'(x_i)C_{ii}, & \text{if } i=j \\ [x_i, x_j; f]C_{ij}, & \text{if } i \neq j \end{cases} \tag{5.60}$$

by (5.56).

For general C^1 functions, f, we can use Lemma 5.5 and convolution with Gaussians of small support. □

One special case of (5.51) has an interesting consequence.

Theorem 5.14 (Lagrange Interpolation) *Let $x_1, \dots, x_n$ be n distinct points in $\mathbb{R}$ and let f be a function defined at the x_j's. Then there is a unique polynomial, of degree at most $n-1$ with*

$$P(x_j) = f(x_j), \qquad j = 1, \dots, n \tag{5.61}$$

Moreover, one has that

$$P(x) = cx^{n-1} + \text{ lower order}; \qquad c = [x_1, x_2, \dots, x_n; f] \tag{5.62}$$

Remarks

1. We will later (Theorem 22.2) provide another proof that relies on the invertibility of Vandermonde matrices.
2. As we'll explain in the Notes to Chapter 22, this result goes back to Newton although the explicit formula (5.63) below is due to Lagrange.
3. The formula (5.63) becomes singular as points become coincident but there is a version called Hermite interpolation that doesn't when f is sufficiently smooth. It is discussed in Chapter 22.

Proof If two polynomials obey (5.61), their difference vanishes at n points and is a polynomial of degree at most $n-1$, so identically zero. This proves uniqueness.

The explicit polynomial

$$P(x; x_1, \dots, x_n; f) = \sum_{j=1}^{n} f(x_j) \frac{\prod_{k \neq j}(x - x_k)}{\prod_{k \neq j}(x_j - x_k)} \tag{5.63}$$

is easily seen to obey (5.61) proving existence. Equation (5.62) is immediate from this formula and (5.51). □

Theorem 5.10 has another consequence:

Theorem 5.15 (Mean value theorem for divided differences) *Let f be a C^{n-1} function on $(a, b) \subset \mathbb{R}$. Let*

$$a < x_1 \leq x_2 \leq \cdots \leq x_n < b \tag{5.64}$$

Then there exists $w \in [x_1, x_n]$ so that

$$[x_1, \dots, x_n; f] = \frac{f^{(n-1)}(w)}{(n-1)!} \tag{5.65}$$

Remark Chapter 22 will have another proof using Theorem 5.14; see Theorem 22.11.

Proof By (5.44),

$$\min_{w \in [x_1, x_n]} \frac{f^{(n-1)}(w)}{(n-1)!} \leq [x_1, \dots, x_n; f] \leq \max_{w \in [x_1, x_n]} \frac{f^{(n-1)}(w)}{(n-1)!} \tag{5.66}$$

since the volume of the simplex is $1/(n-1)!$. By continuity of the derivative, there is a point in $[x_1, x_n]$ where the intermediate value is taken. □

As a first application of divided differences, we note the following analog of the Daleckiĭ–Krein formula

Theorem 5.16 *Let A, C be as in Theorem 5.6 and let $f \in C^2(a, b)$. Then*

$$\left. \frac{d^2}{d\lambda^2} f(A + \lambda C)_{ij} \right|_{\lambda=0} = 2 \sum_k [x_i, x_k, x_j; f]\, C_{ik}\, C_{kj} \tag{5.67}$$

Remarks

1. This method can also be used to give an alternate proof of Theorem 5.6 (the Daleckiĭ–Krein formula).
2. There is also a proof of this using the ideas in our second proof of the Daleckiĭ–Krein formula.

Proof By Proposition 5.4, it suffices to handle the case where f is a polynomial and so the case $f(x) = x^m$, $m = 0, 1, \dots$. In that case,

$$(A + \lambda C)^m = \sum_{\ell=0}^{m} \lambda^\ell \sum_{\substack{q_1,\dots,q_{\ell+1}\geq 0 \\ q_1+\dots+q_{\ell+1}=m-\ell}} A^{q_1} C A^{q_2} C \cdots C A^{q_{\ell+1}}$$

so

$$\frac{d^2}{d\lambda^2}(A + \lambda C)^m \bigg|_{\lambda=0} = 2 \sum_{q_1+q_2+q_3=m-2} A^{q_1} C A^{q_2} C A^{q_3}$$

and therefore

$$\frac{d^2}{d\lambda^2}[(A + \lambda C)^m]_{ij} \bigg|_{\lambda=0} = 2 \sum_{q_1+q_2+q_3=m-2,\, k} x_i^{q_1} C_{ik} x_k^{q_2} C_{kj} x_j^{q_3}$$

On the other hand,

$$[x_1, x_2; x^m] = \frac{x_1^m - x_2^m}{x_1 - x_2} = \sum_{q_1+q_2=m-1} x_1^{q_1} x_2^{q_2}$$

so, by induction

$$[x_1, x_2, x_3; x^m] = \sum_{q_1+q_2+q_3=m-2} x_1^{q_1} x_2^{q_2} x_3^{q_3}$$

proving (5.67) in case $f(x) = x^m$. □

Remark Analogously, if f is C^m, then

$$\frac{1}{m!}\left[\frac{d^m f}{d\lambda^m}(A + \lambda C)\bigg|_{\lambda=0}\right]_{j_1 j_{m+1}} = \sum_{j_2,\dots,j_m=1}^{n} C_{j_1 j_2} \dots C_{j_m j_{m+1}} [x_{j_1}, \dots, x_{j_{m+1}}; f] \tag{5.68}$$

(5.67) suggests that we single out for $x_0, x_1, \dots, x_n \in (a, b)$ the $n \times n$ *Kraus matrix*,

$$K_n(x_0; x_1, \dots, x_n; f)_{ij} = [x_0, x_i, x_j; f]; \qquad 1 \leq i, j \leq n \tag{5.69}$$

(5.67) focuses interest on K_n in cases where $x_0 \in \{x_j\}_{j=1}^n$. Notice that

$$K_n(x_0; x_1, \dots, x_n; f) = L_n(x_1, \dots, x_n; [x_0, \cdot; f]) \tag{5.70}$$

There will also be an analog of B_n. The *Hansen–Tomiyama* matrix is defined by

$$H_n(x;f)_{ij} = \frac{f^{(i+j)}(x)}{(i+j)!}; \qquad 1 \le i, j \le n \tag{5.71}$$

We now return to studying the Dobsch matrix. We will approximate the Dobsch matrix (5.38) with matrices

$$[A_n(x_1,\dots,x_n;f)]_{k\ell} = [x_1,\dots,x_k,\, x_1,\dots,x_\ell;\, f] \tag{5.72}$$

since (5.41) implies $A_n(x_1,\dots,x_n) \to B_n(x_0)$ if $x_1,\dots,x_n \to x_0$. Some authors call A_n the extended Loewner matrix, a name we reserve for another object (see (5.100)). We'll call A_n the *multipoint Loewner matrix*. In this regard, it will be useful to know a priori that (5.72) is, for suitable f, strictly positive.

Lemma 5.17 *Let $d\mu$ be a nontrivial finite measure on $\mathbb{R}\backslash(a,b)$ and define f for $x \in (a,b)$ by*

$$f(x) = \int (y-x)^{-1}\, d\mu(y) \tag{5.73}$$

Then $A_n(x_1,\dots,x_n;f)$ is strictly positive for each $x_1,\dots,x_n \in (a,b)$.

Remarks

1. A measure is nontrivial if its support is not a finite set of points.
2. Below, we will only need one such f. For $b < c < d$, we could take

$$\int_c^d (y-x)^{-1}\, dy = \log\left(\frac{d-x}{c-x}\right) \tag{5.74}$$

Proof By (5.54),

$$[A_n(x_1,\dots,x_n;f)]_{k\ell} = \int \prod_{j=1}^{k} (y-x_j)^{-1} \prod_{j=1}^{\ell} (y-x_j)^{-1}\, d\mu(y) \tag{5.75}$$

Thus,

$$\sum_{k,\ell=1}^{n} A_{k\ell}\bar{\zeta}_k \zeta_\ell = \int |Q(x_1,\dots,x_n;\zeta_1,\dots,\zeta_n;y)|^2\, d\mu(y) \tag{5.76}$$

with

$$Q(y) = \sum_{\ell=1}^{n} \zeta_\ell \prod_{j=1}^{\ell} (y-x_j)^{-1} \tag{5.77}$$

Now, Q is a rational function of y, so its zeros are a finite set if $(\zeta_1, \dots, \zeta_\ell) \neq (0, \dots, 0)$. Since $d\mu$ is nontrivial, $\int |Q|^2 \, d\mu > 0$. □

Theorem 5.18 *Let $f \in \mathcal{M}_n(a, b)$ be C^1. Then for any distinct $x_1, \dots, x_n$ in (a, b), $A_n(x_1, \dots, x_n; f)$ of (5.55) is positive. If f is C^{2n-1}, then the matrix $B_n(x_0; f)$ is positive for all $x_0 \in (a, b)$.*

Remarks

1. We will shortly prove that if $n \geq 2$ and $f \in \mathcal{M}_1$, then f is C^1.
2. We will see in Chapter 6 that the last result has a converse—i.e. if $B_n(x_0; f)$ is positive for all $x_0 \in (a, b)$, then $f \in \mathcal{M}_n(a, b)$. In that chapter, we'll also find a more direct way to go from positive Loewner matrices to positive Dobsch matrices.
3. We'll see in Chapter 8 that the first assertion also has a converse, indeed, for any distinct $x_1, \dots, x_n$, one has that $L_n(x_1, \dots, x_n; f) \geq 0 \iff A_n(x_1, \dots, x_n; f) \geq 0$. See Theorem 8.6.

Proof The second statement follows from the first by (5.41) and taking $x_j = x_0 + j\varepsilon$ with $\varepsilon \downarrow 0$.

In the matrix $L_n(x_1, \dots, x_n; f) = [x_i, x_j; f]$, subtract row 1 from rows $2, 3, \dots, n$, and see

$$\det([x_i, x_j; f]) = \prod_{i=2}^{n} (x_i - x_1) \det(A_{ij}^{(1)})$$

where

$$A_{ij}^{(1)} = \begin{cases} [x_1, x_j; f] & \text{if } i = 1 \\ [x_1, x_i, x_j; f] & \text{if } i \geq 2 \end{cases}$$

Now subtract row 2 from rows $3, \dots, n$ and find

$$\det([x_i, x_j; f]) = \prod_{i=2}^{n} (x_i - x_1) \prod_{j=3}^{n} (x_j - x_2) \det(A_{ij}^{(2)})$$

where

$$A_{ij}^{(2)} = \begin{cases} [x_1, x_j; f] & \text{if } i = 1 \\ [x_1, x_2, x_j; f] & \text{if } i = 2 \\ [x_1, x_2, x_i, x_j; f] & \text{if } i \geq 3 \end{cases}$$

Iterating,

$$\det([x_i, x_j; f]) = \prod_{i<j}(x_i - x_j)\det(A_{ij}^{(n)})$$

where

$$A_{ij}^{(n)} = [x_1, \dots, x_i, x_j; f]$$

Doing the same with the columns,

$$\det([x_i, x_j; f]) = \prod_{i<j}(x_i - x_j)^2 \det(A_n(x_1, \dots, x_n; f)) \tag{5.78}$$

so, by Theorem 5.9, if $f \in \mathcal{M}_n(a, b)$,

$$\det(A_n(x_1, \dots, x_n; f)) \geq 0 \tag{5.79}$$

Now fix f_0 of the form (5.56) (say, f_0 given by (5.74)) and let

$$d_k(\theta) = \det(A_k(x_1, \dots, x_k; (1-\theta)f + \theta f_0))$$

By (5.79), $d_k(\theta) \geq 0$ for $k = 1, \dots, n$ and, by Lemma 5.17, $d_k(1) > 0$. Since each $d_k(\theta)$ is a polynomial in θ, $d_k(\theta) > 0$ for all $\theta \in [0, 1]$ but a finite number, so by Proposition 5.8, $A_n(x_1, \dots, x_n; (1-\theta)f + \theta f_0) > 0$ for all but finitely many θ in $[0, 1]$. Taking such a set of θ going to 0, we see that $A_n(x_1, \dots, x_n; f) \geq 0$. □

Similarly, we have the following:

Theorem 5.19 *Let f be a C^{2n} function on (a, b). Suppose that for all $a < x_1 < x_2 < \cdots < x_n < b$ and $x_0 \in \{x_j\}_{j=1}^n$, we know that $K_n(x_0; x_1, \dots, x_n; f) \geq 0$. Then for all $x_0 \in (a, b)$, we have that $H_n(x_0; f) \geq 0$.*

Remarks

1. This is important because we'll eventually prove (see Theorem 9.2) that the condition on K_n is equivalent to f being convex on $n \times n$ matrices.
2. We'll prove a converse of this in Chapters 6 and 8

Proof By (5.70) and the proof of the last theorem, we see that $A_n(x_1, \dots, x_n; [x_0, \cdot; f]) \geq 0$. If we now let all of $x_1, \dots, x_n$ approach x_0, we see that $H_n(x_0; f) \geq 0$. □

Recall [325, Section 6.2] that any locally integrable, measurable function, f, on (a, b) defines a distribution, that is, a function on $C_0^\infty(a, b)$ via

$$g \mapsto \int_a^b f(x)g(x)\,dx \tag{5.80}$$

and that any distribution T has a distributional derivative dT/dx defined by

$$\frac{dT}{dx}(g) = T(-g') \tag{5.81}$$

Proposition 5.20 *Any $f \in \mathcal{M}_1(a, b)$ is a measurable function. If $f \in \mathcal{M}_n(a, b)$, then $f^{(2n-1)}$, the $(2n-1)$st distribution derivative is positive.*

Remark We say a distribution, T, is *positive* if and only if $g \geq 0 \Rightarrow T(g) \geq 0$.

Proof $\mathcal{M}_1(a, b)$ is monotone functions. Let

$$q(c) = \sup_{x \in \{a,b\}} \{x \mid f(x) < c\}$$

Then

$$f^{-1}((-\infty, c]) = (a, q(c)) \quad \text{or} \quad (a, q(c)]$$

so f is measurable.

Let $f \in \mathcal{M}_n(a, b)$. By Proposition 5.4(b), there exist C^∞ functions f_m in $\mathcal{M}_n(a - \frac{1}{m}, b + \frac{1}{m})$ so that f_m are bounded uniformly on any $[c, d] \subset (a, b)$ and $f_m \to f$ pointwise almost everywhere. It follows that if T_m, T are the distributions for f_m and f, then $T_m(g) \to T(g)$ for each test function g.

By Theorem 5.18, $B_n(x; f_m) \geq 0$ for all m and all $x \in (a, b)$. Since the n, n matrix element of B is $f_m^{(2n-1)}/(2n-1)!$, we see $f_m^{(2n-1)}(x) \geq 0$, so $T_m^{(2n-1)} \geq 0$, so $T^{(2n-1)} \geq 0$. □

This Proposition, Theorem 5.2, and the mean value theorem for divided difference, (5.65), immediately imply that:

Corollary 5.21 *Let $f \in \mathcal{M}_n(a, b)$. Then for $j = 1, \ldots, n$ and any distinct $x_1, \ldots, x_{2j} \in (a, b)$, we have that*

$$[x_1, \ldots, x_{2j}; f] \geq 0 \tag{5.82}$$

Remark The smoothness required for the mean value theorem for divided differences may not hold for $j = n$ but one can, as usual, convolute f with a smooth positive function and take limits to get the case $j = n$.

Corollary 5.22 *Let $f \in \mathcal{M}_n(-1, 1)$, $n \geq 2$ be an odd function (i.e. $f(-x) = -f(x)$). Then f is convex on $(0, 1)$ and concave on $(-1, 0)$.*

Remark This should be distinguished from the fact that if $f \in \mathcal{M}_n(0, \infty)$, $n \geq 2$, then f is concave (see Corollary 14.5)!

Proof By the same approximation argument used in the Proposition, we can suppose that f is C^∞. Then $f^{(3)} \geq 0$. Since oddness implies that $f''(0) = 0$, we see that $\pm f''(x) \geq 0$ for $\pm x \geq 0$. The convexity and concavity follow. □

Lemma 5.23

(a) *If T is a distribution on (a, b) with $T' = 0$, then T is given by the constant function.*
(b) *If T is a distribution on (a, b) with $T^{(k)} = 0$, then T is given by a polynomial.*
(c) *If T is a distribution on (a, b) so $T^{(k)}$ is given by a continuous function, then T is given by a continuous function which is classically C^k.*
(d) *If T is a distribution on (a, b) so $T^{(2)} \geq 0$, then T is given by a continuous function.*

Remark The proof of (d) shows that in that case T is given by a convex function.

Proof (a) Pick a function h in $C_0^\infty(a, b)$ with $\int h(y)\,dy = 1$. For any $g \in C_0^\infty(a, b)$, let $\ell_0(g) = \int g(y)\,dy$. Then $\ell_0(g - \ell_0(g)h) = 0$, so $g - \ell_0(g)h = q'$ where $q(x) = \int_a^x (g - \ell_0(g)h(y))\,dy$ is in C_0^∞. Thus, $T(q') = 0$ implies $T(g) = \ell_0(g)T(h)$, that is, $T = T(h)1$.

(b) By (a), $T^{(k-1)} = c_1 1$. Thus, $(T - c_1 x^{k-1}/(k-1)!)^{(k-1)} = 0$, so $T^{(k-2)} = c_2 + c_1 x$. Iterating, T is a polynomial of degree $k - 1$.

(c) If $T^{(k)} = f$, then with

$$g(x) = \int_a^x dx_1 \int_a^{x_1} dx_2 \ldots \int_a^{x_{k-1}} f(x_k)\,dx_k$$

we have $T^{(k)} - g^{(k)} = 0$, so $T = g +$ polynomial. g is C^k, so T is also.

(d) For each $[c, d]$, pick h in $C_0^\infty(a, b)$ with $h \geq 0$ and $h = 1$ on $[c, d]$. Then $h\|f\|_\infty \pm f \geq 0$ for each $f \in C_0^\infty(a, b)$ supported in $[c, d]$, so $|T^{(2)}(f)| \leq \|f\|_\infty T^{(2)}(h)$, and thus $T^{(2)}$ extends to a positive functional on $C[c, d]$, that is, a measure μ on $[c, d]$. If

$$g(x) = \int_c^x \mu([c, y])\,dy$$

then $g^{(2)}(f) = T^{(2)}(f)$ for $f \in C_0^\infty(c, d)$, and thus $T = g +$ linear function, that is, T is given by a continuous function in each $[c, d]$. □

Distributional results normally only determine a function for a.e. x, and we want results for all x. The key is

Lemma 5.24 *Let g, f be two functions on (a, b) so that g is monotone (i.e., in $\mathcal{M}_1$) and f is continuous. If $g(x) = f(x)$ for a.e. x, then $g(x) = f(x)$ for all x.*

Proof Fix $x_0 \in (a, b)$. By monotonicity, $\lim_{\varepsilon \downarrow 0} g(x_0 \pm \varepsilon) = g_\pm(x_0)$ exists and

$$g_-(x_0) \leq g(x_0) \leq g_+(x_0) \tag{5.83}$$

Since $f = g$ for a.e. x, we can find $x_n^\pm \to x_0$ so $x_n^+ \uparrow$ and $x_n^- \downarrow$ and $g(x_n^\pm) = f(x_n^\pm)$. Thus, $g_\pm(x_0) = f(x_0)$, so by (5.83), $g(x_0) = f(x_0)$. □

We can now prove Theorem 5.2, that any $f \in \mathcal{M}_n(a,b)$ lies in $C^{(2n-3)}$.

Proof of Theorem 5.2 (Proof of Theorem 5.2) Let $f \in \mathcal{M}_n(a,b)$. By Proposition 5.20, $f^{(2n-1)} \geq 0$, the $(2n-1)$st distributional derivative is positive. By Lemma 5.23(d), $f^{(2n-3)}$ is given by a continuous function, so by Lemma 5.23(c), f is equal a.e. to a C^{2n-3} function. By Lemma 5.24, f is equal to this C^{2n-3} function for all, not just a.e., x. □

Next, we want to present a generalization of the positivity of Loewner determinants that is closer to Loewner's original proof of Theorem 5.1. (We will not need this extension in most of this book—indeed, it will only be used in Loewner's proof (Chapter 25).) We will start by analyzing some rank one perturbations. Let A be a finite self-adjoint matrix and let $\varphi \in \mathbb{C}^n$ and let

$$B = A + \langle \varphi, \cdot \rangle \varphi \tag{5.84}$$

We will suppose φ is cyclic for A, that is, A has simple eigenvalues and φ is not orthogonal to any eigenspace. Let

$$F_A(z) = \langle \varphi, (A-z)^{-1}\varphi \rangle \qquad F_B(z) = \langle \varphi, (B-z)^{-1}\varphi \rangle \tag{5.85}$$

Since

$$(A-z)^{-1} = (B-z)^{-1} + (A-z)^{-1}(B-A)(B-z)^{-1} \tag{5.86}$$

we see

$$F_A(z) = F_B(z) + F_A(z)F_B(z) \tag{5.87}$$

or

$$F_B(z) = \frac{F_A(z)}{1+F_A(z)} \tag{5.88}$$

It will be convenient instead to use

$$G_A(z) = 1 + F_A(z) \qquad G_B(z) = 1 + F_B(z) \tag{5.89}$$

so (5.88) says B has eigenvalues at zeros of G_A. Since $\frac{\partial}{\partial z}(\lambda - z)^{-1} = (\lambda - z)^{-2}$, G is monotone in x on $\mathbb{R}$ between poles, so it has exactly one zero between such poles. Since $G \to 1$ as $x \to \pm\infty$, the final zero of B is above the last pole of G, that is, the eigenvalues $\lambda_1 < \lambda_2 < \cdots < \lambda_n$ of A and $\mu_1 < \cdots < \mu_n$ of B obey

$$\lambda_1 < \mu_1 < \lambda_2 < \cdots < \lambda_n < \mu_n \tag{5.90}$$

related, of course, to (5.29). Suppose $\eta^{(1)}, \ldots, \eta^{(n)}$ and $\psi^{(1)}, \ldots, \psi^{(n)}$ are the orthonormal eigenvectors of A and B, respectively. Then

$$G_A(z) = 1 + \sum_{j=1}^{n} \frac{|\langle \eta^{(j)}, \varphi \rangle|^2}{\lambda_j - z} \tag{5.91}$$

This shows that one can obtain the eigenvalues of B by forming (5.91) and finding its zeros. But it also gives us a way to solve the inverse problems given the λ's and μ's obeying (5.90) to form

$$G(z) = \prod_{j=1}^{n} \frac{(\mu_j - z)}{(\lambda_j - z)} \tag{5.92}$$

and use it and (5.91) to find $\langle \eta^{(j)}, \varphi \rangle$. We have:

Proposition 5.25 *Let $\{\lambda_j\}_{j=1}^n$ and $\{\mu_j\}_{j=1}^n$ obey* (5.90)*. Then there exist $n \times n$ self-adjoint matrices A and B obeying*

$$A\eta^{(j)} = \lambda_j \eta^{(j)} \qquad B\psi^{(j)} = \mu_j \psi^{(j)} \tag{5.93}$$

for orthonormal families and a vector φ so that (5.84) *holds and*

$$\langle \varphi, \eta^{(j)} \rangle > 0 \qquad \langle \varphi, \psi^{(j)} \rangle > 0 \tag{5.94}$$

and

$$\det(\langle \eta^{(j)}, \psi^{(j)} \rangle) = 1 \tag{5.95}$$

Proof Let A be given by

$$A = \begin{pmatrix} \lambda_1 & & \\ & \ddots & \\ & & \lambda_n \end{pmatrix} \tag{5.96}$$

and let $\eta^{(j)}$ be the vector with 1 in the j place and zero elsewhere. Form $G(z)$ by (5.92) and note that

$$G(z) = 1 + \sum_{j=1}^{n} \frac{\alpha_j}{\lambda_j - z} \tag{5.97}$$

where

$$\alpha_j = \frac{\prod_{k=1}^{n} (\mu_k - \lambda_j)}{\prod_{k \neq j} (\lambda_k - \lambda_j)} > 0 \tag{5.98}$$

The positivity in (5.98) holds since both numerator and denominator have $(j-1)$ negative terms. Equation (5.97) holds since both sides have the same polar singularities and their difference goes to zero at infinity.

Let

$$\varphi = \sum_{j=1}^{n} \alpha_j^{1/2} \eta^{(j)} \tag{5.99}$$

Thus, with B given by (5.84), $G_A(z) = G(z)$, so the zeros are at the μ_j, which must be the eigenvalues for B.

By (5.88), F_B has a pole at each μ_j, so φ is not orthogonal to any eigenspace of B, and thus we can uniquely pick $\psi^{(j)}$, so (5.93) and (5.94) hold.

It remains to prove (5.95). $\langle \eta^{(j)}, \psi^{(j)} \rangle$ is a real orthogonal matrix, so its determinant is $+1$ or -1. For $y \in [0, 1]$, let $B(y) = A + y\langle \varphi, \cdot \rangle \varphi$ and pick an orthonormal basis $\psi^{(k)}(y)$ for $B(y)$ uniquely by $\langle \varphi, \psi^{(k)}(y) \rangle > 0$. By eigenvalue perturbation theory (see Corollary 5.30), $\psi^{(k)}(y)$ is continuous in y, so

$$\det(\langle \eta^{(j)}, \psi^{(k)}(y) \rangle)$$

is continuous in $[0, 1]$ with values ± 1, so identically 1. □

Following Loewner [209], we define the extended Loewner matrix:

Definition Let f be a real-valued continuous function on (a, b). For $2n$ distinct points $\lambda_1, \dots, \lambda_n, \mu_1, \dots, \mu_n$ in (a, b), we define the *extended Loewner matrix* by

$$(L_n^{(e)}(\lambda_1, \dots, \lambda_n; \mu_1, \dots, \mu_n; f))_{k\ell} = [\lambda_k, \mu_\ell; f] \tag{5.100}$$

Theorem 5.26 *If $a < \lambda_1 < \mu_1 < \lambda_2 < \cdots < \mu_n < b$ and $f \in \mathfrak{M}_n(a, b)$, then*

$$\det(L_n^{(e)}(\lambda_1, \dots, \lambda_n; \mu_1, \dots, \mu_n; f)) \geq 0 \tag{5.101}$$

Remark The proof does not depend on matrix monotonicity on an open interval. With the natural definition of $\mathfrak{M}_n(\Omega)$ for an arbitrary open subset of $\mathbb{R}$ (see Theorem 1.7 and Chapter 35), all we need is $f \in \mathfrak{M}_n(\Omega)$ for some Ω containing all the λ's and μ's.

Proof Construct $A, B, \varphi, \eta^{(j)}, \psi^{(j)}$ as in Proposition 5.25. For any function f, note

$$\begin{aligned} \langle \psi^{(k)}, (f(B) - f(A))\eta^{(\ell)} \rangle &= \langle f(B)\psi^{(k)}, \eta^{(\ell)} \rangle - \langle \psi^{(k)}, f(A)\eta^{(\ell)} \rangle \\ &= (f(\mu_k) - f(\lambda_\ell))\langle \psi^{(k)}, \eta^{(\ell)} \rangle \end{aligned} \tag{5.102}$$

Take $f(x) = x$ and get

$$\langle \psi^{(k)}, \varphi \rangle \langle \varphi, \eta^{(\ell)} \rangle = (\mu_k - \lambda_\ell)\langle \psi^{(k)}, \eta^{(\ell)} \rangle \tag{5.103}$$

and use (5.103) and (5.102) to obtain

$$M_{k\ell} \equiv \langle \psi^{(k)}, (f(B) - f(A))\eta^{(\ell)} \rangle = [\mu_k, \lambda_\ell; f] \langle \psi^{(k)}, \varphi \rangle \langle \varphi, \eta^{(\ell)} \rangle \tag{5.104}$$

Now, if $L_{k\ell} = \langle \eta^{(k)}, \psi^{(\ell)} \rangle$, then

$$(LM)_{k\ell} = \langle \eta^{(k)}, (f(B) - f(A))\eta^{(\ell)} \rangle$$

so since $\det(L) = 1$ (by (5.95)), (5.104) becomes

$$\det(f(B) - f(A)) = \det([\mu_k, \lambda_\ell; f]) \prod_{k,\ell} \langle \psi^{(k)}, \varphi \rangle \langle \varphi, \eta^{(\ell)} \rangle \tag{5.105}$$

Since $\prod_{k,\ell} \langle \psi^{(k)}, \varphi \rangle \langle \varphi, \eta^{(\ell)} \rangle > 0$, we see that if $f(B) - f(A) \geq 0$, then $\det(f(B) - f(A)) \geq 0$ and (5.105) implies (5.101). □

Remark The proof shows that strict positivity of (5.101) is equivalent to strict positivity of $f(B) - f(A)$ (see (5.105)). In particular, if $f, f_1 \in \mathfrak{M}_n(a, b)$ and $\det(L_n^{(e)}(f_1)) > 0$, so is $\det(L_n^{(e)}(f + f_1))$.

Note that, by taking $\mu_k = \lambda_k + \varepsilon$ and $\varepsilon \downarrow 0$, (5.101) implies (5.35) and thus provides a second (actually, the original) proof that (a) ⇒ (c) in Theorem 5.9.

We want next to translate the monotone matrix hypotheses of Theorems 1.7 and 1.8 into results about Loewner matrices.

Theorem 5.27 *Let f obey* (a) *of Theorem 1.8. Then for any $x_1, \dots, x_n$ obeying*

$$a < x_1 < \cdots x_j < b < c < x_{j+1} < \cdots < x_n < d \tag{5.106}$$

for some j, we have

$$L_n(x_1, \dots, x_n; f) \geq 0 \tag{5.107}$$

Proof Given such an $(x_1, \dots, x_n)$, let A be given by (5.3) and let $C \geq 0$. By Corollary 5.29 below, for λ small, $A + \lambda C$ has j eigenvalues in (a, b) and $n - j$ in (c, d). So, by the hypothesis, $\frac{d}{d\lambda} f(A + \lambda C) \geq 0$. By (5.35), $L_n \odot C \geq 0$, so taking $C = (\varphi, \cdot)\varphi$, $L_n \geq 0$. □

Our final topic in this chapter is to indicate the proofs of the results in eigenvalue perturbation theory used earlier in this chapter.

Theorem 5.28 *Let $A(\lambda)$ be an analytic family of $n \times n$ matrices in a neighborhood, N, of $\lambda = 0$ symmetric about $\mathbb{R}$ with*

$$A(\lambda)^* = A(\bar{\lambda}) \tag{5.108}$$

Let $x_1(0)$ be an eigenvalue of $A(0)$ for which the corresponding spectral projection $P_1(0)$ is one-dimensional. Then there exists a connected open set, N_0, symmetric about $\mathbb{R}$, with $0 \in N_0 \subset N$ and a scalar analytic function, $x_1(\lambda)$, and operator-valued analytic function, $P_1(\lambda)$, on N_0 so that

$$\text{(i)} \qquad P_1(\lambda)|_{\lambda=0} = P_1(0) \qquad x_1(\lambda)|_{\lambda=0} = x_1(0) \tag{5.109}$$

$$\text{(ii)} \qquad A(\lambda)P_1(\lambda) = P_1(\lambda)A(\lambda) = x_1(\lambda)A(\lambda) \tag{5.110}$$

$$\text{(iii)} \qquad P_1(\lambda)^2 = P_1(\lambda) \tag{5.111}$$

$$\text{(iv)} \qquad P_1(\lambda)^* = P_1(\bar{\lambda})$$

Remark (5.111) says $P_1(\lambda)$ is a projection. By (5.109) and continuity, P_1 is rank one. By (5.110), $\operatorname{Ran}(P_1)$ are eigenvectors of $A(\lambda)$.

Proof Pick ε so $\{w \mid |w - x_1(0)| \leq 2\varepsilon\}$ contains only one eigenvalue of $A(0)$, namely $x_1(0)$. Since $\{B \mid \det(B) \neq 0\}$ is open and $B \to B^{-1}$ is analytic on this set, there is a disk N_1 about $\lambda = 0$, so $\det(A(\lambda) - w) \neq 0$ if $\lambda \in N_1$ and $\frac{\varepsilon}{2} < |w - x_1(0)| < 2\varepsilon$, and $(A(\lambda) - w)^{-1}$ is jointly analytic there. Define

$$P_1(\lambda) = \frac{1}{2\pi i} \int_{|w-x_1(0)|=\varepsilon} \frac{dw}{w - A(\lambda)} \tag{5.112}$$

$$x_1(\lambda) = \operatorname{Tr}(A_1(\lambda)P_1(\lambda)) \tag{5.113}$$

By diagonalizing $A(\lambda)$ when λ is real, one sees that $P_1(\lambda)$ is precisely the projection onto the eigenspaces whose eigenvalues, x, obey $|w - x| < \varepsilon$. In particular, (5.111) holds for λ real, and so by analytic continuation, for all λ.

Since $A(\lambda)(w - A(\lambda))^{-1} = (w - A(\lambda))^{-1}A(\lambda)$, we have $A(\lambda)P_1(\lambda) = P_1(\lambda)A(\lambda)$, and once we prove that $\dim(\operatorname{Ran}(P_1)) = 1$, it is clear this equals $x_1(\lambda)P_1(\lambda)$. Since $P_1(\lambda)$ is analytic and $\operatorname{Tr}(P_1(\lambda))$ is an integer, this trace is constant, so 1, since (5.109) obviously holds. (iv) is obvious from (5.112) and (5.108). □

Corollary 5.29 *Let $\{A(\lambda) \mid 0 \leq \lambda \leq 1\}$ be a family of self-adjoint $n \times n$ matrices indexed by $[0, 1]$ so that each $A(\lambda)$ has simple eigenvalues and so that $A(\lambda)$ can be analytically continued to a complex neighborhood of $[0, 1]$. Then there exist n real analytic rank one matrices $\{P_j(\lambda)\}_{j=1}^n$ and real analytic functions $\{x_j(\lambda)\}_{j=1}^n$ on $[0, 1]$ so that*

$$A(\lambda) = \sum_{j=1}^{n} x_j(\lambda)P_j(\lambda) \tag{5.114}$$

$$P_j(\lambda)P_k(\lambda) = \delta_{jk}P_j(\lambda) \tag{5.115}$$

Proof By the theorem, these are analytic locally, but the x's and P's obeying (5.114)/(5.115) are uniquely determined up to order for matrices with simple eigenvalues. □

Corollary 5.30 *Under the hypotheses of Corollary 5.29, suppose also there is* φ *with*

$$\langle \varphi, P_j(\lambda)\varphi \rangle \neq 0 \tag{5.116}$$

for all j *and all* λ *in* $[0, 1]$. *Then one can choose real analytic vectors,* $\eta_j(\lambda)$, *with*

$$\eta_j(\lambda) \in \operatorname{Ran}(P_j(\lambda)) \tag{5.117}$$

$$\langle \eta_j(\lambda), \eta_k(\lambda) \rangle = \delta_{jk} \tag{5.118}$$

Moreover, if each $A(\lambda)$ *is real and* φ *is real, the* $\eta_j(\lambda)$ *can be chosen real.*

Remark It is not hard to show this can be done even if (5.116) does not hold, but it is simpler in that case and is all we need.

Proof Take

$$\eta_j = \frac{P_j(\lambda)\varphi}{\langle \varphi, P_j(\lambda)\varphi \rangle^{1/2}}$$

□

Notes and Historical Remarks The idea of reducing matrix monotonicity to positivity of certain matrices and determinants is basic to Loewner's original paper [209] and was extended by his doctoral student Dobsch [76] who, in particular, introduced the matrix we call $B(x_0)$. We have named the matrix after him. We will have more to say about it in Chapter 6. Loewner's proof used what we call the extended Loewner matrix with L coming as a limit.

The terminology for these matrices is not standard. What we call the Loewner matrix, Donoghue [81] calls the Pick matrix, based on work of Pick that we will discuss in Chapters 16–20. We will see that L is a degenerate limit of Pick matrices, but it is sufficiently different that I prefer not using that name. Donoghue does call determinants of our L_n, Loewner determinants. Donoghue's "extended Loewner matrix" is our A, very different from what we call the extended Loewner matrix! Anyhow, the reader needs to be aware that there is no standard naming in the literature.

Divided differences (under a different name) for unequal points go back to Newton [242] (see the discussion in Chapter 22). The name seems to be due to de Morgan [75]. The Genocchi–Hermite formula is named after Genocchi [121–123] and Hermite [154] who had formulae related to (5.44). The Cauchy integral formula approach (5.56) goes back to Frobenius [103]. The mean value theorem for divided differences appeared first in Schwarz [311]. For more on the history, see [269] and see de Boor [72] for more on the formalism. They are so important to the theory, that the second section in Donoghue's book [81] is entitled "Divided Differences". Our treatment here of divided differences is influenced by de Boor [72] and Heinävaara [147]

Bernstein polynomials appeared in Bernstein [25]. As our notation suggests, there is a probabilistic interpretation of the polynomials and of the convergence. Let $v_1, v_2, \ldots$ be independent identically distributed random variables taking only two values 0 and 1. Suppose 1 is taken with probability x and 0 with probability $1-x$ so

$$\mathbb{E}(v_j) = x \tag{5.119}$$

Let

$$s_m = \frac{1}{m}\sum_{\ell=1}^{n} v_\ell \tag{5.120}$$

Then

$$\text{Prob}\left(s_m = \frac{j}{m}\right) = x^j(1-x)^{m-j}\binom{m}{j} \tag{5.121}$$

and

$$\mathbb{E}(f(x)) = B_m(x)$$

the Bernstein polynomial of order m. That $B_m(x) \to f(x)$ is just the law of large numbers. Equation (5.12) is a standard calculation of $\mathbb{E}((s_m - x)^2)$ and the calculation that follows is essentially the Chebyshev inequality proof of the law of large numbers; see [325, Section 7.2]

The Daleckiĭ–Krein formula is due to Ju. L. Daleckiĭ and S. G. Krein [68]. As explained in Daleckiĭ [67] when S. Krein's more famous brother, M. G. Krein, was told of their result, he immediately remarked that it must be connected with Loewner's theorem! Similar formulae are in Loewner [209]

In [69], Ju. L. Daleckiĭ and S. G. Krein developed formulae for higher derivatives which includes Theorem 5.16 and even a Taylor theorem with remainder. They also stated results for operators on Hilbert space including

$$\frac{d}{d\lambda}f(A(\lambda)) = \int_a^b \int_a^b \frac{f(\nu)-f(\mu)}{\nu-\mu} dE_\lambda(\nu)\frac{dA(\lambda)}{d\lambda}dE_\lambda(\mu) \tag{5.122}$$

where $dE_\lambda(\cdot)$ is the spectral measure of the self-adjoint operator $A(\lambda)$. Their result requires strong hypothesis on the regularity of $\lambda \mapsto A(\lambda)$ and of f and making sense of the double spectral integral. [68, 69] are in Russian; [67] has a summary (without proofs) in English.

(5.122) is the starting point for a long series on double spectral integral by Birman–Solomyak [39–41] (the last is a summary in English) with many applications, the study of what happens when $A(\lambda)$ lies in certain trace ideal spaces and applications to Krein spectral shifts. Simon [321] has a proof of one of their spectral shift results that relies on

$$\frac{d}{d\lambda}\mathrm{Tr}(f(A(\lambda)) = \mathrm{Tr}\left(\left(\frac{dA(\lambda)}{d\lambda}\right) f'(A(\lambda))\right) \tag{5.123}$$

One can simplify his proof by noting that (5.123) follows immediately for the finite dimensional case from the Daleckiĭ–Krein formula and one can then extend to the trace class by taking limits of the integral of the derivative to get an integral formula for the trace class case that one can differentiate.

Schur products were discussed by Schur [307] who, in particular, proved Lemma 5.3. They are also called Hadamard products, although Horn's historical notes [168] seem to indicate this is due to an incorrect offhand attribution of von Neumann propagated by Halmos rather than due to actual work of Hadamard!

The Hansen–Tomiyama matrix was defined by them [138, 139] in 2007. They noted Theorem 5.19 and conjectured that a converse holds, a result proven ten years later by Heinävaara as we'll discuss in Chapters 6 and 8. As we remarked, the condition on Kraus matrices in Theorem 5.19 was proven by Kraus (in 1936) to be equivalent to convexity on $n \times n$ matrices; see Chapter 9.

Corollary 5.22 is an observation of Moslehian et al. [227]

Two standard references for eigenvalue perturbation theory are Kato [178] and Reed–Simon [281]; see also [329, Sections 1.4 and 2.3]. The basic papers are Rellich [282], Nagy [338], and Kato [175]. In particular, Rellich proved that if (5.108) holds, then eigenvalues and eigenprojections are analytic in a neighborhood of $N \cap \mathbb{R}$ even if there are degeneracies. Analyticity of the eigenvalues (which we actually use once in this chapter) is easier. All solutions of

$$\det(x - A(\lambda)) = 0 \tag{5.124}$$

are real for λ real, and one shows, first, that the solutions of (5.124) are given by convergent series in $(\lambda - \lambda_0)^{1/p}$ for suitable p, and then that reality of all branches implies analyticity. Analyticity of the eigenprojections is more subtle [178, 281, 329], but not used in this chapter.

In proving Proposition 5.25, we used rank one perturbation theory which has extra features like (5.88). This theory is associated especially with work of Aronszajn [19], Krein [191], and Donoghue [78]. For further discussion of rank one perturbations, see Simon [320]

Chapter 6
Heinävaara's Integral Formula and the Dobsch–Donoghue Theorem

Recall (see (5.38)) that given a C^{2n-1} function, f, on (a, b), we defined its Dobsch matrix at $x_0 \in (a, b)$ by

$$B_n(x_0; f)_{k\ell} = \frac{f^{(k+\ell-1)}(x_0)}{(k+\ell-1)!} \qquad 1 \le k, \ell \le n \tag{6.1}$$

We proved (see Theorem 5.18) that if $f \in \mathcal{M}_n(a, b)$ and $f \in C^{2n-1}$, then $B_n(x_0; f)$ is positive (but not necessarily strictly positive) for each $x_0 \in (a, b)$. We also showed that in general functions in $\mathcal{M}_n(a, b)$ are in C^{2n-3} (and in the next chapter, we'll see there are examples of such functions which are not in C^{2n-2}), so we'll need a replacement for pointwise positivity if f is only C^{2n-3}. Any continuous function, f, on (a, b) defines a distribution and so we can form distributional derivatives (see Simon [325, Chapters 6 and 9] for background on distributions). We define the matrix $B_n(x; f)$ as the distribution given by (6.1) where $f^{(k+\ell-1)}$ is a distributional derivative. We say that $B_n(\cdot; f)$ is positive if, for every $\mathbb{C}^n$-valued function, $\boldsymbol{\varphi}(x)$, a C^∞ function of compact support in (a, b), we have that

$$\sum_{1 \le k, \ell \le n} \int_a^b \frac{f(x)}{(k+\ell+1)!} (-1)^{k+\ell-1} \frac{d^{k+\ell-1}}{dx^{k+\ell-1}} [\overline{\varphi_k(x)} \varphi_\ell(x)]\, dx \ge 0 \tag{6.2}$$

Since any $f \in \mathcal{M}_n(a, b)$ is a distributional limit of C^∞ functions in $\mathcal{M}(a + \epsilon, b - \epsilon)$ (see Proposition 5.4), we conclude that for any $f \in \mathcal{M}_n(a, b)$, $B_n(\cdot; f)$ is a positive distribution.

One main result of this chapter is a converse to Theorem 5.18 which means that positivity of $B_n(x_0; f)$ is equivalent to $f \in \mathcal{M}_n(a, b)$. The second main result is an analog of the next theorem for matrix convex functions. In Chapter 8, we'll provide a simpler proof of these facts (that is not unrelated), so the reader can skip this chapter. We include it for historical reasons and because it is an interesting proof.

B. Simon, *Loewner's Theorem on Monotone Matrix Functions*, Grundlehren der mathematischen Wissenschaften 354, https://doi.org/10.1007/978-3-030-22422-6_6

Theorem 6.1 (The Dobsch–Donoghue Theorem) *Let f be a C^{2n-1} function on (a, b). Then $f \in \mathcal{M}_n(a, b)$ if and only if the matrix $B_n(x_0; f)$ is positive for all $x_0 \in (a, b)$. If f is only a priori continuous, then $f \in \mathcal{M}_n(a, b)$ if and only if the distribution $B(\cdot; f)$ is a positive distribution.*

Remarks

1. Since we already know one direction, we'll only prove after many preliminaries that positivity of B_n implies monotonicity on $n \times n$ matrices.
2. In many ways, this theorem belongs to Chapter 5. It is placed here only because we'll also have an analog for $\mathcal{C}_n$ which relies on the results given in Chapter 9.
3. Without this theorem, it isn't clear how to construct functions in $\mathcal{M}_n$ that aren't already in $\mathcal{M}_\infty$. In the next chapter, using this theorem, we'll easily construct functions $f \in \mathcal{M}_n(a, b)$ that aren't in $\mathcal{M}_{n+1}(a, b)$, let alone $\mathcal{M}_\infty(a, b)$.
4. While the distributional result only assumes that f is a priori continuous, we recall that positivity of the distribution $B_n(\cdot; f)$ implies positivity of the distributional derivative, $f^{(2n-1)}$, and this implies that f is classically C^{2n-3} (see Lemma 5.23).

Before starting the proof, we note a simple corollary

Theorem 6.2 (Donoghue Localization Theorem) *Let $a < c < b < d$ all in $\mathbb{R}$ and let f be a real-valued function on (a, d). If $f \in \mathcal{M}_n(a, b)$ and $f \in \mathcal{M}_n(c, d)$, then $f \in \mathcal{M}_n(a, d)$.*

Remarks

1. For $n = \infty$, this is a simple consequence of Loewner's theorem since if f has a Herglotz continuation from (a, b) and from (c, d), they agree on (c, b) and so are the same and thus f has a Herglotz continuation from (a, d).
2. We'll see this as an immediate consequence of Theorem 6.1. Historically though, the first proof of Theorem 6.1 relied on first proving Theorem 6.2; see the Notes.

Proof The hypothesis and Theorem 6.1 imply that $B_n(x; f)$ is positive for x in $(a, b) \cup (c, d) = (a, d)$. By the other direction in the theorem, $f \in \mathcal{M}_n(a, b)$.

There is a tiny subtlety we brushed under the rug in the last paragraph. If f is C^{n-1}, we are dealing with pointwise positivity and the argument is valid. If however f is only C^{n-3}, then $B_n(x; f)$ is only a positive distribution. There are two ways of fixing this. One can use a partition of unity to show that if a distribution is positive on two overlapping intervals, it is positive on their union. Alternatively, one can convolute with a smooth function of small support, shrinking the intervals somewhat so that they still overlap. □

We now turn to the proof of Theorem 6.1. Because of Theorem 5.18, we need to show that if $B_n(x; f)$ is a positive distribution on (a, b), then the Loewner matrix $L_n(x_1, \ldots, x_n; f)$ is positive for all $a < x_1 < x_2 < \cdots < x_n < b$. By the usual approximation plus adding a small amount of log, we can suppose that f is C^∞ on (a, b) with all derivatives bounded and that B_n is a pointwise strictly positive matrix on all of (a, b). Let X be shorthand for $(x_1, \ldots, x_n)$ as above and write $L_n(x_1, \ldots, x_n; f) \equiv L(X; f)$. Then we will prove

Theorem 6.3 (Heinävaara's First Integral Formula) *Fix $n \geq 2$. Let $f \in C^\infty(a, b)$ and X as above. Then there exists a non-negative continuous function, $I_X(t)$, on $\mathbb{R}$ supported on $[x_1, x_n]$ and a continuous $n \times n$ matrix valued function, $C(t, X)$, so that*

$$L(X; f) = (2n-1) \int_{-\infty}^{\infty} C^*(t, X) B_n(t; f) C(t, X) I_X(t)\, dt \tag{6.3}$$

Remarks

1. This is not a mere existence statement. C and I will be explicit, albeit complicated looking, functions (the Notes in Chapter 8 will explain why they have the form that they do). $C(t, X)$ will be a polynomial in t while $I_X(t)$ will be a piecewise polynomial function of t with changes of polynomial at the x_j. Both functions are f independent.
2. While we write $\int_{-\infty}^{\infty}$, it is really $\int_{x_1}^{x_n}$ so $B_n(t; f)$ is defined there.
3. It is easy to extend this result to C^{2n-1} functions, f.

Proof of Theorem 6.1 Given Theorem 6.3 As already noted, we can suppose that f is C^∞ with bounded derivatives and that $B_n(t; f)$ is strictly positive. Since B_n is positive, so is $C^* BC$. Since $I \geq 0$, the integrand in (6.3) is positive so $L(X; f)$ is a positive matrix. □

We begin the proof of (6.3) by carefully defining C and I. For $1 \leq k \leq n$, define polynomials $g_{k,X}(t, y)$ by

$$g_{k,X}(t, y) = \prod_{\ell \neq k} (1 + y(t - x_\ell)) \tag{6.4}$$

For $1 \leq i, j \leq n$ let C_{ij} be the coefficient of y^{i-1} in $g_{j,X}(t, y)$, i.e.,

$$C(t, X)_{ij} = \sum_{\{x_{a_1}, \ldots, x_{a_{i-1}}\} \in X \setminus \{x_j\}} \prod_{\ell=1}^{i-1} (t - x_{a_\ell}) \tag{6.5}$$

$C(t, X)$ is the $n \times n$ matrix with matrix elements $C(t, X)_{ij}$

We will also need

$$p_X(z) = \prod_{\ell=1}^{n} (z - x_\ell) \tag{6.6}$$

We recall the function $h_z(t) = (z - t)^{-1}$ of (5.53). C enters because of

Proposition 6.4 *For any $X, t \in \mathbb{R}$ and $z \in \mathbb{C} \setminus (\{t\} \cup [x_1, x_n])$, we have that*

$$C^*(t, X)B_n(t, h_z)C(t, X) = \frac{L(X, h_z)p_X(z)^2}{(z-t)^{2n}} \tag{6.7}$$

Proof Use D for the left-hand side of (6.7) and $\langle \cdot, \cdot \rangle_r$ for the bilinear form without a complex conjugate. Since $h_z^{(k)}(t)/k! = (z-t)^{-k-1}$, we have that $B_n(t; h_z)$ is a rank one matrix

$$B_n(t; h_z) = (z-t)^{-2} \langle v(z,t), \cdot \rangle_r v(z,t) \tag{6.8}$$

where $v_j(z,t) = (z-t)^{-(j-1)}$, $j = 1, \dots, n$. Thus, since C has real matrix elements

$$D = (z-t)^{-2} \langle C^*(t, X)v(z,t), \cdot \rangle_r C^*(t, X)v(t,z) \tag{6.9}$$

C was chosen so that

$$\begin{aligned} [C^*(t, X)v(z,t)]_j &= \sum_{i=1}^n C_{ij} v_i(z,t) \\ &= \sum_{i=1}^n C_{ij}(z-t)^{-(i-1)} = g_{j,X}(t, \frac{1}{z-t}) \end{aligned} \tag{6.10}$$

Thus

$$\begin{aligned} D_{ij} &= \frac{g_{i,X}(t, \frac{1}{z-t}) g_{j,X}(t, \frac{1}{z-t})}{(z-t)^2} \\ &= \frac{1}{(z-t)^2} \prod_{\ell \neq i} \left(1 + \frac{t - x_\ell}{z-t}\right) \prod_{\ell \neq j} \left(1 + \frac{t - x_\ell}{z-t}\right) \\ &= \frac{1}{(z-t)^{2n}} \prod_{\ell \neq i} (z - x_\ell) \prod_{\ell \neq j} (z - x_\ell) \\ &= \frac{p_X(z)^2}{(z-t)^{2n}} \frac{1}{z - x_i} \frac{1}{z - x_j} \\ &= [x_i, x_j; h_z] \frac{p_X(z)^2}{(z-t)^{2n}} \end{aligned} \tag{6.11}$$

by the formula (5.54) for $[x_i, x_j; h_z]$. This proves (6.7). □

Next, we begin the construction of I_X by defining residue functionals, R_j, $j = 1, \dots, n$ for rational functions, r, on $\mathbb{C}$ whose only possible poles are at points in X.

We set $\rho = \frac{1}{2} \min_{i \neq j} |x_i - x_j|$ and let

$$R_j(r) = \frac{1}{2\pi i} \oint_{|z-x_j|=\rho} r(z)\, dz \tag{6.12}$$

R_j is, of course, the residue of r at x_j. When we look at $R_j(H)$ where H is a function of several variables, say $H(z, w)$, we'll use $R_j^{(z)}$ to indicate the variable that is integrated over.

Define

$$S_X(z, t) = -\frac{(z-t)^{2n-2}}{p_X(z)^2} \tag{6.13}$$

Given (6.7), it is clear why this is a natural function. Then define

$$I_{j,X}(t) = R_j^{(z)}(S_X(z, t)) \tag{6.14}$$

where the residue is taken in the z variable with t as a parameter. Thus

$$I_{j,X}(t) = -\frac{d}{dz} \frac{(z-t)^{2n-2}}{\prod_{\ell \neq j}(z - x_\ell)^2} \bigg|_{z=x_j} \tag{6.15}$$

so we see that $I_{j,X}$ is a polynomial in t of degree $2n - 2$ with

$$I_{j,X}(x_j) = 0$$

because $2n - 2 \geq 2$ and there is a $(z-t)^{2n-3}$ or $(z-t)^{2n-2}$ in all the terms in the derivative.

Define for $t \in \mathbb{R}$

$$I_X(t) = \sum_{\{j | x_j < t\}} I_{j,X}(t) \tag{6.16}$$

Since $I_{j,X}(x_j) = 0$, this function is continuous which means that $I_X(t)$ is a *spline*, that is a piecewise polynomial, continuous function. Clearly, $I_X(t) = 0$ if $t \leq x_1$. Moreover, we claim that

$$\sum_{j=1}^{n} I_{j,X}(t) = 0 \tag{6.17}$$

for this is the sum of the residues of all the poles of $S_X(z, t)$ which is the integral over z of S_X over any contour surrounding X, for example, a large circle. Since for t fixed and $|z|$ large, we have that $|S_X(z, t)| \leq C(t)|z|^{-2}$, this integral goes to zero as the radius of the circle goes infinity which proves (6.17). Thus $I_X(t) = 0$ also when $t \geq x_n$.

Proposition 6.5 *If* $z \in \mathbb{C} \setminus [x_1, x_n]$, *then*

$$(2n-1)\int_{-\infty}^{\infty} \frac{I_X(t)}{(z-t)^{2n}}\,dt = \frac{1}{p_X(z)^2} \tag{6.18}$$

Proof We compute for $z \notin [x_1, \infty)$:

$$(2n-1)\int_{-\infty}^{\infty} \frac{I_X(t)}{(z-t)^{2n}}\,dt = -(2n-1)\sum_{j=1}^{n} R_j^{(w)}\left(\int_{x_j}^{\infty} \frac{(w-t)^{2n-2}}{p_X(w)^2(z-t)^{2n}}\,dt\right) \tag{6.19}$$

$$= \sum_{j=1}^{n} R_j^{(w)}\left(\frac{\left(\frac{w-x_j}{z-x_j}\right)^{2n-1} - 1}{(w-z)p_X(w)^2}\right) \tag{6.20}$$

$$= -\sum_{j=1}^{n} R_j^{(w)}\left(\frac{1}{(w-z)P_X(w)^2}\right) \tag{6.21}$$

$$= R_{w=z}^{(w)}\left(\frac{1}{(w-z)P_X(w)^2}\right) \tag{6.22}$$

$$= p_X(z)^{-2} \tag{6.23}$$

where we get (6.19) using the definitions (6.16) and (6.14). Note that I_j does not have compact support, but it is a polynomial in t with degree at most $2n-2$, so, because of the $(z-t)^{-2n}$, the integral converges and we can interchange the sum and integral.

We can do the integral over t to get (6.20) by noting that

$$\frac{d}{dt}\left[\frac{(w-t)^{2n-1}}{(z-t)^{2n-1}}\right] = (2n-1)\left(\frac{w-t}{z-t}\right)^{2n-2} \frac{d}{dt}\left(1 + \frac{w-z}{z-t}\right)$$

$$= (w-z)(2n-1)\frac{(w-t)^{2n-2}}{(z-t)^{2n}} \tag{6.24}$$

We then get (6.20) where we slightly extend $R_j^{(w)}$ to allow functions with poles at points not in X (in this case, at w=z). We then get (6.21) by noting that the first term in the fraction does not have a pole at $w = x_j$.

The function $h(w) = 1/(w-z)p_X(w)^2$ is $O(|w|^{-(2n+1)})$ at $w = \infty$, so as in the proof of (6.17), the sum of its residues is 0. This gives us (6.22) where $R_{w=z}^{(w)}$ means the residue at $w = z$. □

The key to application of Heinävaara integral formula is that $I(t) \geq 0$. We'll need a lemma for that

Lemma 6.6 *Fix $w_1, \ldots, w_q \in (0, \infty)$. Let $Q(t)$ be defined for $t \in (-\infty, 0)$ by*

$$Q(t) = \prod_{j=1}^{q} (w_j - t)^{-1} \tag{6.25}$$

Then for all $k \geq 0$

$$Q^{(k)}(t) \geq 0 \tag{6.26}$$

for all $t \in (-\infty, 0)$.

Proof Let $t = -s + \delta t$ for $s > 0$. If $|\delta t| < s$, then

$$\begin{aligned}(w_j - t)^{-1} &= (w_j + s - \delta t)^{-1} \\ &= \sum_{m=0}^{\infty} \frac{(\delta t)^m}{(w_j + s)^{m+1}}\end{aligned} \tag{6.27}$$

by the geometric series formula. Thus $Q(-s+\delta t)$ is a product of series with positive Taylor coefficients and so a series with positive Taylor coefficients. It follows that $Q^{(k)}(-s) > 0$ for all k. □

Theorem 6.7 *$I_X(t) \geq 0$ for all $t \in \mathbb{R}$.*

Proof Let $X + a$ be the set where a is added to each x_j. Then

$$S_{X+a}(z + a, t + a) = S_X(z, t) \Rightarrow I_{X+a}(t + a) = I_X(t) \tag{6.28}$$

This means that by picking $a = -t$, we can, without loss, prove that $I_X(0) \geq 0$ for all X. By continuity of I, we can also suppose that no $x_j = 0$ so we can suppose that

$$x_1 < x_2 < \cdots < x_\ell < 0 < x_{\ell+1} \leq x_n \tag{6.29}$$

Since $S_X(z, 0)$ is $O(z^{-2})$ at infinity, we can replace a curve around $[x_1, x_\ell]$ by the imaginary axis (by closing the contour near infinity), i.e.,

$$I_X(0) = -\frac{1}{2\pi i} \int_{-i\infty}^{i\infty} \frac{z^{2n-2}}{p_X(z)^2}\, dz \tag{6.30}$$

Change variables to $w = \frac{1}{z}$. Then $\frac{dz}{z} = -\frac{dw}{w} \Rightarrow -\frac{dz}{z^2} = dw$. Moreover, under this map, $(i\epsilon, i\epsilon^{-1}) \mapsto (-i\epsilon^{-1}, -i\epsilon)$ and $(-i\epsilon^{-1}, -i\epsilon) \mapsto (i\epsilon, i\epsilon^{-1})$ so the curve from $-i\infty$ to $i\infty$ goes to itself with no change of orientation. Let $\{y_j\} = \{1/x_j\}$ be written in increasing order (so $y_1 = 1/x_\ell, \dots, y_\ell = 1/x_1, y_{\ell+1} = 1/x_n, \dots, y_n = 1/x_{\ell+1}$). Then

$$\frac{z^{2n}}{p_X(z)^2} = \frac{1}{\left[\prod_{j=1}^n w\left(\frac{1}{w} - \frac{1}{y_j}\right)\right]^2} = \frac{\prod_{j=1}^n y_j^2}{p_Y(w)^2}$$

Putting this together, we get

$$I_X(0) = \frac{1}{2\pi i}\int_{-i\infty}^{i\infty} \frac{\prod_{j=1}^n y_j^2\, dw}{p_Y(w)^2} \tag{6.31}$$

Again, we can close the contour at infinity and see that if γ is a bounded closed curve that surrounds $[y_1, y_\ell]$ and if

$$f(w) = \frac{\prod_{j=1}^n y_j^2}{\prod_{j=\ell+1}^n (y_j - w)^2}$$

then

$$I_X(0) = \frac{1}{2\pi i}\oint_\gamma \frac{\prod_{j=1}^n y_j^2}{p_Y(w)^2}\, dw \tag{6.32}$$

$$= \frac{1}{2\pi i}\oint_\gamma \frac{f(w)\, dw}{\prod_{j=1}^\ell (w - y_j)^2} \tag{6.33}$$

$$= [y_1, y_1, y_2, y_2, \dots, y_\ell, y_\ell; f] \tag{6.34}$$

by (5.56). Now use the mean value theorem for divided differences, Theorem 5.15, to see that

$$I_X(0) = \frac{f^{(2\ell-1)}(w)}{(2\ell - 1)!} \tag{6.35}$$

for some $w \in [y_1, y_\ell]$. By Lemma 6.6, $I_X(0) > 0$. □

Remark The proof shows that I is strictly positive on $(x_1, x_n) \setminus \{x_j\}_{j=1}^n$.

Proof of Theorem 6.3 We've already proven that $I_X(t)$ is supported on $[x_1, x_n]$ and is non-negative, so we only need to prove (6.3). Suppose first that $f = h_z$ for $z \in \mathbb{C} \setminus [x_1, x_n]$. Then, by (6.7) and (6.18),

$$\text{RHS of (6.3)} = (2n-1)L(X, h_z)\int_{-\infty}^{\infty} \frac{p_X(z)^2}{(z-t)^{2n}} I_X(t)\, dt \tag{6.36}$$

$$= L(X, h_z) \tag{6.37}$$

proving (6.3) when $f = h_z$. If f is analytic in a neighborhood of (a, b), then we can pick a simple closed curve, γ, surrounding $[x_1, x_n]$ inside the domain of f and so that the inside of γ is also in the domain of analyticity of f. Then the Cauchy integral formula can be written as

$$f(\cdot) = \frac{1}{2\pi i} \oint_\gamma h_z(\cdot) f(z)\, dz \tag{6.38}$$

so we can integrate (6.3) for the special case of all h_z to get it for all f analytic in a neighborhood of (a, b).

If $f \in C^\infty(a, b)$, we can convolute with Gaussians to get a family of entire functions, f_n so that $f_n^{(\ell)} \to f^{(\ell)}$ for each ℓ uniformly in a neighborhood of $[x_1, x_n]$ and so prove (6.3) for general C^∞ f. □

This completes our discussion of the Dobsch–Donoghue theorem. We turn to an analog for matrix convex functions where we will prove

Theorem 6.8 (Heinävaara's Theorem) *Fix n. Let $f \in C^{2n}(a, b)$. Then the following are equivalent*

(a) *The Kraus matrix, $K_n(x_0; x_1, \ldots, x_n; f)$, of* (5.69) *is positive for all $x_0, x_1, \ldots, x_n \in (a, b)$.*
(b) *The Kraus matrix, $K_n(x_0; x_1, \ldots, x_n; f)$, of* (5.69) *is positive for all $x_0, x_1, \ldots, x_n \in (a, b)$ with $x_0 \in \{x_j\}_{j=1}^n$.*
(c) *The Hansen–Tomiyama matrix, $H_n(x; f)$, of* (5.71) *is positive for all $x \in (a, b)$.*

Remark We'll eventually see (Theorem 9.2) that (b) (and therefore all three) are equivalent to f being an $n \times n$ matrix convex function.

The proof we give in this chapter of this theorem will rely on an integral representation analogous to (6.3). We will fix x_0 and $X = \{x_1, \ldots, x_n\}$ with $x_j \in (a, b)$; $j = 0, 1, \ldots, n$. In place of $S_X(z, t)$ given by (6.13) we define

$$T_{x_0, X}(z, t) = -\frac{(z - t)^{2n-1}}{(z - x_0) p_X(t)^2} \tag{6.39}$$

and instead of $I_{j,X}$ given by (6.14), we define

$$J_{j, x_0, X}(t) = R_j^{(z)}(T_{x_0, X}(z, t)) \tag{6.40}$$

where now $j = 0$ is included. And, of course,

$$J_{x_0,X}(t) = \sum_{\substack{\{j|x_j<t\} \\ 0\le j\le n}} J_{j,x_0,X}(t) \tag{6.41}$$

Analogous to Propositions 6.4, 6.5, and Theorem 6.7, we have that

Theorem 6.9 *Fix n. Let $f \in C^{2n}(a,b)$. Then*

(i) *We have for any $z \in \mathbb{C} \setminus (a,b)$ that*

$$C^*(t,X)H_n(t,h_z)C(t,X) = \frac{K(x_0;X;h_z)p_X(z)^2(z-x_0)}{(z-t)^{2n+1}} \tag{6.42}$$

(ii) $$2n \int_{-\infty}^{\infty} \frac{J_{x_0,X}(t)}{(z-t)^{2n+1}}\, dt = \frac{1}{(z-x_0)p_X(z)^2} \tag{6.43}$$

(iii) *For all $t \in \mathbb{R}$, $J_{x_0,X}(t) \geq 0$* (6.44)

Proof

(i) $h_z^{(\ell)}(x) = (z-x)^{-\ell-1}$ and $[x_1,\dots,x_n;h_x]$ is given by (5.54). Thus

$$H_n(t;h_z) = (z-t)^{-1}B_n(t;h_z) \tag{6.45}$$

$$K(x_0;X;h_z) = (z-x_0)^{-1}L_n(X;h_z) \tag{6.46}$$

so (6.7) implies (6.42).

(ii) A formula like (6.19) holds but with $2n-1$ replaced by $2n$ and on the right $\frac{(w-t)^{2n-2}}{p_X(w)^2(z-t)^{2n}}$ is replaced by $\frac{(w-t)^{2n-1}}{p_X(w)^2(w-x_0)(z-t)^{2n+1}}$. This leads to a formula like (6.21) but with

$$-\sum_{j=1}^{n} R_j^{(w)}\left(\frac{1}{(w-z)(w-x_0)P_X(w)^2}\right) \tag{6.47}$$

which by the zero sum of residues including the one at $w=z$ leads to (6.43).

(iii) As in the proof of Theorem 6.7, we can take $t=0$ and we can suppose $x_0 > 0$ (otherwise, replace x_j by $-x_j$). In (6.30), $z^{2n-2}/p_X(z)^2$ is replaced by $z^{2n-1}/(z-x_0)p_X(z)^2$ so that (6.31) becomes

$$\frac{1}{2\pi i}\int_{-i\infty}^{i\infty} \frac{dw}{(y_0-w)p_Y(w)^2} \tag{6.48}$$

which is positive by the same argument.

□

Given the above results, one easily gets the following by proving it first when $f = h_z$, then for functions analytic in a neighborhood of $[a, b]$, and then, in general, as we did in the proof of (6.3):

Theorem 6.10 (Heinävaara's Second Integral Formula) *Fix n. Let $f \in C^{2n}(a, b)$. Then for $\{x_0\} \cup X \subset (a, b)$, we have that:*

$$K(x_0; X; f) = 2n \int_{-\infty}^{\infty} C^*(t, X) H_n(t; f) C(t, X) J_{x_0, X}(t)\, dt \tag{6.49}$$

where $J_{x_0, X} \geq 0$.

Proof of Theorem 6.8 (a)⇒(b) is trivial.

(b)⇒(c) is Theorem 5.19.

(c)⇒(a) is immediate from (6.49). □

Notes and Historical Remarks The Dobsch–Donoghue Theorem is named after Dobsch [76] who proved the half that $f \in \mathcal{M}_n \Rightarrow B_n(\cdot, ; f) \geq 0$ in 1937 and Donoghue who proved the more subtle half in his book [81] in 1974. His proof is long, quite involved and not easy to understand.

Donoghue proved Theorems 6.1 and 6.2 in the opposite order we do. Namely, he used Loewner's interpolation machinery to prove Theorem 6.2. Then he showed that if $B_n(x_0, f) > 0$, there is an interval, I_{x_0}, containing x_0 so that $f \in \mathcal{M}_n(I_{x_0})$. In this way he could use Theorem 6.2 to prove Theorem 6.1 when B_n is strictly positive (which can be arranged by adding a small amount of a log).

Loewner stated the localization theorem (Theorem 6.2) in his initial paper [209] saying it was easy to prove, but he gave no details. It took 45 years for the first proof to appear and an additional 35 years for the truly simple proofs of Heinävaara [146, 148] (see also Chapter 8 below).

As we noted in the Notes to Chapter 5, (a)⇒(c) in Theorem 6.8 is due to Hansen–Tomiyama [138], who conjectured the other direction held (and so, given results of Kraus we discuss in Chapter 9 the analogs of Theorems 6.1 and 6.2 for matrix convex functions). Heinävaara [146] proved this conjecture and this chapter follows that work. Chapter 8 will present his second proof of these results.

Chapter 7
$\mathcal{M}_{n+1} \neq \mathcal{M}_n$

In this chapter, we'll present explicit examples that show that $\mathcal{M}_{n+1}(a, b)$ is strictly smaller than $\mathcal{M}_n(a, b)$ and that show that $2n - 3$ in Theorem 5.2 is optimal.

Example 7.1 In Proposition 24.2, we will prove that for f a Herglotz function, $L^{(e)}(f)$ is strictly positive if and only if its Herglotz representation in the form (3.39) has at least n points in the support of the representing measure. To prove the analog for Dobsch matrices, we need to compute $B_n(0; f)$ for $f(z) = -(z - z_0)^{-1}$. For small z, we can write

$$-(z - z_0)^{-1} = z_0^{-1}\left(1 - \frac{z}{z_0}\right)^{-1} = \sum_{n=0}^{\infty} z_0^{-1-n} z^n \tag{7.1}$$

to see that $f^{(n)}(0)/n! = z_0^{-1-n}$. We conclude that for $f(z) = -(z - z_0)^{-1}$ with $z_0 \in \mathbb{R} \setminus \{0\}$, $B_n(0; f)$ is the positive rank one operator

$$B_n(0; f)_{k\ell} = v(z_0)_k v(z_0)_\ell \qquad v(z_0)_k = z_0^{-k} \tag{7.2}$$

Notice if $f(z) = z$, then (7.2) holds with $v(\infty)_k = \delta_{k1}$. $\square$

Proposition 7.2 *Let f be a Herglotz function which is real and analytic on $(a, b) \subset \mathbb{R}$. Then for $x_0 \in (a, b)$, $B_n(x_0; f)$ is strictly positive if and only if the Herglotz representation,* (3.39)*, has μ with at least n points in its support.*

Proof As in the proof of Proposition 24.2, we need to show that if $\{z_j\}_{j=1}^n$ are n distinct points in $\mathbb{R} \cup \{\infty\} \setminus \{0\}$, then $\{v(z_j)\}_{j=1}^n$ span $\mathbb{C}^n$. This is equivalent to the $n \times n$ matrix with columns $\{v(z_j)\}_{j=1}^n$ having non-zero determinant. If no z_j is ∞, this determinant is $\prod_{j=1}^n z_j^{-1}$ times a non-vanishing $n \times n$ Vandermonde determinant. If $z_1 = \infty$, it is $\prod_{j=2}^n z_j^{-2}$ times a non-vanishing $(n - 1) \times (n - 1)$ Vandermonde determinant. $\square$

B. Simon, *Loewner's Theorem on Monotone Matrix Functions*, Grundlehren der mathematischen Wissenschaften 354, https://doi.org/10.1007/978-3-030-22422-6_7

Example 7.3 We will construct $f \in \mathcal{M}_n(a, b)$, which is not a C^{2n-1} function. In particular, $f \notin \mathcal{M}_{n+1}(a, b)$. Start with g which has $B_n(0; g)$ strictly positive, for example $g(z) = -\sum_{j=1}^{n}(z + j)^{-1}$. Define

$$f_-(x) = \sum_{j=1}^{2n-1} \frac{g^{(j)}(0)}{j!} x^j \qquad f_+(x) = f_-(x) + \frac{x^{2n-1}}{(2n-1)!} \tag{7.3}$$

Since $B_n(0; f_-) = B_n(0; g)$, we have that $B_n(0, f_-)$ is strictly positive. Since $B_n(0; f_+) = B_n(0; f_-) + [(2n-1)!]^{-1}\delta^{(nn)}$ where $\delta^{(nn)}$ is the matrix with zeros in all places except for a 1 in the nn place, we see that $B_n(0, f_+)$ is strictly positive also. By continuity of $B_n(x_0; h)$ in x_0 for h which is C^∞, for some $\epsilon > 0$, $B_n(x; f_\pm)$ is strictly positive for $|x| < \epsilon$.

Define

$$f(x) = f_\pm(x) \text{ if } 0 \le \pm x < \epsilon \tag{7.4}$$

Then $B_n(x; f)$ is strictly positive for $0 < |x| < \epsilon$. $B(x; f)$ has a simple jump discontinuity at $x = 0$, which doesn't impact the positivity of $B(\cdot; f)$ as a distribution on $(-\epsilon, \epsilon)$. We conclude from Theorem 6.1 that $f \in \mathcal{M}_n(-\epsilon, \epsilon)$. Notice that f is C^∞ away from 0 but that $\lim_{y \downarrow 0} f^{(2n-1)}(y) - f^{(2n-1)}(-y) = 1$ so f is not $C^{(2n-1)}$ as claimed.

If we want (a, b), a given proper open interval of $\mathbb{R}$ instead of $(-\epsilon, \epsilon)$, we can use conformal mapping as in Theorem 2.4. □

Example 7.4 We'll construct $f \in \mathcal{M}_n(a, b)$ with f not C^{2n-2} showing that the $2n - 3$ of Theorem 5.2 is optimal. Pick g as in the last example. Then we can find $\alpha > 0$ so that if α is added to all the $g^{2n-2}(0)$ terms in B, it is still strictly positive. Now define f_- as in (7.3) but f_+ by

$$f_+(x) = f_-(x) + \frac{\alpha x^{2n-2}}{(2n-2)!} \tag{7.5}$$

Again, $B_n(0; f_\pm)$ is strictly positive and we can find $\epsilon > 0$, so $B_n(x; f_\pm) > 0$ if $|x| < \epsilon$. Define f by (7.4). Then $B_n(x; f_\pm) > 0$ if $0 < |x| < \epsilon$, but f is not C^{2n-2}. $f^{(2n-2)}$ has a jump discontinuity at 0, so the distributional derivative $f^{(2n-1)}$ has a delta function term which contributes a term to the distribution $B_n(\cdot; f)$ but since $\alpha > 0$, it doesn't change the positivity. Thus, $f \in \mathcal{M}_n(-\epsilon, \epsilon)$ but not C^{2n-2} and so not in $\mathcal{M}_{n+1}$. By conformal mapping, we can replace $(-\epsilon, \epsilon)$ by any given proper interval (a, b). In particular, there is $f \in \mathcal{M}_n(-1, 1)$, which is C^{2n-3} but not C^{2n-2}. □

Example 7.5 The examples above are not smooth. Moreover, that f is not in $\mathcal{M}_{n+1}$ is at a single point; one can show for small ϵ, f is in both $\mathcal{M}_{(0,\epsilon)}$ and $\mathcal{M}_{(-\epsilon,0)}$. There are simple examples which are smooth; indeed there are polynomials for which the failure to lie in $\mathcal{M}_{n+1}$ is not at a single point! Fix n. Let

$$f(x) = x + \frac{1}{3}x^3 + \cdots + \frac{1}{2n-1}x^{2n-1} \tag{7.6}$$

By looking at the power series for $\log(1+x)$, one sees that

$$g(x) \equiv x + \frac{1}{3}x^3 + \cdots + \frac{1}{2j-1}x^{2j-1} + \cdots = \frac{1}{2}\log\left(\frac{1+x}{1-x}\right) \tag{7.7}$$

By Theorem 4.1 (or a simple check that it is Herglotz), $g \in \mathcal{M}_\infty(-1, 1)$. Since it is not rational, Proposition 7.2 implies that $B_n(0; g)$ is strictly positive. Since up to order $2n-1$ f and g have the same derivatives at $x = 0$, $B_n(0; f)$ is also strictly positive.

The lower 3×3 block of $B_{n+1}(0; f)$ is

$$\begin{pmatrix} a(x) & b(x) & (2n-1)^{-1} \\ b(x) & (2n-1)^{-1} & 0 \\ (2n-1)^{-1} & 0 & 0 \end{pmatrix}$$

where $a(x)$ and $b(x)$ are irrelevant since the zero elements imply the determinant of this matrix is $-(2n-1)^{-3}$ coming from the main anti-diagonal. Thus we conclude that $B_{n+1}(x; f)$ is not positive for any x.

It follows by continuity of B_n that $f \in \mathcal{M}_n(-\epsilon, \epsilon)$ for some small ϵ but that $f \notin \mathcal{M}_{n+1}(a, b)$ for any $a < b$. □

Example 7.6 We will construct $f \in \mathcal{M}_n(a, b)$ so that for some $c \in (a, b)$, $f \restriction (c, b)$ is a rational function of degree $n-1$ (so for $x \in (c, b)$, $\det(B_n(x; f)) = 0$) but $f \restriction (a, c)$ has B_n strictly positive. This shows that B_n can fail to be strictly positive on part of the domain of definition of f without that holding everywhere. Let g be a rational Herglotz function of degree $n-1$ on $(-1, \infty)$, e.g., $g(z) = -\sum_{j=1}^{n-1}(z+j)^{-1}$. Let $f_+(z) = g(z)$ and

$$f_-(z) = \frac{z^{2n-1}}{(2n-1)!} + \sum_{j=0}^{2n-1} \frac{g^j(0)}{j!} z^j$$

and $f(z) = f_\pm(z)$ for $\pm z \geq 0$. As above, for some $\epsilon > 0$, $f \in \mathcal{M}_{(-\epsilon,\infty)}$. $f \restriction (0, \infty)$ is rational Herglotz and $f \restriction (-\epsilon, 0)$ has $B_n > 0$. □

Notes and Historical Remarks The idea behind Example 7.3 is given in Donoghue's book [81, pgs 83–84] but without being very specific and without an explicit example. He refers to adding a singular monotone function to $f^{(2n-2)}$ without saying what to choose (in our example, we make the simplest choice adding $x\chi_{(0,\infty)}(x)$ to $f^{(2n-2)}$). Also, his statement that $\mathcal{M}_{n+1} \neq \mathcal{M}_n$ is not so explicit (he says "Unfortunately, we have had to invoke the theorem that $\mathcal{M}_n$ is a local property without having given its proof. In the absence of that theorem, we would not yet be sure that the classes $\mathcal{M}_n(a, b)$ were all distinct as n increases"; his proof of the local theorem is only later in the book).

Because Donoghue's discussion was so murky, Hansen–Ji–Tomiyama [134] wrote a paper claiming the first explicit examples with $\mathcal{M}_{n+1} \neq \mathcal{M}_n$. They have the lovely example in Example 7.5. For further discussion of this and related examples, see Hansen–Tomiyama [138], Osaka–Silvestrov–Tomiyama [252], and Osaka [250]. In Example 7.5, we saw that $B_n(x; f)$ was not positive for any x because there was a principle submatrix which was a Hankel matrix whose determinant was negative since came from a single antidiagonal product of strictly positive numbers. As noted in [252], the same argument implies that if f is a polynomial with $\deg(f) \leq 2n - 2$, then $f \notin \mathcal{M}_n(a, b)$ for all $a < b$.

Chapter 8
Heinävaara's Second Proof of the Dobsch–Donoghue Theorem

In this chapter, we provide a really short, simple, and elegant proof due to Heinävaara of Theorem 6.1. It will be a direct consequence of the mean value theorem for divided differences. Below, $\mathcal{P}_n$ will denote the real polynomials of degree at most n, a real vector space of dimension $n + 1$.

Proposition 8.1 *Fix k. Let f be a C^k function on $(a, b) \subset \mathbb{R}$.*

(a) *If $f^{(k)}(x) \geq 0$ for all $x \in (a, b)$, then $[y_1, \dots, y_{k+1}; f] \geq 0$ for any choice of $k + 1$ y's in (a, b).*
(b) *If, for every $x \in (a, b)$, there exists $\{y_j^{(m)}\}_{j=1,m=1}^{k+1,\infty}$ so that as $m \to \infty$, we have that $y_j^{(m)} \to x$ and $[y_1^{(m)}, \dots, y_{k+1}^{(m)}; f] \geq 0$, then $f^{(k)}(x) \geq 0$ for all $x \in (a, b)$.*

Remarks

1. If $g^{(k)}(x) \geq 0$ for all x, then g is called a *k-tone function*.
2. The y's need not be distinct.

Proof (a) is an immediate consequence of Theorem 5.15, the mean value theorem for divided difference. (b) follows from continuity of $[y_1, \dots, y_{k+1}; f]$ in its arguments, even at coincidence points, and (5.41). □

Proposition 8.2 *Let f be $C^{(2n-1)}$ near x_0. Let $B_n(x_0; f)$ be the Dobsch matrix at x_0. Then $B_n(x_0; f)$ is a positive matrix if and only if for every $q \in \mathcal{P}_{n-1}$, we have that $(q^2 f)^{(2n-1)}(x_0) \geq 0$*

Proof By translation covariance, we can suppose that $x_0 = 0$. By Leibniz' rule for higher derivatives of products

B. Simon, *Loewner's Theorem on Monotone Matrix Functions*, Grundlehren der mathematischen Wissenschaften 354, https://doi.org/10.1007/978-3-030-22422-6_8

$$\frac{(q^2 f)^{(2n-1)}(0)}{(2n-1)!} = \sum_{i,j \mid i+j \leq 2n-1} \frac{q^{(i)}(0)}{i!} \frac{f^{(2n-1-i-j)}(0)}{(2n-1-i-j)!} \frac{q^{(j)}(0)}{j!} \tag{8.1}$$

$$= \sum_{i,j=0}^{n-1} \frac{q^{(i)}(0)}{i!} \frac{f^{(2n-1-i-j)}(0)}{(2n-1-i-j)!} \frac{q^{(j)}(0)}{j!} \tag{8.2}$$

$$= \sum_{k,\ell=1}^{n} \frac{q^{(n-k)}(0)}{(n-k)!} \frac{f^{(k+\ell-1)}(0)}{(k+\ell-1)!} \frac{q^{(n-\ell)}(0)}{(n-\ell)!} \tag{8.3}$$

$$= \sum_{k.\ell=1}^{n} c_k c_\ell (B_n(0; f))_{k\ell} \tag{8.4}$$

where $q(x) = \sum_{k=0}^{n-1} c_{n-k} x^k$. To get (8.2), we note that $q^{(i)} = 0$ if $i \geq n$ and that if $i, j \leq n-1$, then $i + j \leq 2n-1$. Equation (8.3) uses $k = n-i$, $\ell = n-j$. Since q runs through all of $\mathcal{P}_{n-1}$ as $(c_1, \dots, c_n)$ runs through $\mathbb{R}^n$, we get the result. □

Proposition 8.3 *Let f be C^1 on (a, b) and $a < x_1 < \cdots < x_n < b$ n distinct points. Then the Loewner matrix, $L_n(x_1, \dots, x_n; f)$, is positive if and only if for every $q \in \mathcal{P}_{n-1}$, we have that*

$$[x_1, x_1, x_2, x_2, \dots, x_n, x_n; q^2 f] \geq 0 \tag{8.5}$$

Proof As usual by adding a function with strictly positive Loewner matrix if need be, we can approximate f with analytic functions and suppose that f is analytic in a simply connected neighborhood of $[x_1, x_n]$. Let C be a contour that surrounds $[x_1, x_n]$. Then, by (5.56),

$$\sum_{i,j=1}^{n} c_i c_j [x_i, x_j; f] = \frac{1}{2\pi i} \oint \sum_{i,j=1}^{n} c_i c_j \frac{f(z)}{(z-x_i)(z-x_j)} dz \tag{8.6}$$

$$= \frac{1}{2\pi i} \oint \left(\sum_{j=1}^{n} \frac{c_j}{z-x_j} \right)^2 f(z)\, dz \tag{8.7}$$

$$= \frac{1}{2\pi i} \oint \frac{q(z)^2}{(z-x_1)^2 \dots (z-x_n)^2} f(z)\, dz \tag{8.8}$$

$$= [x_1, x_1, x_2, x_2, \dots, x_n, x_n; q^2 f] \tag{8.9}$$

by (5.56) again. Here

$$q(z) = \sum_{i=1}^{n} c_i \prod_{j \neq i} (z - x_j) \tag{8.10}$$

Such a q obeys $q(x_i) = c_i \prod_{j\neq i}(x_i - x_j)$ so $q \equiv 0 \Rightarrow c_i \equiv 0$. Thus the n functions $\{\prod_{j\neq i}(z - x_j)\}_{i=1}^n$ are linearly independent and so a basis for $\mathcal{P}_{n-1}$. Thus, as $(c_1, \dots, c_n)$ runs through all of $\mathbb{R}^n$, q runs through all of $\mathcal{P}_{n-1}$. Equation (8.9) implies the result. □

Proof of Theorem 6.1

$$\begin{aligned} f \in \mathcal{M}_n(a,b) &\iff \forall_{a<x_1<\dots<x_n} L(x_1,\dots,x_n; f) \geq 0 \text{ by Theorem 5.1} \\ &\iff \forall_{q\in\mathcal{P}_{n-1}} \forall_{a<x_1<\dots<x_n} [x_1, x_1, \dots, x_n, x_n; q^2 f] \geq 0 \\ &\qquad\qquad \text{by Proposition 8.3} \\ &\iff \forall_{q\in\mathcal{P}_{n-1}} \forall_{x\in(a,b)} (q^2 f)^{(2n-1)} \geq 0 \\ &\qquad\qquad \text{by Proposition 8.1 with } k = 2n-1 \\ &\iff \forall_{x\in(a,b)} B_n(x; f) \geq 0 \text{ by Proposition 8.2} \end{aligned}$$

□

We now turn to the second proof of Heinävaara's Theorem, Theorem 6.8.

Proposition 8.4 *Let f be a C^{2n} function near x_0. Let $H_n(x_0; f)$ be the Hansen–Tomiyama matrix at x_0. Then $H_n(x_0; f)$ is a positive matrix if and only if, for all $q \in \mathcal{P}_{n-1}$, we have that $(q^2 f)^{(2n)} \geq 0$*

Proof We do the same calculation that led to (8.4) but for $(q^2 f)^{(2n)}/(2n)!$ rather than $(q^2 f)^{(2n-1)}/(2n-1)!$. In (8.3), we get $f^{k+\ell}(0)/(k+\ell)!$ instead of $f^{k+\ell-1}(0)/(k+\ell-1)!$. □

Remark The reason that $f^{(m)}$ enters only for $m \geq 2$ is that $\deg(p) \leq n-1$, so in the analog of (8.2), i and j are at most $n-1$ and $2n-(n-1)-(n-1) = 2$.

Proposition 8.5 *Let f be a C^2 function on $(a,b) \subset \mathbb{R}$ and let $a < x_1 < \dots < x_n < b$ be n distinct points. Let $x_0 \in (a,b)$ (which may be one of the x_j's). Then the Kraus matrix $K(x_0; x_1, \dots, x_n; f)$ is positive if and only if for all $q \in \mathcal{P}_{n-1}$, we have that*

$$[x_0, x_1, x_1, x_2, x_2, \dots, x_n, x_n; q^2 f] \geq 0 \tag{8.11}$$

Remark This holds if some of the x_j are coincident so long as f is sufficiently smooth.

Proof In (8.6), $1/(z-x_i)(z-x_j)$ is replaced by $1/(z-x_i)(z-x_j)(z-x_0)$, so in (8.7) and (8.8), $f(z)$ is replaced by $f(z)/(z-x_0)$. □

Second Proof of Theorem 6.8 Since f is C^{2n}, the Kraus matrix is continuous in $x_1, \dots, x_n$ even if there are coincidences, so in checking we can suppose the x_j are distinct.

(c)⇒(a) is just the mean value theorem for divided differences given the last two propositions.

(a)⇒(b) is trivial.

(b)⇒(c) follows by making the points coincident and the last two propositions. □

Finally, following Heinävaara[148], we prove

Theorem 8.6 *Let f be a C^1 function on (a, b) and $x_1, \ldots, x_n$ distinct points in (a,b). Then*

$$L_n(x_1, \ldots, x_n; f) \geq 0 \iff A_n(x_1, \ldots, x_n; f) \geq 0 \tag{8.12}$$

Remark We proved ⇒ in Chapter 5 (see Theorem 5.18).

Proof (Heinävaara[148]) By the usual trick of adding $\epsilon \log(x - a + 1)$ and convoluting with Gaussians and taking limits, it suffices to prove the result for f analytic in a neighborhood of (a, b).

By (8.7)

$$L_n(x_1, \ldots, x_n; f) \geq 0 \iff \frac{1}{2\pi i} \oint \left(\sum_{j=1}^{n} \frac{c_j}{z - x_j} \right)^2 f(z)\, dz \geq 0 \tag{8.13}$$

By a similar calculation

$$A_n(x_1, \ldots, x_n; f) \geq 0 \iff \frac{1}{2\pi i} \oint \left(\sum_{j=1}^{n} \frac{d_j}{(z - x_1) \ldots (z - x_j)} \right)^2 f(z)\, dz \geq 0 \tag{8.14}$$

The lemma below completes the proof. □

Lemma 8.7 *Fix $x_1, \ldots, x_n$ distinct points in (a, b). The real span of $\{\frac{1}{z-x_j}\}_{j=1}^n$ and $\{\frac{1}{(z-x_1)\ldots(z-x_j)}\}_{j=1}^n$ are the same.*

Proof Let V be the span of $\{\frac{1}{z-x_j}\}_{j=1}^n$. By a partial fraction expansion, each function $\frac{1}{(z-x_1)\ldots(z-x_j)}$ lies in V, indeed by computing residues

$$\frac{1}{(z - x_1) \ldots (z - x_j)} = \sum_{\ell=1}^{j} \frac{1}{\prod_{k \neq \ell} (x_\ell - x_k)} \frac{1}{z - x_\ell} \tag{8.15}$$

Thus by counting dimensions, it suffices to prove that $\{\frac{1}{(z-x_1)\ldots(z-x_j)}\}_{j=1}^n$ are independent. So suppose that

$$\sum_{j=1}^{n} \frac{d_j}{(z - x_1)\dots(z - x_j)} = 0 \tag{8.16}$$

Then, since near ∞, $g(z) = d_1/z + O(1/z^2)$, we have that $d_1 = 0$. By looking at successive Taylor coefficients at infinity, we conclude each $d_j = 0$. □

Notes and Historical Remarks This proof is from Heinävaara [148] who has pointed out that one can understand the relation of this proof to the one in Chapter 6 by using the Peano kernel which we discuss in the Notes to Chapter 22. His function $I_{\{x_1,\dots,x_n\}}(t)$ of (6.16) is just the Peano kernel for $[x_1, x_1, \dots, x_n, x_n; f]$ (see (22.37)). When there are coincident points as there are in this case, the formula is more complicated than (22.38). As in (5.52), we get derivatives as in (6.15). The matrix $C(t, X)$ is just the change of basis matrix from $\{(x - t)^k\}_{k=0}^{n-1}$ to $\{\prod_{j \neq k}(x - x_j)\}_{k=1}^{n}$. If we evaluate $\langle \varphi, L(X; f)\varphi \rangle$ with $\varphi = \{c_j\}_{j=1}^{n}$, then (6.3) is just the Peano formula for $[x_1, x_1, \dots, x_n, x_n; q^2 f]$ with q given by (8.10).

The equivalence (8.12) for all n-tuples follows from Donoghue's localization theorem. The pointwise result is a new one of Heinävaara[148].

Chapter 9
Convexity, I: The Theorem of Bendat–Kraus–Sherman–Uchiyama

Recall that

Definition A real-valued function on (a, b) is called (resp. *concave*) if for all $x, y \in (a, b)$ and $\theta \in [0, 1]$,

$$f(\theta x + (1 - \theta)y) \leq \theta f(x) + (1 - \theta)f(y) \tag{9.1}$$

(resp. $\geq$).

To distinguish from matrix convex functions, defined below, we'll sometimes use *scalar convex* for this notion. It is not hard to show (see reference in the Notes) that f is scalar convex (resp. concave) if and only if it is continuous and the distributional derivative $f'' \geq 0$ (resp. $f'' \leq 0$). There is a related condition that doesn't involve distributional derivatives: f is convex, if and only if its second divided difference $[x_1, x_2, x_3; f]$ is non-negative.

This is the first of five chapters on matrix convex functions. $f : [a, b] \to \mathbb{R}$ is called convex on n-matrices, written $f \in \mathcal{C}_n(a, b)$, if and only if, for all self-adjoint $n \times n$ matrices A, B with eigenvalues in (a, b), we have that

$$f(\theta A + (1 - \theta)B) \leq \theta f(A) + (1 - \theta)f(B) \tag{9.2}$$

for all $\theta \in [0, 1]$. We have that $\mathcal{C}_n \subset \mathcal{C}_{n-1}$ for the same reason as the analog of monotone functions. We set $\mathcal{C}_\infty(a, b) = \cap_n \mathcal{C}_n(a, b)$. $f \in \mathcal{C}_\infty(a, b)$ is called *matrix convex* or *operator convex*isoperator convex function. We will normally not use operator convex except for "strongly operator convex" where we choose to follow the literature on that class. As for the monotone function case (see Theorem 2.7), if $f \in \mathcal{C}_\infty(a, b)$, (9.2) holds for all (bounded) self-adjoint operators on a Hilbert space with spectrum in (a, b). If f obeys (9.2) with the sign reversed, we say that f is *matrix concave* or *operator concave*. Since "convex" and "concave" start with the

B. Simon, *Loewner's Theorem on Monotone Matrix Functions*, Grundlehren der mathematischen Wissenschaften 354, https://doi.org/10.1007/978-3-030-22422-6_9

same letter, we will not introduce a separate symbol for matrix concave functions but instead write $-f \in \mathcal{C}_n(a, b)$.

For scalar C^2 functions, f is convex if and only f' is monotone increasing. This chapter will discuss the matrix analog of this fact, which is not the obvious generalization. The next chapter will discuss the remarkable fact that if f is a non-negative function on $(0, \infty)$, then $f \in \mathcal{M}_\infty(0, \infty) \iff -f \in \mathcal{C}_\infty(0, \infty)$! The third chapter discusses operator analogs of Jensen's inequality. The fourth relates matrix convexity to Loewner matrices. The fifth discusses a stronger notion than matrix convexity.

A first guess might be that $f \in \mathcal{C}_\infty(a, b) \iff f' \in \mathcal{M}_\infty(a, b)$ but this is false. We'll see soon that $f' \in \mathcal{M}_\infty(a, b) \Rightarrow f \in \mathcal{C}_\infty(a, b)$, but there is an example later in the chapter with $f \in \mathcal{C}_\infty(a, b)$ but f' not in $\mathcal{M}_\infty(a, b)$. Instead we'll prove the following: If f is a real-valued function on (a, b) and $y \in (a, b)$, we define

$$f_y^{(1)}(x) \equiv [x, y; f] = \begin{cases} \frac{f(x)-f(y)}{x-y}, & \text{if } x \neq y \\ f'(x), & \text{if } x = y \end{cases} \tag{9.3}$$

Theorem 9.1 (Bendat–Kraus–Sherman–Uchiyama Theorem) *The following are equivalent for a C^1 function, f, on $(a, b) \subset \mathbb{R}$:*

(a) $f \in \mathcal{C}_\infty(a, b)$
(b) *For all* $y \in (a, b)$, $f_y^{(1)} \in \mathcal{M}_\infty(a, b)$
(c) *For one* $y \in (a, b)$, $f_y^{(1)} \in \mathcal{M}_\infty(a, b)$

Since some convexity results will play a role in some of the proofs of Loewner's theorem, it is important to keep track of where exactly one uses Loewner's theorem in the arguments. We will freely use the equivalence of $\mathcal{M}_n$ and positivity of the associated Loewner matrix. In the above $(a) \Leftrightarrow (b)$, indeed the related results for $n \times n$ matrices, will *not* use Loewner's theorem. The proof we will give of $(c) \Rightarrow (b)$ will (although we'll have a degenerate special case that doesn't use Loewner's theorem). This is the only use of Loewner's theorem in the first four chapters (except for a side remark in the fourth chapter). The fifth chapter relies in many places on Loewner's theorem. We also note that the equivalence of $\mathcal{M}_n$ and positivity of the associated Loewner matrix is only used in this and the fourth chapter and not in the next two.

Theorem 9.2 *Let f be a C^2 function on $(a, b) \subset \mathbb{R}$. Fix n. The following are equivalent:*

(a) $f \in \mathcal{C}_n(a, b)$
(b) *For all* $y \in (a, b)$, $f_y^{(1)} \in \mathcal{M}_n(a, b)$
(c) *For all* $y_0, y_1, \ldots, y_\ell \in (a, b)$; $\ell = 1, \ldots, n$, *the* $\ell \times \ell$ *Kraus matrix* $K(y_0; y_1, \ldots, y_\ell; f)$ *is a positive matrix.*
(d) *For all distinct* $y_1, \ldots, y_n \in (a, b)$ *and* $y_0 \in \{y_1, \ldots, y_n\}$, *the* $n \times n$ *Kraus matrix* $K(y_0; y_1, \ldots, y_n; f)$ *is a positive matrix.*
(e) *For all* $x \in (a, b)$, *the Hansen–Tomiyama matrix,* $H_n(x; f)$ *is a positive matrix.*

Remarks

1. This implies (a) $\Leftrightarrow$ (b) in Theorem 9.1.
2. We will prove that for $n \geq 2$, any $f \in \mathfrak{C}_n(a, b)$ is C^2, in fact it is C^{2n-2}.
3. The proof shows that we could add a sixth condition. Namely the condition (9.2) need only hold for pairs with $A \leq B$.
4. The Hansen–Tomiyama matrix is defined in (5.71).

Proof The equivalence of (c), (d), (e) is Heinävaara's Theorem (Theorem 6.8). Theorem 5.9 shows that $(b) \iff (c)$. Thus we are left with proving that $(a) \iff (d)$. We begin by noting that $f \in \mathfrak{C}_n(a, b)$ is equivalent to $g_{A,B,\varphi}(\lambda) = \langle\varphi, f(A + \lambda(B - A))\varphi\rangle$ being scalar convex for all $\varphi \in \mathbb{C}^n$, $\lambda \in [0, 1]$ and self-adjoint A, B with $\sigma(A), \sigma(B) \subset (a, b)$. Since f is C^2, so is g. Such a function is convex if and only if its second derivative is positive.

A little thought shows that, therefore, $f \in \mathfrak{C}_n(a, b)$ if and only if $\frac{d^2}{d\lambda^2}\langle\varphi, f(A + \lambda C)\varphi\rangle\Big|_{\lambda=0} \geq 0$ for all $\varphi \in \mathbb{C}^n$ and self-adjoint A, C with $\sigma(A) \subset (a, b)$. Thus shifting to a basis in which A is diagonal (and then taking limits restricting to the case where A has simple spectrum) we can use (5.67) to see that

$$f \in \mathfrak{C}_n(a, b) \iff \forall_{a<x_1<\cdots<x_n<b} \forall_{\substack{\{C_{k\ell}\}_{k,\ell=1}^n \\ C_{k\ell}=\overline{C_{\ell k}}}} \forall\{\varphi_j\}_{j=1}^n \sum_{m,k,\ell=1}^{n} \bar{\varphi}_k\varphi_\ell C_{km}C_{m\ell}[x_k, x_m, x_\ell; f] \geq 0 \tag{9.4}$$

(d)$\Rightarrow$(a) Let $K^{(m)}$ be the matrix with $K_{k\ell}^{(m)} = [x_m, x_k, x_\ell; f]$ and let $\psi^{(m)} \in \mathbb{C}^n$ be given by $\psi_\ell^{(m)} = C_{m\ell}\varphi_\ell$. Since $C_{km} = \overline{C_{mk}}$, the right side of (9.4) is $\sum_{m=1}^n \langle\psi^{(m)}, K^{(m)}\psi^{(m)}\rangle$. (d) says each $K^{(m)} \geq 0$, so RHS of (9.4) holds and thus $f \in \mathfrak{C}_n(a, b)$.

(a)$\Rightarrow$(d) By (a), we have the right side of (9.4) and we need to show that for each m_0 and $\eta \in \mathbb{C}^n$, we have that

$$\langle\eta, K^{(m_0)}\eta\rangle \geq 0 \tag{9.5}$$

Fix $\{y_j\}_{j=1}^n$ with each $y_j > 0$. Pick $C_{k\ell} = y_k y_\ell$ and $\varphi_\ell = y_\ell^{-1}\eta_\ell$. Then the right side of (9.4) says that

$$\sum_{m,k,\ell=1}^{n} \bar{\eta}_k\eta_\ell y_m^2[x_k, x_m, x_\ell; f] \geq 0 \tag{9.6}$$

Now take $y_{m_0} = 1$ and $y_m \downarrow 0$ for $m \neq m_0$ to get (9.5). $\square$

Just as Theorem 6.1 implies Theorem 6.2, the equivalence (a) and (e) immediately implies

Theorem 9.3 (Heinävaara's Localization Theorem) *Let $a < c < b < d$ all in $\mathbb{R}$ and let f be a real-valued function on (a, b). If $f \in \mathfrak{C}_n(a, b)$ and $f \in \mathfrak{C}_n(c, d)$, then $f \in \mathfrak{C}_n(a, d)$.*

Corollary 9.4 *Any $f \in \mathfrak{C}_n(a, b)$ has $f^{(2j)} \geq 0$ as a distribution for $j = 1, 2, \ldots, n$ and is a C^{2n-2} function.*

Proof By a standard limiting argument, $H_n(x; f)$ is a positive distribution since it can be approximated by C^∞ functions with positive Kraus matrices. Thus $f^{(2j)} \geq 0$, $j = 1, 2, \ldots, n$ are positive distributions, so by Lemma 5.23 (c) and (d), f is equal a.e. to a C^{2n-2} function. Since f is scalar convex, it is continuous, so f is equal to the C^{2n-2} function everywhere. □

As in the argument to go from Proposition 5.20 to Corollary 5.21, Corollary 9.4 implies that

Corollary 9.5 *If $f \in \mathfrak{C}_n(a, b)$ and $j = 1, \ldots, n$, then for any distinct $x_1, \ldots, x_{2j+1}$, we have that*

$$[x_1, \ldots, x_{2j+1}; f] \geq 0 \tag{9.7}$$

Corollary 9.6 $\mathfrak{C}_n(a, b) \neq \mathfrak{C}_{n+1}(a, b)$. *Indeed, there exists $f \in \mathfrak{C}_n(a, b)$ which is not C^{2n-1} (and so not in $\mathfrak{C}_{n+1}$). Moreover, C^{2n-2} in Corollary 9.4 is optimal.*

Proof As in Example 7.1, one sees that if $c < 0$, then $f(x) = (x - c)^{-1}$ has $H_n(0; f) \geq 0$ (so $f \in \mathfrak{C}_\infty(c, \infty)$) and $H_n(0; f)$ is rank 1. It follows that $g(x) = \sum_{j=1}^n (x + j)^{-1}$ has $H_n(0, g)$ strictly positive. By following Example 7.4, one finds f with $H_n(x; f) \geq 0$ for $|x| < \epsilon$, which is not C^{2n-1}. Scaling and translation proves the result for any finite (a, b). □

Corollary 9.7 *Let f be a C^1 function on (a, b). Fix n. Then, $f' \in \mathfrak{M}_n(a, b) \Rightarrow f \in \mathfrak{C}_n(a, b)$.*

Proof Immediate from

$$f_y^{(1)}(x) = \int_0^1 f'((1 - s)y + sx)\, ds \tag{9.8}$$

□

Remark We'll see later (see Example 9.12) that there exist $f \in \mathfrak{C}_\infty$ with $f' \notin \mathfrak{M}_n$ for all $n \geq 2$.

Theorem 9.8 *Let $0 \in (a, b) \subset \mathbb{R}$. Let f be a C^∞ function on (a, b) so that $f_0^{(1)} \in \mathfrak{M}_\infty(a, b)$. Then, for all $y \in (a, b)$, $f_y^{(1)} \in \mathfrak{M}_\infty(a, b)$.*

Remarks

1. By translation invariance, this completes the proof of Theorem 9.2.
2. This uses Loewner's theorem.

3. The proof extends with some minor changes to allow $y = a$ when $\lim_{x\downarrow a} f(x)$ is finite; see below.

Proof Without loss, we can suppose that $f(0) = 0$ (since constants don't change $f_y^{(1)}$) and that $|a| + |b| < \infty$ (since $\mathcal{M}_\infty(a,\infty) = \cap_{b<\infty}\mathcal{M}_\infty(a,b)$). In that case $f_0^{(1)}$ obeys a representation like (1.13). There is a positive measure $d\nu$ on $[a^{-1}, b^{-1}]$ and $\beta \in [0,\infty)$ so that

$$f(x) = \beta x + \int_{a^{-1}}^{b^{-1}} \frac{x^2}{1-\lambda x}\, d\nu(\lambda) \tag{9.9}$$

By a direct calculation,

$$\frac{1}{x-y}\left[\frac{x^2}{1-\lambda x} - \frac{y^2}{1-\lambda y}\right] = \frac{x+y-\lambda xy}{(1-\lambda x)(1-\lambda y)} \tag{9.10}$$

We have that $z \in \mathbb{C}_+$, $y \in (a,b)$, $\lambda \in [a^{-1}, b^{-1}]$

$$\begin{aligned}\operatorname{Im}\left[\frac{z+y-\lambda zy}{(1-\lambda z)(1-\lambda y)}\right] &= \frac{\operatorname{Im}[(z+y-\lambda zy)(1-\lambda\bar{z})]}{|1-z\lambda|^2(1-\lambda y)} \\ &= \frac{\operatorname{Im} z}{|1-z\lambda|^2(1-\lambda y)} \geq 0\end{aligned}$$

so $\operatorname{Im} f_y^{(1)}(z) > 0$ on $\mathbb{C}_+$ and $f_y^{(1)} \in \mathcal{M}_\infty(a,b)$. □

Here is an explicit version of the end point result:

Theorem 9.9 *Let f be a C^∞ function on $(0,b)$ with $b \in (0,\infty)\cup\{\infty\}$ so that $\lim_{t\downarrow 0} f(t) = 0$. Then f is matrix convex on $(0,b)$ if and only if $g(t) \equiv f(t)/t$ is matrix monotone on $(0,b)$.*

Proof If f is matrix convex, then $f_y^{(1)}$ is matrix monotone by Theorem 9.1 for all $y \in (0,b)$. Taking $y \downarrow 0$, we see that g is matrix monotone.

If g is matrix monotone, f has a representation of the form (9.9) where a^{-1} is replaced by $-\infty$ and b^{-1} is 0 if $b = \infty$. The same calculation shows that $f_y^{(1)}$ is a Herglotz function. Thus, by Theorem 9.1, f is concave. □

Here is another proof of Theorem 9.9 that does not rely on Loewner's theorem (via the integral representation, (9.9)) and so could be used in a part of proof of it. Moreover, we can track the n dependence. The proof relies on ideas that will recur in Chapter 11:

Proposition 9.10

(a) *Let $f \in \mathcal{C}_{2n}(0,b)$ for $b \in (0,\infty)\cup\{\infty\}$ with $f(0) \equiv \lim_{x\downarrow 0} f(x) = 0$. Let A, X be $n\times n$ matrices with $\|A\| \leq 1$ and X self-adjoint with spectrum in $(0,b)$. Then*

$$f(A^*XA) \le A^* f(X)A \tag{9.11}$$

(b) *Suppose f is a real-valued, continuous function on* $[0, b)$ *so that* (9.11) *holds for all* $2n \times 2n$ *matrices,* A, X *where* A *is an orthogonal projection and* X *self-adjoint with spectrum in* $(0, b)$*. Then* $f \in \mathfrak{C}_n(0, b)$.

Remark When $n = \infty$ and $A = P$, a projection, (9.11) goes back to Davis [70] who proved an equivalence to concavity. We discuss this further in Chapter 11.

Proof

(a) By a simple continuity argument, in the convexity inequality we can suppose that the matrices have spectrum in $[0, b)$ (i.e., 0 eigenvalue allowed). Define

$$B = \sqrt{1 - AA^*}, \qquad C = \sqrt{1 - A^*A} \tag{9.12}$$

and $2n \times 2n$ matrices, U and V by

$$U = \begin{pmatrix} A & B \\ C & -A^* \end{pmatrix}, \qquad V = \begin{pmatrix} A & -B \\ -C & -A^* \end{pmatrix} \tag{9.13}$$

Note first that $BA = A^*C$ (for example, by expanding the square roots in a power series), that $U^*U = V^*V = \mathbf{1}$ (so that U and V are unitary), and finally that

$$U^* \begin{pmatrix} X & 0 \\ 0 & 0 \end{pmatrix} U = \begin{pmatrix} A^*XA & A^*XB \\ B^*XA & B^*XB \end{pmatrix} \tag{9.14}$$

$$V^* \begin{pmatrix} X & 0 \\ 0 & 0 \end{pmatrix} V = \begin{pmatrix} A^*XA & -A^*XB \\ -B^*XA & B^*XB \end{pmatrix}$$

Thus, by convexity on $2n \times 2n$ matrices

$$\begin{aligned} \begin{pmatrix} f(A^*XA) & 0 \\ 0 & f(B^*XB) \end{pmatrix} &= f\begin{pmatrix} A^*XA & 0 \\ 0 & B^*XB \end{pmatrix} \\ &= f\left(\tfrac{1}{2}U^*\begin{pmatrix} X & 0 \\ 0 & 0 \end{pmatrix}U + \tfrac{1}{2}V^*\begin{pmatrix} X & 0 \\ 0 & 0 \end{pmatrix}V\right) \\ &\le \tfrac{1}{2}f\left(U^*\begin{pmatrix} X & 0 \\ 0 & 0 \end{pmatrix}U\right) + \tfrac{1}{2}f\left(V^*\begin{pmatrix} X & 0 \\ 0 & 0 \end{pmatrix}V\right) \\ &= \tfrac{1}{2}U^*\begin{pmatrix} f(X) & 0 \\ 0 & f(0) \end{pmatrix}U + \tfrac{1}{2}V^*\begin{pmatrix} f(X) & 0 \\ 0 & f(0) \end{pmatrix}V \\ &= \begin{pmatrix} A^*f(X)A & 0 \\ 0 & B^*f(X)B \end{pmatrix} \end{aligned}$$

proving (9.11)

(b) Given Y, Z, two $n \times n$ matrices and $\theta \in [0, 1]$, let $\mathbf{1}$ be an $n \times n$ identity matrix and

$$U = \begin{pmatrix} \sqrt{\theta}\mathbf{1} & \sqrt{1-\theta}\mathbf{1} \\ -\sqrt{1-\theta}\mathbf{1} & \sqrt{\theta}\mathbf{1} \end{pmatrix} \quad X = \begin{pmatrix} Y & 0 \\ 0 & Z \end{pmatrix} \quad A = \begin{pmatrix} 1 & 0 \\ 0 & 0 \end{pmatrix} \tag{9.15}$$

Then U is unitary and

$$f(A^*U^*XUA)_{11} = f(\theta Y + (1-\theta)Z) \tag{9.16}$$

while

$$\begin{aligned} [A^* f(U^*XU)A]_{11} &= [A^*U^* f(X)UA]_{11} \\ &= \theta f(Y) + (1-\theta) f(Z) \end{aligned} \tag{9.17}$$

so (9.11) implies concavity.

□

Theorem 9.11 *Let f be a continuous function on $[0, b)$ with $f(0) = 0$. Define g on $(0, b)$ by $g(t) = f(t)/t$.*

(a) *If $f \in \mathcal{C}_{n+1}(0, b)$, then $g \in \mathcal{M}_n(0, b)$*
(b) *If $g \in \mathcal{M}_{2n}(0, b)$, then $f \in \mathcal{C}_n(0, b)$*
In particular, $f \in \mathcal{C}_\infty(0, b) \iff g \in \mathcal{M}_\infty(0, b)$.

Remark The proof of (b) relies on the second half of Proposition 9.10. There is a proof of a weaker form of (a) (with $\mathcal{C}_{n+1}(0, b)$ replaced by $\mathcal{C}_{2n}(0, b)$) that uses the other half of Proposition 9.10. For let $Y \leq X$, both $n \times n$ matrices with spectrum in $(0, b)$, so both invertible. Let $A = X^{-1/2}Y^{1/2}$ so $A^*XA = Y$. If $\psi = X^{1/2}\varphi$, then

$$\|A^*\psi\|^2 = \langle \varphi, Y\varphi \rangle \leq \langle \varphi, X\varphi \rangle = \|\psi\|^2$$

Thus $\|A\| = \|A^*\| \leq 1$. Since $f \in \mathcal{C}_{2n}(0, b)$, by (9.11),

$$f(Y) = f(A^*XA) \leq A^* f(X)A \tag{9.18}$$

Multiplying by $Y^{-1/2}$ on left and right (which preserves operator inequalities), we see that

$$Y^{-1} f(Y) \leq X^{-1} f(X) \Rightarrow g(X) \leq g(Y) \tag{9.19}$$

so g is monotone.

Proof

(a) By Theorem 9.2, if $f \in \mathcal{C}_{n+1}(0, b)$, then for all $y \in (0, b)$, $f_y^{(1)} \in \mathcal{M}_n(0, b)$. Taking $y \downarrow 0$, we see that $g \in \mathcal{M}_n(0, b)$.

(b) Let $g \in \mathcal{M}_{2n}(0, b)$, X a $2n \times 2n$ self-adjoint matrix with eigenvalues in $(0, b)$, and P a $2n \times 2n$ orthogonal projection onto a subspace of $\mathbb{C}^{2n}$. For some $\epsilon_0 > 0$, $(1+\epsilon_0)\|X\| < b$, so for $0 < \epsilon < \epsilon_0$, $X^{1/2}(\epsilon\mathbf{1} + P)X^{1/2}$ has spectrum in $(0, b)$. Since $\epsilon\mathbf{1} + P \leq (1+\epsilon)\mathbf{1}$, we have that $X^{1/2}(\epsilon\mathbf{1} + P)X^{1/2} \leq (1+\epsilon)X \Rightarrow g(X^{1/2}(\epsilon\mathbf{1} + P)X^{1/2}) \leq g((1+\epsilon)X)$. Thus

$$X^{-1/2}(\epsilon\mathbf{1}+P)^{-1/2}X^{-1/2}f(X^{1/2}(\epsilon\mathbf{1}+P)X^{1/2}) \leq (1+\epsilon)^{-1}X^{-1}f((1+\epsilon)X)) \tag{9.20}$$

Multiply on the left by $PX(\epsilon\mathbf{1} + P)X^{1/2}$ and on the right by $X^{1/2}(\epsilon\mathbf{1} + P)XP$ and, using the continuity of f, take $\epsilon \downarrow 0$ to find that

$$PX^{1/2}f(X^{1/2}PX^{1/2})X^{1/2}PXP \leq PXPf(X)PXP \tag{9.21}$$

We claim for any continuous function, h, on $[0, b]$, one has that

$$h(X^{1/2}PX^{1/2})X^{1/2}P = X^{1/2}Ph(PXP) \tag{9.22}$$

for this is trivial for polynomials which are dense. Once one has this, it holds for continuous functions on $[0, b)$ by a continuity argument. Thus

$$\begin{aligned} PXP\,Pf(PXP)P\,PXP &\leq PXP\,f(X)\,PXP \\ &= PXP\,Pf(X)P\,PXP \end{aligned} \tag{9.23}$$

On both sides, all operators act on Ran P and on this space PXP is invertible, so

$$Pf(PXP)P \leq Pf(X)P \tag{9.24}$$

Since $f(0) = 0$, $f(PXP) = Pf(PXP)P$, proving (9.11) when A is a projection. By Proposition 9.10(b), $f \in \mathcal{C}_n(0, b)$. □

Example 9.12 Let

$$f(x) = \frac{x}{1-x} \tag{9.25}$$

as a function on $(-1, 1)$. Since $g(x) \equiv f_0^{(1)}(x) = (1-x)^{-1}$ is in $\mathcal{M}_\infty(-1, 1)$, we see that $f \in \mathcal{C}_\infty(-1, 1)$ on account of Theorem 9.2. Since $f(x) = -1 + (1-x)^{-1}$, we have that

$$f'(z) = \frac{1}{(1-z)^2} \tag{9.26}$$

f' is a scalar monotone function on $(-1,1)$ (as it must be since f is scalar convex), but it is not in $\mathcal{M}_\infty(-1, 1)$ since

$$\operatorname{Im} f'(x+iy) = \frac{2y(1-x)}{[(1-x)^2+y^2]^2}$$

has both signs in $\mathbb{C}_+$. Indeed, by looking at Example 1.2, one sees without recourse to Loewner's theorem that f' is not even in $\mathcal{M}_2(-1,1)$. This is the promised example where f is matrix convex but f' is not matrix monotone. Another (related) example is $f(x) = x^\alpha; -1 \le \alpha < 0$ where f is matrix convex (see the next example) but $f'(x)$ is not matrix monotone (see the remarks after Theorem 4.1).

Similar to (9.25) is $x \mapsto x/(1+x)$ which is matrix concave on $(-1,\infty)$ (since $x \mapsto -(1+x)$ is matrix monotone). This is related to the matrix concavity (in two variables) of the harmonic mean (see Example 37.8). □

Example 9.13 We saw in Theorem 4.1 that $x \mapsto x^\alpha \in \mathcal{M}_\infty(0,\infty)$ if and only if $0 \le \alpha \le 1$ and $x \mapsto -x^\alpha \in \mathcal{M}_\infty(0,\infty)$ if and only if $-1 \le \alpha \le 0$ (indeed, if α is not in this range, the function is not even in $\mathcal{M}_2(0,\infty)$). By Theorem 9.9, when $\alpha > 0$, $f \in \mathcal{C}_\infty(0,\infty)$ if and only if $x^{-1}f(x) \in \mathcal{M}_\infty(0,\infty)$. It follows that $x \mapsto x^\alpha$ is matrix convex if and only if $1 \le \alpha \le 2$ and matrix concave if and only if $0 \le \alpha \le 1$ (and neither matrix convex nor matrix convex if $\alpha \ge 0$ is not in this range).

$\alpha < 0$ is more subtle. We'll see below in Example 10.4 that, $x \mapsto x^\alpha$ is matrix convex if $-1 \le \alpha \le 0$. On the other hand, using that when $\alpha < -1$, $f(z) = z^\alpha$ has half rays in the upper half plane where $|\operatorname{Im} f(z)|$ and $|\operatorname{Re} f(z)|$ are comparable and where (for different rays) $\operatorname{Im} f(z)$ goes to in one case ∞ and in the other $-\infty$, one can see that $(f(z) - f(1))/(z-1)$ is neither a Herglotz function nor the negative of a Herglotz function, so for such α, $x \mapsto x^\alpha$ is neither matrix convex nor matrix concave.

Summarizing, $x \mapsto x^\alpha$ is matrix convex if $\alpha \in [-1,0] \cup [1,2]$, matrix concave if $\alpha \in [0,1]$ and neither for other α. This should be compared to when $x \mapsto x^\alpha$ is scalar convex or scalar convex. Since that depends on positivity or negativity of the second derivative, all that matters is the sign of $\alpha(\alpha-1)$ so this map is scalar convex if $\alpha \ge 1$ or $\alpha \le 0$ and scalar concave if $\alpha \in [0,1]$. □

Example 9.14 If $f_s(x)$ is matrix convex, in general $h_s(x) \equiv df_s(x)/ds|_{s=s_0}$ will not be because in forming $(f_s(x) - f_{s_0}(x))/(s-s_0)$, we have a difference of convex functions which is neither convex nor concave in general. However, if f_{s_0} is affine, the derivative will be convex (resp. concave) if f_s is convex for $s \in (s_0, s_0+\epsilon]$ (resp. $[s_0-\epsilon, s_0)$). In the last example, we have that $x \mapsto -x^\alpha$ is affine at $\alpha = 0$ or 1 and convex in between. It follows that $x \mapsto \log(x)$ is matrix concave and $x \mapsto x\log(x)$ is matrix convex, in each case on $[0,\infty)$ □

Example 9.15 Let $f(x) = \frac{\pi\tan(\pi x)}{x}$ on $(-\frac{1}{2},\frac{1}{2})$ with $f(0) = \pi^2$. Then by (4.7), one has that

$$z^{-1}(f(z) - \pi^2) = \sum_{n=0}^{\infty} \frac{2z}{((n+\frac{1}{2})^2)[(n+\frac{1}{2})^2 - z^2]} \tag{9.27}$$

which is matrix monotone by (4.8), so $f \in \mathcal{C}_\infty(-\frac{1}{2}, \frac{1}{2})$. □

By combining Theorem 9.1 with the general representation formula (3.52) for functions in $\mathcal{M}_\infty(a, b)$, one sees that $g \in \mathcal{C}_\infty(a, b)$ if and only if there are finite measures, $\mu_\pm$ on $J_\pm$ given by (3.48) with (3.51)

$$\begin{aligned} g(z) =& \alpha z^2 + \beta z + \gamma + \int_b^\infty \frac{(z - x_0)^2}{(y - z)(y - x_0)^2}\, d\mu_+(y) \\ &+ \int_{-\infty}^a \frac{(z - x_0)^2}{(z - y)(y - x_0)^2}\, d\mu_-(y) \end{aligned} \tag{9.28}$$

and where $\alpha \geq 0$, $\beta, \gamma \in \mathbb{R}$. Since $(y - x_0)^2 - (y - z)(y + z - 2x_0) = (z - x_0)^2$, we can rewrite this as

$$\begin{aligned} g(z) = \alpha z^2 + \beta z + \gamma + &\int_b^\infty \left[\frac{1}{y - z} - \frac{y + z - 2x_0}{(y - x_0)^2} \right] d\mu_+(y) \\ + &\int_{-\infty}^a \left[\frac{1}{z - y} + \frac{y + z - 2x_0}{(y - x_0)^2} \right] d\mu_-(y) \end{aligned} \tag{9.29}$$

Instead of (3.52), one can use some of the various alternate forms discussed in Chapter 3, e.g., (3.55) or (3.68).

Notes and Historical Remarks For more information on scalar convex and scalar concave functions, see Simon [324].

In 1936, Kraus [189], a student of Loewner, initiated the study of matrix convex functions, f, in terms of the matrix $[\lambda_0, \lambda_i, \lambda_j; f]$. He proved the equivalence of the conditions (a) and (d) in Theorem 9.2. (b) $\Longleftrightarrow$ (c) in that Theorem is just Loewner's criterion for $f \in \mathcal{M}_n$ in terms of Loewner matrices. (a)$\Rightarrow$(e) in the Theorem is from Hansen–Tomiyama [138] (see Theorem 5.19). In an early draft of this book, the best I could do (if $(x)_n$ means condition (x) for a given n) is show that $(a)_{n+1} \Rightarrow (c)_n \Rightarrow (d)_n \Rightarrow (a)_n$. The full equivalence is a recent result of Heinävaara. That (e)$\Rightarrow$(a) and Theorem 9.3 are conjectures of Hansen–Tomiyama [138] proven by Heinävaara [146, 148].

In 1955, Bendat–Sherman [22] related this to $f_y^{(1)} \in \mathcal{M}_\infty$, wrote down an integral representation and noted Example 9.12. They proved for $(a, b) = (-1, 1)$ that $f_0^{(1)} \in \mathcal{M}_\infty(-1, 1) \Rightarrow f_y^{(1)} \in \mathcal{M}_\infty(-1, 1)$ for all $y \in (-1, 1)$. But their special point had to be the midpoint. That there was an argument for a general point seems to have only been written down in 2010 by Uchiyama [350] although at least one expert I consulted said he has known it for many years.

Theorem 9.9 was proven and heavily exploited by Hansen–Pedersen [136] in their proof of Loewner's theorem (see Chapter 28). Of course, since this is part of a proof of Loewner's theorem, they can't use Loewner's theorem as we do to prove it (!) but they proved finite n results. They had Theorem 9.11(b) but instead of part(a) had the result in the remark that required $f \in \mathcal{C}_{2n}(0, b)$. That $\mathcal{C}_{n+1}$ suffices was noted in Hiai–Sano [158].

That $x \mapsto -x \log x$ is in $-\mathcal{C}_\infty[0, 1]$ was first proven by Umegaki [355]. It plays an important role in quantum statistical mechanics and in quantum information theory.

Chapter 10
Convexity, II: Concavity and Monotonicity

This chapter will first provide a remarkable equivalence between matrix concavity and matrix monotonicity for positive functions not even hinted at in the scalar case. Then we'll discuss a connection between matrix convexity and Loewner matrices. As a warmup we claim

Proposition 10.1 *If g is scalar concave and non-negative on $(0, \infty)$, then g is scalar monotone increasing.*

Remarks

1. Below, this will be generalized to higher n.
2. $g(x) = x^\alpha, \alpha \geq 1$ shows that positivity (or at least being bounded from below) is essential for this result.

Proof If g is concave and $x < y < z$, then (9.1) (with the $\geq$ sign for resp. concave functions) implies that

$$g(z) \leq \frac{z-x}{y-x}\, g(y) - \frac{z-y}{y-x}\, g(x) \tag{10.1}$$

In particular, for x, y fixed,

$$\limsup_{z\to\infty} \frac{g(z)}{z} \leq \frac{g(y)-g(x)}{y-x} \tag{10.2}$$

Thus, if $g(y) < g(x)$, then $\limsup_{z\to\infty} g(z)/z < 0$. It follows that if g is a non-negative concave function on (α, ∞), then $g(y) \geq g(x)$ for all $y > x > \alpha$. □

As we've seen (see Corollary 14.5), if f is scalar concave on $(0, \infty)$ and positive, then it is scalar monotone increasing, but many positive scalar monotone functions are not concave, for example $f(x) = x^2$. Of course, this function, while scalar monotone, is not matrix monotone. In this chapter, we'll prove:

B. Simon, *Loewner's Theorem on Monotone Matrix Functions*, Grundlehren der mathematischen Wissenschaften 354, https://doi.org/10.1007/978-3-030-22422-6_10

Theorem 10.2

(a) $-f \in \mathcal{C}_n(0, \infty)$ *and* $f \geq 0 \Rightarrow f \in \mathcal{M}_n(0, \infty)$
(b) $f \in \mathcal{M}_{2n}(0, \infty) \Rightarrow -f \in \mathcal{C}_n(0, \infty)$

In particular, if $f \geq 0$*, then* $-f \in \mathcal{C}_\infty(0, \infty) \Leftrightarrow f \in \mathcal{M}_\infty(0, \infty)$

Remark Some simple examples show that all the conditions are needed. The function $f(x) = -x^2$ lies in $-\mathcal{C}_\infty(0, \infty)$ (see Example 9.13) but is not even scalar monotone showing that $f \geq 0$ is needed in (a). The function $f(x) = 1 - x^2$ is positive on $(0, 1)$ and has $-f \in \mathcal{C}_\infty(0, 1)$ but is not even scalar monotone showing that (a) requires a semi-infinite interval. Also $f(x) = (2 - x)^{-1}$ lies in $\mathcal{M}_\infty(0, 1)$ (see Proposition 2.2; note that $0 < A \leq B < 1 \Rightarrow (2 - B) \leq (2 - A) \Rightarrow (2 - A)^{-1} \leq (2 - B)^{-1}$ by (2.5)) but is strictly scalar convex (even matrix convex by the next example) and so in no $-\mathcal{C}_n(0, 1)$ showing that (b) also needs a semi-infinite interval.

Corollary 10.3 *Let* $n \geq 2$*. If* $f \in \mathcal{M}_{2n}(0, \infty)$*, then for* $j = 1, \ldots, n$*,* $f^{(2j)}(x) \leq 0$ *on* $(0, \infty)$*. In particular, if* $f \in \mathcal{M}_\infty(0, \infty)$*, then for all* $j \geq 1$*, we have that*

$$(-1)^{j+1} f^{(j)}(x) \geq 0 \tag{10.3}$$

Remarks

1. Equation (10.3) also follows from Loewner's theorem and the Herglotz representation but we emphasize that this proof does not use Loewner's theorem.
2. We'll eventually show (see Theorem 26.1 and the Notes to Chapter 26) that (10.3) implies that f is real analytic on $(0, \infty)$ with an analytic continuation to the right half plane.

Proof By Corollary 9.4 and Theorem 10.2(b), $f^{(2j)} \leq 0$ in distributional sense. Since $f \in C^{4n-3}$, it holds pointwise. Equation (10.3) for j odd follows from Proposition 5.20. □

Example 10.4 Since Theorem 10.2(b) doesn't depend on the sign of f and $x \mapsto -x^\alpha$ is matrix monotone for $-1 \leq \alpha \leq 0$ (see the Remark after Theorem 4.1), we conclude that $x \mapsto x^\alpha$ is matrix convex for $-1 \leq \alpha \leq 0$. □

Proof of Theorem 10.2 (Only (a); for part (b), see Theorem 10.6) Suppose that $-f \in \mathcal{C}_n(0, \infty)$, $f \geq 0$ and $0 \leq A \leq B$ for two $n \times n$ matrices. Write

$$\begin{aligned} B &= A + \epsilon(\epsilon^{-1}(B - A)) \\ &= (1 + \epsilon)\left[\frac{1}{1+\epsilon}A + \frac{\epsilon}{1+\epsilon}\epsilon^{-1}(B - A)\right] \end{aligned} \tag{10.4}$$

Since $\frac{1}{1+\epsilon} + \frac{\epsilon}{1+\epsilon} = 1$, concavity of f implies that

$$f\left(\frac{1}{1+\epsilon}B\right) \geq \frac{1}{1+\epsilon}f(A) + \frac{\epsilon}{1+\epsilon}f(\epsilon^{-1}(B-A))$$
$$\geq \frac{1}{1+\epsilon}f(A)$$

since $f \geq 0$. Taking $\epsilon \downarrow 0$, we see that $f(B) \geq f(A)$ so f is monotone on $n \times n$ matrices. □

Recall that if P is an orthogonal projection on a Hilbert space, $\mathcal{H}$, and $A \in \mathcal{L}(\mathcal{H})$, we defined in Chapter 2 $A_P \in \mathcal{L}(P\mathcal{H})$ by $A_P\varphi = PA\varphi$ for all $\varphi \in P\mathcal{H}$ (i.e., $A_P = PAP \restriction P\mathcal{H}$). The following is not unconnected to the Jensen's inequality of the next chapter.

Proposition 10.5 *Let $A \in \mathcal{L}(\mathcal{H})$ for some (perhaps finite dimensional) Hilbert space, $\mathcal{H}$, with $A \geq 0$. Suppose that f is continuous and monotone on bounded positive operators on $\mathcal{H}$. Let P be an orthogonal projection in $\mathcal{L}(\mathcal{H})$. Then*

$$f(A_P) \geq f(A)_P \tag{10.5}$$

Proof Writing $\mathcal{H} = P\mathcal{H} \oplus (1-P)\mathcal{H}$, we can decompose

$$A = \begin{pmatrix} a_{11} & a_{12} \\ a_{21} & a_{22} \end{pmatrix} \tag{10.6}$$

where, for example $A_P = a_{11}$. Given $\epsilon > 0$, let $T_\epsilon = \begin{pmatrix} \epsilon^{1/2} & 0 \\ 0 & -\epsilon^{-1/2} \end{pmatrix}$. Then

$$T_\epsilon^* A T_\epsilon = \begin{pmatrix} \epsilon a_{11} & -a_{12} \\ -a_{21} & \epsilon^{-1} a_{22} \end{pmatrix} \tag{10.7}$$

is ≥ 0 if $A \geq 0$. Thus

$$A \leq A + T_\epsilon^* A T_\epsilon = \begin{pmatrix} (1+\epsilon)a_{11} & 0 \\ 0 & (1+\epsilon^{-1})a_{22} \end{pmatrix} \tag{10.8}$$

Since f is monotone

$$f(A) \leq \begin{pmatrix} f((1+\epsilon)a_{11}) & 0 \\ 0 & f((1+\epsilon^{-1})a_{22}) \end{pmatrix} \tag{10.9}$$

Since $B \leq C \Rightarrow B_P \leq C_P$, we conclude that

$$f(A)_P \leq f((1+\epsilon)A_P) \tag{10.10}$$

Taking $\epsilon \downarrow 0$ yields (10.5). □

The following completes the proof of Theorem 10.4.

Theorem 10.6 *If $f \in \mathcal{M}_{2n}(0,\infty)$, then $-f \in \mathcal{C}_n(0,\infty)$*

Proof Write $\mathbb{C}^{2n} = \mathbb{C}^n \oplus \mathbb{C}^n$ with P the projection onto $\mathbb{C}^n \oplus \{0\}$. Given $A, B \geq 0$ and $\theta \in [0,1]$, define

$$U_\theta = \begin{pmatrix} \sqrt{\theta}\,\mathbb{1} & \sqrt{1-\theta}\,\mathbb{1} \\ -\sqrt{1-\theta}\,\mathbb{1} & \sqrt{\theta}\,\mathbb{1} \end{pmatrix} \tag{10.11}$$

which is unitary on $\mathbb{C}^{2n}$.

By a simple calculation, $\left[U_\theta \begin{pmatrix} A & 0 \\ 0 & B \end{pmatrix} U_\theta^{-1}\right]_{11} = \theta A + (1-\theta)B$. Therefore, by Proposition 10.5

$$\begin{aligned} f(\theta A + (1-\theta)B) &\geq \left[f\left(U_\theta \begin{pmatrix} A & 0 \\ 0 & B \end{pmatrix} U_\theta^{-1}\right)\right]_{11} \\ &= \left[U_\theta \begin{pmatrix} f(A) & 0 \\ 0 & f(B) \end{pmatrix} U_\theta^{-1}\right]_{11} \\ &= \theta f(A) + (1-\theta) f(B) \end{aligned}$$

proving that f is concave on positive matrices on $\mathbb{C}^n$. □

Corollary 10.7 **($\equiv$ Theorem 1.1; see also Corollary 14.3)** *Any $f \in \mathcal{M}_2(-\infty,\infty)$ has the form*

$$f(x) = ax + b \tag{10.12}$$

for $a \geq 0$.

Proof By Theorem 10.6, f is scalar concave on each (a,∞) and so on $(-\infty,\infty)$. But $-f(-x)$ is also in $\mathcal{M}_2(-\infty,\infty)$, so $-f(-x)$ is scalar concave, and thus $f(x)$ is scalar convex. Any function which is both scalar convex and scalar concave is linear. Monotonicity implies $a \geq 0$. □

Since some proofs of the hard part of Loewner's theorem rely on the finite n aspects of Theorem 10.2, it is important to have the finite n results. But if one has Loewner's theorem, there is an easy, direct proof of the "in particular" assertion at the end of that theorem, that is one can prove

Theorem 10.8 . *Let $f \geq 0$ be a function on $(0,\infty)$. Then $f \in \mathcal{M}_\infty(0,\infty) \iff -f \in \mathcal{C}_\infty(0,\infty)$.*

Proof We compare the representations (3.52) for $f \in \mathcal{M}_\infty(0,\infty)$ and (9.29) for $g = -f \in \mathcal{C}_\infty(0,\infty)$. For $\mathcal{M}_\infty(0,\infty)$, we take $x_0 = 1, b = \infty, u = -y$, and $d\rho(u) = d\nu_-(-u)/(1+u)$ to see that

$$f(z) = \alpha z + \gamma + \int_0^\infty \left[-\frac{1}{z+u} + \frac{1}{u+1} \right] d\rho(u) \tag{10.13}$$

with $\alpha \geq 0$, $\gamma \in \mathbb{R}$ and

$$\int_0^\infty (1+y^2)^{-1}\, d\rho(u) < \infty \tag{10.14}$$

For $-\mathcal{C}_\infty(0,\infty)$, we take $x_0 = 1$, $b = \infty$, $u = -y$, and $d\xi(u) = d\mu_-(-u)$ and get a representation

$$f(z) = \eta z^2 + \beta z + \kappa + \int_0^\infty \left[-\frac{1}{z+u} + \frac{u-z+2}{(u+1)^2} \right] d\xi(u) \tag{10.15}$$

with $\eta \leq 0$, $\beta, \kappa \in \mathbb{R}$ and

$$\int_0^\infty (1+u^2)^{-3/2} < \infty \tag{10.16}$$

The integrand in (10.15) is $-\frac{(z-1)^2}{(z+u)(u+1)^2}$ since $(u-z+2)(z+u)-(u+1)^2 = -(z-1)^2$. Thus (10.16)$\Rightarrow$ integral in (10.15) is $o(z^2)$, so

$$\lim_{x\to+\infty} \frac{f(x)}{x} = \begin{cases} -\infty, & \text{if } \eta < 0 \\ \beta - \int \frac{d\xi(u)}{(u+1)^2}, & \text{if } \eta = 0 \end{cases}$$

so if f obeys (10.15) and $f \geq 0$, we must have that $\eta = 0$ and $\alpha \equiv \beta - \int \frac{d\xi(u)}{(u+1)^2} \geq 0$. In particular, $\int \frac{d\xi(u)}{(u+1)^2} < \infty$. Write $u - z + 2 = (u+1) - (-z-1)$ and see that if $f \in \mathcal{C}_\infty(0,\infty)$ with $f \geq 0$, then f has the form (10.13) where ρ obeys (10.14). Conversely, any function of the form (10.13) can be rewritten in the form (10.15). □

The following remarkable result is related to these ideas.

Theorem 10.9 *Let f be a real-valued function on (a,b), a finite open interval. Suppose that f lies in $\mathcal{M}_\infty(a,b) \cap [-\mathcal{C}_\infty(a,b)]$ (i.e., is both matrix monotone and matrix concave). Then f has an analytic continuation to (a,∞) which lies in $\mathcal{M}_\infty(a,\infty) \cap [-\mathcal{C}_\infty(a,\infty)]$.*

Proof f has a representation of the form (3.52) and $-f$ one of the form (9.29). Pick $x_0 = \frac{1}{2}(a+b)$. On (x_0,∞), we have that

$$\lim_{\epsilon\downarrow 0} \pi \operatorname{Im} f(x+i\epsilon)\, dx = \frac{d\nu_+(x)}{x-x_0} \tag{10.17}$$

$$\lim_{\epsilon \downarrow 0} -\pi \operatorname{Im} f(x + i\epsilon)\, dx = d\mu_+(x) \tag{10.18}$$

in the sense of weak convergence. Since ν_+ and μ_+ are both positive, we conclude that both are zero. Equation (3.52) with $\nu_+ = 0$ gives a function with real analytic values on (a, ∞). This function is matrix monotone by the easy half of Loewner's theorem. By Theorem 10.8, $-f \in \mathcal{C}_\infty(a, \infty)$.

□

If we combine Theorems 10.2(b) and 9.11(a), we see that if $f \in \mathcal{M}_{2n+2}(0, \infty)$ is continuous at 0 with $f(0) = 0$ and $g(t) = f(t)/t$, then $-g \in \mathcal{M}_n(0, \infty)$. In fact more is true—we can go from $2n$ instead of $2n + 2$:

Theorem 10.10 *Let $f \in \mathcal{M}_{2n}(0, \infty)$ be continuous at 0 with $f(0) = 0$ and let $g(t) = f(t)/t$. Then $-g \in \mathcal{M}_n(0, \infty)$. In particular, $f \in \mathcal{M}_\infty(0, \infty) \Rightarrow -g \in \mathcal{M}_\infty(0, \infty)$.*

Remark If $f(t) = t^\alpha$ with $\alpha > 1$, then $-g$ is not even scalar monotone, so this shows once again that such f's are not in $\mathcal{M}_\infty(0, \infty)$, not even in $\mathcal{M}_2(0, b)$ for any b.

Proof This will combine calculations in the proofs of Proposition 9.10 and the remark to Theorem 9.11.

Let $\|A\| \leq 1$ for an $n \times n$ matrix A. Let $U = \left(\begin{smallmatrix} A & B \\ C & -A^* \end{smallmatrix}\right)$ with B, C given by (9.12). By (9.14), for an $n \times n$ matrix, Z,

$$\left[U \left(\begin{smallmatrix} Z & 0 \\ 0 & 0 \end{smallmatrix}\right) U^{-1}\right]_{11} = A^* Z A \tag{10.19}$$

By Proposition 10.5, for any $n \times n$ self-adjoint matrix, X, with costive spectrum, $f\left(\left[U \left(\begin{smallmatrix} X & 0 \\ 0 & 0 \end{smallmatrix}\right) U^{-1}\right]_{11}\right) \geq \left[U f\left(\left(\begin{smallmatrix} X & 0 \\ 0 & 0 \end{smallmatrix}\right)\right) U^{-1}\right]_{11}$ which, by (10.19) says that $f(A^* X A) \geq A^* f(X) A$. If $0 < Y \leq X$, take $A = X^{-1/2} Y^{1/2}$ to get $g(Y) \geq g(X)$, i.e., that $-g$ is monotone. □

We now want to consider three maps of functions $f : (0, \infty) \to (0, \infty)$ to other such functions:

$$f^\sharp(t) = t f(t)^{-1}; \qquad f^\flat(t) = f(t^{-1})^{-1}; \qquad f^*(t) = t f(t^{-1}) \tag{10.20}$$

Notice that

$$f^{\sharp\sharp} = f; \quad f^{\flat\flat} = f; \quad f^{**} = f; \quad (f^\flat)^\sharp = f^* \tag{10.21}$$

Now let $\mathcal{P}$ be the set of strictly positive functions in $\mathcal{M}_\infty(0, \infty)$

Theorem 10.11 *Each of the maps $f \mapsto f^\sharp$, $f \mapsto f^\flat$, and $f \mapsto f^*$ is a bijection of $\mathcal{P}$ to itself.*

Proof Because of the first three equations of (10.21), it suffices to prove that each of the maps takes $\mathcal{P}$ to $\mathcal{P}$. By the last of those equations, we need only prove this for the first two of the three maps.

By (2.5), $A \mapsto A^{-1}$ is order reversing on strictly positive matrices, so we conclude that $f^{\flat}$ is order preserving if f is order preserving.

Let $f^{\natural}(t) = f(t)/t$. By Theorem 10.10, if f is order preserving on positive matrices, $f^{\natural}$ is order reversing. Since $A \mapsto A^{-1}$ is also order reversing, $f^{\sharp} = (f^{\natural})^{-1}$ is order preserving. □

Another theorem of the same genre that depends on Loewner's theorem

Theorem 10.12 *Let $f \in \mathcal{M}_\infty(0,\infty)$ with $f \geq 0$ and not identically zero. Then $h(t) \equiv tf(t)$ is a bijection of $[0,\infty)$ to itself and its compositional inverse, g, lies in $\mathcal{M}_\infty(0,\infty)$.*

Remark Compositional inverses of functions in $\mathcal{M}_\infty(0,\infty)$ may not seem very interesting but we'll see in Chapter 43 that they are.

Proof Since f is scalar strictly monotone, continuous, and strictly positive on $(0,\infty)$, h is strictly monotone and continuous with $h(0) = 0$ and $\lim_{t\uparrow\infty} h(t) = \infty$, so h is a bijection as claimed.

Suppose first that

$$\int \lambda\, d\mu(\lambda) < \infty \Rightarrow \int \frac{\lambda}{1+\lambda}(\lambda\, d\mu(\lambda)) < \infty$$

so we can use $\lambda\, d\mu(\lambda)$ as a measure in (3.68). Using that representation for f, we see that

$$h(t) = \gamma t + \beta t^2 - q(t) \tag{10.22}$$

where

$$\gamma = \alpha + \int_0^\infty \lambda\, d\mu(\lambda); \qquad q(t) = \int_0^\infty \frac{\lambda t}{\lambda + t} \lambda\, d\mu(\lambda) \tag{10.23}$$

by writing

$$\frac{\lambda t^2}{\lambda + t} = \lambda t - \frac{\lambda^2 t}{\lambda + t}$$

In particular, by (3.68), $q \in \mathcal{M}_\infty(0,\infty)$.

Note that $\gamma \geq 0$, $\beta \geq 0$ and since f is not identically zero, at least one of them is strictly positive. By (10.22), g is the unique positive solution of

$$\gamma g(t) + \beta g(t)^2 = q(g(t)) + t \tag{10.24}$$

Define $\{g_n(t)\}_{n=0}^{\infty}$ inductively by $g_0(t) \equiv 0$ and $g_{n+1}(t)$ is the unique positive solution of

$$\gamma g_{n+1}(t) + \beta g_{n+1}(t)^2 = q(g_n(t)) + t \tag{10.25}$$

so that

$$g_{n+1}(t) = \begin{cases} \frac{-\gamma+\sqrt{\gamma^2+4\beta(q(g_n(t))+t)}}{2\beta}, & \text{if } \beta > 0 \\ \frac{q(g_n(t))+t}{\gamma}, & \text{if } \beta = 0 \end{cases} \tag{10.26}$$

We first claim inductively that each g_n lies in $\mathcal{M}_\infty(0, \infty)$. For this is trivial if $n = 0$ and if $g_n \in \mathcal{M}_\infty(0, \infty)$, so is g_{n+1} since both q and the square root function lie in $\mathcal{M}_\infty(0, \infty)$ and the composition of positive functions in $\mathcal{M}_\infty(0, \infty)$ is in $\mathcal{M}_\infty(0, \infty)$.

Next, we claim inductively that $g_n \leq g$. Again this is trivial if $n = 0$. Moreover, if $g_n \leq g$, then $4q(g_n(t)) + t \leq q(g(t)) + t$ since q is monotone. Thus, $\gamma g_{n+1}(t)+\beta g_{n+1}(t)^2 \leq \gamma g(t)+\beta g(t)^2$. Since $x \mapsto \gamma x+\beta x^2$ is strictly monotone, we conclude that $g_{n+1} \leq g$.

We also claim inductively that $g_n \leq g_{n+1}$. Again, this is trivial for $n = 0$ so suppose it is true for $n = m - 1$. $g_{n-1} \leq g_n$, (10.25), and the scalar monotonicity of q imply that $\gamma g_n + \beta g_n^2 \leq \gamma g_{n+1} + \beta g_{n+1}^2$. Since $x \mapsto \gamma x + \beta x^2$ is strictly monotone, we conclude that $g_n \leq g_{n+1}$.

Thus g_n has a pointwise limit, g_∞ which obeys (10.24) and so equals g. Therefore $g \in \mathcal{M}_\infty(0, \infty)$.

Given an arbitrary $f \in \mathcal{M}_\infty(0, \infty)$, let $d\mu$ be the measure in (3.68). Define

$$f_n(t) = \alpha + \beta t + \int_0^n \frac{\lambda t}{\lambda + t} d\mu(t) \tag{10.27}$$

By the special case above, each $g_n \in \mathcal{M}_\infty(0, \infty)$. It is easy to see that $g_n \to g$ pointwise, so $g \in \mathcal{M}_\infty(0, \infty)$. □

Example 10.13 Let

$$h(t) = t^p \log(t + 1); \quad p \geq 1 \tag{10.28}$$

and

$$\tilde{h}(t) = t[\log(1 + t)]^{1/p} \tag{10.29}$$

Since $x \mapsto x^{1/p}$ and $x \mapsto \log(1+x)$ are both in $\mathcal{M}_\infty(0, \infty)$, so is their composition and Theorem 10.12 implies that $\tilde{h}$ is the compositional inverse to a function in $\mathcal{M}_\infty(0, \infty)$. Composing this with $x \mapsto x^p$ which is also the inverse of a function in $\mathcal{M}_\infty(0, \infty)$, we see that h is the compositional inverse of a function in $\mathcal{M}_\infty(0, \infty)$. □

Suppose that $h(t)$ is a matrix convex function on $(0, \infty)$ which is a bijection of $(0, \infty)$ to itself (and so scalar monotone). $f(t) = h(t)/t$ is matrix monotone by Theorem 9.9 and $f \geq 0$, so by Theorem 10.12 can be rephrased:

Theorem 10.14 *Let h be a bijection of $(0, \infty)$ to itself with $h'(t) > 0$ on $(0, \infty)$ and so that $-h \in \mathcal{C}_\infty(0, \infty)$. Then the compositional inverse, g, of h (i.e., the function with $g(h(t)) = h(g(t)) = t$) lies in $\mathcal{M}_\infty(0, \infty)$.*

The following, proven in Chapter 12, is a result that extends this last theorem:

Theorem 10.15 ($\equiv$ Theorem 12.16) *Let h be a C^1 function on an interval $(a, b) \subset \mathbb{R}$ (a or b can be infinite) with $h'(t) > 0$ for all $t \in (a, b)$. Suppose that for some $c \in (a, b)$, we have that*

$$f(x) = \frac{h(x) - h(c)}{x - c} \tag{10.30}$$

lies in $-\mathcal{C}_n(a, b)$. Then g, the compositional inverse of h, lies in $\mathcal{M}_n(h(a), h(b))$.

To see the connection, we note that

(Second) Proof of Theorem 10.14 given Theorem 10.15 By Theorem 9.1, if $h \in C_\infty(0, \infty)$, then f given by (10.30) is in $\mathcal{M}_\infty(0, \infty)$ for any c. By Theorem 10.2(b), $f \in -\mathcal{C}(0, \infty)$. Since $h((0, \infty)) = (0, \infty)$ by the bijection hypothesis, Theorem 10.15 implies that $g \in \mathcal{M}_\infty(0, \infty)$. $\square$

We will prove Theorem 10.15 in Chapter 12.

Notes and Historical Remarks In 1982, Hansen–Pedersen [136] proved results that imply that if $f \in \mathcal{M}_{4n}(0, \infty)$, then $-f \in \mathcal{C}_n(0, \infty)$ and they proved for $f \geq 0$, that $-f \in \mathcal{C}_n(0, \infty) \Rightarrow f \in \mathcal{M}_n(0, \infty)$. In particular, they had the results for $n = \infty$. The first published place that the $\mathcal{M}_{2n}$ result appears in Mathias [218]. The proof we give for Theorem 10.6 appeared in print first in Hansen [131]; he repeated the argument in [133]. Earlier, in his unpublished 1989 note, Boutet de Monvel [45] not only had the $\mathcal{M}_{2n}$ result (apparently unaware of the earlier work of Hansen–Pedersen [136]) but also the proof we give here.

Theorem 10.10 is implicit in [136] and made explicit in Hiai–Sana[158]. Theorem 10.12 and Example 10.13 are from Ando [16]. Theorem 10.15 when $n = \infty$ is due to Pedersen–Uchiyama [263]; see the Notes to Chapter 12.

The idea of looking at block matrices is very powerful and often used; see, for example, the books of Bhatia [33, 34] and Hiai–Petz[157]. Bhatia says that the pioneering work on this technique is Ando [14, 15]. While Ando undoubtedly popularized the method and showed its power, it certainly appeared earlier (e.g., Connes [63]). Related ideas appeared in von Neumann's work in 1930s. We used it in Chapter 4, will use it in the next chapter, Chapter 13 and in Part III. In particular, the ideas discussed in Chapter 13 under the rubric Schur complements are very powerful.

Theorem 10.9 appears in Brown–Uchiyama [50] who say they suspect it is known to some people but they haven't seen a proof. While it is not stated explicitly, the result is used and proven in Šmul'jan [331] in 1965.

Hansen [133] has emphasized the use of the maps $f \mapsto f^\sharp$ and $f \mapsto f^*$ and that paper has Theorem 10.11. They will play a significant role in Chapter 28.

There has been discussion of Theorem 10.8 when $f \leq 0$, equivalently, for $f \geq 0$, of the relation of the conditions $f \in \mathcal{C}_\infty(0, \infty)$ and $-f \in \mathcal{M}_\infty(0, \infty)$. By Theorem 10.2 (b), we have for $f \geq 0$ that $-f \in \mathcal{M}_\infty(0, \infty) \Rightarrow f \in \mathcal{C}_\infty(0, \infty)$. The example $f(t) = t^2$ (essentially in the Remark after Theorem 10.2) shows that the opposite implication does not hold in general. Hiai et al. [156] have shown that if $f \geq 0$ obeys

$$tf(t) = f(t^{-1}) \tag{10.31}$$

for all $t \in (0, \infty)$, then $-f \in \mathcal{M}_\infty(0, \infty) \iff f \in \mathcal{C}_\infty(0, \infty)$. The obvious example where (10.31) holds is $f(t) = t^{-1/2}$ but there are others, for example, $f(t) = (1 + t)^{-1}$ or $f(t) = 1 + t^{-1}$. If g obeys (10.31), is matrix monotone and normalized by $g(t) = 1$, then $f(t) = 1/g(t)$ obeys (37.19) and $(A, B) \mapsto A^{1/2} f(A^{-1/2} BA^{-1/2}) A^{1/2}$ is an operator mean in the sense of Chapter 36 and is symmetric (i.e., invariant under the interchange of A and B). This is discussed further in Chapter 37.

Chapter 11
Convexity, III: Hansen–Jensen–Pedersen (HJP) Inequality

Jensen's inequality in its original form says that if f is a scalar convex function (on an open convex set, A, of a vector space, V) and if $\{x_j\}_{j=1}^n \subset A$; $0 \leq \theta_j \leq 1$, $j = 1, \ldots, n$ with $\sum_{j=1}^n \theta_j = 1$, then $f(\sum_{j=1}^n \theta_j x_j) \leq \sum_{j=1}^n \theta_j f(x_j)$. This follows by an easy induction from the definition. More generally, if μ is a probability measure on A, then $f(\int x\, d\mu) \leq \int f(x)\, d\mu$ which one can prove by approximating with point measures (or by other means; see [325, Section 5.3]). Recall that given an orthogonal projection P on a Hilbert space, $\mathcal{H}$, and $A \in \mathcal{L}(\mathcal{H})$, we defined $A_P \in \mathcal{L}(\text{Ran}(P))$ by $A_P = PAP \restriction \text{Ran}(P)$. Here we'll prove the following operator version:

Theorem 11.1 (Hansen–Jensen–Pedersen Theorem) *Let f be a real-valued continuous function on an open interval, $I \subset \mathbb{R}$. Let $\mathcal{H}$ be an infinite dimensional (separable) Hilbert space. Then the following are equivalent:*

(a) *f is matrix convex*
(b) *For each n and each n-tuple of bounded self-adjoint operators, $\{x_j\}_{j=1}^n$ with spectra in I and every n-tuple of bounded operators, $\{a_j\}_{j=1}^n$ with $\sum_{j=1}^n a_j^* a_j = \mathbb{1}$, we have that*

$$f\left(\sum_{j=1}^n a_j^* x_j a_j\right) \leq \sum_{j=1}^n a_j^* f(x_j) a_j \tag{11.1}$$

(c) *For every isometry v on $\mathcal{H}$, and every self-adjoint operator, x, with spectrum in I*

$$f(v^* x v) \leq v^* f(x) v \tag{11.2}$$

(d) *For any self-adjoint projection, P and any self-adjoint operator, x, with spectrum in I*

B. Simon, *Loewner's Theorem on Monotone Matrix Functions*, Grundlehren der mathematischen Wissenschaften 354, https://doi.org/10.1007/978-3-030-22422-6_11

$$f(x_P) \leq f(x)_P \tag{11.3}$$

Remarks

1. x is a self-adjoint operator with spectrum in a closed interval, J, if and only if $\langle \varphi, x\varphi \rangle \in J$ for any unit vector φ. If $\varphi \in P\mathcal{H}$, then $\langle \varphi, x_P\varphi \rangle = \langle \varphi, x\varphi \rangle$ so we see that x_P also has its spectrum in J. Alternatively, if $J = (a, b)$, then x has its spectrum in J if and only if $a\mathbf{1} \leq x \leq b\mathbf{1}$ and this implies that $aP \leq PxP \leq bP$ which restricted to $P\mathcal{H}$ says that x_P has spectrum in J.
2. Equation (11.3) is called the *Davis condition*. Davis [71] calls it the Sherman property because he proved the equivalence of (a) and (d) in [70] following a suggestion of Sherman.

The key implication (a) $\Rightarrow$ (b) is the hardest step in the proof and requires some preliminary machinery to which we now turn. Given $\mathcal{H}$, a Hilbert space, and $m \in \mathbb{Z}_+$, $\mathcal{H}^{(m)}$ is the direct sum $\mathcal{H}_1 \oplus \cdots \oplus \mathcal{H}_m$ where each $\mathcal{H}_j$ is a copy of $\mathcal{H}$. $A \in \mathcal{L}(\mathcal{H}^{(m)})$ will be written $(a_{ij})_{i,j=1}^m$, where a_{ij} is a bounded map from $\mathcal{H}_j$ to $\mathcal{H}_i$, (i.e., $[A(\varphi_1, \ldots, \varphi_m)^t]_i = \sum_{j=1}^m a_{ij}\,\varphi_j$).

Proposition 11.2 *Let f be a matrix convex function on an open interval, $I \subset \mathbb{R}$. Let $X \in \mathcal{L}(\mathcal{H}^{(m)})$ be a self-adjoint operator with spectrum in I and let U be a unitary map in $\mathcal{L}(\mathcal{H}^{(m)})$. Then*

$$f([U^*XU]_{mm}) \leq [U^*f(X)U]_{mm} \tag{11.4}$$

Proof $A \in \mathcal{L}(\mathcal{H}^{(m)})$ is called *diagonal* if $a_{ij} = 0$ for $i \neq j$ and $\mathrm{diag}(A)$ is the diagonal operator obtained from a general A by replacing all a_{ij} by 0 for $i \neq j$. Let $\omega = e^{2\pi i/m}$ be a primitive mth root of unity so that $\frac{1}{m}\sum_{j=1}^m (\omega^\ell)^j = \delta_{\ell 0}$ if $|\ell| < m$. Let W be the diagonal unitary matrix with $W_{jj} = \omega^j \mathbb{1}$, $j = 1, \ldots, m$ so $(W^*AW)_{ij} = \omega^{j-i}a_{ij}$ which implies that

$$\frac{1}{m}\sum_{j=1}^m (W^j)^* A W^j = \mathrm{diag}(A) \tag{11.5}$$

Thus

$$\begin{aligned} f\left([U^*XU]_{mm}\right) &= f\left([\mathrm{diag}(U^*XU)]_{mm}\right) \\ &= f(\mathrm{diag}(U^*XU))_{mm} &(11.6) \\ &= f\left(\frac{1}{m}\sum_{j=1}^m (W^j)^* U^*XUW^j\right)_{mm} &(11.7) \\ &\leq \frac{1}{m}\sum_{j=1}^m f((W^j)^*U^*XUW^j)_{mm} &(11.8) \end{aligned}$$

$$= \frac{1}{m}\sum_{j=1}^{m}[(W^j)^* U^* f(X) U W^j]_{mm} \tag{11.9}$$

$$= [U^* f(X) U]_{mm} \tag{11.10}$$

proving (11.4)

In the above, (11.6) uses $f(A)_{ii} = f(a_{ii})$ for diagonal A and (11.7) uses (11.5). (11.8) uses matrix convexity of f and (11.9) the unitary covariance of $A \mapsto f(A)$, i.e., that $f(V^*AV) = V^* f(A) V$ for any unitary, V. Finally (11.10) uses (11.5) again. □

Proposition 11.3 *Let V be an isometry of $\mathcal{K}_2$ to $\mathcal{K}_1$, two Hilbert spaces. Let $U : \mathcal{K} = \mathcal{K}_1 \oplus \mathcal{K}_2 \to \mathcal{K}$ by*

$$U = \begin{pmatrix} P & V \\ V^* & 0 \end{pmatrix} \tag{11.11}$$

where P is the projection onto the orthogonal complement of Ran V. *Then U is unitary.*

Remarks

1. VV^* is the projection onto Ran V so $P = \mathbb{1} - VV^*$. In particular, for $\varphi_1 \in \mathcal{K}_1$:

$$\|P\varphi_1\|^2 + \|V^*\varphi_1\|^2 = \|\varphi_1\|^2 \tag{11.12}$$

2. If $\mathcal{K}_1 = \mathcal{K}_2$, we can conjugate with $\left(\begin{smallmatrix} 0 & \mathbb{1} \\ \mathbb{1} & 0 \end{smallmatrix}\right)$ to get $\widetilde{U} = \left(\begin{smallmatrix} V^* & P \\ 0 & V \end{smallmatrix}\right)$ with $\widetilde{U} \restriction \mathcal{K}_2 = V$, so $\widetilde{U}$ is a unitary dilation in the language of the Notes of Chapter 19, indeed the minimal dilation. It is interesting to find explicit bases that show if V is right shift on $\ell^2(\mathbb{Z}_+)$, then $\widetilde{U}$ is the two sided shift on $\ell^2(\mathbb{Z})$.

Proof Since $U = U^*$, it suffices to show that $U^*U = \mathbb{1}$, i.e., $\|U\left(\begin{smallmatrix} \varphi_1 \\ \varphi_2 \end{smallmatrix}\right)\| = \|\left(\begin{smallmatrix} \varphi_1 \\ \varphi_2 \end{smallmatrix}\right)\|$. We compute

$$\begin{aligned} \left\| U \begin{pmatrix} \varphi_1 \\ \varphi_2 \end{pmatrix} \right\|^2 &= \left\| \begin{pmatrix} P\varphi_1 + V\varphi_2 \\ V^*\varphi_1 \end{pmatrix} \right\|^2 \\ &= \|P\varphi_1 + V\varphi_2\|^2 + \|V^*\varphi_1\|^2 \\ &= \|P\varphi_1\|^2 + \|V\varphi_2\|^2 + \|V^*\varphi_1\|^2 \qquad (\text{since } V\varphi_2 \perp \text{Ran } P) \\ &= \|\varphi_1\|^2 + \|\varphi_2\|^2 \qquad (\text{by (11.12)}) \end{aligned}$$

□

Proposition 11.4 *Let $\mathcal{H}$ be a Hilbert space and $\{a_j\}_{j=1}^n$ bounded operators with*

$$\sum_{j=1}^{n} a_j^* a_j = \mathbb{1} \tag{11.13}$$

Then with $m = n + 1$ and $a_m = 0$, there is a unitary map $U \in \mathcal{L}(\mathcal{H}^{(m)})$ so that $U_{jm} = a_j$.

Proof Let $\mathcal{K}_2 = \mathcal{H}, \mathcal{K}_1 = \mathcal{H}^{(n)}$ so $\mathcal{K} \equiv \mathcal{K}_1 \oplus \mathcal{K}_2 = \mathcal{H}^{(m)}$. $V : \mathcal{K}_2 \to \mathcal{K}_1$ by $(V\varphi)_j = a_j\varphi$ is an isometry by (11.13), so the U of (11.11) is the required unitary. □

Proof of Theorem 11.1 We'll show (a) ⇒ (b) ⇒ (c) ⇒ (d) ⇒ (a).

(a) ⇒ (b) Given $\{a_j\}_{j=1}^n$ obeying (11.13), let U be the unitary on $\mathcal{H}^{(m)}$ (where $m = n + 1$) constructed in Proposition 11.4. Then for any diagonal $Y \in \mathcal{L}(\mathcal{H}^{(m)})$,

$$(U^*YU)_{mm} = \sum_{j=1}^{n} a_j^* y_{jj} a_j \tag{11.14}$$

Let X be the diagonal matrix with $x_{jj} = x_j$, $j = 1, \dots, n$, and $x_{mm} = \lambda\mathbb{1}$ for some $\lambda \in I$. Then, by (11.4) and (11.14)

$$\begin{aligned} f\left(\sum_{j=1}^{n} a_j^* x_j a_j\right) &= f([U^*XU]_{mm}) \\ &\le [U^* f(X) U]_{mm} \\ &= \sum_{j=1}^{n} a_j^* f(x_j) a_j \end{aligned}$$

(b) ⇒ (c) Set $n = 1$

(c) ⇒ (d) Suppose first that $\operatorname{Ran} P$ has infinite dimension. Then there is an isometry, $V : \mathcal{H} \to \mathcal{H}$, with $\operatorname{Ran} V = \operatorname{Ran} P$. In particular, $VV^* = P$.

We claim that for any continuous real-valued function, g, on I and self-adjoint, x, on $\mathcal{H}$ with spectrum in I, we have that

$$[Vg(V^*xV)V^*]_P = g(x_P) \tag{11.15}$$

Since any such g is a uniform limit of polynomials, we need only prove that for $g(x) = x^\ell$. In that case

$$\begin{aligned}[Vg(V^*xV)V^*]_P &= [V(V^*xV)^\ell V^*]_P \\ &= [(VV^*xVV^*)^\ell]_P \\ &= [(PxP)^\ell]_P = (x_P)^\ell\end{aligned}$$

verifying (11.15).

Thus if f obeys (3)

$$\begin{aligned}f(x_P) &= [Vf(V^*xV)V^*]_P \\ &\leq [VV^*f(x)VV^*]_P \\ &= [Pf(x)P]_P \\ &= f(x)_P\end{aligned}$$

proving (11.3).

If Ran P is finite dimensional, let $\widetilde{\mathcal{H}} = \mathcal{H} \otimes \ell^2(\mathbb{Z})$, $\tilde{x} = x \otimes \mathbb{1}$, $\widetilde{P} = P \otimes \mathbb{1}$ so Ran $\widetilde{P}$ is infinite dimensional. Noting that $\tilde{x}_{\widetilde{P}} = x_P \otimes \mathbb{1}$ and $f(\tilde{x}) = f(x) \otimes \mathbb{1}$, we see that (11.3) for $\tilde{x}$, $\widetilde{P}$ implies it for x, P.

$\underline{(d) \Rightarrow (a)}$ The argument is essentially the same as in Theorem 10.6. Given $A, B \in \mathcal{L}(\mathcal{H})$, with spectrum in I, let $U_\theta \in \mathcal{L}(\mathcal{H} \oplus \mathcal{H})$ be given by (10.11), $X = \left(\begin{smallmatrix} A & 0 \\ 0 & B \end{smallmatrix}\right)$, $P = \left(\begin{smallmatrix} \mathbb{1} & 0 \\ 0 & 0 \end{smallmatrix}\right)$. Then $[U_\theta^* Y U_\theta]_P = \theta Y_{11} + (1-\theta) Y_{22}$, for any diagonal Y, so by (11.3)

$$\begin{aligned}f(\theta A + (1-\theta)B) &= f([U_\theta^* X U_\theta]_P) \\ &\leq [f(U_\theta^* X U_\theta)]_P \\ &= [U_\theta^* f(X) U_\theta]_P \\ &= \theta f(A) + (1-\theta) f(B)\end{aligned}$$

proving that f is matrix convex. □

Theorem 11.5 (Contractive HJP Inequality) *Let $0 \in I \subset \mathbb{R}$ with I an open interval and let f be a matrix convex function on I with $f(0) \leq 0$. Suppose that $\{a_j\}_{j=1}^n$ is an n-tuple of bounded operators with $\sum_{j=1}^n a_j^* a_j \leq \mathbb{1}$ and that $\{x_j\}_{j=1}^n$ is an n-tuple of bounded self-adjoint operators with spectrum in I. Then* (11.1) *holds.*

Proof Let $a_{n+1} \equiv (\mathbb{1} - \sum_{j=1}^n a_j^* a_j)^{1/2}$, $x_{n+1} = 0$. By Theorem 11.1

$$
\begin{aligned}
f\left(\sum_{j=1}^{n} a_j^* x_j a_j\right) &= f\left(\sum_{j=1}^{n+1} a_j^* x_j a_j\right) \\
&\leq \sum_{j=1}^{n+1} a_j^* f(x_j) a_j \\
&\leq \sum_{j=1}^{n} a_j^* f(x_j) a_j
\end{aligned}
$$

since $a_{n+1}^* f(0) a_{n+1} \leq 0$. □

Notes and Historical Remarks Jensen's 1906 paper [171] caused analysts to take notice of the subject of convex functions and included the discrete version of his inequality (for the scalar case). For a time, this inequality was a cancelation stamp of the Danish Post Office. In 1982, Hansen–Pedersen [136], as part of their proof of Loewner's theorem (see Chapter 28), proved Theorem 11.5. They realized this result could be improved and proved Theorem 11.1 in 2003 [137]. Our proofs follow theirs. Much earlier, in a prescient paper, Davis [70] proved (1) $\Longleftrightarrow$ (4) in Theorem 11.1. Davis included a corollary for certain maps of general C^*-algebras which was generalized by Choi [59]. It is possible to deduce (1) $\Rightarrow$ (2) from these general C^* results so that some have called what we call the Hansen–Jensen–Pedersen theorem, the Choi–Davis–Jensen inequality. Davis [71] proved a finite n relation, namely the analogs of the two halves of Proposition 9.10 for A replaced by a projection. The two papers of Davis [70, 71] had considerable impact.

Related to the ideas of this chapter, Osaka–Tomiyama [253, 254] prove that a continuous function, f on $[0, 1)$ has $f(t)/t$ in $\mathcal{M}_n(0, 1)$ if and only if for every $n \times n$ matrix, A, with spectrum in $[0, 1)$ and every contraction, C, on $\mathbb{C}^n$, one has that $f(C^*AC) \leq C^* f(A) C$. Also related, Hansen et. al [135] have proven maps defined on all self-adjoint operators, A with spectrum in (a, b) with values in the self-adjoint operators, that obey a bound like (11.1) for $n = 2$, are given by $A \mapsto f(A)$ where f is a matrix convex function on (a, b).

Chapter 12
Convexity, IV: Bhatia–Hiai–Sano (BHS) Theorem

In Chapter 5, given a C^1 function, f, on $(a, b) \subset \mathbb{R}$, we defined the Loewner matrix by

$$(L_k^{(f)}(x_1, \dots, x_k))_{ij} = \begin{cases} \frac{f(x_i)-f(x_j)}{x_i-x_j} & i \neq j \\ f'(x_i) & i = j \end{cases} \tag{12.1}$$

for $a < x_1 < \cdots < x_k < b$. We showed (Theorem 5.1) that $f \in \mathcal{M}_n(a, b) \iff L_n^{(f)} \geq 0$ in the sense that $\sum_{i,j} \bar{\varphi}_i L_{ij} \varphi_j \geq 0$ for all φ. Later (see Chapter 21), we'll see that $L^{(f)}$ is strictly positive (in that the sum is > 0 if $\varphi \not\equiv 0$) if f is in $\mathcal{M}_\infty$ and is not a rational function. In this chapter, we'll relate matrix convexity of general (not just negative) functions on $(0, \infty)$ to Loewner matrices.

By Theorems 5.1 and 10.2 if $f \leq 0$, it is matrix convex if and only if all $L^{(f)}$ are negative. As we'll see later, in a certain sense, knowing about all positive matrix convex functions will tell us about all matrix convex functions but this is not true about knowing about all negative functions so there may be new phenomena. The negative result can't be true for all matrix convex f's. For given a negative such f, let $g(x) = f(x) + \alpha x$, which is also matrix convex. But $L^{(g)} = L^{(f)} + \alpha Y$ where $Y_{ij} \equiv 1$. If $\mathbf{1} = (1, 1, \dots, 1)$, then

$$\langle \mathbf{1}, L_k^{(g)} \mathbf{1} \rangle = \langle \mathbf{1}, L_k^{(f)} \mathbf{1} \rangle + \alpha k^2$$

which is positive if α is large and positive. Thus $L^{(g)}$ is not negative but it is cnd where:

Definition Let Q be a real symmetric $n \times n$ matrix. We say that Q is *conditionally positive* (*cpd*) if and only if

B. Simon, *Loewner's Theorem on Monotone Matrix Functions*, Grundlehren der mathematischen Wissenschaften 354, https://doi.org/10.1007/978-3-030-22422-6_12

$$\sum_{i=1}^{n} x_i = 0;\ x_i \in \mathbb{R} \Rightarrow \sum_{i,j=1}^{n} x_i Q_{ij} x_j \geq 0 \tag{12.2}$$

If (12.2) holds with ≥ 0 replaced by ≤ 0, we say that Q is *conditionally negative* (*cnd*). If ≥ 0 is replaced by $= 0$, we say that Q is *conditionally zero* (*cz*).

Remarks

1. Often one looks at complex Hermitian matrices (replacing x_i by $\bar{x}_i$), but the only cases of interest here are real matrices and the analysis of cz matrices is simpler in that case.
2. One might guess that only multiples of Y are cz, but, as we'll see, that's wrong.
3. See the Notes for some other situations where cpd and cnd matrices arise.

We will denote the set of functions, f on $(0, \infty)$ so that for all $0 < x_1 < \cdots < x_n < \infty$, $L_n^{(f)}(x_1, \ldots, x_n)$ is cnd by $\mathcal{B}_n$. We let $\mathcal{B}_\infty = \cap_{n=1}^{\infty} \mathcal{B}_n$.

If M is an $n \times n$ matrix and $\widetilde{M}$ is the $(n-1) \times (n-1)$ matrix with $\widetilde{M}_{ij} = M_{ij}$ for $1 \leq i, j \leq n-1$, then M cnd $\Rightarrow \widetilde{M}$ cnd, for given $\tilde{x} \in \mathbb{R}^{n-1}$ with $\sum_{j=1}^{n-1} \tilde{x}_j = 0$, let $x_j = \tilde{x}_j$ for $j = 1, 2, \ldots, n-1$ and $x_n = 0$ and note that $\langle \tilde{x}, \widetilde{M}\tilde{x} \rangle = \langle x, Mx \rangle$. Thus $\mathcal{B}_{n-1} \supset \mathcal{B}_n$.

Here is the main result of this chapter:

Theorem 12.1 (Bhatia–Hiai–Sano Theorem) *Let f be a C^1 function on* $(0, \infty)$.

(i) $f \in \mathcal{C}_{2n+1}(0, \infty) \Rightarrow \liminf_{t\to\infty} f(t)/t > -\infty$ *and* $f \in \mathcal{B}_n$
(ii) $\liminf_{t\to\infty} f(t)/t > -\infty$ *and* $f \in \mathcal{B}_{4n+1} \Rightarrow f \in \mathcal{C}_n(0, \infty)$.

In particular, $f \in \mathcal{C}_\infty(0, \infty) \iff \liminf_{t\to\infty} f(t)/t > -\infty$ *and* $f \in \mathcal{B}_\infty$.

The "in particular" result is due to Bhatia–Sano [37]. Their proof of (ii) for $n = \infty$ is essentially a finite n argument made explicit by Hiai–Sano [158] but their proof of (i) for $n = \infty$ uses Loewner's theorem (see below) and doesn't have a direct finite n version. Hiai–Sano had to find a new approach for the finite n version of (i). We'll also see below that in (ii), one cannot drop the condition that $\liminf_{t\to\infty} f(t)/t > -\infty$.

That positive matrix convex functions tell us about general matrix convex functions are seen by

Proposition 12.2 *Let f be a scalar concave function on $(0, \infty)$. Let $a > 0$ and define*

$$h_a(x) \equiv f(x+a) - f(a) - (x-a)f'(a) \tag{12.3}$$

Then f is in $\mathcal{C}_n(0, \infty)$ if and only if each h_a is and $h_a(x) \geq 0$ for all x, so, in particular,

$$\liminf_{t\to\infty} f(t)/t > -\infty \tag{12.4}$$

Thus, to prove Theorem 12.1(i), it suffices to prove it in the case where f is C^1 on $[0, \infty)$ with

$$f(0) = f'(0) = 0 \quad and \quad f(x) > 0 \text{ on } (0, \infty) \tag{12.5}$$

Proof Any scalar convex function is larger than its tangent at any point (see [325, Section 5.3]), so $h_a \geq 0$. Equation (12.4) follows from $h_a \geq 0$ which implies that $\liminf_{t\to\infty} f(t)/t \geq f'(a)$ for any a.

For any $\epsilon > 0$, $h_{a,\epsilon}(x) \equiv h_a(x) + \epsilon x^2$ is matrix convex if f is and it is C^1 on $[0, \infty)$ with (12.5). Since

$$\langle\varphi, L^{(f)}(x_1, \ldots, x_n)\varphi\rangle = \lim_{\epsilon\downarrow 0} \langle\varphi, L^{(h_{a,\epsilon})}(x_1 - a, \ldots, x_n - a)\varphi\rangle \tag{12.6}$$

if $\sum_{j=1}^n \varphi_j = 0$, $x_j > a$, cnd of $L^{(h_{a,\epsilon})}$ implies it for $L^{(f)}$ (in fact, as we'll see, one has (12.6) without the limit). □

One can also restrict part (ii) to positive f's.

Proposition 12.3 *It suffices to prove Theorem 12.1 (ii) for f C^1 on $[0, \infty)$ with $f(0) = 0$, $f'(0) > 0$, $f(x) > 0$ on $(0, \infty)$.*

Proof Since f is matrix concave if and only if each $f(x + a), a > 0$ is, we can suppose that f is C^1 on $[0, \infty)$. If also $\liminf_{t\to\infty} f(t)/t > -\infty$, then $\inf_t f(t)/t \equiv -b > -\infty$. Let $g(x) = f(x) + (b + 1)x$. Then $\lim_{x\to\infty} g(x) = \infty$, so $-a \equiv \inf_t g(t) > -\infty$ and there is x_0 with $g(x_0) = -a$ and $g(x_0) > a$ for all $x > x_0$. Let $-c \equiv \inf_{0\leq x\leq x_0} g'(x) \leq 0$ (since $g'(x_0) = 0$) and $k(x) = g(x) + (c + 1)x + a$. Then $k'(x) \geq 1$ on $(0, x_0)$, so on that interval, $k(0) < k(x)$. Moreover, $k(0) = (c+1)x_0$ so $k(x) > k(0)$ on $[x_0, \infty)$ also. If $h(x) = k(x) - k(0)$, then $h(x) > 0$ on $(0, \infty)$, $h(0) = 0$, $h'(0) > 0$ and, since f and h differ by an affine function, $f \in \mathcal{B}_m \Rightarrow h \in \mathcal{B}_m$. □

Example 12.4 Fix $\mathbf{a} \in \mathbb{R}^n$. Let

$$B_{ij} = a_i + a_j \tag{12.7}$$

if $\mathbf{1}$ is the vector with all 1's, then

$$B = \langle\mathbf{a}, \cdot\rangle\mathbf{1} + \langle\mathbf{1}, \cdot\rangle\mathbf{a} \tag{12.8}$$

so

$$\langle\mathbf{1}, \varphi\rangle = 0 \quad \Rightarrow \quad \langle\varphi, B\varphi\rangle = 0 \tag{12.9}$$

i.e., B is a cz matrix. We'll see next they are the only ones. □

Proposition 12.5 *Let B be a real symmetric matrix. Then B is cz if and only if B has the form* (12.7) *for some* $\mathbf{a} \in \mathbb{R}^n$. *Let f be a function on $(a, b) \subset \mathbb{R}$. Then $L_2^{(f)}(x_1, x_2)$ is cz for all $a < x_1 < x_2 < b$ if and only if $f(x)$ is a quadratic polynomial and, if f is such a polynomial, then $L_n^{(f)}$ is cz for all $a < x_1 < \cdots < x_n < b$.*

Proof We've already seen that if B is of the form (12.7), then it is cz. On the other hand, if B is cz and $i \neq j$, then $\langle(\delta_i - \delta_j), B(\delta_i - \delta_j)\rangle = 0$, so $b_{ij} = \frac{1}{2}(b_{ii} + b_{jj})$ which is (12.7) if we let $a_i = \frac{1}{2}b_{ii}$. If t^α is the function $f(x) = x^\alpha$, then

$$(L_n^{(t)})_{ij} = 1; \qquad (L_n^{(t^2)})_{ij} = x_i + x_j \tag{12.10}$$

so any quadratic polynomial has a cz Loewner matrix.

For the other direction, suppose first that $0 \in (a, b)$ and $f(0) = f'(0) = 0$. If $x \neq 0$, then

$$L_2^{(f)}(0, x) = \begin{pmatrix} 0 & f(x)/x \\ f(x)/x & f'(x) \end{pmatrix} \tag{12.11}$$

so if $L_2^{(f)}(0, x)$ is cn, we have, looking at $(1 \;\; -1)L_2^{(f)}\begin{pmatrix} 1 \\ -1 \end{pmatrix}$ that

$$f'(x) = 2f(x)/x \tag{12.12}$$

a first order ODE on $(0, b)$ and on $(0, a)$, whose only solutions are $f(x) = c_{\pm}x^2$. Doing the same for $f(x) - f(0) - f'(0)x$, we see that

$$f(x) = \begin{cases} c_+x^2 + \beta x + \alpha, & \text{if} \quad x \in (0, b) \\ c_-x^2 + \beta x + \alpha, & \text{if} \quad x \in (a, 0) \end{cases}$$

for suitable $\alpha, \beta \in \mathbb{R}$. In particular, f is C^∞ on $(a, b) \setminus \{0\}$. But 0 is an arbitrary point, so f is C^∞ at 0 also. Thus $c_+ = c_-$ and f is a quadratic polynomial. □

Example 12.6 Let $f(x) = -x^2$ on $(0, \infty)$. Then $L_n^{(f)}$ is cz and so cnd for any $0 < x_1 < x_2 < \cdots < x_n$ while f is not scalar convex so certainly not matrix convex; indeed, it is strictly matrix concave. This shows that the condition $\liminf_{t\to\infty} f(t)/t > -\infty$ cannot be dropped from Theorem 12.1 (ii).

One might argue that cz is special and perhaps strictly cnd would suffice. But by the results in Chapter 24, $f(x) = -x^{1/2}$ has strictly negative Loewner matrices so $f(x) = -x^{1/2} - x^2$ has strictly cnd Loewner matrices and is certainly not matrix convex. □

Example 12.7 Let $(a, b) \subset \mathbb{R}$ be a finite interval. Given a C^1 function, f, on (a, b), define

$$\tilde{f}(x) = f(a + b - x) \tag{12.13}$$

which is also a C^1 function on (a, b). It is easy to see that f is matrix convex, if and only if $\tilde{f}$ is matrix convex. On the other hand, if $a < x_1 < \cdots < x_n < b$ and $\tilde{x}_j = a + b - x_{n+1-j}$, then $a < \tilde{x}_1 < \cdots < \tilde{x}_n < b$ and

$$L_n^{(\tilde{f})}(\tilde{x}_1, \dots, \tilde{x}_n) = -L_n^{(f)}(x_1, \dots, x_n) \tag{12.14}$$

Unless $L^{(f)}$ is cz, it can't be that both $L^{(f)}$ and $L^{(\tilde{f})}$ are cnd, so there will be lots of matrix convex functions on (a, b) with $L^{(f)}$ not cnd. For example, if $(a, b) = (0, 1)$ and $f(x) = -x^{1/2}$, then $L^{(f)}$ is strictly negative, so if $g(x) = -(1 - x)^{1/2}$, then g is matrix convex while $L^{(g)}$ is strictly positive and certainly not cnd. □

Example 12.8 Let $\alpha > 0$ and $f(x) = x^2(x + \alpha)^{-1}$. Then for $x \neq y$,

$$\frac{f(x) - f(y)}{x - y} = 1 - \frac{\alpha^2}{(x + \alpha)(y + \alpha)} \tag{12.15}$$

If ψ is the vector with components $\psi_j = 1/(x_j + \alpha)$ for $0 < x_1 < \cdots < x_n < \infty$, then, for any φ and $C_{ij} = 1/(x_i + \alpha)(x_j + \alpha)$

$$\langle \varphi, C\varphi \rangle = |\langle \varphi, \psi \rangle|^2$$

so C is positive which means that each $L^{(f)}$ is cnd. By rewriting (9.9), any matrix convex function on $(0, \infty)$ which is C^1 at 0 has a representation

$$f(x) = a + bx + cx^2 + \int \frac{x^2}{x + \alpha} d\mu(\alpha) \tag{12.16}$$

where μ is a measure on $(0, \infty)$ with $\int (1+\alpha)^{-1} d\mu(\alpha) < \infty$. Since x, x^2, and $(x + \alpha)^{-1}$ all lead to cnd Loewner matrices, so does f proving that $f \in \mathcal{C}_\infty(0, \infty) \Rightarrow f \in \mathcal{B}_\infty$. This is the original proof of this fact. It relies on Loewner's theorem. □

We need two lemmas in preparation for the proof of Theorem 12.1:

Lemma 12.9 *Let f be a C^1 strictly positive function on $(0, \infty)$ and let*

$$g(x) = \frac{x^2}{f(x)} \tag{12.17}$$

Then for $0 < x_1 < \cdots < x_n < \infty$,

$$L_n^{(f)}(x_1, \dots, x_n)_{ij} = -\frac{f(x_i)}{x_i} L_n^{(g)}(x_1, \dots, x_n)_{ij} \frac{f(x_j)}{x_j} + \frac{f(x_i)}{x_i} + \frac{f(x_j)}{x_j} \tag{12.18}$$

Proof For $s \neq t$ in $(0, \infty)$, we compute:

$$\frac{1}{s-t}\left[\frac{s^2}{f(s)} - \frac{t^2}{f(t)}\right] = \frac{1}{s-t}\left[\frac{s^2 - st}{f(s)} - \frac{t^2 - st}{f(t)}\right] + \frac{st}{s-t}\left[\frac{1}{f(s)} - \frac{1}{f(t)}\right]$$
$$= \frac{s}{f(s)} + \frac{t}{f(t)} - \frac{st}{f(s)f(t)}\left[\frac{f(s)-f(t)}{s-t}\right] \quad (12.19)$$

Multiplying by $\frac{f(s)}{s}\frac{f(t)}{t}$ yields (12.18). □

Lemma 12.10 *Let $\{A_{ij}\}_{1 \le i,j \le n}$ be an $n \times n$ cnd matrix. Define for $1 \le i, j \le n-1$*

$$B_{ij} = A_{ij} + A_{nn} - A_{in} - A_{nj} \quad (12.20)$$

Then the $(n-1) \times (n-1)$ matrix $\{B_{ij}\}_{1 \le i,j \le n-1}$ is negative.

Proof Given $\mathbf{x} \in \mathbb{R}^{n-1}$, define $\mathbf{y} \in \mathbb{R}^n$ by

$$y_j = x_j,\ j = 1, \dots, n-1; \qquad y_n = -\sum_{j=1}^{n-1} x_j \quad (12.21)$$

Since $\sum_{j=1}^n y_j = 0$, $\langle \mathbf{y}, A\mathbf{y}\rangle \le 0$. But $\langle \mathbf{y}, A\mathbf{y}\rangle = \langle \mathbf{x}, B\mathbf{x}\rangle$. □

Proof of Theorem 12.1

(i) By Proposition 12.2, we can suppose that f is C^1 on $[0, \infty)$, $f(0) = f'(0) = 0$ and $f(x) > 0$ on $(0, \infty)$. If $f \in \mathcal{C}_{2n+1}(0, \infty)$, then by Theorem 9.11, $f(x)x^{-1} \in \mathcal{M}_{2n}(0, \infty)$. Thus, by Theorem 10.11, $x/(f(x)x^{-1}) = x^2/f(x)$ is in $\mathcal{M}_n(0, \infty)$. Thus the first term on right side of (12.18) is negative. Since the second term is cz, the left side is cnd, i.e., $f \in \mathcal{B}_n$.

(ii) By Proposition 12.3, we can suppose that $f(0) = 0$, $f(x) > 0$ on $(0, \infty)$, and $f'(0) > 0$. By Lemma 12.10, by taking x_0 as the "$(4n+1)$st coordinate" and letting $x_0 \downarrow 0$, we see that for $0 < x_1 < \dots < x_{4n} < \infty$

$$\frac{f(x_i) - f(x_j)}{x_i - x_j} - \frac{f(x_i)}{x_i} - \frac{f(x_j)}{x_j} + f'(0) \quad (12.22)$$

is a negative $4n \times 4n$ matrix. Since $f'(0) > 0$, the $f'(0)$ term (constant in (ij)) gives a positive matrix so, if it is dropped, the matrix is still negative.

By (12.18), $-\frac{f(x_i)}{x_i} L_{4n}^{(g)}(x_1, \dots, x_{4n})_{ij} \frac{f(x_j}{x_j}$ is negative so, since $f(x_i) \neq 0$, $L_{4n}^{(g)}$ is positive. By Theorem 5.1, $x^2/f(x)$ is in $\mathcal{M}_{4n}(0, \infty)$. By Theorem 10.11, $x/(x^2/f(x)) = f(x)/x$ is in $\mathcal{M}_{2n}(0, \infty)$. By Theorem 9.11, $f \in \mathcal{C}_n(0, \infty)$. □

If A is an $n \times n$ real symmetric matrix, P a projection onto an $n-1$-dimensional subspace and $B \equiv PAP \restriction \operatorname{Ran} P$, then the eigenvalues of B interlace those of A;

in particular, if $B \leq 0$, then A has at most one strictly positive eigenvalue. Thus, if A is cnd, it has either one or zero strictly positive eigenvalues. We end this chapter by asking about strictly positive eigenvalues of Loewner matrices of matrix convex functions.

Example 12.11 Let B be a 2×2 cz matrix so B_{ij} is given by (12.7). Suppose that we have for some i, j that $a_i \neq a_j$. Then

$$\det \begin{pmatrix} B_{ii} & B_{ij} \\ B_{ji} & B_{jj} \end{pmatrix} = 4a_i a_j - (a_i + a_j)^2$$

$$= -(a_i - a_j)^2 < 0$$

so that 2×2 matrix has one strictly positive and one strictly negative eigenvalue. By the interlacing fact, we conclude that B has exactly one strictly positive and one strictly negative eigenvalue. In particular the matrix convex function cx^2 $(c > 0)$ has a positive eigenvalue. This is typical: □

Theorem 12.12 *Let f be C^1 on $[0, \infty)$ and in $\mathcal{C}_k(0, \infty)$. Then:*

(i) If $k = n$ and $f(x) \leq 0$ for all x, then $L_n^{(f)}(x_1, \ldots, x_n)$ always has no positive eigenvalues.
(ii) If $k = 2n + 1$, $f(0) = 0$ and $f(x) > 0$ for $x > 0$, then each $L_n^{(f)}(x_1, \ldots, x_n)$ has a single positive eigenvalue.

Proof

(i) If $f \leq 0$, then $-f$ is a positive matrix concave function. By Theorem 10.2, $-f \in \mathcal{M}_n(0, \infty)$, so $L_n^{(f)}$ is negative and thus it has no positive eigenvalues.
(ii) By Theorem 12.1, $L_n^{(f)}$ is cnd. Since f is scalar convex with $f(0) = 0$ and $f(x) > 0$ for $x > 0$, we have that $f'(x) > 0$ for such x. Thus $\text{Tr}(L_n^{(f)}) > 0$, so $L_n^{(f)}$ cannot have only non-positive eigenvalues.

□

Example 12.13 Let $f(x) = x^2 - \sqrt{x}$ which is matrix convex. By scaling, it is easy to see that for small t, $L_2^{(f)}(t, 2t)$ is very close to $L_2^{(-\sqrt{x})}(t, 2t)$ and for large t to $L_2^{(x^2)}(t, 2t)$. The first is strictly negative and the second has a positive eigenvalue. Thus, in this case, we see that the possibilities for the positive eigenvalue of $L_f^{(n)}(x_1, \ldots, x_n)$ can depend on the exact values of x_j. To get a definite value, we need to specify the sign of f. □

We finally turn to the ideas behind the proof of Theorem 10.15 which involve cnd matrices.

Lemma 12.14 *Let A be a Hermitian cpd finite matrix. Then the matrix, B, with matrix elements*

$$b_{ij} = e^{a_{ij}} \tag{12.23}$$

is strictly positive.

Remark B is sometimes called the *Hadamard exponential* in line with the strange name Hadamard product (see the Notes to Chapter 5). We'll call it the *Schur exponential*. Similarly, if $a_{ij} \geq 0$ for all i, j, the matrices with matrix elements a_{ij}^{α} ($\alpha > 0$) are called Hadamard powers and if each $a_{ij} > 0$, the matrix with elements a_{ij}^{-1} is called the Hadamard inverse. We'll use *Schur powers* and *Schur inverse*.

Proof Let φ be the vector with all components 1 and P the orthogonal projection onto φ, so the matrix with all components $1/n$. Let

$$C = (\mathbf{1} - P)A(\mathbf{1} - P) \tag{12.24}$$

so $C \geq 0$ by hypothesis. Then

$$(AP)_{ij} = \sum_{k=1}^{n} \frac{1}{n} a_{ik} \equiv x_i \tag{12.25}$$

is independent of j. By $A^* = A$, we see that

$$(PA)_{ij} = \bar{x}_j \tag{12.26}$$

is independent of i. Similarly

$$(PAP)_{ij} = \frac{1}{n^2} \sum_{k\ell} a_{k\ell} \equiv \alpha \tag{12.27}$$

is constant and real.

Let

$$y_i = x_i - \frac{1}{2}\alpha \tag{12.28}$$

Then (12.24)–(12.28) imply that

$$a_{ij} = c_{ij} + y_i + \bar{y}_j \tag{12.29}$$

By Lemma 5.3, the Schur exponential $S \equiv \{e^{c_{ij}}\}_{1 \leq i,j \leq n}$ of C is strictly positive. Let D be the diagonal matrix with $d_{jj} = e^{y_j}$. Then (12.29) implies that

$$B = DSD^* > 0$$

since $S > 0$. □

Proposition 12.15 *Let A have strictly positive matrix elements and be cnd. Then its Schur inverse (i.e., B with $b_{ij} = a_{ij}^{-1}$) is strictly positive.*

Remark The same proof shows that b is infinitely divisible, that is for each $\alpha > 0$, the matrix with matrix elements $a_{ij}^{-\alpha}$ is positive.

Proof Let C_α be the matrix with

$$(C_\alpha)_{ij} = e^{-\alpha a_{ij}} \tag{12.30}$$

Since $-\alpha A$ is cpd, by Lemma 12.14, C_α is strictly positive. Thus, since $\int_0^\infty e^{-\alpha x}\, d\alpha = x^{-1}$, we have that

$$B = \int_0^\infty C_\alpha \, d\alpha \tag{12.31}$$

is positive. □

Theorem 12.16 (≡ Theorem 10.15) *Let h be a C^1 function on an interval $(a, b) \subset \mathbb{R}$ (a or b can be infinite) with $h'(t) > 0$ for all $t \in (a, b)$. Suppose that for some $c \in (a, b)$, we have that*

$$f(x) = \frac{h(x) - h(c)}{x - c} \tag{12.32}$$

lies in $-\mathcal{C}_n(a, b)$. Then g, the compositional inverse of h, lies in $\mathcal{M}_n(h(a), h(b))$.

Proof Let

$$q(x) = \frac{f(x) - f(c)}{x - c} \tag{12.33}$$

Then by a calculation that we defer (initially for $x \neq y$ and then taking limits for $x = y \neq c$)

$$\frac{h(x) - h(y)}{x - y} = f(x) + f(y) - f(c) + (x - c)\frac{q(x) - q(y)}{x - y}(y - c) \tag{12.34}$$

By hypothesis, $-f$ is in $\mathcal{C}_n(a, b)$. It follows from Theorem 9.2 that $-q \in \mathcal{M}_n(a, b)$. Thus for any $a < x_1 < \cdots < x_n < b$, (with no $x_j = c$), the matrix $L(x_1, \ldots, x_n; q)$ is negative. If D is the diagonal matrix with $d_{jj} = x_j - c$ and $a_i = f(x_i) - \frac{1}{2} f(c)$, $b_{ij} = a_i + a_j$, then (12.34) says that

$$L(x_1, \ldots, x_n; h) = B + DL(x_1, \ldots, x_n; q)D \tag{12.35}$$

Note that $L(x_1, \ldots, x_n; q) \leq 0 \Rightarrow DL(x_1, \ldots, x_n; q)D \leq 0$. By the calculation in Example 12.4, B is cz. It follows that $L(x_1, \ldots, x_n; h)$ is cnd. Since $h'(t) > 0$, this matrix has strictly positive elements. It follows from Proposition 12.15 that the matrix with

$$M(x_1, \ldots, x_n; h)_{ij} \equiv \begin{cases} \frac{x_i - x_j}{h(x_i) - h(x_j)}, & \text{if } i \neq j \\ \frac{1}{h'(x_i)}, & \text{if } i = j \end{cases} \tag{12.36}$$

is a positive matrix. h is bijection of (a, b) to $(h(a), h(b))$. Suppose that $h(a) < y_1 < \cdots < y_n < h(b)$ with no $y_j = h(c)$. Let $x_j = g(y_j)$ so $h(x_j) = y_j$ and, by the chain rule $1/h'(x_j) = g'(y_j)$. Then

$$L(y_1, \ldots, y_n; g) = M(x_1, \ldots, x_n; h)$$

so the Loewner matrix of g is positive. Taking limits, this remains true if some y_j is $h(c)$. Thus $g \in \mathcal{M}_\infty(h(a), h(b))$.

We turn finally to the verification of (12.34). Substituting

$$h(x) = h(c) + f(x)(x - c); \;\; f(x) = f(c) + q(x)(x - c) \tag{12.37}$$

we see that

$$\frac{h(x) - h(y)}{x - y} = \frac{f(x)(x - c))}{x - y} - \frac{f(y)(y - c))}{x - y} \tag{12.38}$$

$$= f(x) + f(y) + \frac{f(x)(y - c)}{x - y} - \frac{f(y)(x - c)}{x - y} \tag{12.39}$$

$$= f(x) + f(y) - f(c) + \frac{q(x)(x - c)(y - c) - q(y)(x - c)(y - c)}{x - y} \tag{12.40}$$

which proves (12.34). We get (12.39) from (12.38) using $(x - c) = (x - y) + (y - c)$ and $(y - c) = -(x - y) + (x + c)$. We then get (12.40) by using (12.37) and $(y - c) - (x - c) = -(x - y)$. □

Notes and Historical Remarks Conditionally positive matrices (and functions—see below) enter in many parts of analysis. We say a L^∞ function, f, on $\mathbb{R}^\nu$ is conditionally positive if $\varphi \in L^1$, real and $\int \varphi(x) d^\nu x = 0 \Rightarrow \int\int \varphi(x) f(x - y)\varphi(y)\, d^\nu x\, d^\nu y \geq 0$. On $\mathbb{R}^2$, $\log |x|^{-1}$ is (strictly) cpd and this implies uniqueness of equilibrium measures in two-dimensional potential theory (see [325, Section 6.9, Problem 6]). In one dimension, $f(x) = -|x|$ is cpd and this is useful in the theory of weakly coupled bound states in quantum theory [316].

More generally, $K(x, y)$ in L^∞ is called a cpd kernel if $\int\int \varphi(x) K(x, y) \varphi(y)\, d^\nu x\, d^\nu y \geq 0$ whenever $\varphi \in L^1$ and $\int \varphi(x) dx = 0$. Notice that by an approximation argument, a C^∞ function $f \in \mathcal{B}_\infty$ if and only if $L^{(f)}(x, y) = (f(x) - f(y))/(x - y)$ is cpd. Cpd kernels and the closely related notion of infinitely divisible kernels (see below) are important in probability theory, see, for example, Parthasarathy–Schmidt [257].

Donoghue's book [81] has a chapter on cpd kernels which he calls *almost positive kernels*. Donoghue discusses it because the first place that cpd matrices appear in Loewner theory is in a 1966 paper of Loewner [212] in the context of schlicht functions (see below), not matrix convex functions.

Lemma 12.14 and the proof we give are due to Loewner [212] while Proposition 12.15 is due to Horn [166]. Horn's work was based on his thesis done at Stanford under the direction of Loewner and Don Spencer. His idea that the results for $e^{-\alpha a_{ij}}$ extend to $f(a_{ij})$ where f is the Laplace transform of a positive measure includes many functions, f.

Theorem 12.16 is due to Pedersen–Uchiyama [263] when $n = \infty$. The fact that, because of Heinävaara's part of Theorem 9.2, the results hold for finite n is a new observation here but is immediate from their proof (used here) and Heinävaara's result.

The results of the first part of this chapter are based on four papers by Sano and coauthors [37, 38, 158, 159] and one by Uchiyama [350]. The original idea and the "in particular" part of Theorem 12.1 (with a much strong condition than $\liminf_{t\to\infty} f(t)/t > -\infty$) is in Bhatia–Sano [37] who used Lemmas 12.9 and 12.10 in their arguments. Uchiyama [350] found the correct $\liminf_{t\to\infty} f(t)/t > -\infty$ condition. The finite n results are from Hiai–Sano [158]. References [38] and [159] have further results.

There has been some interest in the eigenvalues of the Loewner matrices for the function $f(x) = x^r$ for $x \in (0, \infty)$ and $r > 0$. As we saw in Chapter 5, if $0 < r < 1$, all eigenvalues are strictly positive and Bhatia–Holbrook [36] proved that if $1 < r < 2$, the Loewner matrices have one strictly positive eigenvalue and the rest strictly negative consistent with the fact that f is matrix convex but not matrix monotone decreasing and the results of this chapter. Bhatia et al. [35] have proven a conjecture in [36] that for r positive and non-integral, one has that if k is an integer larger than 1, $2k - 1 < r < 2k$, the $n \times n$ Loewner matrices for $n > r$ have exactly k strictly negative eigenvalues and the remaining strictly positive and if $2k < r < 2k + 1$ for the same n there are exactly k strictly positive eigenvalues and the rest strictly negative. There also results for integral r (where the Loewner matrices have a zero eigenvalue) and for smaller values of n.

We mention three further themes related to the subject of this chapter. There is another set of equivalences, first found when $n = \infty$ in [37] with finite n results in [158]:

$$f \in \mathcal{C}_{n+1}(0, \infty) \Rightarrow \limsup_{t \downarrow 0} tf(t) \geq 0 \,\&\, -tf \in \mathcal{B}_n$$

$$\limsup_{t \downarrow 0} tf(t) \geq 0 \,\&\, -tf \in \mathcal{B}_n \Rightarrow f \in \mathcal{C}_n(0, \infty)$$

(of course, the papers don't say $-tf \in \mathcal{B}_n$ but that $L_n^{(tf)}$ is cpd). The example $f(t) = -t^{-1}$ shows that the condition $\limsup_{t \downarrow 0} tf(t) \geq 0$ is necessary and $f(t) = -1 + t^2$ that the lim sup needs to involve tf and not just f.

Secondly, one can look at the *Kwong matrix*:

$$\left[K_n^{(f)}(t_1, \dots, t_n)\right]_{ij} = \frac{f(t_i) + f(t_j)}{t_i + t_j}$$

named after Kwong [194] who proved that if f is matrix monotone, then all $K_f^{(n)}$ are positive matrices; see also [37].

Finally, we note a connection to whether a matrix monotone function, $f \in \mathcal{M}_\infty(0, \infty)$ defines a $1-1$ map of $\mathbb{C}_+$ to $\mathbb{C}_+$. A matrix $A = \{a_{ij}\}_{1 \leq i,j \leq n}$ is called *infinitely divisible* if and only if $a_{ij} > 0$ and for all $r > 0$, the matrix with matrix elements $(a_{ij})^r$ (the Schur power) defines a positive matrix. It is not hard to show this is true if and only if the matrix with matrix elements $\log(a_{ij})$ is cpd. Horn [167] proved that a matrix monotone function has an analytic continuation which is $1-1$ on $\mathbb{C}_+$ if and only if $L_n^{(f)}$ is always infinitely divisible. This is further discussed in Donoghue [81, Chapter XVII], Loewner [210, 212], Fitzgerald [99], and Bhatia–Sano [37].

Chapter 13
Convexity, V: Strongly Operator Convex Functions

Let f be a real-valued function on $(a, b) \subset \mathbb{R}$. Recall the Davis condition, (11.3), that for all finite self-adjoint matrices, A, and all self-adjoint projections, P, one has that

$$Pf(PAP)P \le Pf(A)P \tag{13.1}$$

(if $0 \notin (a, b)$, then $f(PAP)$ won't be defined since $PAP \restriction (\mathbf{1} - P)\mathcal{H} = 0$, but because $P(f(0)(\mathbf{1} - P))P = 0$, independently of $f(0)$, we can assign any value to $f(0)$ without it changing the truth of (13.1); equivalently, the left side looks at $A_P = PAP$ as an operator on $P\mathcal{H}$ and defines $f(A_P)$ as an operator on $P\mathcal{H}$ and views $Pf(PAP)P$ as $f(A_P) \oplus 0$ under the $P\mathcal{H} \oplus (\mathbf{1} - P)\mathcal{H}$ decomposition). Davis proved that (13.1) was equivalent to f being matrix convex.

We say that f is *strongly operator convex* (SOC for short) and write $f \in \mathcal{S}_\infty(a, b)$ if and only if for all finite matrices, A, and all self-adjoint projections, P, one has that

$$Pf(PAP)P \le f(A) \tag{13.2}$$

To agree with the small number of papers on the subject we use "operator convex" rather than strongly matrix convex. Two things are immediate

Proposition 13.1

(a) *If* $f \in \mathcal{S}_\infty(a, b)$*, then* $f(x) \ge 0$ *for all* $x \in (a, b)$.
(b) $f \in \mathcal{S}_\infty(a, b) \Rightarrow f \in \mathcal{C}_\infty(a, b)$.

Remark Later (Lemma 13.7), we'll prove that either $f \equiv 0$ or else $f(x) > 0$ for all $x \in (a, b)$.

B. Simon, *Loewner's Theorem on Monotone Matrix Functions*, Grundlehren der mathematischen Wissenschaften 354, https://doi.org/10.1007/978-3-030-22422-6_13

Proof

(a) Let $x \in (a, b)$, $A = x\mathbf{1}$. Then (13.2) for $A \iff Pf(x)P \leq f(x) \iff f(x)(\mathbf{1} - P) \geq 0 \iff f(x) \geq 0$ (if $P \neq \mathbf{1}$).
(b) (13.2)⇒(13.1) by multiplying on the left and right by P. Thus, this is just a restatement of Davis' result, (1) $\iff$ (4) in Theorem 11.1.

□

Example 13.2 Let $f(x) \equiv x$ on a non-empty interval (a, b). Pick $c \in (a, b)$ and let

$$A = \begin{pmatrix} c & \epsilon \\ \epsilon & c \end{pmatrix} \tag{13.3}$$

as a 2×2 matrix. For ϵ small and strictly positive, A has spectrum in (a, b). Let $P = \begin{pmatrix} 1 & 0 \\ 0 & 0 \end{pmatrix}$. Then

$$f(A) - Pf(PAP)P = \begin{pmatrix} 0 & \epsilon \\ \epsilon & c \end{pmatrix} \equiv B \tag{13.4}$$

If $\varphi = \begin{pmatrix} 1 \\ -\delta \end{pmatrix}$, then $\langle \varphi, B\varphi \rangle = c\delta^2 - 2\epsilon\delta < 0$ for $0 < \delta < 2\epsilon/c$ so (13.2) fails and f is not SOC. □

This example is somewhat surprising and might suggest no function is SOC. But we are heading towards showing that $f(x) = x^{-1}$ lies in $\mathcal{S}_\infty(0, \infty)$. We'll want to consider block matrix representations for $\mathcal{H} = P\mathcal{H} \oplus (\mathbf{1} - P)\mathcal{H}$ or $\mathcal{H} = (\mathbf{1} - P)\mathcal{H} \oplus P\mathcal{H}$. $M \in \mathcal{L}(\mathcal{H})$ with $M^* = M$ can then be written

$$M = \begin{pmatrix} X & Z \\ Z^* & Y \end{pmatrix} \tag{13.5}$$

where, in the first case, $X \in \mathcal{L}(P\mathcal{H})$, $Y \in \mathcal{L}((\mathbf{1} - P)\mathcal{H})$, $Z : (\mathbf{1} - P)\mathcal{H} \to P\mathcal{H}$ with $X = X^*$, $Y = Y^*$. Given such an M with Y invertible, we define the *Schur complement* of Y (sometime written M/Y) as

$$S = X - ZY^{-1}Z^* \tag{13.6}$$

Let

$$L = \begin{pmatrix} \mathbf{1} & 0 \\ -Y^{-1}Z^* & \mathbf{1} \end{pmatrix} \text{ so } L^{-1} = \begin{pmatrix} \mathbf{1} & 0 \\ Y^{-1}Z^* & \mathbf{1} \end{pmatrix} \tag{13.7}$$

S arises via a simple calculation

$$L^*ML = L^* \begin{pmatrix} S & Z \\ 0 & Y \end{pmatrix} = \begin{pmatrix} S & 0 \\ 0 & Y \end{pmatrix} \tag{13.8}$$

This has three immediate consequences:

Theorem 13.3 *Let M have the form* (13.5). *Then*

(a) *M is strictly positive if and only if both S and Y are strictly positive.*
(b) *Let Y be invertible. Then M is invertible if and only if S is invertible.*
(c) *If M and Y are invertible, then*

$$\begin{aligned} M^{-1} &= L \begin{pmatrix} S^{-1} & 0 \\ 0 & Y^{-1} \end{pmatrix} L^* \\ &= \begin{pmatrix} S^{-1} & -S^{-1}ZY^{-1} \\ -Y^{-1}Z^*S^{-1} & Y^{-1} + Y^{-1}Z^*S^{-1}ZY^{-1} \end{pmatrix} \end{aligned} \tag{13.9}$$

Proof

(a) If $M > 0$, then $Y > 0$ and so invertible. We can then define L and see that $L^*ML > 0$, which implies that $S > 0$. Conversely, if $S, Y > 0$, then Y is invertible so (13.8) holds and $\begin{pmatrix} S & 0 \\ 0 & Y \end{pmatrix} > 0 \Rightarrow M > 0$.
(b) Since Y is invertible, (13.8) holds which shows the desired result.
(c) is immediate from (13.8) which says that $M = (L^*)^{-1} \begin{pmatrix} S & 0 \\ 0 & Y \end{pmatrix} L^{-1}$.

□

Example 13.4 Let $f(x) = x^{-1}$ on $(0, \infty)$. Given $M > 0$ and P, decomposed as in (13.5) according to a $\mathcal{H} = (\mathbf{1} - P)\mathcal{H} \oplus P\mathcal{H}$ decomposition, we have that $PMP = Y$ so

$$\begin{aligned} f(M) - Pf(PMP)P &= M^{-1} - PY^{-1}P \\ &= \begin{pmatrix} S^{-1} & -S^{-1}ZY^{-1} \\ -Y^{-1}Z^*S^{-1} & Y^{-1}Z^*S^{-1}ZY^{-1} \end{pmatrix} &\quad (13.10) \\ &\equiv \begin{pmatrix} A & B \\ B^* & C \end{pmatrix} &\quad (13.11) \end{aligned}$$

where (13.10) comes from (13.9). Notice that $M > 0 \Rightarrow S, Y > 0$, so we can do all these calculations. Notice also that

$$\begin{aligned} BC^{-1}B^* &= S^{-1}ZY^{-1}(YZ^{-1}S(Z^*)^{-1}Y)Y^{-1}Z^*S^{-1} \\ &= S^{-1} = A \end{aligned} \tag{13.12}$$

so $A - BC^{-1}B^* = 0$, which by a slight extension of Theorem 13.3 implies that $\begin{pmatrix} A & B \\ B^* & C \end{pmatrix} \geq 0$. If Z is not invertible, then neither is C, so we can't do the above calculation but in (13.10) we can then replace Z by $Z + \epsilon\mathbf{1}$, which is invertible for ϵ small, get positivity, and then take ϵ to 0. This shows that f is SOC. We summarize in the theorem below. □

Theorem 13.5 *The function $f(x) = x^{-1}$ on $(0, \infty)$ lies in $\mathcal{S}_\infty(0, \infty)$.*

The argument also proves the next interesting maximal characterization which we won't need here but will in Chapter 35. Recall that in Chapter 11 we defined A_P by $A_P = PAP \restriction \text{Ran}(P)$. If $B \in \mathcal{L}(\text{Ran } P)$, we use $B \oplus 0$ for its extension to an operator on $\mathcal{H}$ via $(B \oplus 0)\varphi = BP\varphi$, i.e., $\begin{pmatrix} B & 0 \\ 0 & 0 \end{pmatrix}$ in the block language we have been using.

Theorem 13.6 *Let $N \geq 0$ be an invertible operator on a Hilbert space, $\mathcal{H}$, and P a self-adjoint projection. Define*

$$N^\sharp = [(N^{-1})_P]^{-1} \tag{13.13}$$

Then $Q \in \mathcal{L}(\text{Ran } P)$ has

$$Q \oplus 0 \leq N \iff Q \leq N^\sharp \tag{13.14}$$

Remarks

1. By Theorem 13.3(c), N^{-1} positive and invertible implies that $(N^{-1})_P$ is invertible, so $N^\sharp$ is a well defined strictly positive operator.
2. In Chapter 36, we'll say a lot more about maximal positive Y's for which $\begin{pmatrix} Y & 0 \\ 0 & 0 \end{pmatrix} \leq \begin{pmatrix} A & B \\ B^* & C \end{pmatrix}$ for explicit examples of A, B, C.

Proof Write $M = N^{-1}$ and use (13.11), noting that, the $f(x) = x^{-1}$, $Pf(PMP)P$ is exactly $N^\sharp$. Thus

$$N - Q \oplus 0 = \begin{pmatrix} A + N^\sharp - Q & B \\ B^* & C \end{pmatrix} \tag{13.15}$$

By Theorem 13.3(a) and the calculation in (13.12), we see that

$$N - Q \oplus 0 \geq 0 \iff N^\sharp - Q \geq 0 \tag{13.16}$$

which is (13.14). □

We will need the following:

Lemma 13.7 *If $g \in \mathcal{S}_\infty(a, b)$, then either $g \equiv 0$ or else $g(t) > 0$ for all $t \in (a, b)$.*

Proof Fix $g \in \mathcal{S}_\infty(a, b)$ and $t, s \in (a, b)$. Let $X = \begin{pmatrix} t & 0 \\ 0 & s \end{pmatrix}$ as a 2×2 matrix, let U be the 2×2 unitary matrix

$$U = \frac{1}{\sqrt{2}} \begin{pmatrix} 1 & -1 \\ 1 & 1 \end{pmatrix}$$

and $P = \begin{pmatrix} 1 & 0 \\ 0 & 0 \end{pmatrix}$. Note that

$$U^* \begin{pmatrix} a & 0 \\ 0 & b \end{pmatrix} U = \begin{pmatrix} \frac{a+b}{2} & \frac{b-a}{2} \\ \frac{b-a}{2} & \frac{a+b}{2} \end{pmatrix} \tag{13.17}$$

Thus,

$$Pg(PU^*XUP)P = \begin{pmatrix} g(\frac{t+s}{2}) & 0 \\ 0 & 0 \end{pmatrix} \equiv M \tag{13.18}$$

while

$$g(U^*XU) = Ug(X)U = U^* \begin{pmatrix} g(t) & 0 \\ 0 & g(s) \end{pmatrix} U$$

$$= \begin{pmatrix} \frac{g(t)+g(s)}{2} & \frac{g(s)-g(t)}{2} \\ \frac{g(s)-g(t)}{2} & \frac{g(t)+g(s)}{2} \end{pmatrix} \equiv N \tag{13.19}$$

If $g(t) = 0$ and $\varphi = \begin{pmatrix} 1 \\ -1 \end{pmatrix}$, then $\langle \varphi, N\varphi \rangle = 0$ while $\langle \varphi, M\varphi \rangle = g(\frac{t+s}{2})$. By hypothesis, $M \leq N$, so it follows that $g(t) = 0 \Rightarrow g(\frac{t+s}{2}) = 0$ for all $s \in (a, b)$ so g vanishes on the interval $(\frac{a+t}{2}, \frac{t+b}{2})$ that goes half way from t to the edges. Iterating we go $3/4$ of the way to the edges, etc., so we conclude that if g ever vanishes, then g is identically zero. □

Here is the big theorem on SOC functions:

Theorem 13.8 *Let $J = (a, b) \subset \mathbb{R}$. Let f be a real-valued C^1 function on J. Pick $x_0 \in J$ and let g be a second C^1 function on J related to f by*

$$f(x) = \begin{cases} \frac{g(x)-g(x_0)}{x-x_0}, & \text{if } x \neq x_0 \\ g'(x_0), & \text{if } x = x_0 \end{cases} \tag{13.20}$$

Then the following are equivalent

(a) $f \in \mathcal{S}_\infty(J)$
(b) *either* $f \equiv 0$ *or* $f > 0$ *and* $-1/f \in \mathcal{C}_\infty(J)$
(c) $g \in \mathcal{M}_\infty(J)$
(d) f *has an integral representation of the form*

$$f(x) = c + \int_{-\infty}^{a} (x-y)^{-1}\, d\xi_-(y) + \int_b^\infty (y-x)^{-1}\, d\xi_+(y) \tag{13.21}$$

where $c \geq 0$ *and* $\xi_\pm$ *are positive measures obeying*

$$\int (1+|y|)^{-1}\, d\xi_\pm(y) < \infty \tag{13.22}$$

Remark This can be viewed either as a theorem about a given g in which case f is defined by (13.20) or else about a given function f in which case g is defined by (13.20) with any choice for $g(x_0)$. One shouldn't worry much about the smoothness assumptions since any of the conditions imply that f and g are real analytic on J.

Proof We'll show that (a)⇒(b)⇒(c)⇒(d)⇒(a).

(a)⇒(b) If $f \equiv 0$, then (b) holds. If not, by the Lemma, (a)⇒ $f > 0$. Let $M = \begin{pmatrix} X & Z \\ Z^* & Y \end{pmatrix}$. If (a) holds we know that

$$\begin{pmatrix} f(X) & 0 \\ 0 & 0 \end{pmatrix} \leq f(M) \equiv \begin{pmatrix} A & C \\ C^* & B \end{pmatrix} \tag{13.23}$$

By the Davis criterion, we need to prove that

$$-f(X)^{-1} \leq -Pf(M)^{-1}P \tag{13.24}$$

Let $f(X) = Q$. By (13.9), we thus need to prove that

$$\begin{pmatrix} A-Q & C \\ C^* & B \end{pmatrix} \geq 0 \Rightarrow -Q^{-1} \leq -(A - CB^{-1}C^*)^{-1} \tag{13.25}$$

By Theorem 13.3 (a), LHS of (13.25) ⇒ $A - Q - CB^{-1}C^* \geq 0 \Rightarrow Q \leq A - CB^{-1}C^*$ ⇒ RHS of (13.25) since $x \mapsto -x^{-1}$ is in $\mathcal{M}_\infty(0,\infty)$ and Q (and therefore also $A - CB^{-1}C^*$) are positive since $f > 0$.

(b)⇒(c) If $f \equiv 0$, g is constant, so in $\mathcal{M}_\infty$. If not, since $f > 0$, we have that $g'(x_0) > 0$. Let $h(x) = 1/f(x)$ so that $h(x_0) = 1/g'(x_0)$. We see that for $x \neq x_0$:

$$\frac{h(x) - h(x_0)}{x - x_0} = \frac{\frac{x-x_0}{g(x)-g(x_0)} - \frac{1}{g'(x_0)}}{x - x_0}$$
$$= \frac{1}{g(x) - g(x_0)} - \frac{1}{g'(x_0)(x - x_0)} \tag{13.26}$$

or

$$q(x) \equiv -\left[\frac{1}{g(x) - g(x_0)}\right]$$
$$= \frac{(-h)(x) - (-h)(x_0)}{x - x_0} - \frac{1}{g'(x_0)(x - x_0)} \tag{13.27}$$

Since we are supposing that $-h \in \mathcal{C}_\infty(a, b)$, by Theorem 9.1, the first term on the right of (13.27) lies in $\mathcal{M}_\infty(a, b)$. The function $x \mapsto -(x - x_0)^{-1}$ lies in both $\mathcal{M}_\infty(-\infty, x_0)$ and $\mathcal{M}_\infty(x_0, \infty)$ so we see that q lies in $\mathcal{M}_\infty(a, x_0) \cap \mathcal{M}_\infty(x_0, b)$. It follows that q has an analytic extension to $\mathbb{C}_+$ and $\mathbb{C}_-$ which is Herglotz there and analytic on $(a, b) \setminus \{x_0\}$. q has a simple pole at x_0.

Thus, $g(z) = -1/q(z) + g(x_0)$ is Herglotz and analytic on $(a, b) \setminus \{x_0\}$. Since q has a pole at x_0, g has a zero and so is analytic on (a, b). Thus, by the easy half of Loewner's theorem, $g \in \mathcal{M}_\infty(a, b)$.

(c)$\Rightarrow$(d) By (3.52) (with f replaced by g), we have that

$$f(x) = \alpha + \int_b^\infty \frac{1}{y - x} \frac{d\nu_+(y)}{(y - x_0)^2} + \int_{-\infty}^a \frac{1}{y - x} \frac{d\nu_-(y)}{(y - x_0)^2} \tag{13.28}$$

This has the form of (13.21) with $d\xi_\pm(y) = d\nu_\pm(y)/(y - x_0)^2$. By (3.51), $d\xi_\pm$ obey (13.22)

(d)$\Rightarrow$(a) By Theorem 13.5, if $y < a$, then $x \mapsto (x - y)^{-1} \in \mathcal{S}_\infty(a, b)$ and by using the elementary fact that $f(x) \in \mathcal{S}_\infty(c, d) \iff f(-x) \in \mathcal{S}_\infty(-d, -c)$, one sees that if $y > b$, then $x \mapsto (y - x)^{-1} \in \mathcal{S}_\infty(a, b)$. Positive combinations and limits of functions in $\mathcal{S}_\infty$ lie in $\mathcal{S}_\infty$ so functions with the integral representation (13.22) lie in $\mathcal{S}_\infty$. □

Example 13.9 In example 9.15, we saw that $\tan x/x$ is matrix convex on $[\pi/2, \pi/2]$ and in Example 4.2 that $\tan x$ is in $\mathcal{M}_\infty(-\pi/2, \pi/2)$. By theorem 13.8, we see that in fact $\tan x/x$ is not merely in $\mathcal{C}_\infty(-\pi/2, \pi/2)$; it is in $\mathcal{S}_\infty(-\pi/2, \pi/2)$.

Example 13.10 Let α be real and let $f(x) = x^\alpha$ as a function on $(0, \infty)$. Consider $h(x) = x^{-\alpha} = f(x)^{-1}$. Theorem 13.8 says that $f \in \mathcal{S}_\infty(0, a) \iff -h \in \mathcal{C}_\infty(0, a)$. By Example 9.13, h is matrix concave if and only if $0 \le -\alpha \le 1$. Thus $x^\alpha \in \mathcal{S}_\infty(0, \infty) \iff -1 \le \alpha \le 0$, extending Theorem 13.5. For $-1 < \alpha \le 0$, one can also use the fact that $x^{\alpha+1} \in \mathcal{M}_\infty(0, \infty)$ and $x^\alpha = \lim_{t \downarrow 0}(x - t)^{-1}[x^{\alpha+1} - t^{\alpha+1}]$. □

There are several ways to use these ideas to generate additional monotone functions from a given one. Recall divided differences defined by (5.39)/(5.40).

Theorem 13.11

(a) *Let* $f \in \mathcal{M}_\infty(a, b)$. *Then for any* $x_1, \ldots, x_{2\ell} \in (a, b)$

$$g(x) = [x, x_1, \ldots, x_{2\ell}; f] \tag{13.29}$$

also lies in $\mathcal{M}_\infty(a, b)$

(b) *Pick* $x_1, x_2 \in (a, b)$. *Let* $f = f_0, f_1(x) = [x, x_1; f_0], f_2(x) = -f_1(x)^{-1}, f_3(x) = [x, x_2; f_2]$. *If* $f_0 \in \mathcal{M}_\infty(a, b)$, *then* $f_1 \in \mathcal{S}_\infty(a, b)$, $f_2 \in \mathcal{C}_\infty(a, b)$, $f_3 \in \mathcal{M}_\infty(a, b)$.

Remarks

1. One can iterate the (b) process. It is easy to see it only ends after finitely many steps if f_0 is a rational function.
2. While we will use SOC to prove (a), there is a direct proof by Loewner's theorem using (5.54). If $y \in \mathbb{R} \setminus (a, b)$, then

$$[x, x_1, \ldots, x_{2\ell}; (y - \cdot)^{-1}] = (y - x)^{-1} \prod_{j=1}^{2\ell} (y - x_j)^{-1} \tag{13.30}$$

and for such y, the product is always positive since 2ℓ is even.

Proof

(a) By induction, we need only handle $\ell = 1$. In that case $[x, x_1; f] \in \mathcal{S}_\infty(a, b)$ by Theorem 13.6 and so in $\mathcal{C}_\infty(a, b)$. Then by Theorem 9.1, $[x, x_1, x_2; f] \in \mathcal{M}_\infty(a, b)$.
(b) By Theorem 13.6, $f_1 \in \mathcal{S}_\infty(a, b)$ and $f_2 \in \mathcal{C}_\infty(a, b)$. Then by Theorem 9.1, $f_3 \in \mathcal{M}_\infty(a, b)$.

□

There are finite matrix variants of a weak version of this last theorem:

Theorem 13.12 *Given a real-valued function* f *on* $(a, b) \subset \mathbb{R}$ *and* $y \in (a, b)$, *let* $f_y^{(1)}(x) \equiv [x, y; f]$. *Then*

(a) *If* $f \in \mathcal{C}_n(a, b)$, *then* $f_y^{(1)} \in \mathcal{M}_n(a, b)$.
(b) *If* $f \in \mathcal{M}_n(a, b)$, *then* $f_y^{(1)} \in \mathcal{C}_{n-1}(a, b)$.

Remarks

1. If f is the function in Example 7.4 and $y = 0$, then $f_y^{(1)}$ is not C^{2n-3} so it cannot lie in $\mathcal{C}_n(a, b)$ whose functions are in C^{2n-2}. This shows that in (b), one cannot replace $\mathcal{C}_{n-1}$ by $\mathcal{C}_n$.

2. There is an important difference between the two assertions. The proof (see Theorem 6.8) shows that if $f_y^{(1)} \in \mathcal{M}_n(a,b)$ for all $y \in (a,b)$, then $f \in \mathcal{C}_n(a,b)$ but one throws something away in the proof of (b) ($\mathcal{P}_{n-2}$ instead of $\mathcal{P}_{n-1}$) and it seems likely to me that there are $f \notin \mathcal{M}_n(a,b)$ for which $f_y^{(1)} \in \mathcal{C}_{n-1}(a,b)$ for all $y \in (a,b)$

Proof

(a) is just (a)⇒(b) in Theorem 9.2 (which is due to Heinävaara).
(b) This follows from Heinävaara's method in Chapter 8. By Proposition 8.5 (and the equivalence of $g \in \mathcal{C}_m(a,b)$ to the positivity of Kraus matrices), $f_y^{(1)} \in \mathcal{C}_{n-1}(a,b) \iff [y, x_0, x_1, x_1, \dots, x_{n-1}, x_{n-1}; q^2 f] \geq 0$ for all choices of distinct $x_1, \dots, x_{n-1} \in (a,b)$, all $x_0 \in (a,b)$ and $q \in \mathcal{P}_{n-2}$. By the mean value theorem for divided differences, this positivity is implied by $(q^2 f)^{(2n-1)} \geq 0$ for all $q \in \mathcal{P}_{n-2}$ but, by Proposition 8.2 and Theorem 6.1, this is true for all $q \in \mathcal{P}_{n-1}$ and so for all $q \in \mathcal{P}_{n-2}$.

□

Notes and Historical Remarks The theory of SOC functions was originally developed in 1988 by Brown [48] who used some deep results in the theory of C^*-algebras. Earlier Krein [190], had proven (as part of a different result) that for $C > 0$, one has that $(PCP)^{-1} \oplus 0 \leq C^{-1}$, i.e., that $x \mapsto x^{-1}$ is SOC on $(0, \infty)$.

Brown considered the theory in general C^*-algebras while we only consider finite matrices although, as in Theorem 2.7, one can easily extend our results to bounded operators on a Hilbert space (see Theorem 2.7). Twenty-five years later, he reexamined the results in two papers [49, 50], one of them joint with Uchiyama. The proofs were made more direct and simpler. Our presentation here borrows ideas from these papers. In particular, Theorem 13.11 is from Brown–Uchiyama [50]. The equivalence of conditions (a), (b), and (d) in Theorem 13.8 is from Brown [48, 49], while condition (c) is from Brown–Uchiyama [50]. References [49, 50] discuss a number of other operator inequalities related to SOC.

The method of Schur complements was named by Haynsworth [144, 145] after work of Schur [308]. In that paper, Schur proved that if A, B, C, D are $n \times n$ matrices so that A and C commute, then

$$\det \begin{pmatrix} A & B \\ C & D \end{pmatrix} = \det(AD - BC)$$

In proving that he used

$$\begin{pmatrix} A^{-1} & 0 \\ -CA^{-1} & \mathbf{1} \end{pmatrix} \begin{pmatrix} A & B \\ C & D \end{pmatrix} = \begin{pmatrix} \mathbf{1} & A^{-1}B \\ 0 & D - CA^{-1}B \end{pmatrix}$$

so Schur complements appeared in a context close to (13.8). The inversion formula (13.9) first appeared in a 1937 paper of Banachiewicz [21]. Puntanen–Styan [277] have a detailed history of these ideas noting that there are precursors in the work of Laplace and Sylvester. Zhang [370] and Gohberg [126] are collections of articles on the Schur complement.

In the physics literature, Schur complements are called the method of Feshbach projections after its 1958 rediscovery by the theoretical physicist Herman Feshbach [97].

Theorem 13.6 is due to Krein [190].

We have used the symbol $\mathcal{S}_\infty$ in analogy to $\mathcal{M}_\infty$ because we assume the bound for all finite matrices. We never formally defined $\mathcal{S}_n$ although there is a natural definition. It would be interesting to have a finite n version of Theorem 13.8 analogous to the finite n results in Theorem 10.2. Theorem 13.12 (b), which is new here, is a weak version of finite n variant of (c)$\Rightarrow$(a) in Theorem 13.8.

Chapter 14
2 x 2 Matrices: The Donoghue and Hansen–Tomiyama Theorems

Loewner's theorem provides a simple characterization of $\mathcal{M}_\infty(a, b)$ but it is not so simple to describe which functions are in a general $\mathcal{M}_n(a, b)$. However, there are complete, succinct descriptions of $\mathcal{M}_2(a, b)$ and $\mathcal{C}_2(a, b)$ that we will describe in this chapter. We'll prove below that

Theorem 14.1 (Donoghue's Theorem) *A nonconstant real-valued function f on $(a, b) \subset \mathbb{R}$ is in $\mathcal{M}_2(a, b)$ if and only if*

(a) *f is C^1 and $f'(x) > 0$ for all $x \in (a, b)$.*
(b) *$(f')^{-1/2}$ is concave.*

Remark We emphasize that this implies that a function in some $\mathcal{M}_n(a, b)$, $n \geq 2$ is either constant or has a strictly positive derivative. This is something we'll use often and without comment throughout the rest of this book. It doesn't require the full theorem above (see Proposition 14.7): If a 2×2 matrix is positive and has a zero element on diagonal, it is easy to see that the off-diagonal element is also zero. Thus looking at the Loewner matrix, $L_2(x_0, y; f)$, one sees that $f'(x_0) = 0 \Rightarrow \forall_y\, f(y) = f(x_0)$.

Theorem 14.2 (Hansen–Tomiyama Theorem) *A real function f on $(a, b) \subset \mathbb{R}$ which is not affine is in $\mathcal{C}_2(a, b)$ if and only if*

(1) *f is C^2 and $f''(x) > 0$ for all $x \in (a, b)$.*
(2) *$(f'')^{-1/3}$ is concave.*

The reason for the simple description is that in this 2×2 case, one can reduce positivity of the 2×2 Dobsch and Hansen–Tomiyama matrices to positivity of their determinants which gives differential inequalities that one can rephrase as positivity of suitable second derivatives. We will prove these two theorems below in a series of propositions. Here is a corollary:

Corollary 14.3 ($\equiv$ Theorem 1.1; See also Corollary 10.7) *If $f \in \mathcal{M}_2(-\infty, \infty)$, then*

B. Simon, *Loewner's Theorem on Monotone Matrix Functions*, Grundlehren der mathematischen Wissenschaften 354, https://doi.org/10.1007/978-3-030-22422-6_14

$$f(x) = ax + b \tag{14.1}$$

for some $a \geq 0, \in \mathbb{R}$.

Proof As we saw in the proof of Proposition 10.1, if g is a non-negative scalar concave function on (α, ∞), then $g(y) \geq g(x)$ for all $y > x > \alpha$. Similarly, if g is a non-negative concave function on $(-\infty, \beta)$, $g(y) \geq g(x)$ for all $y < x < \beta$. It follows that a non-negative concave function on $(-\infty, \infty)$ is constant!

By Theorem 14.1, if $f \in \mathcal{M}_2(-\infty, \infty)$, then $(f')^{-1/2}$ is constant, that is, f' is constant. So (14.1) holds. □

Similarly, Theorem 14.2 implies that

Corollary 14.4 *If* $f \in \mathcal{C}_2(-\infty, \infty)$, *then*

$$f(x) = ax^2 + bx + c \tag{14.2}$$

for some $a \geq 0, \in \mathbb{R}$.

The following is a Corollary of Theorem 14.1 (it is, of course, a special case of Theorem 10.2):

Corollary 14.5 *If* $f \in \mathcal{M}_2(0, \infty)$, *then* f *is scalar concave.*

Proof By Theorem 14.1, $g = (f')^{-1/2}$ is concave and non-negative, so, by Proposition 10.1, $g = (f')^{-1/2}$ is monotone increasing. Thus $f' = g^{-2}$ is monotone decreasing, so f is concave. □

Note that by prior results (Theorem 5.2 and Corollary 9.4), any $f \in \mathcal{M}_2(a, b)$ is C^1 and any $f \in \mathcal{C}_2(a, b)$ is C^2.

Here is another proof of Corollaries 14.3 and 14.5, due to Heinävaara[149], which doesn't go through Theorem 14.1 or even Loewner matrices but relies directly on the definition of matrix monotonicity. We won't even use the consequence of Loewner matrices about continuity and won't suppose continuity but prove it directly.

Theorem 14.6 (Heinävaara [149]) *Suppose that* $f \in \mathcal{M}_2(0, \infty)$. *Then* f *is concave. If* $f \in \mathcal{M}_2(\mathbb{R})$, *then* f *is affine, i.e.* (14.1) *holds.*

Proof Because we don't want to suppose a priori that f is continuous, we need some preliminaries. By looking at constant matrices, we know that f is scalar monotone, so for any $t \in (0, \infty)$, $f^{\pm}(t) = \lim_{\varepsilon \downarrow 0} f(t \pm \varepsilon)$ exists. Moreover, by scalar monotonicity we have that

$$\lim_{s \uparrow t} f^+(s) = \lim_{s \uparrow t} f^-(s) = f^-(t) \tag{14.3}$$

Moreover, by monotonicity, we have that $f^-(t) \leq f(t) \leq f^+(t)$, so by (14.3), it suffices to prove that $f^-(t) \geq f^+(t)$ to conclude that f is continuous at t.

We'll need the matrix $U = \begin{pmatrix} \frac{1}{\sqrt{2}} & \frac{1}{\sqrt{2}} \\ \frac{1}{\sqrt{2}} & -\frac{1}{\sqrt{2}} \end{pmatrix}$ which is unitary and obeys

$$U \begin{pmatrix} a & 0 \\ 0 & b \end{pmatrix} U^{-1} = \begin{pmatrix} \frac{a+b}{2} & \frac{a-b}{2} \\ \frac{a-b}{2} & \frac{a+b}{2} \end{pmatrix} \tag{14.4}$$

We claim that, for any $\alpha \in \mathbb{R}$ and $\delta > 0$, we have that $\begin{pmatrix} \alpha^2\delta & 0 \\ 0 & \delta^{-1} \end{pmatrix} \geq \begin{pmatrix} 0 & \alpha \\ \alpha & 0 \end{pmatrix}$ since the difference has positive trace and determinant 0. Thus, by (14.4), for any $y \in \mathbb{R}$, we have that

$$\begin{aligned} \begin{pmatrix} \alpha^2\delta + y & 0 \\ 0 & \delta^{-1} + y \end{pmatrix} &\geq \begin{pmatrix} y & \alpha \\ \alpha & y \end{pmatrix} \\ &= U \begin{pmatrix} y+\alpha & 0 \\ 0 & y-\alpha \end{pmatrix} U^{-1} \end{aligned} \tag{14.5}$$

It follows that if $f \in \mathcal{M}_2(0, \infty)$ and $y \pm \alpha > 0$, then

$$\begin{pmatrix} f(\alpha^2\delta + y) & 0 \\ 0 & f(\delta^{-1} + y) \end{pmatrix} \geq U \begin{pmatrix} f(y+\alpha) & 0 \\ 0 & f(y-\alpha) \end{pmatrix} U^{-1} \tag{14.6}$$

so looking at the 11 matrix elements, we have that

$$f(\alpha^2\delta + y) \geq \frac{f(y+\alpha) + f(y-\alpha)}{2} \tag{14.7}$$

Taking $\delta \downarrow 0$, we see that

$$0 < \alpha < y \Rightarrow f^+(y) \geq \frac{f(y+\alpha) + f(y-\alpha)}{2} \tag{14.8}$$

Fix $x > 0$. For n large, we have $3/n < x$ so we can take $y = x - 1/n$ and $\alpha = 2/n$ in (14.8) and see that

$$f^+(x - 1/n) \geq \frac{f(x+1/n) + f(x-3/n)}{2} \tag{14.9}$$

Taking $n \uparrow \infty$ and using (14.3), we conclude that $f^-(x) \geq f^+(x)$ proving continuity. Once we have continuity, we conclude that f is concave by (14.8). If $f \in \mathcal{M}_2(\mathbb{R})$, we see that both f and $x \mapsto -f(-x)$ are concave, so f is affine. □

We begin the proof of Theorem 14.1 with

Proposition 14.7

(1) *If* $f \in \mathcal{M}_2(a,b)$ *and* $f'(x_0) = 0$ *for some* $x_0 \in (a,b)$, *then* f *is constant.*
(2) *If* $f \in \mathcal{C}_2(a,b)$ *and* $f''(x_0) = 0$ *for some* $x_0 \in (a,b)$, *then* f *is affine.*

Proof

(1) By the positivity of $\det(L_2(x_0, x_1; f))$, we have

$$\left| \frac{f(x_0) - f(x_1)}{x_0 - x_1} \right|^2 \leq f'(x_0) f'(x_1) \tag{14.10}$$

so $f'(x_0) = 0$ implies $f(x_0) = f(x_1)$, that is, f is constant.
(2) Let $g(x) = [x, x_0; f]$. By Theorem 9.2, $g \in \mathcal{M}_2(a,b)$ and $g'(x_0) = \frac{1}{2} f''(x_0) = 0$, so by (1), g is constant. It follows that $f(x) = f(x_0) + g(x)(x - x_0)$ is affine.

□

Proposition 14.8 *Let* A *be a* 2×2 *self-adjoint matrix with matrix elements* $\{a_{ij}\}_{1 \leq i,j \leq 2}$. *Suppose that* $a_{22} > 0$. *Then*

$$A \geq 0 \iff \mathrm{Tr}(A) > 0,\ \det(A) \geq 0 \tag{14.11}$$

Proof Let λ_1, λ_2 be the eigenvalues of A.

$\Rightarrow$ $A \geq 0 \Rightarrow \lambda_1, \lambda_2 \geq 0$. Moreover, at least one of them is non-zero since $\lambda_1 = \lambda_2 = 0 \Rightarrow A = 0 \Rightarrow a_{22} = 0$. Thus $\mathrm{Tr}(A) = \lambda_1 + \lambda_2 > 0$ and $\det(A) = \lambda_1 \lambda_2 \geq 0$.

$\Leftarrow$ If $\det(A) > 0$, then either both of λ_1 and λ_2 are positive or both are negative. In the latter case, $\mathrm{Tr}(A) < 0$ so $\mathrm{Tr}(A) > 0 \Rightarrow \lambda_1, \lambda_2 > 0 \Rightarrow A > 0$.

If $\det(A) = 0$, then one eigenvalue is 0 and the other is $\mathrm{Tr}(A)$, so $\mathrm{Tr}(A) > 0 \Rightarrow \lambda_1, \lambda_2 \geq 0 \Rightarrow A \geq 0$. □

Suppose now that f is C^3. Then $B_2(x_0; f)$ is a function and we can compute

$$\det(B_2(x_0; f)) = \det \begin{pmatrix} f'(x_0) & \frac{f''(x_0)}{2} \\ \frac{f''(x_0)}{2} & \frac{f'''(x_0)}{6} \end{pmatrix} \tag{14.12}$$

$$= \frac{1}{6} f'''(x_0) f'(x_0) - \frac{f''(x_0)^2}{4} \tag{14.13}$$

$$= -\frac{1}{3} (f')^{5/2} [(f')^{-1/2}]''(x_0) \tag{14.14}$$

As an aside we note that this quantity is also $\frac{f'(x)^2}{6}\,(S_x f)$ where

$$S_x f = \left(\frac{f''}{f'}\right)' - \frac{1}{2}\left(\frac{f''}{f'}\right)^2 = \frac{f'''}{f'} - \frac{3}{2}\left(\frac{f''}{f'}\right)^2 \tag{14.15}$$

is the Schwarzian derivative (see the Notes).

Proof of Theorem 14.1 Suppose first that f is C^3. Since f is supposed to be nonconstant and monotone, by Proposition 14.7, $f'(x_0) > 0$ for all $x_0 \in (a, b)$, so by Theorem 6.1, Proposition 14.8, and the above calculation we have that

$$\begin{aligned} f \in \mathcal{M}_2(a,b) &\iff \forall_{x_0 \in (a,b)} B_2(x_0; f) \geq 0 \\ &\iff \forall_{x_0 \in (a,b)} \det(B_2(x_0; f)) \geq 0 \\ &\iff \forall_{x_0 \in (a,b)} [f'^{-1/2}]''(x_0) \leq 0 \iff f'\prime - 1/2\text{is concave} \end{aligned}$$

If f is not assumed in C^3, we can approximate f with $f_n = j_n * f$ where j_n is smooth, non-negative, with integral 1 and supported in $(-1/n, 1/n)$ and note that $f \in \mathcal{M}_2(a,b) \iff f_n \in \mathcal{M}_2(a+1/n, b-1/n) \iff (f_n')^{-1/2}$ is concave on (a, b). Since, in either direction, f is C^1, we have that $f_n' \to f'$ pointwise (and uniformly on compact subsets of (a, b)) which completes the proof. □

To prove Theorem 14.2, we make a calculation similar to (14.14) letting $g \equiv f''$

$$\det(H_2(x_0; f)) = \det \begin{pmatrix} \frac{g(x_0)}{2} & \frac{g'(x_0)}{6} \\ \frac{g'(x_0)}{6} & \frac{g''(x_0)}{24} \end{pmatrix} \tag{14.16}$$

$$= \frac{1}{48} g''(x_0) g(x_0) - \frac{g'(x_0)^2}{36} \tag{14.17}$$

$$= -\frac{1}{16} g^{7/3} [(g)^{-1/3}]''(x_0) \tag{14.18}$$

Proof of Theorem 14.2 Proposition 14.8, Theorem 6.8, Theorem 9.2, and (14.18) yield a proof of this theorem by following the steps in the proof of Theorem 14.1. □

Calculations like those above (for $n = 1, 2$) show that

$$\det \begin{pmatrix} \frac{h}{n!} & \frac{h'}{(n+!)!} \\ \frac{h'}{(n+1)!} & \frac{h''}{(n+2)!} \end{pmatrix} = -\frac{1}{(n+2)n!^2} h^{2+1/(n+1)} (h^{-1/(n+1)})'' \tag{14.19}$$

Picking $h = f^{(n)}$ and looking at suitable 2×2 diagonal submatrices of the Dobsch and Hansen–Tomiyama matrices, we find necessary (but far from sufficient) conditions on $\mathcal{M}_n$ and $\mathcal{C}_n$.

Theorem 14.9

(1) *If $f \in \mathcal{M}_n(a, b)$, $n \geq 2$, then for $j = 1, \ldots, n-1$, we have that $(f^{(2j-1)})^{-1/2j}$ is concave.*

(2) *If $f \in \mathcal{C}_n(a, b)$, $n \geq 2$, then for $j = 1, \ldots, n-1$, we have that $(f^{(2j)})^{-1/(2j+1)}$ is concave.*

We want to note one simple consequence of Theorem 14.1 that will be needed in Chapter 31.

Proposition 14.10 *If $f \in \mathcal{M}_2(a,b)$, then f' is convex. In particular, for any $x \in (a,b)$ and all $0 < c < \min(x-a, b-x)$, we have that $\frac{f(x+c)-f(x-c)}{2c}$ is monotone decreasing to $f'(x)$ as $c \downarrow 0$.*

Remark Alternatively, we know that $f''' \geq 0$ which implies convexity of f' by the proof of Lemma 5.23.

Proof Since $f'^{-1/2}$ is concave and so is $y \mapsto y^{-2}$, we get the claimed convexity. Thus $\frac{1}{2}[f'(x+q) + f'(x-q)]$ is decreasing in q so its average for $q \in (0,c)$ is decreasing in c. But this average is $\frac{f(x+c)-f(x-c)}{2c}$. □

Notes and Historical Remarks Theorem 14.1 appeared first in Donoghue [81] who says it is due to Dobsch [76], but I can't find any discussion of 2 × 2 matrices in [76]! (Loewner [209] does discuss the 2 × 2 case but doesn't have anything definitive). So I have decided to call it Donoghue's theorem. Interesting enough, while Donoghue goes from monotonicity to concavity of $(f')^{-1/2}$ by using the Dobsch matrix, even though he proves the converse half of the Dobsch–Donoghue theorem elsewhere in his book, he doesn't use that in his proof of the converse direction of Theorem 14.1 as we do. Rather he argues that if $x_1 < x_2$ and

$$a = f'(x_1)^{-1/2} \qquad b = f'(x_2)^{-1/2} \tag{14.20}$$

concavity of $(f')^{-1/2}$ says

$$f'(\theta x_1 + (1-\theta)x_2) \leq (\theta a + (1-\theta)b)^{-2} \tag{14.21}$$

By a direct calculation,

$$\int_0^1 \frac{d\theta}{(\theta a + (1-\theta)b)^2} = (ab)^{-1} \tag{14.22}$$

and

$$\frac{f(x_2) - f(x_1)}{x_2 - x_1} = \int_0^1 d\theta \, f'(\theta x_1 + (1-\theta)x_2) \tag{14.23}$$

(14.25), (14.22), and (14.23) imply (14.10) which is the positivity of $\det(L_2(x_1, x_2; f)$. Since $f' > 0 \Rightarrow \mathrm{Tr}(L_2(x_1, x_2; f)) > 0$, we see the positivity of the Loewner matrix which implies monotonicity.

Theorem 14.2 is from Hansen–Tomiyama [138]. They did not have the converse of their result that concavity of f implies positivity of H_2 (i.e., Heinävaaras's proof of (c)⇒(b) in Theorem 6.8) so they needed a different and rather involved proof that $(f'')^{-1/3}$ concave implies that f is in $\mathcal{C}_2$. Because of Heinävaara's work, we can avoid this part of their argument.

While Corollaries 14.3 and 14.4 are simple consequences of Theorem 14.1, I am not aware of it having been noted before the 2009 paper of Hansen–Tomiyama [139].

The "direct" proof we give of Corollaries 14.5 and 14.3 as Theorem 14.6 and its proof is due to Heinävaara[149]. He has a proof that only requires that $f \in \mathcal{M}_2(\Omega)$ for an open set Ω that is unbounded above and below (where $\mathcal{M}_2(\Omega)$ is defined in Chapter 35). This extension requires ideas from Theorem 35.9.

That $\det(B_2)$ is essentially a Schwarzian derivative was pointed out to me by Jeremy Schiff. The Schwarzian derivative (14.15) was introduced by Schwarz [310] in work on fractional linear transformations. It has two important properties in that regard:

$$g(x) = \frac{\alpha f(x) + \beta}{\gamma f(x) + \delta} \Rightarrow S_x f = S_x g \tag{14.24}$$

and

$$S_x f = 0 \text{ on } (a, b) \Rightarrow f(x) = \frac{\alpha x + \beta}{\gamma x + \delta} \text{ on } \quad (a, b) \tag{14.25}$$

Schwarzian derivatives enter in many places in complex analysis and dynamical systems [28, 51, 62, 128, 188, 199, 234, 235, 249, 299]. An excellent survey is Osgood [255]. It is interesting that they also occur here.

Chapter 15
Quadratic Interpolation: The Foiaş–Lions Theorem

In this chapter, we'll begin by considering a mathematically interesting problem that seems unconnected to the subject of matrix monotone functions. The punch line of the chapter is that a special case of the problem is identical to the question of monotone matrix functions! Let X be a complex vector space and $\|\cdot\|_1$ and $\|\cdot\|_2$ two norms on X. We'll mainly consider finite dimensional X but in the infinite dimensional case, one normally demands that the two norms are compatible in the sense that if $\|x_n\|_1 \to 0$ and $\{x_n\}_{n=1}^\infty$ is Cauchy in $\|\cdot\|_2$, then $\|x_n\|_2 \to 0$. For now, we use X_j for X with the norm, $\|\cdot\|_j$.

A third norm, $\|\cdot\|_*$, on X is said to be an *exact interpolation* norm (for $(\|\cdot\|_1, \|\cdot\|_2)$ if $T : X \to X$ linear, $\|Tx\|_1 \le \|x\|_1$ and $\|Tx\|_2 \le \|x\|_2$ implies that $\|Tx\|_* \le \|x\|_*$. By replacing T by $T/\max(\|T\|_{\mathcal{L}(X_1)}, \|T\|_{\mathcal{L}(X_2)})$, one sees that this is equivalent to $T \in \mathcal{L}(X_1) \cap \mathcal{L}(X_2)$ which implies that $T \in \mathcal{L}(X_*)$ and $\|T\|_{\mathcal{L}(X_*)} \le \max(\|T\|_{\mathcal{L}(X_1)}, \|T\|_{\mathcal{L}(X_2)})$.

In the infinite dimensional case, one often forms the completions, X_1 and X_2 of X in the norms $\|\cdot\|_1$ and $\|\cdot\|_2$ so that one talks about two Banach spaces, X_1 and X_2, where $X_1 \cap X_2$ is dense in each X_j. One discusses maps $T \in \mathcal{L}(X_1)$ and $\mathcal{L}(X_2)$ which agree on $X_1 \cap X_2$.

Suppose now that all norms are quadratic, i.e. come from inner products. Thus we can think of X_1 as a Hilbert space with $\|\varphi\|_1 = \langle \varphi, \varphi \rangle^{1/2}$ and X_2 has the norm $\|\varphi\|_2 = \langle \varphi, A\varphi \rangle^{1/2}$ for some $A \ge 0$. In the infinite dimensional case, one allows A to be unbounded—consistency then implies that the quadratic form of A is closable (see [329, Chapter 7]). But we'll restrict to $\mathcal{H} = \mathbb{C}^n$ for a while. We now ask for which positive operators, B, is $\|\varphi\|_B \equiv \langle \varphi, B\varphi \rangle^{1/2}$ an exact interpolation norm. Note that $\|T\varphi\|_C \le \|\varphi\|_C \Leftrightarrow \langle \varphi, T^*CT\varphi \rangle \le \langle \varphi, C\varphi \rangle \Leftrightarrow T^*CT \le C$. We have thus proven:

Proposition 15.1 *Let $\|\cdot\|, \|\cdot\|_B$, and $\|\cdot\|_A$ be three quadratic norms on $\mathbb{C}^n$. Then $\|\cdot\|_B$ is an exact interpolation for $(\|\cdot\|, \|\cdot\|_A) \Leftrightarrow$*

$$T^*T \le 1 \;\&\; T^*AT \le A \Rightarrow T^*BT \le B \tag{15.1}$$

B. Simon, *Loewner's Theorem on Monotone Matrix Functions*, Grundlehren der mathematischen Wissenschaften 354, https://doi.org/10.1007/978-3-030-22422-6_15

We next head towards a proof that if (15.1) holds, then $B = h(A)$ for some function, h.

Proposition 15.2 *Let E be an orthogonal projection on a Hilbert space, $\mathcal{H}$ and $A \geq 0$, a bounded operator. Then*

$$EAE \leq A \Leftrightarrow EA = AE \tag{15.2}$$

Proof $\Leftarrow$ $E \leq \mathbf{1}$, so $A^{1/2}EA^{1/2} \leq A$. Since A commutes with E, it computes with $A^{1/2}$ (see [329, Section 2.4]). Thus, since $E^2 = E$, we have that $A^{1/2}EA^{1/2} = AE = EAE$.

$\Rightarrow$ Let $\varphi \in \operatorname{Ran} E$, $\psi \in \operatorname{Ran}(\mathbf{1} - E)$. Let $\alpha \in \mathbb{C}$. Then $EAE \leq A \Rightarrow \langle \varphi, A\varphi \rangle \leq \langle \varphi + \alpha\psi, A(\varphi + \alpha\psi) \rangle = \langle \varphi, A\varphi \rangle + 2\operatorname{Re}(\alpha\langle \varphi, A\psi \rangle) + |\alpha|^2 \|\psi\|^2$. Thus, subtracting the two sides and dividing by α and taking $\alpha \downarrow 0$, we see that for all $e^{i\theta} \in \partial\mathbb{D}$, $\operatorname{Re}(\langle e^{i\theta}\varphi, A\psi \rangle) \geq 0 \Rightarrow \langle \varphi, A\psi \rangle = 0$. Therefore, $EA(\mathbf{1} - E) = 0 \Rightarrow EA = EAE \Rightarrow AE = EA$. □

Lemma 15.3 *Let $A, B \in \mathcal{L}(\mathbb{C}^n)$, both non-negative. Suppose that all orthogonal projections, E, which commute with A commute with B. Then, for some function, h, we have that $B = h(A)$.*

Remark This is also true on an infinite dimensional Hilbert space, $\mathcal{H}$. The proof in that case relies on the von Neumann double commutant theorem (see [329, Section 5.4]).

Proof Let α be an eigenvalue of A and P_α the orthogonal projection onto $\{\varphi \mid A\varphi = \alpha\varphi\}$. Since $AP_\alpha = P_\alpha A$, we have that $BP_\alpha = P_\alpha B$, i.e. B leaves $P_\alpha[\mathbb{C}^n]$ fixed. If $\varphi \in \operatorname{Ran} P_\alpha$, then E, the rank one projection onto multiples of φ commutes with A and so with B. Thus $(\mathbf{1} - E)BE = 0$ which means φ is an eigenvector of B. If φ, ψ, and $\varphi + \psi$ are all eigenvectors of B, the eigenvalues are the same. Thus there is a number $h(\alpha)$ so that $B\varphi = h(\alpha)\varphi$ for all $\varphi \in \operatorname{Ran} P_\alpha$. Therefore $A = h(B)$. □

Now suppose that (15.1) holds for fixed A, B, and all T. If E is a projection, then, by Proposition 15.2 the LHS of (15.1) holds for $T = E$, $\Longleftrightarrow$ $EA = AE$. Therefore, we conclude that $EA = AE \Rightarrow EB = BE$ so, by Lemma 15.3, B is a function of A. Thus, to solve the exact interpolation problem for quadratic norms, we focus on functions, h which obey

$$T^*T \leq \mathbf{1}, \quad T^*AT \leq A \Rightarrow T^*h(A)T \leq h(A) \tag{15.3}$$

Functions, h, that obey (15.3) for all pairs of $n \times n$ matrices A, T with $A \geq 0$ are called *interpolation functions* (of order n). The set of such functions is denoted by $\mathcal{I}_n$. Clearly, $\mathcal{I}_{n+1} \subset \mathcal{I}_n$. We define $\mathcal{I}_\infty = \cap_{n=1}^{\infty} \mathcal{I}_n$. The main result of this chapter is

Theorem 15.4 (Foiaş–Lions Theorem) $\mathcal{I}_\infty = \mathcal{M}_\infty(0, \infty)$, *the matrix monotone functions on* $(0, \infty)$.

It then follows easily from Loewner's theorem (see the remark after Theorem 15.12) that if $h \in \mathcal{I}_\infty$, then (15.3) holds also in the infinite dimensional case.

Proposition 15.5 $\mathcal{M}_{2n}(0, \infty) \subset \mathcal{I}_n$

Proof Let $A \geq 0, T$ be in $\mathcal{L}(\mathbb{C}^n)$ with LHS of (15.3). Let $S = \sqrt{\mathbf{1} - T^*T}$ and $Q = \sqrt{\mathbf{1} - TT^*}$. Since $T(T^*T)^n = (TT^*)^n T$ and the square root is given by a convergent power series [329, Theorem 2.4.4], one has that $TS = QT$ and so $ST^* = T^*Q$. Define $U = \begin{pmatrix} T & Q \\ -S & T \end{pmatrix}$ which is easily seen to obey $U^*U = \mathbf{1}$, so since U acts on the finite dimensional space $\mathbb{C}^{2n}$, U is unitary. Let $h \in \mathcal{M}_{2n}(0, \infty)$. Let $P = \begin{pmatrix} \mathbf{1} & 0 \\ 0 & 0 \end{pmatrix}$, the projection onto the first $\mathbb{C}^n$ and let $C = \begin{pmatrix} A & 0 \\ 0 & 0 \end{pmatrix}$. By Proposition 10.5,

$$P(U^*h(C)U)P = Ph(U^*CU)P \leq h(P(U^*CU)P) \tag{15.4}$$

since U is unitary. Thus

$$T^*h(A)T \leq h(T^*AT) \tag{15.5}$$

Since $T^*AT \leq A$, monotonicity of h implies that

$$h(T^*AT) \leq h(A) \tag{15.6}$$

Thus LHS of (15.3) and $h \in \mathcal{M}_{2n}(0, \infty) \Rightarrow$ RHS of (15.3), i.e. $h \in \mathcal{I}_n$. □

As a Corollary of the proof, we have that

Proposition 15.6 *If h is in $\mathcal{M}_\infty(0, \infty)$, then* (15.3) *holds for bounded operators on an infinite dimensional Hilbert space.*

Proof By Theorem 2.7, h is monotone on bounded operators on an infinite dimensional Hilbert space. The proof of Proposition 15.5 then implies the result. □

Proposition 15.7 $\mathcal{I}_{2n} \subset \mathcal{M}_n(0, \infty)$

Proof Let $h \in \mathcal{I}_{2n}$. Let $A, B \in \mathcal{L}(\mathbb{C}^n)$ with $A \leq B$. Let $C = \begin{pmatrix} A & 0 \\ 0 & B \end{pmatrix}$ and $T = \begin{pmatrix} 0 & \mathbf{1} \\ B^{-1/2}A^{1/2} & 0 \end{pmatrix}$. Then $T^*T = \begin{pmatrix} A^{1/2}B^{-1}A^{1/2} & 0 \\ 0 & \mathbf{1} \end{pmatrix} \leq \mathbf{1}$ and $T^*CT = \begin{pmatrix} A & 0 \\ 0 & A \end{pmatrix} \leq C$. Since $h \in \mathcal{I}_{2n}$, $T^*h(C)T \leq h(C)$, i.e. $\begin{pmatrix} A^{1/2}B^{1/2}h(B)B^{1/2}A^{1/2} & 0 \\ 0 & h(A) \end{pmatrix} \leq \begin{pmatrix} h(A) & 0 \\ 0 & h(B) \end{pmatrix}$. Thus, $h(A) \leq h(B)$ and $B^{-1}h(B) \leq A^{-1}h(A)$. □

As a Corollary of the proof and Proposition 15.5:

Corollary 15.8 $h \in \mathcal{M}_{4n}(0, \infty) \Rightarrow (z \mapsto -z^{-1}h(z) \in \mathcal{M}_n(0, \infty))$

Remarks

1. Thus $h \in \mathcal{M}_\infty(0,\infty) \Rightarrow (z \mapsto -z^{-1}h(z) \in \mathcal{M}_\infty(0,\infty))$. This follows also from Loewner's theorem, (3.62), and the facts that $z \mapsto -z^{-1}$ and $z \mapsto -(\lambda + (1-\lambda)z)^{-1}$ for $0 < \lambda \leq 1$ are matrix monotone.
2. Propositions 15.5 and 15.7 imply Theorem 15.4.

Next, we want to note an infinitesimal form of (15.3). Given a bounded operator, B, on a Hilbert space, we define

$$\operatorname{Re} B = \tfrac{1}{2}(B + B^*) \tag{15.7}$$

Theorem 15.9 *Fix a function, h, on* $(0,\infty)$. *Suppose that* (15.3) *holds for all bounded operators T and A on a Hilbert space with* $A \geq 0$. *Then for all bounded B and* $A \geq 0$

$$\operatorname{Re} B \geq 0 \text{ and } \operatorname{Re} AB \geq 0 \Rightarrow \operatorname{Re} h(A)B \geq 0 \tag{15.8}$$

Proof Let $T_t = e^{-tB}$ defined via a convergent power series. Then for any bounded operator C

$$\frac{d}{dt}\langle T_t\varphi, CT_t\varphi\rangle = -\langle T_t\varphi, (CB + B^*C)T_t\varphi\rangle \tag{15.9}$$

so

$$T_t^* C T_t \leq C \text{ for all } t \geq 0 \iff \operatorname{Re}(CB) \geq 0 \tag{15.10}$$

Thus, if (15.3) holds, LHS of (15.8) $\Rightarrow$ LHS of (15.3) for T_t, all $t \geq 0 \Rightarrow$ RHS of (15.3) for all $t \geq 0$, $\Rightarrow$ RHS of (15.8). □

We want to conclude with some further remarks including the fact that Loewner's theorem implies that any $h \in \mathcal{M}_\infty(0,\infty)$ is an interpolation function, even in the infinite dimensional case. There is another proof that rather than Proposition 15.6 depends on an important way to construct quadratic interpolation norms.

Given two norms, $\|\cdot\|_1$ and $\|\cdot\|_2$, on a space, X, define

$$K_2(t,\varphi) = \inf_{\psi_1+\psi_2=\varphi} \left(\|\psi_1\|_1^2 + t\|\psi_2\|_2^2\right)^{1/2} \tag{15.11}$$

for $t \in (0,\infty)$ and $\varphi \in X$. We have that

Theorem 15.10 *Fix* $t \in (0,\infty)$. *Then* $K_2(t,\cdot)$ *is a* $(\|\cdot\|_1, \|\cdot\|_2)$ *exact interpolation norm.*

Remark This proof also holds when X is an infinite dimensional space.

Proof Since

$$\|\eta_{11}+\eta_{12}\|_1^2+t\|\eta_{21}+\eta_{22}\|_2^2 \leq \left[\left(\|\eta_{11}\|_1^2+t\|\eta_{21}\|_1^2\right)^{1/2}+\left(\|\eta_{12}\|_2^2+t\|\eta_{22}\|_2^2\right)^{1/2}\right]^2$$

(for by the Schwarz inequality, $(ac + bd) \leq (a^2 + b^2)^{1/2}(c^2 + d^2)^{1/2}$ and take $a = \|\eta_{11}\|_1, b = t^{1/2}\|\eta_{21}\|_2, c = \|\eta_{12}\|_1, andd = t^{1/2}\|\eta_{22}\|_2$), we see that $K_2(t, \cdot)$ is a norm by first taking the inf on the left and then the inf on the right.

If $\|T\psi\|_1 \leq \|\psi\|_1$ and $\|T\psi\|_2 \leq \|\psi\|_2$, then

$$\|T\psi_1\|_1^2 + t\|T\psi_2\|_2^2 \leq \|\psi_1\|_1^2 + t\|\psi_2\|_2^2 \tag{15.12}$$

Since those $\eta_1 = T\psi_1$ and $\eta_2 = T\psi_2$ with $\psi_1 + \psi_2 = \varphi$ are a subset of all η_1, η_2 with $\eta_1 + \eta_2 = T\varphi$, and we are taking an inf,

$$K_2(t, T\varphi) \leq \|\psi_1\|_1^2 + t\|\psi_2\|_2^2 \tag{15.13}$$

for all ψ_1, ψ_2 with $\psi_1 + \psi_2 = \varphi$, taking an inf again, we see that

$$K_2(t, T\varphi) \leq K_2(t, \varphi) \tag{15.14}$$

showing that $K_2(t, \cdot)$ is an exact interpolation norm. □

Remarkably, if $\|\cdot\|_1$ and $\|\cdot\|_2$ are quadratic norms, not only are each $K_2(t, \cdot)$ quadratic norms but there is an explicit formula for K_2. So suppose that $\mathcal{H}$ is a Hilbert space and $A \geq 0$ a positive operator on $\mathcal{H}$ with

$$\|\varphi\|_1 = \langle\varphi, \varphi\rangle^{1/2}, \qquad \|\varphi\|_2 = \langle\varphi, A\varphi\rangle^{1/2} \tag{15.15}$$

Fix $t > 0$ and $\varphi \in \mathcal{H}$. Formally the function

$$F(\psi) = \|\varphi - \psi\|^2 + t\langle\psi, A\psi\rangle \tag{15.16}$$

is a function of ψ and $\bar{\psi}$, so we can use $\frac{\partial}{\partial\psi}, \frac{\partial}{\partial\bar{\psi}}$ notation (see [326, Section 2.1]) and find that

$$\frac{\partial F}{\partial\bar{\psi}} = -(\varphi - \psi) + tA\psi$$

solved by $\psi_0 = (\mathbf{1} + tA)^{-1}\varphi, \varphi - \psi_0 = tA(\mathbf{1} + tA)^{-1}\varphi$.

$$F(\psi_0) = \left\langle\varphi, \frac{t^2A^2}{(\mathbf{1}+tA)^2} + \frac{tA}{(\mathbf{1}+tA)^2}\varphi\right\rangle \tag{15.17}$$

$$= \left\langle\varphi, \frac{tA}{\mathbf{1}+tA}\varphi\right\rangle \tag{15.18}$$

Rather than try to justify this calculation in the infinite dimensional case, we use it for motivation:

Theorem 15.11 *If* $\|\cdot\|_1$, $\|\cdot\|_2$ *are given by* (15.15), *then*

$$K_2(t,\varphi) = \left\langle \varphi, \frac{tA}{\mathbf{1}+tA}\varphi \right\rangle^{1/2} \tag{15.19}$$

In particular,

$$\lim_{t\to\infty} K_2(t,\varphi) = \|\varphi\|_1, \qquad \lim_{t\downarrow 0} t^{-1/2} K_2(t,\varphi) = \|\varphi\|_2 \tag{15.20}$$

Proof Given t and φ, let $\psi_0 = (\mathbf{1}+tA)^{-1}\varphi$ so that

$$\varphi - \psi_0 = tA\psi_0 \tag{15.21}$$

That means the cross term (i.e., O(η) terms) in $F(\psi_0+\eta)$ vanishes so

$$F(\psi_0+\eta) = F(\psi_0) + \Big[\|\eta\|^2 + t\langle \eta, A\eta\rangle\Big] \tag{15.22}$$

Since the term in brackets is strictly positive if $\eta \neq 0$, it follows that ψ_0 minimizes F and by (15.17)/(15.18), we get (15.19).

As $t \to \infty$, $\frac{tA}{1+tA} \to \mathbf{1}$, so we get the first equation in (15.20). Similarly as $t \downarrow 0$, $\frac{A}{1+tA} \to A$, so we get the second equation in (15.20). □

Theorem 15.12 (Donoghue's Exact Interpolation Theorem) *Let* $A \geq 0$ *and* K_2 *be given by* (15.19). *If* $h \in \mathfrak{M}_\infty(0,\infty) = \mathfrak{I}_\infty$, *then for some measure,* $d\rho_h$, *we have that*

$$\langle \varphi, h(A)\varphi\rangle = \int \left(1+\frac{1}{t}\right) K_2(t,\varphi)^2 \, d\rho_h(t) \tag{15.23}$$

Proof The proof relies on Loewner's theorem. By that theorem and (3.67),

$$\begin{aligned} \langle \varphi, h(A)\varphi\rangle &= \int (1+t) \left\langle \varphi, \frac{A}{A+t\mathbf{1}}\varphi\right\rangle d\rho_h(t) \\ &= \int (1+t^{-1}) \left\langle \varphi, \frac{tA}{A+t\mathbf{1}}\varphi\right\rangle d\rho_h(t) \\ &= \text{RHS of (15.23)} \end{aligned} \tag{15.24}$$

□

Remark Since $K_2(t,\cdot)$ is an exact interpolation norm, we see that if $h \in \mathfrak{I}_\infty$, then $\langle\,\cdot\,, h(A)\,\cdot\,\rangle^{1/2}$ is an exact interpolation norm even in the infinite dimensional case.

While we proved Theorem 15.12 using Loewner's theorem, the calculations leading to (15.24) are reversible, so if one proves (15.23) for any $h \in \mathcal{I}_\infty$, then using $\mathcal{M}_\infty(0,\infty) = \mathcal{I}_\infty$, we get a new proof of Loewner's Theorem. This is the strategy of Chapter 30.

Notes and Historical Remarks There is a huge literature on interpolation spaces and norms, for example, three books [23, 24, 346]. The complex method (see [325, Section 6.6] and [326, Section 5.2] for discussion including history) of Calderón and Lions abstracts the Riesz–Thorin Theorem. The real method in part abstracts the Marcinkiewicz interpolation theorem (see [328, Section 6.1]). The pioneers include Lions and Peetre. In particular, Peetre [264, 265] introduced his K-functional

$$K_1(\varphi, t) = \inf_{\psi_1+\psi_2=\varphi} \|\psi_1\|_1 + t\|\psi_2\|_2$$

and proved it was an exact interpolation norm. The K_2 of this chapter is clearly an analog.

The Foiaş–Lions theorem (Theorem 15.4) is from their paper [102] with further development by Donoghue[79], who in particular, had Theorem 15.12.

The arguments we give to prove that $\mathcal{M}_{2n}(0,\infty) \subset \mathcal{I}_n$ and $\mathcal{I}_{2n} \subset \mathcal{M}_n(0,\infty)$ are from Boutet de Monvel [45]. Donoghue [79] had $\mathcal{I}_{2n} \subset \mathcal{M}_n(0,\infty)$ using $T = \left(\begin{smallmatrix} 0 & 0 \\ 1 & 0 \end{smallmatrix}\right)$ which shows $f(A) \le f(B)$. Boutet's choice of $T = \left(\begin{smallmatrix} 0 & 1 \\ B^{-1/2}A^{1/2} & 0 \end{smallmatrix}\right)$ also yields the $B^{-1}f(B) \le A^{-1}f(A)$ result. Boutet de Monvel [45] also has the infinitesimal form of being an interpolation norm. Since the exponential map from $n \times n$ complex matrices to invertible matrices is surjective, for finite dimensional spaces, the infinitesimal form has a converse. Alternatively, in Chapter 34, we'll see the converse via a proof of Loewner's theorem.

Ameur [6–8] has the stronger $\mathcal{M}_{n+1}(0,\infty) \subset \mathcal{I}_{2n}$, improved to $\mathcal{M}_{n+1}(0,\infty) \subset \mathcal{I}_{2n+1}$ in [10](so $\mathcal{M}_{n+1}(0,\infty) \subset \mathcal{I}_{2n+1} \subset \mathcal{I}_{2n} \subset \mathcal{M}_n(0,\infty)$), but this stronger result relies on Theorem 25.1 which itself immediately implies the hard part of Loewner's theorem as we'll see in Corollary 25.2. So, in essence, this stronger inclusion relies on Loewner's theorem while Boutet's result does not.

Here is the argument from [10]: Given Corollary 25.3 and $f \in \mathcal{M}_{n+1}$ and given A, a matrix on $\mathbb{C}^{2n+1}$, we let $\{\lambda_j\}_{j=1}^{2n+1}$ be the eigenvalues. Let h be the Herglotz function analytic on $\mathbb{C} \setminus (-\infty, 0]$ guaranteed by Corollary 25.3 with $h(\lambda_j) = f(\lambda_j)$, $j = 1, \ldots, 2n+1$. Since $f(A) = h(A)$ and h has a Herglotz representation, $h(A)$ has a representation of the form (15.23) which implies that $\langle \cdot, f(A) \cdot \rangle^{1/2}$ is an exact interpolation norm.

Theorem 15.9 is from Boutet de Monvel [45]. While it was written in 1989, it was unpublished until this book. Independently, Uchiyama [347] proved the special case where B is self-adjoint and $h(t) = t^\alpha$; $0 < \alpha < 1$. His proof uses an integral representation and so holds for any h which is matrix monotone on $(0,\infty)$ so long as B is self-adjoint. Our application of this theorem in Chapter 34 is for a non-self-adjoint B.

Part II
Proofs of the Hard Direction

This part presents eleven distinct proofs of the hard direction of Loewner's theorem, that is, a real-valued function $f \in \mathcal{M}_\infty(-1, 1)$ obeys an integral representation of the form (1.13), equivalently, given the Herglotz representation theorem that f is the restriction to $(-1, 1)$ of a function analytic in a neighborhood of $\mathbb{C}_+ \cup (-1, 1)$ with non-negative imaginary values in $\mathbb{C}_+$. It will sometimes be more convenient to consider non-negative $f \in \mathcal{M}_\infty(0, \infty)$ in which case the integral representation takes the form

$$f(x) = \int_0^1 \frac{x}{\lambda + (1 - \lambda)x}\, d\nu(\lambda) \tag{p2.1}$$

for $x > 0$, and some positive measure $d\nu$ on $[0, 1]$. Notice, despite the naive fact that it seems as if the integrand vanishes at $x = 0$, in fact, one has that $\lim_{x \downarrow 0} f(x) = \nu(\{0\})$.

That we only need this for non-negative f's follows by noting that if $g \in \mathcal{M}_\infty(0, \infty)$, then for $a > 0$, $f(x) = g(x + a) - g(a)$ is a non-negative function in $\mathcal{M}_\infty(0, \infty)$, so (p2.1) implies the necessary analyticity for g (by then taking $a \downarrow 0$). The remark following Lemma 2.3 shows that the analyticity result for $\mathcal{M}_\infty(-1, 1)$ or $\mathcal{M}_\infty(0, \infty)$ implies it for all (a, b).

Here are the proofs that we will present:

(1) **Degenerate Pick's Theorem Proof** (*New Here*) This proof is in Chapter 20 with background on Pick's theorem in Chapters 16–18 and 24. Pick's theorem studies the question of when there is a Herglotz function, f, on $\mathbb{C}_+$ with $f(z_j) = w_j$ for specified points $\{z_j, w_j\}_{j=1}^n$ in $\mathbb{C}_+$ and states this is true exactly when certain matrices are positive. The hard part of Loewner's theorem is a kind of degenerate version of the hard part of Pick's theorem where the z's and w's are moved into $\mathbb{R}$. This proof moves them slightly into $\mathbb{C}_+$ and takes limits. There is a proof of Pick's theorem using Hilbert space methods due to Korányi and Sz.-Nagy (exposed in Chapter 17) and if you combine that proof with this proof of Loewner's theorem, you get a proof very close to Korányi's

proof so, in one sense, this is a variant of Korányi's proof (see (4) below). But there are other proofs of Pick's theorem than the one of Korányi and Sz.-Nagy, so this really is a distinct proof. We note that Pick was Loewner's thesis advisor so Loewner certainly knew this connection and may have been motivated by analogy in finding his theorem.

(2) **Rational Interpolation Proof** (*Loewner*) This proof, the original one, is presented in Chapter 25 with background on rational interpolation in Chapters 21–24.

(3) **Moment Problem Proof** (*Bendat–Sherman*) This proof is presented in Chapter 26. If f has a representation of the form (1.13), then f is C^∞ and

$$\frac{f^{(n+1)}(0)}{(n+1)!} = \int_{-1}^{1} \lambda^n \, d\nu(-\lambda) \tag{p2.2}$$

so one can hope to find $d\nu$ by solving the moment problem. Once one has done that, one knows that the right side of (1.13) is just $\sum_{n=0}^{\infty} \frac{f^{(n+1)}(0)x^n}{(n+1)!}$. So, we need to know that f is analytic in $\mathbb{D}$ to complete the proof. Here we are saved by a variant of a remarkable theorem of Bernstein which says if f on $(-1, 1)$ has $(-1)^n f^{(n)}(x) \geq 0$ for all $x \in (-1, 1)$, then f has an analytic continuation to $\mathbb{D}$! As we've seen (see Theorem 5.2) any $f \in \mathcal{M}_\infty(-1, 1)$ is C^∞ and (see Theorem 5.18 and the fact that, by (5.38), $f^{(2n-1)}(x)$ is a diagonal matrix element of a positive matrix) $f^{(2n-1)}(x) \geq 0$. This is enough to use the Bernstein's theorem ideas to get analyticity of $f \in \mathcal{M}_\infty(-1, 1)$ on $\mathbb{D}$. Positivity of the Dobsch matrix also provides conditions that guarantee that the moment problem is solvable. Bendat–Sherman used the Hamburger moment problem (seeking measures on $(-\infty, \infty)$), while here we use the simpler-to-control Hausdorff moment problem (seeking measures on $(-1, 1)$). This chapter is self-contained in the sense that we provide proofs of the solubility of the Hausdorff moment problem and of the improved Bernstein's theorem that we need.

(4) **Hilbert Space Proof** (*Korányi*) This proof is presented in Chapter 27. Since the Loewner matrix is positive, we can use it to define an inner product, $< ., . >_L$. One then constructs a vector ϕ and self-adjoint operator, A, so that

$$< \phi, (A - x)^{-1} \phi >_L = f(x) \tag{p2.3}$$

and obtains the measure for the Loewner representation as a spectral measure for operator A and vector ϕ. Korányi demanded (p2.3) for all $x \in (-1, 1)$ and required the theory of self-adjoint extensions and the spectral theorem for unbounded operators. Instead, our presentation constructs $n \times n$ matrices, point measures, and takes limits as $n \to \infty$. This variant of his proof avoids some technical issues with unbounded operators.

(5) **Krein–Milman Proof** (*Hansen–Pedersen*) This proof is presented in Chapter 28. Many of the classical integral representation theorems of analysis can

be viewed as special cases of a strong form of the Krein–Milman that holds when the set of extreme points are closed; see Choquet [61] or Simon [324]. This is true of Loewner's theorem also—one needs to set up a topology in which $\mathcal{M}_\infty(a, b)$ (or some convenient normalized subset) is compact and then identify the extreme points. We provide a variant of the original Hansen–Pedersen proof that borrows from improvements of Boutet de Monvel, Bhatia, and, most importantly, a later paper of Hansen (see the Notes to Chapter 28). One disadvantage of this proof is that analyticity of matrix monotone functions is unexplained—the extreme points are uniformly analytic, so their integrals are. But analyticity is something that one observes "just happens to hold" for the extreme points.

(6) **Hahn–Banach Proof** (*Sparr*) This proof is presented in Chapter 29. Sparr's idea is to define a particular class of continuous functions on $[-\infty, 1]$, and use a given $f \in \mathcal{M}_\infty(0, \infty)$ to define a positive functional on such functions. He then appeals to a variant of M. Riesz' version of the Hahn–Banach theorem (see [325, Theorem 5.5.8]) to extend to a positive functional on all continuous functions and thereby get a measure which provides the required integral representation. In fact, we prove that Sparr's set of functions is dense, so positivity allows extension without appealing to the Hahn–Banach theorem. Therefore, it might be better to call this the Riesz–Markov proof, except the Riesz–Markov theorem is behind many of the other proofs.

(7) **Interpolation Theory Proof** (*Ameur*) This proof is presented in Chapter 30. It relies on the fact that matrix monotone functions are also interpolation functions. By the careful construction of certain operators which obey the hypothesis of what an interpolation function has to do, it allows the construction of a positive measure.

(8) **Continued Fraction Proof** (*Wigner–von Neumann*) This proof is presented in Chapter 31. This proof uses a key lemma that if f is a C^1 function on $(-1, 1)$ with $f(0) = 0$ and if

$$g(x) = -\frac{1}{f(x)} + \frac{1}{f'(0)x} \tag{p2.4}$$

then $f \in \mathcal{M}_n(-1, 1) \Rightarrow g \in \mathcal{M}_{n-1}(-1, 1)$. They prove this lemma by looking at the Loewner matrices for f and g. For $f \in \mathcal{M}_\infty(-1, 1)$, one can iterate and get a continued fraction expansion. They then use a variant of Markov's method for convergence of continued fractions to get convergence of the truncated fractions to a function obeying the required integral representation.

(9) **Multipoint Continued Fraction Proof** (*New Here*) This proof is presented in Chapter 32. It is a small variant of the Wigner–von Neumann proof. The truncated continued fraction for that proof has more and more derivatives at 0 that agree with those of f and one needs some estimates that, while not hard, are also not trivial. This proof does the procedure at more and more dyadic rationals and so gets functions with an integral representation that provide f

at more and more points so weak convergence and an integral representation for the limitx are easy.

(10) **Hardy Space Proof** (*Rosenblum–Rovnyak*) This proof is presented in Chapter 33. It depends on the fact that operators on the Hardy space $H^2(\mathbb{D})$ that commute with multiplication by z are multiplication by $H^\infty(\mathbb{D})$ functions. That result is proven in Chapter 19.

(11) **Mellin Transform Proof** (*Boutet de Monvel*; Previously Unpublished) This proof is presented in Chapter 34. It depends on writing one version of the Herglotz representation theorem on $(0, \infty)$ as a convolution on the multiplicative group of positive reals of a measure with the function $x/(1+x)$. Mellin transform which is Fourier transform on this group lets one construct a measure using Bochner's theorem. The proof relies on the infinitesimal version of when a function is an interpolation function.

The proofs (5), (6), (7), and (11) do not explicitly rely on positive of Loewner matrices for all n but the other 7 proofs do (the proof (5) uses the fact that if f is matrix monotone and nonconstant, then $f'(t) > 0$ for all t; while there may be other proofs, we prove this by using positivity of 2×2 Loewner matrices).

Chapter 35 discusses versions of Loewner's theorem when (a, b) is replaced by a general open set. In particular, it proves the perhaps surprising result that if $a < b < c < d$ and $f \in \mathcal{M}_\infty((a, b) \cup (c, d))$, then f has an analytic continuation to all of (a, d) and so f is the restriction of a function in $\mathcal{M}_\infty(a, d)$.

Chapter 16
Pick Interpolation, I: The Basics

In 1916, Pick asked and answered the following question: Given indexed collections $\{z_j\}_{j \in I}$ in $\mathbb{D}$ and $\{w_j\}_{j \in I}$ in $\{w \mid \operatorname{Re} w > 0\}$ with the z_j distinct, when does there exist an analytic function F on $\mathbb{D}$ with $\operatorname{Re} F > 0$ so that for all $j \in I$,

$$F(z_j) = w_j \tag{16.1}$$

This is the first of five chapters on Pick's theorems. In this chapter, we will first state the theorems in three versions: for Carathéodory functions on $\mathbb{D}$, for Schur functions on $\mathbb{D}$, and for Herglotz functions on $\mathbb{C}_+$. We prove the conditions are necessary and that the three theorems are "equivalent." Finally, we will reduce to the case where I is finite using a compactness result. In the next three chapters, we will provide the proofs of the hard part of the theorem: one using the spectral theorem for finite Hermitian matrices, the second using continued fractions, and the third using Hardy spaces. The fifth chapter deduces the hard part of Loewner's theorem as a limit of Pick's theorem!

This is something of an aside on our main theme, but it is a relevant aside. That is not clear now but will be shortly.

In Chapter 3, we used the same symbol f for Herglotz and Carathéodory functions. In this chapter, for clarity as we jump back and forth, we will use

$$G \text{ for a Herglotz function,} \qquad G: \mathbb{C}_+ \to \mathbb{C}_+ \tag{16.2}$$

$$F \text{ for a function,} \qquad F: \mathbb{D} \to \{z \mid \operatorname{Re} z > 0\} \tag{16.3}$$

$$f \text{ for a function,} \qquad f: \mathbb{D} \to \mathbb{D} \tag{16.4}$$

F's of (16.3) are "essentially" *Carathéodory functions*; the latter are normalized by $F(0) = 1$. We will define *Schur functions* as analytic maps of $\mathbb{D} \to \overline{\mathbb{D}}$, that is, analytic with

B. Simon, *Loewner's Theorem on Monotone Matrix Functions*, Grundlehren der mathematischen Wissenschaften 354, https://doi.org/10.1007/978-3-030-22422-6_16

$$\sup_{z\in\mathbb{D}} |f(z)| \le 1 \tag{16.5}$$

the unit ball in $\mathbb{H}^\infty$. We will call the functions in (16.4) *strong Schur functions*. The Schur functions that are not strong are precisely of the form

$$f(z) = e^{i\theta} \in \partial\mathbb{D} \tag{16.6}$$

for a fixed θ and all z, that is, constants of magnitude 1. We will see that the Schur functions are a natural compactification of (16.4).

We will look at the Pick problem for (16.2) first. By (3.33) and (3.34) for $z, \zeta \in \mathbb{C}_+$,

$$\frac{G(z) - \overline{G(\zeta)}}{z - \bar{\zeta}} = A + \int \frac{1+x^2}{(x-z)(x-\bar{\zeta})}\, d\mu(x) \tag{16.7}$$

Thus, if

$$G(z_j) = w_j \tag{16.8}$$

we have for each finite $F \subset I$,

$$\sum_{j,k\in F} \bar{\alpha}_j \alpha_k \frac{w_j - \bar{w}_k}{z_j - \bar{z}_k} = A\left|\sum_{j\in F} \alpha_j\right|^2 + \int (1+x^2)\left|\sum_{j\in F} \frac{\alpha_j}{x - z_j}\right|^2 d\mu(x) \tag{16.9}$$

We thus have half of

Theorem 16.1 (Pick's Theorem for Herglotz Functions) *Let* $\{z_j\}_{j\in I}$, $\{w_j\}_{j\in I}$ *be indexed sets in* $\mathbb{C}_+$ *with the* z_j *distinct. Then* (16.8) *holds for some Herglotz function, G, if and only if the* Pick matrix

$$M_{jk} = \frac{w_j - \bar{w}_k}{z_j - \bar{z}_k} \tag{16.10}$$

is positive, that is, $\sum_{j,k} \bar{\alpha}_j \alpha_k M_{jk} \ge 0$ *for any* $\{\alpha_j\}_{j\in I}$, *which is non-zero only for finitely many* j.

We have proven positivity of (16.10) is necessary for (16.8) to have a solution. Equation (16.9) also lets us answer when M is strictly positive for I finite.

Theorem 16.2 *If* G *obeys* (3.33) *and* (16.8) *for* $\{z_i\}_{i=1}^n$, *if* M *is given by* (16.10), *and if* $\ker(M) \neq 0$, *then either* $A = 0$ *and* $\mathrm{supp}(d\mu)$ *is at most* $(n-1)$ *points or else* $A \neq 0$ *and* $\mathrm{supp}(d\mu)$ *is at most* $(n-2)$ *points. Conversely, if either* $A = 0$ *and* $\mathrm{supp}(d\mu)$ *is at most* $(n-1)$ *points or else* $A \neq 0$ *and* $\mathrm{supp}(d\rho)$ *is at most* $(n-2)$ *points, then* $\ker(M) \neq 0$.

Remarks

1. If we think of $A \neq 0$ meaning $\mu(\{\infty\}) \neq 0$ (as discussed in Chapter 3), then this says M is strictly positive if and only if $\text{supp}(d\mu)$ (in $\mathbb{R} \cup \{\infty\}$) has at least n points in its support.
2. One can refine this theorem (by using our proof below) to show that $\dim(\ker(M)) = j > 0$ if and only if the number of points in $\text{supp}(d\mu)$ (in $\mathbb{R} \cup \{\infty\}$) is exactly $n - j$.

Proof One can write

$$\sum_{j=1}^{n} \frac{\alpha_j}{x - z_j} = \frac{Q(x)}{\prod_j (x - z_j)} \tag{16.11}$$

where

$$Q(x) = \left(\sum_{j=1}^{n} \alpha_j\right) x^{n-1} + \text{lower order} \tag{16.12}$$

Moreover, if $Q \equiv 0$, each pole on the left of (16.11) has zero residue, so $\alpha_j = 0$, that is, $(\alpha_1, \dots, \alpha_n) \neq (0, \dots, 0)$ implies Q has at most $n - 1$ zeros, and if $\sum_{j=1}^{n} \alpha_j = 0$, at most $n - 2$ zeros. This plus (16.9), which shows $\sum \bar{\alpha}_j \alpha_k M_{jk} = 0$ implies that $d\mu$ is supported on the zeros of Q, yield the first sentence in the theorem.

For the converse, note that if $d\mu$ is supported on $\{x_k\}_{k=1}^{\ell}$, then (RHS of (16.9) $=$ 0) $\Leftrightarrow Q(x_k) = 0$ for all x_k, and if $A \neq 0$ also $\sum_{j=1}^{n} \alpha_j = 0$ which is ℓ (or if $A \neq 0$, $\ell + 1$) linear conditions on α. If $A = 0$, the kernel of this map is of dimension at least $n - \ell$ (and if $A \neq 0$, $n - (\ell + 1)$). This proves the second, converse part of the theorem. □

The "hard part" of Theorem 16.1 is to show $\mu \geq 0$ is sufficient. The connection to Loewner's theorem is now clear. If $x_1, \dots, x_n \in (0, 1)$ are distinct, and f is a possible matrix monotone function and $z_j = x_j + i\varepsilon$ and

$$w_j = f(x_j) + i\varepsilon f'(x_j) \tag{16.13}$$

then the limit in (16.10) is $L_n(x_1, \dots, x_n; f)$! So the Loewner matrix is the limit of Pick matrices.

To pass from $\mathbb{C}_+$ to $\mathbb{D}$, it is natural to use the map from $\mathbb{D}$ to $\mathbb{C}_+$,

$$T(z) = i\left[\frac{1 + z}{1 - z}\right] \tag{16.14}$$

Notice that

$$T(z) - \overline{T(\zeta)} = \frac{2i(1 - z\bar{\zeta})}{(1 - z)(1 - \bar{\zeta})} \tag{16.15}$$

Since M_{jk} is a positive matrix if and only if $N_{ij} \equiv \bar{a}_j M_{jk} a_k$ is a positive matrix for all vectors a_j non-vanishing, (16.15) suggests the replacement for $z - \bar{\zeta}$ when $z, \zeta \in \mathbb{D}$ is $1 - z\bar{\zeta}$. This motivates the original form of Pick's theorem:

Theorem 16.3 (Pick's Theorem for Carathéodory Functions) *Let $\{z_j\}_{j\in I}$ and $\{w_j\}_{j\in I}$ be indexed sets in $\mathbb{D}$ and $\{z \mid \operatorname{Re} z > 0\}$ with the z_j distinct. Then there is an analytic function $F\colon \mathbb{D} \to \{z \mid \operatorname{Re} z > 0\}$ with*

$$F(z_j) = w_j \tag{16.16}$$

if and only if the Pick matrix

$$M_{jk}^{(\mathrm{C})} = \frac{w_j + \bar{w}_k}{1 - z_j \bar{z}_k} \tag{16.17}$$

is positive.

Theorem 16.4 (Pick's Theorem for Schur Functions) *Let $\{z_j\}_{j\in I}$ and $\{w_j\}_{j\in I}$ be indexed sets in $\mathbb{D}$ with the z_j distinct. Then there is an analytic function $f\colon \mathbb{D} \to \mathbb{D}$ so*

$$f(z_j) = w_j \tag{16.18}$$

if and only if the Pick matrix

$$M_{jk}^{(\mathrm{S})} = \frac{1 - w_j \bar{w}_k}{1 - z_j \bar{z}_k} \tag{16.19}$$

is positive.

We note first that if $\{z_j\}_{j\in I} \subset \mathbb{C}_+$, $\{w_j\}_{j\in I} \subset \mathbb{C}_+$, $\{\zeta_j\}_{j\in I} \subset \mathbb{D}$, and $\{\omega_j\}_{j\in I} \subset \{z \mid \operatorname{Re} z > 0\}$, and if

$$z_j = T(\zeta_j) \qquad w_j = i\omega_j \tag{16.20}$$

with T given by (16.14), then, by (16.15),

$$\frac{w_j - \bar{w}_k}{z_j - \bar{z}_k} = \frac{1}{2} \frac{\omega_j + \omega_k}{1 - \zeta_j \bar{\zeta}_k} (1 - \zeta_j)(1 - \bar{\zeta}_k) \tag{16.21}$$

Thus

$$M_{jk}(z; w) = \tfrac{1}{2}(1 - \zeta_j)(1 - \bar{\zeta}_k) M_{jk}^{(\mathrm{C})}(\zeta; \omega) \tag{16.22}$$

and one side of (16.22) is positive if and only if the other side is. Thus,

Theorem 16.5 *Either half of Theorem 16.1 is equivalent to the corresponding half of Theorem 16.3. In particular, since we have proven necessity of* $M_{ij} \geq 0$ *in Theorem 16.1, we have necessity of* $M_{ij}^{(\mathrm{C})} \geq 0$ *in Theorem 16.3.*

Similarly, if $\{z_j\}_{j\in I} \subset \mathbb{C}_+$, $\{w_j\}_{j\in I} \subset \mathbb{C}$, $\{\zeta_j\}_{j\in I} \subset \mathbb{D}$, and $\{\omega_j\}_{j\in I} \subset \mathbb{D}$, and if

$$z_j = T(\zeta_j) \qquad w_j = T(\omega_j) \tag{16.23}$$

with T given by (16.14), then, by (16.20),

$$\frac{w_j - \bar{w}_k}{z_j - z_k} = \left(\frac{1 - \omega_j\bar{\omega}_k}{1 - \zeta_j\bar{\zeta}_k}\right)\frac{(1-\zeta_j)(1-\bar{\zeta}_k)}{(1-\omega_j)(1-\bar{\omega}_j)} \tag{16.24}$$

and thus $M_{jk}(z; w)$ is positive if and only if $M_{jk}^{(\mathrm{S})}(\zeta; \omega)$ is positive. Thus,

Theorem 16.6 *Either half of Theorem 16.1 is equivalent to the corresponding half of Theorem 16.4. In particular, since we have proven necessity of* $M_{ij} \geq 0$ *in Theorem 16.1, we have necessity of* $M_{ij}^{(\mathrm{S})} \geq 0$ *in Theorem 16.4.*

Thus, as we choose, we can prove the hard half of any of these theorems. Finally, we turn to reducing to the case where I is finite. The key is a compactness result which follows from:

Theorem 16.7 *For each* $\zeta_0 \in \mathbb{D}$ *and* $w_0 \in \mathbb{C}_+$,

$$\{F \mid F \text{ analytic on } \mathbb{D},\ \operatorname{Re} F > 0,\ F(\zeta_0) = -iw_0\} \tag{16.25}$$

is uniformly bounded on compact subsets of $\mathbb{D}$. *For each* $w_0, z_0 \in \mathbb{C}_+$,

$$\{G \mid G \text{ analytic on } \mathbb{C}_+,\ \operatorname{Im} G > 0,\ G(z_0) = w_0\} \tag{16.26}$$

is uniformly bounded on compact subsets of $\mathbb{D}$.

Proof If T is given by (16.14), then

$$\mathcal{P}(G) = -iG \circ T \tag{16.27}$$

is a bijection of G's in (16.26) to F's in (16.25) if $\zeta_0 = T^{-1}(z_0)$. Thus, the result for (16.25) implies it for (16.26).

For $z_0 \in \mathbb{D}$, the map

$$T_{z_0}(z) = \frac{z - z_0}{1 - \bar{z}_0 z} \tag{16.28}$$

maps $\mathbb{D}$ to $\mathbb{D}$ (since $T_{z_0}(e^{i\theta}) = e^{i\theta}\frac{w}{\bar{w}}$ with $w = 1 - e^{-i\theta}z_0$, it maps $\partial\mathbb{D}$ to $\partial\mathbb{D}$ and $T_{z_0}(0) \in \mathbb{D}$) and

$$T_{z_0}(z_0) = 0 \qquad T_{z_0}(0) = -z_0 \tag{16.29}$$

By a direct calculation,

$$T_{z_0}^{-1} = T_{-z_0} \tag{16.30}$$

If $\mathcal{C}_{\zeta_0,w_0}$ is the set (16.25), then

$$F \to \frac{((F \circ T_{-\zeta_0}) + i \operatorname{Re} w_0)}{\operatorname{Im} w_0}$$

maps $\mathcal{C}_{\zeta_0,w_0}$ to $\mathcal{C}_{0,-i}$, so it suffices to prove (16.25) is bounded from the case $\mathcal{C}_{0,-i}$ of Carathéodory functions, that is, $F(0) = 1$.

The Schur lemma says that if $f : \mathbb{D} \to \mathbb{D}$ and $f(0) = 0$, then $g(z) = f(z)/z$ also maps $\mathbb{D} \to \overline{\mathbb{D}}$ (see the Notes). By this fact,

$$F(z) = \frac{1 + zf(z)}{1 - zf(z)} \tag{16.31}$$

sets up a one–one correspondence between $F \in \mathcal{C}_{0,-i}$ and Schur functions. In particular, since $|f(z)| \leq 1$,

$$|F(z)| \leq \frac{1 + |z|}{1 - |z|} \tag{16.32}$$

which is uniformly bounded on any compact subset of $\mathbb{D}$. □

Montel's theorem says that if $\Omega \subset \mathbb{C}$ is an open set and if $K_n \subset \Omega$ is a family of compact subsets with $\cup K_n = \Omega$, then for any sequence $C_n \in (0, \infty)$, $\{f \mid f \text{ analytic on } \Omega;\ \sup_{z \in K_n} |f(z)| \leq C_n\}$ is compact in the topology of uniform convergence on each K_n. Moreover this topology is the same as pointwise convergence (see the Notes). Thus,

Theorem 16.8 *In the topology of uniform convergence on compacts, the set of Schur functions is compact as are the sets* (16.25) *and* (16.26).

Remark By the proof, in place of $G(z_0) = w_0$, we can show compactness if $G(z_0) \in C$ a compact subset of $\mathbb{C}_+$.

Theorem 16.9 *To prove the sufficiency of $M \geq 0$ (resp. $M^{(C)} \geq 0$, $M^{(S)} \geq 0$) in Theorem 16.1 (resp. Theorem 16.3, Theorem 16.4), it suffices to prove it for the case where I is finite.*

Proof Fix $j_0 \in I$ and note $\{G \mid G(z_{j_0}) = w_{j_0}\}$ is compact. For each finite $F \subset I$ with $j_0 \in F$, we can find G_F so $G_F(z_j) = w_j$ for $j \in F$. The set of finite F's in I under inclusion is a directed set and $\{G_F\}_{F \text{ finite} \subset I}$ is a net so, by compactness, there is a limit point G (see the Notes), which it is easy to see solves the problem for I. □

Since we will have proofs of Loewner's theorem that go through approximation by Herglotz functions, the following compactness result related to the above results is of interest. Notice that fixing $G(z_0)$ for $z_0 \in \mathbb{C}_+$ fixes two real numbers, $\operatorname{Re} G(z_0)$ and $\operatorname{Im} G(z_0)$. Since those Herglotz functions, f, analytic in (a, b) which we are interested in are real on (a, b), to fix two real numbers we need to fix f at two points.

Theorem 16.10 *Let $a < c < d < b$ where a can be $-\infty$ or b can be $+\infty$. Let $\mathcal{N}$ be the set of Herglotz functions, f, with an analytic continuation to $\Omega = (\mathbb{C}\backslash\mathbb{R}) \cup (a, b)$ so that f is real on (a, b) and $f(c) = 0$, $f(d) = 1$. Then $\mathcal{N}$ is compact in the topology of uniform convergence on compact subsets of Ω, equivalently of pointwise convergence.*

Remark If $\alpha < \beta$ and

$$g(c) = \alpha \qquad g(d) = \beta \tag{16.33}$$

then $f(x) = (g(x) - \alpha)/(\beta - \alpha)$ is in $\mathcal{N}$ if and only if (16.33) holds, so there is no loss in fixing these values.

Proof If $(a, b) = \mathbb{R}$, the only functions are linear, and there is nothing to prove since $\mathcal{N}$ has a single function! If $(a, b) \neq \mathbb{R}$, then there is a conformal map that takes $\mathbb{C}_+$ to $\mathbb{C}_+$, $a \mapsto -1$, $b \mapsto 1$, and $c \mapsto 0$. Thus, with x_0 the image of d, functions in $\mathcal{N}$ obey

$$f(x) = \int_{-1}^{1} \frac{x}{1 + \lambda x}\, d\rho(x) \tag{16.34}$$

where

$$\int_{-1}^{1} \frac{1}{1 + \lambda x_0}\, d\rho(x) = \frac{1}{x_0} \tag{16.35}$$

so

$$\begin{aligned} \int_{-1}^{1} \rho &\leq \sup_{\lambda \in [-1,1]} |1 + \lambda x_0| \int_{-1}^{1} \frac{d\rho}{1 + \lambda x_0} \\ &\leq \left[\frac{1 + x_0}{x_0}\right] \end{aligned} \tag{16.36}$$

The set of measures with this bound is compact and those obeying (16.35) is closed. □

In the same way, one proves

Theorem 16.11 *Let $a < c < b$ where a can be $-\infty$ and b can be $+\infty$. Let $\mathcal{N}$ be the set of Herglotz functions, f, with an analytic continuation to $\Omega = \mathbb{C}\backslash\mathbb{R} \cup (a,b)$ so that f is real on (a,b) and $f(c) = 0$, $f'(c) = 1$. Then $\mathcal{N}$ is compact in the topology of uniform convergence on compact subsets of Ω, equivalently of pointwise convergence.*

The Loewner matrix is a degenerate form of the Pick matrix (16.10). This is the idea behind the proof of the hard part of Loewner's theorem in Chapter 20 but it also gives a quick proof of the easy half of Loewner's theorem without needing the explicit form of the Herglotz representation. (It is true that our proof of the easy half of Pick's theorem used a Herglotz representation but there are other proofs; see, e.g., Chapter 19.)

Proof of (c)⇒(a) in Theorems 1.3 and 1.6 If f is Herglotz, then for any n and any $z_1, \dots, z_n$ in $\mathbb{C}_+$, we have that the $n \times n$ matrix $M(z_1, \dots, z_n; f) = \{(\overline{f(z_i)} - f(z_j))/(\overline{z_i} - z_j)\}_{1 \le i,j \le n}$ is a positive matrix. Given $x_1, \dots, x_n$ distinct in (a,b), one sees that $L(x_1, \dots, x_n; f) = \lim_{\epsilon \downarrow 0} M(x_1 + i\epsilon, \dots, x_n + i\epsilon; f)$ so the Loewner matrices are all positive which means that f is matrix monotone. □

Notes and Historical Remarks Pick's theorem in the form of Theorem 16.3 is due to Pick [273]. Independently, Nevanlinna [240] proved Theorem 16.1. Agler–McCarthy [2] have a whole book on extensions, including to multivariable situations using an approach of Sarason [294].

Of course, (16.30) says T_{z_0} commutes with T_{-z_0}. It is not hard to see that T_{z_0} commutes with T_{w_0} if and only if z_0 and w_0 lie on a line through 0. In general, $T_{z_0} \circ T_{z_1}$ is not $T_{-T_{z_0}(-z_1)}$ as you might guess from (16.29), but rather, it is $z \mapsto T_{-T_{z_0}(-z_1)}(e^{i\theta} z)$ where θ is a function of z_0, z_1.

The Schwarz lemma, while very powerful, is easy to prove. If $f(0) = 0$ and $|f(z)| \le 1$ on $\mathbb{D}$, then for each $R < 1$ with $g(z) = f(z)/z$, we have

$$\sup_{|z| \le R} |g(z)| \le R^{-1}$$

by the maximum principle. Taking $R \uparrow 1$, we see $\sup_{z\in\mathbb{D}} |g| \le 1$.

Montel's theorem is also simple (see, e.g., Titchmarsh [344] or Simon [326, Section 6.2]). For by Cauchy estimates, covering K_n by the interiors of suitable K_ℓ's, we get bounds not only on f but on $d^m f/dz^m$ on each K_n, bounds good enough to sum the Taylor zeros in a neighborhood of any point in Ω. A subsequence argument then easily establishes compactness.

For a discussion of nets, see Kelley [181]. If I is countable, one can use sequences.

Related to Pick's theorem is the following result of Hindmarsh [161] which appeared in his thesis done under the direction of Loewner:

Theorem 16.12 (Hindmarsh [161]) *If f is a continuous function of $\mathbb{C}_+$ to $\mathbb{C}_+$ so that for any three distinct points in $\mathbb{C}_+$,*

$$M_{ij} = \frac{f(z_i) - \overline{f(z_j)}}{z_i - \bar{z}_j}$$

is a positive 3×3 matrix, then f is analytic.

By Pick's theorem, this result holds if each $n \times n$ matrix generated by any n points is positive. So this refines Pick's theorem. 2×2 matrices are not sufficient, as can be seen by taking $f(z) = i \operatorname{Im} z$. The key to the proof is to take sequences of triples $(z_1, z_1 + \varepsilon, z_1 + i\varepsilon)$ and by taking $\varepsilon \downarrow 0$ to get the Cauchy–Riemann equations. For details, see Hindmarsh [161] or Donoghue [81].

Chapter 17
Pick Interpolation, II: Hilbert Space Proof

Our goal here is to prove the following part of Theorem 16.1 which, by the results of the last chapter, completes the proofs of Theorems 16.1, 16.3, and 16.4.

Theorem 17.1 *Let* $\{z_j\}_{j=1}^n$ *and* $\{w_j\}_{j=1}^n$ *be points in* $\mathbb{C}_+$ *so that the* z_j *are distinct and* M_{jk} *given by* (16.10) *is positive. Then there exist* $A \geq 0$ *and a measure* μ *with at most* n *points in its support* ($n-1$ *points if* $A > 0$) *so that*

$$w_j = Az_j + \int \frac{d\mu(x)}{x - z_j} \tag{17.1}$$

Remarks

1. This is slightly stronger than Theorem 16.1 in that we specify that the number of points in $\text{supp}(d\mu)$ is at most n.
2. We will prove that if M is strictly positive, then one can take $A = 0$. By Theorem 16.2, $M > 0$ implies $\#(\text{supp}(d\mu)) \geq n$, so if $M > 0$, (17.1) holds with $d\mu$ having precisely n points and no fewer.

   Below we will prove:

Theorem 17.2 *If* M *is strictly positive, then*

$$w_j = \int \frac{d\mu(x)}{x - z_j} \tag{17.2}$$

where μ *has* n *points in its support.*

Proof of Theorem 17.1 given Theorem 17.2 Suppose M is positive but not strictly positive. The function $\log(z)$ analytically continued from $(0, \infty)$ to $\mathbb{C}\backslash(-\infty, 0]$ has (essentially (16.7))

B. Simon, *Loewner's Theorem on Monotone Matrix Functions*, Grundlehren der mathematischen Wissenschaften 354, https://doi.org/10.1007/978-3-030-22422-6_17

$$\frac{\log(z) - \log(\bar{\zeta})}{z - \bar{\zeta}} = \int_{-\infty}^{0} \frac{dx}{(x - z)(x - \bar{\zeta})} \tag{17.3}$$

so, by (16.8), the matrix

$$M_{jk}^{(\varepsilon)} = M_{jk} + \frac{\varepsilon[\log(z_j) - \overline{\log(z_k)}]}{z_j - \bar{z}_k} \tag{17.4}$$

is strictly positive for $\varepsilon > 0$, so there exist measures $d\mu^{(\varepsilon)}$ with

$$w_j + \varepsilon \log(z_j) = \int \frac{d\mu^{(\varepsilon)}(x)}{x - z_j} \tag{17.5}$$

By the remark after Theorem 16.8, the set of Herglotz functions with $|f(z_1) - w_1| \leq \mathrm{Im}(w_1)/2$ is compact, so with

$$G^{(\varepsilon)}(z) = \int \frac{d\mu^{(\varepsilon)}(x)}{x - z}$$

find a limiting Herglotz function $G(z)$ solving $G(z_j) = w_j$.

By Theorem 16.2, since $\ker(M) \neq 0$, G has a representation

$$G(z) = Az + \int \frac{d\mu(x)}{x - z} \tag{17.6}$$

where either $A > 0$ and μ has at most $(n - 2)$ points in its support or $A = 0$ and μ has at most $n - 1$ points in its support. □

Here is the intuition behind the construction below. Suppose we find an $n \times n$ symmetric matrix A and vector φ in $\mathbb{C}^n$ so that

$$w_j = \langle \varphi, (A - z_j)^{-1} \varphi \rangle \tag{17.7}$$

Let $x_1, \ldots, x_n$ be the eigenvalues of A and $\psi_1, \ldots, \psi_n$ an orthonormal basis with $A\psi_j = x\psi_j$. Then

$$w_j = \sum_{\ell=1}^{n} \frac{|\langle \varphi, \psi_\ell \rangle|^2}{x_\ell - z} \tag{17.8}$$

which is precisely of the form (17.2), and the result is proven.

Continuing the supposition on A existing so (17.7) holds, let

$$\eta_j = (A - z_j)^{-1} \varphi \tag{17.9}$$

Since

$$(A - \bar{z}_j)^{-1}(A - z_k)^{-1} = \frac{[(A - z_k)^{-1} - (A - \bar{z}_j)^{-1}]}{(z_k - \bar{z}_j)} \tag{17.10}$$

we have

$$\langle \eta_j, \eta_k \rangle = \frac{w_k - \bar{w}_j}{z_k - \bar{z}_j} = M_{kj} \tag{17.11}$$

Since M is strictly positive, the η's are a basis, so φ is uniquely determined by

$$\langle \varphi, \eta_j \rangle = w_j \tag{17.12}$$

and A is determined by (from (17.9) $\Rightarrow (A - z_k)\eta_k = \varphi$)

$$\begin{aligned} \langle \eta_j, A\eta_k \rangle &= z_k \langle \eta_j, \eta_k \rangle + \langle \eta_j, \varphi \rangle \\ &= z_k \left(\frac{w_k - \bar{w}_j}{z_k - \bar{z}_j} \right) + \bar{w}_j \\ &= \frac{z_k w_k - \bar{z}_j \bar{w}_j}{z_k - \bar{z}_j} \end{aligned} \tag{17.13}$$

The argument below will turn this around. These calculations explain why we define φ and A as we do!

Proof of Theorem 17.2 Let M_{jk} be given by (16.10). Let η_j be the vector in $\mathbb{C}^n$ with

$$(\eta_j)_k = \delta_{jk} \tag{17.14}$$

Define an inner product on $\mathbb{C}^n$ by

$$\langle \alpha, \beta \rangle = \sum \bar{\alpha}_j M_{kj} \beta_k \tag{17.15}$$

which is a strictly positive inner product since $M^t > 0$. Thus (17.11) holds.

Since $M^t > 0$, we can define any vector in $\mathbb{C}^n$ by its inner products with the η_j. Explicitly,

$$\psi = \sum \ell_j(\psi)\eta_j \tag{17.16}$$

where

$$\ell_j(\psi) = \sum_k (M^{-1})_{jk} \langle \eta_k, \psi \rangle \tag{17.17}$$

So we pick φ so that (17.12) holds. This uniquely determines φ.

Define a matrix, A, by (17.13). Since the η's are a basis (albeit, not an orthonormal basis), this determines A uniquely as a matrix on $\mathbb{C}^n$. Moreover, by the uniqueness, A is self-adjoint if and only if

$$\langle A\eta_j, \eta_k\rangle = \langle A^*\eta_j, \eta_k\rangle = \langle \eta_j, A\eta_k\rangle$$

that is, if and only if

$$\overline{\langle \eta_k, A\eta_j\rangle} = \langle \eta_j, A\eta_k\rangle \tag{17.18}$$

Thus, $A = A^*$ if and only if the matrix elements in the nonorthonormal basis η_j are a self-adjoint matrix. Equation (17.13) has this property, that is,

$$\overline{\left(\frac{z_j w_j - \bar{z}_k \bar{w}_k}{z_j - \bar{z}_k}\right)} = \frac{z_k w_k - \bar{z}_j \bar{w}_j}{z_k - \bar{z}_j}$$

so $A = A^*$.

Moreover,

$$\begin{aligned}\langle \eta_j, (A - z_k)\eta_k\rangle &= \frac{z_k w_k - \bar{z}_j \bar{w}_j}{z_k - \bar{z}_j} - z_k \frac{(w_k - \bar{w}_j)}{z_k - \bar{z}_j} \\ &= \bar{w}_j \\ &= \langle \eta_j, \varphi\rangle\end{aligned}$$

and thus,

$$(A - z_k)\eta_k = \varphi \tag{17.19}$$

It follows that (recall $\operatorname{Im} z_k \neq 0$ and A is self-adjoint, so $A - z_k$ is invertible) $\eta_k = (A - z_k)^{-1}\varphi$, so

$$w_j = \langle \varphi, \eta_j\rangle = \langle \varphi, (A - z_j)^{-1}\varphi\rangle \tag{17.20}$$

As explained before (17.8), this implies (17.7), which proves (17.2). □

Notes and Historical Remarks The proof in this chapter is essentially that exposed in Chapters X–XI in Donoghue [81], which in turn is motivated by Sz.-Nagy and Korányi [342]. Donoghue uses the language of reproducing kernels, which seems to me to obscure the situation. Moreover, these authors deal directly with infinite sets using the spectral theorem rather than using compactness to reduce to the finite dimensional case.

There is an awkwardness in the above proof when M is not strictly positive. We added something small and strictly positive, found f, took a limit, and then used Theorem 16.3 to prove the measure could be taken to have rank(M) points rather than dim(M). It is more natural if rank(M) $<$ dim(M) to instead still use M to define an inner product $\langle \cdot, \cdot \rangle_M$ and consider the Hilbert space $\mathcal{H} = M/\{\varphi \mid \langle \varphi, \varphi \rangle_M = 0\}$. This space has dimension rank(M), so if we can define A on $\mathcal{H}$, we get a measure with only rank(M) points.

Alas, I do not see any way to directly prove that $\langle \varphi, \varphi \rangle_M = 0 \Rightarrow \langle \psi, A\varphi \rangle_M = 0$, and so no direct way to follow this strategy for finitely many points in $\mathbb{C}_+$. Indeed, Sz.-Nagy and Korányi [342] prove a theorem for $\mathbb{C}_+$ by considering a set of points, S, which is required to have 0 as a nontangential limit point! They can then handle degenerate M's.

However, Sz.-Nagy and Korányi [342] do have an argument that works on the quotient space in the Carathéodory case (i.e., Theorem 16.3) if one of the points is 0 (and then by conformal transformation, one gets Theorem 16.3). Let us describe their argument: $\{z_j\}_{j=1}^n$ are distinct points in $\mathbb{D}$ with $z_n = 0$ and $\{w_j\}_{j=1}^n$ in $\mathbb{C}_+$ so that the matrix $M^{(C)}$ of (16.17) is positive but not necessarily strictly positive. For $\alpha \in \mathbb{C}^n$, define

$$\langle \alpha, \alpha \rangle_M = \sum_{j,k=1}^{n} \bar{\alpha}_j \alpha_k M_{kj}^{(C)} \tag{17.21}$$

Define $V\colon \mathbb{C}^{n-1} \to \mathbb{C}^n$ (where, by $\mathbb{C}^{n-1}$, we mean $\{\alpha \in \mathbb{C}^n \mid \alpha_n = 0\}$) by

$$V\eta_j = \frac{\eta_j - \eta_n}{z_j} \tag{17.22}$$

for $j = 1, \dots, n-1$. Then for $1 \le j, k \le n-1$,

$$\begin{aligned}
\langle V\eta_j, V\eta_k \rangle_M &= \frac{1}{\bar{z}_j z_k} \left[\langle \eta_j, \eta_k \rangle_M + \langle \eta_n, \eta_n \rangle_M - \langle \eta_j, \eta_n \rangle_M - \langle \eta_n, \eta_k \rangle_M \right] \\
&= \frac{1}{\bar{z}_j z_k} \left[\frac{w_k + \bar{w}_j}{1 - z_k \bar{z}_j} + (w_n + \bar{w}_n) - (\bar{w}_j + w_n) - (\bar{w}_n + w_k) \right] \\
&= \frac{w_k + \bar{w}_j}{\bar{z}_j z_k} \left[\frac{1}{1 - z_k \bar{z}_j} - 1 \right] \\
&= \frac{w_k + \bar{w}_j}{1 - z_k \bar{z}_j} \\
&= \langle \eta_j, \eta_k \rangle_M
\end{aligned} \tag{17.23}$$

Thus,

$$\langle \varphi, \varphi \rangle_M = 0 \Rightarrow \langle V\varphi, V\varphi \rangle_M = 0 \tag{17.24}$$

Let $\pi : \mathbb{C}^n \to \mathbb{C}^\ell$ where $\ell = \text{rank}(M)$ is the canonical projection of $\mathbb{C}^n$ to $\mathbb{C}^n/\{\varphi \mid \langle\varphi, \varphi\rangle_M = 0\}$, then (17.24) says that we can define a map $V_1 : \pi[\mathbb{C}^{n-1}] \to \mathbb{C}^\ell$ by

$$V_1\pi(u) = \pi(Vu) \tag{17.25}$$

If $\pi[\mathbb{C}^{n-1}] = \mathbb{C}^\ell$, take $\tilde{V} = V_1$. Otherwise, one can extend V_1 to a unitary map on $\mathbb{C}^\ell$ (the codimension of $\pi[\mathbb{C}^{n-1}]$ is either 0 or 1). Sz.-Nagy and Korányi deal with infinitely many z_j's (with $z_0 = 0$) and so their $\tilde{V}$ may only be a contraction.

Now,

$$(1 - z_k V)\eta_k = \eta_k - (\eta_k - \eta_n) = \eta_n \tag{17.26}$$

so

$$(1 - z_k \tilde{V})\pi(\eta_k) = \pi(\eta_n) \tag{17.27}$$

Since $|z_k| < 1$ and $\|\tilde{V}\| = 1$, $1 - z_k\tilde{V}$ is invertible, and (17.27) implies

$$\pi(\eta_k) = (1 - z\tilde{V})^{-1}\pi(\eta_n) \tag{17.28}$$

It follows that $k \leq n - 1$,

$$\begin{aligned}
\langle\pi(\eta_n), (1 + z_k\tilde{V})&(1 - z_k\tilde{V})^{-1}\pi(\eta_n)\rangle \\
&= \langle\pi(\eta_n), (1 + z_k\tilde{V})\pi(\eta_k)\rangle \\
&= \langle\pi(\eta_n), [2 - (1 - z_k\tilde{V})]\pi(\eta_k)\rangle \\
&= \langle\pi(\eta_n), 2\pi(\eta_k)\rangle - \langle\pi(\eta_n), \pi(\eta_n)\rangle && (17.29) \\
&= 2\langle\eta_n, \eta_k\rangle_M - \langle\eta_n, \eta_n\rangle_M \\
&= 2(w_k + \bar{w}_n) - (w_n + \bar{w}_n) \\
&= 2w_k - 2i\,\text{Im}(w_n) && (17.30)
\end{aligned}$$

Equation (17.29) follows from (17.26). Equation (17.30) also holds for $k = n$ since ($z_n = 0$)

$$\langle\pi(\eta_n), \pi(\eta_n)\rangle = w_n + \bar{w}_n = 2w_n - 2i\,\text{Im}(w_n)$$

Since $\tilde{V}$ is unitary, by the spectral theorem, there is a point measure on $\partial\mathbb{D}$ with at most ℓ pure points with

$$\tfrac{1}{2}\langle\pi(\eta_n), (1 + z\tilde{V})(1 - z\tilde{V})^{-1}\pi(\eta_n)\rangle = \int \frac{e^{i\theta} + z}{e^{i\theta} - z}\, d\mu(\theta)$$

By (17.30),

$$w_k = i\,\mathrm{Im}(w_n) + \int \frac{e^{i\theta} + z_k}{e^{i\theta} - z_k}\, d\mu(\theta)$$

is the required representation.

Chapter 18
Pick Interpolation, III: Continued Fraction Proof

In this chapter, we will prove the following which provides the missing part (given Chapter 16) of Theorem 16.4 and thereby completes (again) the proofs of Theorems 16.1, 16.3, and 16.4:

Theorem 18.1 *Let* $\{z_j\}_{j=1}^n$ *and* $\{w_j\}_{j=1}^n$ *be collections of points in* $\mathbb{D}$ *with the* z_j *distinct and so that* $M_{jk}^{(\mathrm{S})} \geq 0$. *Then there is a Blaschke product,* $B(z)$, *of order at most* n *so*

$$B(z_j) = w_j \tag{18.1}$$

Remarks

1. $M^{(\mathrm{S})}$ is given by (16.19).
2. If $\mathrm{Ran}(M^{(\mathrm{S})}) = n - \ell$, one can show that B has exactly order $n - \ell$. In particular, if $M^{(\mathrm{S})}$ is strictly positive, B has order n.

In the above, we define a Blaschke product of order n by

$$B(z) = e^{i\theta_0} \prod_{j=1}^{n} T_{z_j}(z) \tag{18.2}$$

for $z_1, \ldots, z_n \in \mathbb{D}$, where

$$T_{z_j}(z) = \frac{z - z_j}{1 - \bar{z}_j z} \tag{18.3}$$

as discussed in (16.28). Since each $T_{z_j} : \mathbb{D} \to \mathbb{D}$, B maps $\mathbb{D}$ to $\mathbb{D}$, and so Theorem 18.1 solves the Schur function interpolation problem. This is *not* quite the usual definition of Blaschke product which is defined, when no z_j is zero to be positive at $z = 0$. That choice is convenient for dealing with infinite Blaschke

B. Simon, *Loewner's Theorem on Monotone Matrix Functions*, Grundlehren der mathematischen Wissenschaften 354, https://doi.org/10.1007/978-3-030-22422-6_18

products (see [326, Section 9.9]); our extra phase factor is necessary for Lemma 18.2 to hold.

The strategy of the proof will be to use induction in n. Assuming the result for $n-1$ points and that $z_n = w_n = 0$, we will show that we can take $B_n(z) = zB_{n-1}(z)$ where B_{n-1} is a suitable Blaschke product of order at most $n-1$. The full result will follow by using the T_z's to move (z_n, w_n) to $(0, 0)$. One part will be to prove that these maps preserve positivity of $M^{(S)}$. We will also need to know that Blaschke products are invariant under such transformation—and it is to that we first turn.

Lemma 18.2 *Let f be a function analytic in a neighborhood of $\overline{\mathbb{D}}$ that obeys*

$$|z| = 1 \Rightarrow |f(z)| = 1 \tag{18.4}$$

(i) *If f is non-vanishing in $\mathbb{D}$, then $f(z) = e^{i\theta_0}$ for some real θ_0.*
(ii) *If f has n zeros in $\mathbb{D}$, then f is a Blaschke product of order n.*

Proof

(i) Suppose f is analytic and non-vanishing in $\{z \mid |z| < R\}$ for $R > 1$, define g on $\{z \mid |z| > R^{-1}\}$ by

$$g(z) = [\,\overline{f(1/\bar{z})}\,]^{-1} \tag{18.5}$$

Since f is non-vanishing in $\mathbb{D}$, g is analytic in $\{z \mid |z| > R^{-1}\}$. By (18.4) $f(z) = g(z)$ for $|z| = 1$, so g provides an analytic continuation of f to $\mathbb{C}$. By (18.5), $\lim_{z\to\infty} g(z) = 1/\overline{f(0)}$. So this entire function is bounded, thus a constant which, by (18.4), must be $e^{i\theta_0}$.

(ii) Let $z_1, \dots, z_n$ be the zeros of f (counting multiplicity). Then

$$\frac{f(z)}{\prod_{j=1}^{n} T_{z_j}(z)} \tag{18.6}$$

is analytic in a neighborhood of $\overline{\mathbb{D}}$, non-vanishing, and obeys (18.4). By (i), this ratio is $e^{i\theta_0}$, that is, f is a Blaschke product.

□

Proposition 18.3 *If B is a Blaschke product of order n and $z_0 \in \mathbb{D}$, then $B \circ T_{z_0}$ and $T_{z_0} \circ B$ are both also finite Blaschke products of order n.*

Proof Since T_{z_0} maps an analytic function in a neighborhood of $\overline{\mathbb{D}}$ into such a neighborhood, it takes functions obeying (18.4) into themselves, either for $B \circ T_{z_0}$ or for $T_{z_0} \circ B$. Since T_{z_0} is a bijection, it preserves the number of zeros. By (ii) of Lemma 18.2, $B \circ T_{z_0}$ and $T_{z_0} \circ B$ are Blaschke products of the same order as B. □

We need an analog of this result for the $M^{(\mathrm{S})}$ matrix:

Proposition 18.4 *Suppose $\{z_j\}_{j=1}^n$ and $\{w_j\}_{j=1}^n$ lie in $\mathbb{D}$ with the z's distinct and so that $M^{(\mathrm{S})}(\{z_j\}_{j=1}^n, \{w_j\}_{j=1}^n)$ is positive. Let $\zeta_0, \zeta_1 \in \mathbb{D}$ and let*

$$\tilde{z}_j = T_{\zeta_0}(z_j) \qquad \tilde{w}_j = T_{\zeta_1}(w_j) \tag{18.7}$$

Then $M^{(\mathrm{S})}(\{\tilde{z}_j\}_{j=1}^n, \{\tilde{w}_j\}_{j=1}^n)$ is also positive.

Proof We havc, by direct calculation,

$$1 - T_{\zeta_0}(z_0)\overline{T_{\zeta_0}(z_1)} = \frac{(1-|\zeta_0|^2)(1-z_0\bar{z}_1)}{(1-z_0\bar{\zeta}_0)(1-\bar{z}_1\zeta_0)} \tag{18.8}$$

Thus,

$$M_{k\ell}^{(\mathrm{S})}(\{\tilde{z}_j\}_{j=1}^n, \{\tilde{w}_j\}_{j=1}^n) = a_k \bar{a}_\ell M_{kl}^{(\mathrm{S})}(\{z_j\}_{j=1}^n, \{w_j\}_{j=1}^n) \tag{18.9}$$

where

$$a_k = (1-|\zeta_1|^2)^{1/2}(1-|\zeta_0|^2)^{-1/2}(1-\bar{\zeta}_1 w_k)^{-1}(1-\bar{\zeta}_0 z_k) \tag{18.10}$$

Positivity of one $M^{(\mathrm{S})}$ implies positivity of the other. □

The final preliminary step is an analog of the Schwarz lemma for this context:

Proposition 18.5 *Let $\{z_j\}_{j=1}^n$ and $\{w_j\}_{j=1}^n$ obey the hypotheses of Theorem 18.1. Suppose that $w_n = z_n = 0$. Then either*

(i) *For some $e^{i\theta} \in \partial\mathbb{D}$, $w_j = e^{i\theta} z_j$, $j = 1, \dots, n-1$; or*
(ii) *For $j = 1, \dots, n-1$, $|w_j| < |z_j|$ and the $(n-1)\times(n-1)$ matrix*

$$M^{(\mathrm{S})}(\{z_j\}_{j=1}^{n-1}, \{w_j/z_j\}_{j=1}^{n-1}) \tag{18.11}$$

is positive.

Proof The matrix $M^{(\mathrm{S})}(\{z_j\}_{j=1}^n, \{w_j\}_{j=1}^n)$ has the form

$$M = \begin{pmatrix} & & & 1 \\ & \widetilde{M} & & 1 \\ & & & \vdots \\ 1 & 1 & \dots & 1 \end{pmatrix} \tag{18.12}$$

that is, the last row and column are all 1's. Given an $n-1$ vector $\alpha_1, \dots, \alpha_{n-1}$, take the n component vector with

$$\alpha_n = -\sum_{j=1}^{n-1} \alpha_j \tag{18.13}$$

Then, by (18.12),

$$\sum_{j,k=1}^{n} M_{jk}\bar{\alpha}_j\alpha_k = \sum_{j,k=1}^{n-1} \widetilde{M}_{jk}\bar{\alpha}_j\alpha_k - 2\left|\sum_{j=1}^{n-1}\alpha_j\right|^2 + \left|\sum_{j=1}^{n-1}\alpha_j\right|^2$$

$$= \sum_{j,k=1}^{n-1} (\widetilde{M}_{jk} - 1)\bar{\alpha}_j\alpha_k \tag{18.14}$$

Thus, the $n-1 \times n-1$ matrix N, given by

$$N_{jk} = \widetilde{M}_{jk} - 1 \tag{18.15}$$

is positive.

In particular, $N_{jj} \geq 0$, so

$$\frac{1-|w_j|^2}{1-|z_j|^2} \geq 1 \tag{18.16}$$

which implies that

$$|w_j|^2 \leq |z_j|^2 \tag{18.17}$$

If $|w_{j_0}| = |z_{j_0}|$ for some j_0, then $w_{j_0} = e^{i\theta}z_{j_0}$ and $N_{j_0j_0} = 0$. Since $N \geq 0$, $N_{j_0k} = 0$ for all j, that is,

$$1 - w_{j_0}\bar{w}_k = 1 - z_{j_0}\bar{z}_k \tag{18.18}$$

which implies $w_k = e^{i\theta}z_k$ for all k, and we are in case (i).

If $|w_j| < |z_j|$ for all j, set $\tilde{w}_j = w_j/z_j \in \mathbb{D}$. Then

$$1 - \tilde{w}_j\overline{\tilde{w}_k} = (z_j\bar{z}_k)^{-1}(z_j\bar{z}_k - w_j\bar{w}_k) \tag{18.19}$$

$$= (z_j\bar{z}_k)^{-1}[(1 - z_j\bar{z}_k)N_{jk}] \tag{18.20}$$

since

$$N_{jk} = \frac{z_j\bar{z}_k - w_j\bar{w}_k}{1 - z_j\bar{z}_k} \tag{18.21}$$

It follows that

$$M_{k\ell}^{(S)}(\{z_j\}_{j=1}^{n-1}, \{\tilde{w}_j\}_{j=1}^{n}) = z_k^{-1}\bar{z}_\ell^{-1} N_{k\ell}$$

is positive. □

Proof of Theorem 18.1 We use induction on n. For $n = 1$, we note that if $f = T_{-w_1} \circ T_{z_1}$, then $f(z_1) = w_1$. Thus, suppose we can construct f for $n-1$ points.

Given n points, let $\{\tilde{z}_j\}_{j=1}^{n}$, $\{\tilde{w}_j\}_{j=1}^{n}$ be

$$\tilde{z}_j = T_{z_n}(z_j) \qquad \tilde{w}_j = T_{w_n}(w_j) \tag{18.22}$$

If we find a Blaschke product so that $g(\tilde{z}_j) = \tilde{w}_j$, then $B = T_{-w_n} \circ g \circ T_{z_n}$ obeys (18.3) and is, by Proposition 18.3, also a Blaschke product. Note that, by Proposition 18.4, $M^{(S)}(\{\tilde{z}_j\}, \{\tilde{w}_j\})$ is positive. Thus, we need only prove the result in the special case $z_n = w_n = 0$.

In that case, we can use Proposition 18.5. If (i) holds, $B(z) = e^{i\theta}z$ solves (18.1). If (ii) holds, by Proposition 18.5 and the induction hypothesis, we can find a Blaschke product, g, of order at most $n-1$ so

$$g(z_j) = \frac{w_j}{z_j} \qquad j = 1, \dots, n-1 \tag{18.23}$$

Then

$$B(z) = zg(z)$$

solves (18.1). □

Notes and Historical Remarks The proof in this chapter is due to D. Marshall. A variant appears in his paper [217] and the proof, essentially in the form we give, appears in Garnett [119], where it is attributed to Marshall's unpublished thesis.

As we will see, there is a connection to the earlier construction of Wigner–von Neumann [363] discussed in Chapter 31 and to the Schur algorithm discussed next.

Schur [308, 309] found a natural way to parameterize what we now call Schur functions. If $f : \mathbb{D} \to \mathbb{D}$, let $\gamma_0 = f(0)$. Define f_1 by

$$f(z) = \frac{\gamma_0 + zf_1}{1 + \bar{\gamma}_0 z f_1} = T_{-\gamma_0}(zf_1) \tag{18.24}$$

Then, by the Schwarz lemma, $f_1 : \mathbb{D} \to \overline{\mathbb{D}}$. If f_1 is a constant in $\partial\mathbb{D}$, set $\gamma_1 = f_1(0)$ and stop. Otherwise, set $\gamma_1 = f_1(0) \in \mathbb{D}$ and iterate using (18.24). In this way, every Schur function is associated to either a finite sequence $(\gamma_0, \dots, \gamma_m)$ with $|\gamma_m| = 1$ and $|\gamma_j| < 1$ for $j = 0, \dots, m-1$, or else to an infinite sequence with

$|\gamma_j| < 1$. Schur proved this map was a bijection and the finite sequence precisely corresponded to Blaschke products.

Wall [358] and Geronimus [124] noted that the Schur algorithm corresponded to a set of continued fraction expansions. Marshall's proof is essentially a multipoint Schur algorithm and so also is a kind of continued fraction expansion.

One can ask about the relation of finiteness in this chapter (i.e., finite Blaschke products) and in the last chapter (i.e., supp$(d\mu)$ is finite). It is the following. Let f, F, and $d\mu$ be related by

$$F(z) = \frac{1 + zf(z)}{1 - zf(z)} = \int \frac{e^{i\theta} + z}{e^{i\theta} - z}\, d\mu(\theta) \tag{18.25}$$

Then $d\mu$ has exactly N points in its support if and only if f is a Blaschke product of order $N - 1$.

Chapter 19
Pick Interpolation, IV: Commutant Lifting Proof

In this chapter, we'll present a proof of Pick's theorem, Theorem 16.4, due to Sarason [294], that relies on the Hardy spaces H^2 and H^∞. It is also a motivation for the Rosenblum–Rovnyak proof [289] of the hard part of Loewner's theorem in Chapter 33.

Pick's theorem asserts the existence of a function $f \in H^\infty$ with $f(z_j) = w_j$ and $\|f\|_\infty \leq 1$. A key fact underlying the approach of this chapter: if a bounded operator, $A : H^2 \mapsto H^2$, commutes with M_z, multiplication by z, then there exists $f \in H^\infty$ so that $Ag(z) = f(z)g(z)$. This will be proven in Theorem 19.3 below.

For Pick's theorem, we'll not construct A directly—rather, we'll look at the n-dimensional space which is the orthogonal complement, K, of $\{g \in H^2 | g(z_j) = 0, j = 1, \dots, n\}$. For now, we denote the orthogonal projection onto K by P. K will have a natural algebraic (but not orthonormal) basis, $\{s_j\}_{j=1}^n$, of those functions in H^2 with $\langle s_j, g\rangle = g(z_j)$. The operator X on K given by $Xs_j = \bar{w}_j s_j$ will turn out to commute with $(PM_zP)^*$.

The key to proving Pick's theorem will be to show there exists G on H^2 commuting with M_z so that $PGP = X^*$ and so that $\|G\|_{\mathcal{L}(H^2)} = \|X\|_{\mathcal{L}(K)}$. It is this lifting of the commutant that gives the name to this approach. As we'll discuss in the Notes, this can be viewed as a special case of a much more general theorem. If G is multiplication by f, we'll prove that $f(z_j) = w_j$ so that $\|X\|_{\mathcal{L}(K)} \leq 1$ will be the sufficient (and necessary) condition to solve the Pick problem for Schur functions.

After presenting this solution, we'll use the machinery to prove a very general abstract theorem that will be the basis of the proof of Loewner's theorem in Chapter 33. In many ways, our discussion here is taking the scenic route. The machinery we'll present relies on lovely theorems of Beurling and of Nehari. While the general abstract theorem we state at the end will need this full machinery, as we'll remark, our application in Chapter 33 only involves a special case where all that is needed is Theorem 19.3 and we could avoid using either the Beurling's or the Nehari's theorems. Our proof of Pick's theorem uses the very special case of

B. Simon, *Loewner's Theorem on Monotone Matrix Functions*, Grundlehren der mathematischen Wissenschaften 354, https://doi.org/10.1007/978-3-030-22422-6_19

K described above, where, as we'll see (the proof of Theorem 16.4 later in this chapter), Beurling's theorem is merely a simple calculation. So neither application requires the Beurling's theorem, but its proof is easy and the result illuminating. Since we have Beurling's theorem, adding to the scenic nature of this chapter, we'll end with an application to prove the Titchmarsh and Lions convolution theorems.

We begin with some elementary facts about H^2 and H^∞ (see the books [86, 185] or chapters in [119, 179, 290], [328, Chapter 5] on H^p spaces for more):

(1) Since $\{e^{in\theta}\}_{n=-\infty}^{\infty}$ is an orthonormal basis for $L^2(\partial\mathbb{D}, \frac{d\theta}{2\pi})$, the Fourier transform (aka Fourier series in this case), $f \mapsto f_n^\# = \langle e^{in\cdot}, f\rangle$, sets up a correspondence between L^2 and $\ell^2(\mathbb{Z}) = \{\{a_n\}_{n=-\infty}^{\infty} | \sum_{n=-\infty}^{\infty} |a_n|^2 < \infty\}$.

(2) The Hardy space, $H^2(\mathbb{D}) = \{f \in L^2 | f_n^\# = 0 \text{ for } n < \infty\}$. It is isomorphic under Fourier transform to $\{\{a_n\}_{n=0}^{\infty} | \sum_{n=0}^{\infty} |a_n|^2 < \infty\}$. Given such an f we can define, for each $z \in \mathbb{D}$, $f(z) = \sum_{n=0}^{\infty} a_n z^n$ which converges uniformly on $\{z \,|\, |z| < r\}$ for each $r < 1$ since $\|\{z^n\}_{n=0}^{\infty}\|_{\ell^2(\mathbb{N})}^2 = \sum_{n=0}^{\infty} |z|^{2n} = (1-|z|^2)^{-1}$. Thus $H^2(\mathbb{D})$ is naturally associated to a vector space of analytic functions on $\mathbb{D}$.

(3) If f is analytic on $\mathbb{D}$, we have that $f(z) = \sum_{n=0}^{\infty} a_n z^n$ and

$$\sum_{n=0}^{\infty} |a_n|^2 = \lim_{r\uparrow 1} \sum_{n=0}^{\infty} |a_n|^2 r^{2n} = \lim_{r\uparrow 1} \int |f(re^{i\theta})|^2 \frac{d\theta}{2\pi} \tag{19.1}$$

where one can take either $\lim_{r\uparrow 1}$ or $\sup_{r<1}$. Thus, we can view $H^2(\mathbb{D})$ either as a subset of $L^2(\partial\mathbb{D})$ or as the set of functions, f, analytic on $\mathbb{D}$ with

$$\|f\|^2 \equiv \lim_{r\uparrow 1} \int |f(re^{i\theta})|^2 \frac{d\theta}{2\pi} = \sup_{r<1} \int |f(re^{i\theta})|^2 \frac{d\theta}{2\pi} < \infty \tag{19.2}$$

(4) If $f \in H^2$, it is easy to see that with $f_r(e^{i\theta}) \equiv f(re^{i\theta})$, we have that

$$\|f_r - f\|_{L^2} \to 0 \text{ as } r \uparrow 1. \tag{19.3}$$

This let's one extend the Poisson formula for harmonic functions on $\mathbb{D}$ which have a continuous extension to $\overline{\mathbb{D}}$ to f's in $H^2(\mathbb{D})$. That is, if

$$P(z, e^{i\varphi}) = \frac{1 - |z|^2}{1 + |z|^2 - 2\,\mathrm{Re}(ze^{-i\varphi})} \tag{19.4}$$

so $P(re^{i\theta}, e^{i\varphi}) = P_r(\theta, \varphi)$ given by (3.9), then, for any $f \in H^2(\mathbb{D})$

$$f(z) = \int P(z, e^{i\theta}) f(e^{i\theta}) \frac{d\theta}{2\pi} \tag{19.5}$$

which shows how to go from the representation of f as an L^2 function on the boundary of the circle to its representation as an analytic function on $\mathbb{D}$. The Poisson kernel is not analytic in z because of the real part—but because $f \in H^2$, not just L^2, the function in (19.5) is analytic. In the other direction, standard maximal function technology (see [328, Chapter 2]) lets one prove for any $f \in H^2(\mathbb{D})$, we have that for a.e. θ

$$f(e^{i\theta}) = \lim_{r\uparrow 1} f(re^{i\theta}) \tag{19.6}$$

which provides the way to go from the analytic representation to the L^2 representative.

(5) As usual, $L^\infty(\partial\mathbb{D}, \frac{d\theta}{2\pi})$ is the subset of those functions in $L^2(\partial\mathbb{D}, \frac{d\theta}{2\pi})$ with

$$\|f\|_\infty = \operatorname{esssup}\{|f(e^{i\theta})\mid e^{i\theta} \in \partial\mathbb{D}\} < \infty \tag{19.7}$$

If $f \in L^\infty$ and $g \in L^2$, then $fg \in L^2$ so $g \mapsto fg$ is an everywhere defined operator, M_f, on L^2 which it is easy to see is a bounded operator with norm (in $\mathcal{L}(L^2)$):

$$\|M_f\| = \|f\|_\infty \tag{19.8}$$

To see this note that $\|fg\|_2 \le \|f\|_\infty\|g\|_2$ to get that $\|M_f\| \le \|f\|_\infty$. On the other hand, for any rational φ and rational $\epsilon > 0$, if $S_{\varphi,\epsilon} = \{z \mid \operatorname{Re}(f(z)e^{-i\varphi}) > \|f\|_\infty - \epsilon\}$ and if $g_{\varphi,\epsilon}$ is its characteristic function, then we have that $\|M_f g_{\varphi,\epsilon}\| \ge (\|f\|_\infty - \epsilon)\|g_{\varphi,\epsilon}\|$. Therefore, $g_{\varphi,\epsilon} \ne 0 \Rightarrow \|M_f\| \ge \|f\|_\infty - \epsilon$. By definition of $\|\cdot\|_\infty$, for any $\epsilon > 0$, there is a rational φ with $|S_{\varphi,\epsilon}| > 0$ proving that $\|M_f\| \ge \|f\|_\infty$. Below we'll see that $\{M_f \mid f \in L^\infty\}$ is precisely the set of operators in $\mathcal{L}(L^2)$ commuting with M_z.

(6) $H^\infty(\mathbb{D}) \equiv H^2(\mathbb{D}) \cap L^\infty(\partial\mathbb{D}, \frac{d\theta}{2\pi})$ by which we intend those analytic functions on $\mathbb{D}$ in H^2 whose boundary values are in L^∞. From the Poisson formula, we see that such a function is bounded on $\mathbb{D}$ by $\sup_{e^{i\theta}\in\partial\mathbb{D}} |f(e^{i\theta})| = \|f\|_\infty$. On the other hand if f is a bounded analytic function its boundary values are a.e. bounded by $\sup_{z\in\mathbb{D}} |f(z)|$. We thus see that $H^\infty(\mathbb{D})$ is the space of bounded analytic functions with norm

$$\|f\|_{L^\infty(\partial\mathbb{D})} = \sup_{z\in\mathbb{D}} |f(z)| \tag{19.9}$$

If $f \in H^2$, then $f(r\cdot) \in H^\infty$, so by (19.3), H^∞ is dense in H^2 (alternatively, the Taylor polynomials, which are bounded, converge!).

(7) If $f \in H^\infty$ and $g \in H^2$, then $fg \in H^2$, so M_f restricts to a map of H^2 to itself that we'll denote by $\widetilde{M}_f$ (but only for a while—shortly, we'll drop the tilde). Since $H^2 \subset L^2$, $\|\widetilde{M}_f\| \le \|M_f\|$. On the other hand, if $g \in H^2$, we have

that $\|M_f z^{-n} g\| = \|z^{-n} M_f g\| = \|M_f g\| \le \|\widetilde{M}_f\| \|g\| = \|\widetilde{M}_f\| \|z^{-n} g\|$. Since $\cup_{n=0}^{\infty} z^{-n} H^2$ is dense in L^2, we conclude that $\|\widetilde{M}_f\| = \|M_f\|$ so we henceforth drop the tilde. Thus (19.8) holds where now M_f is the operator on H^2 and $\|f\|_\infty$ is the H^∞-norm. Below we'll see that $\{M_f \mid f \in H^\infty\}$ is precisely the set of operators in $\mathcal{L}(H^2)$ commuting with M_z

(8) We next need to consider infinite Blaschke products. To get convergence we can't use the factors of the form (18.3) since even at $z = 0$, there is no reason for the phases of the $-z_j$ to converge. Thus we define single Blaschke factors by:

$$b(z, w) = \begin{cases} z & \text{if } w = 0 \\ \frac{|w|}{w} \frac{w-z}{1-\bar{w}z}, & \text{otherwise} \end{cases} \tag{19.10}$$

This function is $|w|$ at $z = 0$ so for absolute convergence of an infinite product of $b(z, z_j)$ at $z = 0$, it suffices that $\sum_{j=1}^{\infty}(1 - |z_j|) < \infty$. In fact, it is easy to see that:

$$|1 - b(z, w)| \le \frac{(1 - |w|)(1 + |z|)}{1 - |z|} \tag{19.11}$$

from which we conclude that

$$\sum_{j=1}^{\infty}(1 - |z_j|) < \infty \Rightarrow$$

$$B(z, \{z_j\}_{j=1}^{\infty}) = \prod_{j=1}^{\infty} b(z, z_j) \text{ converges} \tag{19.12}$$

and defines a function analytic on $\mathbb{D}$ vanishing exactly at the z_j's. This function is bounded by 1 in magnitude so in H^∞. Thus, it has boundary values on $\partial\mathbb{D}$ and one can show that for a.e. θ, $|B(e^{i\theta})| = 1$. Conversely, if $H^1(\mathbb{D})$ is the set of functions analytic in $\mathbb{D}$ with

$$\|f\|_1 \equiv \lim_{r \uparrow 1} \int |f(re^{i\theta})| \frac{d\theta}{2\pi} = \sup_{r<1} \int |f(re^{i\theta})| \frac{d\theta}{2\pi} < \infty \tag{19.13}$$

then Jensen's formula [326, Section 9.8] implies that if z_j is a counting of the zeros of a function f counting multiplicity, then

$$f \in H^1(\mathbb{D}) \Rightarrow \sum_{j=1}^{\infty}(1 - |z_j|) < \infty \tag{19.14}$$

(indeed this is true for H^p with $p < 1$ and even for the so called Nevanlinna class (see [326, Section 9.9]). The point, of course, is that this implies if $f \in H^1$, one can form the Blaschke product of its zeros.

(9) It is a basic result of Riesz that if $f \in H^1(\mathbb{D})$ and B is the Blaschke product of its zeros, then $g \equiv f/B$ is in $H^1(\mathbb{D})$ also and $\|g\|_1 = \|f\|_1$. Taking $h_1 = g^{1/2}$ and $h_2 = g^{1/2}B$, this implies that $f = h_1 h_2$ where $h_1, h_2 \in H^2(\mathbb{D})$ with $\|h_1\|_2 = \|h_2\|_2 = \|f\|_1$ This proof shows that if $H_0^p \equiv \{f \in H^p \mid f(0) = 0\}$ for $p = 1, 2$, then we get all $f \in H_0^1$ by taking $h_1 \in H^2$ and $h_2 \in H_0^2$.

(10) Among multiplication operators, M_f, on H^2 considered above, especially interesting are the isometries. This happens if and only if $f \in H^\infty$ and $|f(e^{i\theta})| = 1$ for a.e. θ. Such functions are called *inner functions*. We won't need it for most results in this chapter but the application to prove the Titchmarsh and Lions convolution theorems will require a structure theorem for such functions. A *singular inner function* is one of the form:

$$I_\nu(z) = \exp\left(-\int \frac{e^{i\theta} + z}{e^{i\theta} - z} d\nu(\theta)\right) \tag{19.15}$$

where $d\nu$ is a measure singular with respect to $\frac{d\theta}{2\pi}$. The simplest example of such a function is $\exp[-(1+z)/(1-z)]$. Every inner function has a unique decomposition $f = e^{i\psi} B_S I_\nu$ where B_S is the Blaschke product of its set of zeros, $S \subset \mathbb{D}$, I_ν is a singular inner function and $e^{i\psi} \in \mathbb{D}$. It is easy to see that if f_1 and f_2 are inner functions, we have that $\forall_{z\in\mathbb{D}} |f_1(z)| \le |f_2(z)| \iff f_1 = f_2 f_3$ for an inner function f_3. Moreover if $f_1 = e^{i\psi_1} B_{S_1} I_{\nu_1}$ $f_2 = e^{i\psi_2} B_{S_2} I_{\nu_2}$ then this happens if and only if $S_2 \subset S_1$ and $0 \le \nu_2 \le \nu_1$.

(11) For some of the material at the end of this chapter (but not needed in the rest of the book), we need some information on $H^2(\mathbb{C}_+)$ and $H^\infty(\mathbb{C}_+)$ where $\mathbb{C}_+ = \{z \in \mathbb{C} \mid \operatorname{Im}(z) > 0\}$. A key role is played by the analytic bijections $\boldsymbol{\zeta} : \mathbb{C}_+ \to \mathbb{D}$, $\mathbf{z} : \mathbb{D} \to \mathbb{C}_+$ given by:

$$\boldsymbol{\zeta}(z) = \frac{i - z}{i + z}; \quad \mathbf{z}(\zeta) = i\frac{1 - \zeta}{1 + \zeta} \tag{19.16}$$

which are inverse to each other and extend analytically to maps between $\mathbb{R}$ and $\partial\mathbb{D}$ by

$$\boldsymbol{\zeta}\left(\tan(\tfrac{\theta}{2})\right) = e^{i\theta}; \quad \mathbf{z}(e^{i\theta}) = \tan(\tfrac{\theta}{2}) \tag{19.17}$$

$$\mathbf{z}(1) = 0, \quad \mathbf{z}(0) = i, \quad \mathbf{z}(-1) = \infty$$

By computing Jacobians, one sees that under $\mathbf{z} : \partial\mathbb{D} \to \mathbb{R}$,

$$\frac{d\theta}{2\pi} = \frac{1}{\pi}\frac{dx}{1 + x^2} \tag{19.18}$$

(12) As in the case of the disk, $H^2(\mathbb{C}_+)$ can be defined in terms of vanishing of its Fourier transform or as boundary values of functions with some L^2 bounds. So, initially, we define it as those functions, $f \in L^2(\mathbb{R})$ whose Fourier transform:

$$\hat{f}(x) = (2\pi)^{-1/2} \int_{\mathbb{R}} f(k) e^{-ikx} dk \tag{19.19}$$

(one dimensional dx) has $\operatorname{supp}(\hat{f}) \subset [0, \infty)$. Then

$$f(z) = (2\pi)^{-1/2} \int \hat{f}(x) e^{izx} dx \tag{19.20}$$

gives a convergent integral for $\operatorname{Im} z > 0$ since $|e^{izx}| = e^{-x \operatorname{Im} z} \in L^2(0, \infty)$. It thus defines an analytic function on $\mathbb{C}_+$. For $y > 0$, the Plancherel theorem shows that

$$\int |f(k+iy)|^2 dk = \int e^{-2yx} |\hat{f}(x)|^2 dx \tag{19.21}$$

so

$$\|f\|_2^2 = \sup_{0<y<\infty} \int |f(k+iy)|^2 dk = \lim_{y \downarrow 0} \int |f(k+iy)|^2 dk$$

The Poisson formula takes the form

$$f(k+iy) = \frac{1}{\pi} \int_{-\infty}^{\infty} f(\ell) \frac{y}{(k-\ell)^2 + y^2} d\ell \tag{19.22}$$

which can be used to prove that for each $f \in H^2(\mathbb{C}_+)$ that, as $y \downarrow 0$, $f(k + iy) \to f(k)$ for a.e. $k \in \mathbb{R}$ and that $\lim_{y \downarrow 0} \|f(\cdot + iy) - f(\cdot)\| = 0$.

Define $U : H^2(\mathbb{D}) \to H^2(\mathbb{C}_+)$ by

$$(Uf)(z) = \pi^{-1/2} (z+i)^{-1} f(\zeta(z)) \tag{19.23}$$

As a map from $L^2(\partial\mathbb{D})$ to $L^2(\mathbb{R})$, this is unitary since

$$|\pi^{-1/2}(x+i)^{-1}|^2 dx = \frac{d\theta}{2\pi}$$

By approximating by polynomials, one can see that boundary values and functions are both mapped by (19.23). Thus, for example, functions, $f(z) \in H^2(\mathbb{C}_+)$ have an expression

$$f(z) = (z+i)^{-1} \sum_{n=0}^{\infty} a_n \boldsymbol{\zeta}(z)^n \tag{19.24}$$

(13) $H^\infty(\mathbb{C}_+)$ is the set of bounded analytic functions on $\mathbb{C}_+$. If $f \in H^\infty(\mathbb{C}_+)$, then $g(\zeta) \equiv f(\mathbf{z}(\zeta)) \in H^\infty(\mathbb{D})$ which implies that $\lim_{y\downarrow 0} f(k+iy)$ exists for a.e. k. As in the case of $\mathbb{D}$, we say that $f \in H^\infty(\mathbb{C}_+)$ is *inner* if and only if $|f(k+i0)| = 1$ for a.e. k.

This completes the rapid review of H^p spaces.

Let P_+ be the orthogonal projection of $L^2(\partial\mathbb{D}, \frac{d\theta}{2\pi})$ onto $H^2(\mathbb{D})$. Let $P_- \equiv 1 - P_+$. Then P_- is the projection onto $\{f \in L^2(\partial\mathbb{D}, \frac{d\theta}{2\pi}) \mid f_n^\# = 0 \text{ if } n \geq 0\}$. By associating such an f with $f(z) = \sum_{n=1}^\infty f_{-n}^\# \bar{z}^n$, we see that Ran $P_- \equiv H_-^2$ is the set of boundary values of antianalytic functions (i.e. analytic in $\bar{z}$, f, on $\mathbb{D}$ with $f(0) = 0$.

Given $f \in L^\infty$, the *Toeplitz operator*, T_f, associated to f is $P_+ M_f P_+$ viewed as an operator on $H^2(\mathbb{D})$. More generally, if $V \subset H^2(\mathbb{D})$ is a closed subspace and P_V is the projection of $L^2(\partial\mathbb{D}, \frac{d\theta}{2\pi})$ onto V, then for any $f \in L^\infty$, $T_f^V \equiv P_V T_f P_V \in \mathcal{L}(V)$ is the *Toeplitz operator for* V. f is called the *symbol* of T_f or T_f^V. If $M_z V \subset V$, then polynomial approximation implies that $M_f V \subset V$ for all $f \in H^\infty$. This fact implies that for such V's, we have that for all $g \in L^\infty$, $f \in H^\infty$, $T_g^V T_f^V = T_{gf}^V$. In particular if $f, g \in H^\infty$, we have that $T_g^V T_f^V = T_f^V T_g^V$.

For $f \in L^\infty$, the *Hankel operator*, H_f, associated to f is $P_- M_f P_+$ viewed as an operator from H^2 to H_-^2. f is called the *symbol* of H_f.

A *Toeplitz matrix* is a matrix, $T(\{a_n\}_{n=0}^\infty)$, on $\ell^2(\mathbb{N})$ with $T_{ij} = a_{j-i}$. If $e_n \equiv e^{in\theta} \in L^2(\partial\mathbb{D}, \frac{d\theta}{2\pi})$, then $\langle e_n, T_f e_m \rangle = T_{mn}(\{f_k^\#\}_{k=0}^\infty)$ for $n, m = 0, 1, \cdots$. A *Hankel matrix* is a matrix, $H(\{a_n\}_{n=1}^\infty)$, on $\ell^2(\mathbb{Z}_+)$ with $H_{ij} = a_{i+j-1}$. For $n, m = 1, 2, \cdots$ then $\langle e_{-n}, H_f e_{m-1} \rangle = H_{nm}(\{f_{-k}^\#\}_{k=1}^\infty)$. Since $\|M_f\| \leq \|f\|_\infty$, we have that $\|T_f\| \leq \|f\|_\infty$ and $\|H_f\| \leq \|f\|_\infty$.

Theorem 19.1 (Nehari's Theorem) *Let* $\{a_n\}_{n=1}^\infty$ *be a bounded sequence. Then* $H(a)$ *is a bounded operator from* H^2 *to* H_-^2 *if and only if* $\exists_{\varphi \in L^\infty}$ *so that* $\varphi_{-n}^\# = a_n$ *for all* $n \geq 1$. *Moreover*

$$\|H(a)\| = \inf\{\|\varphi\|_\infty \mid \varphi \in L^\infty,\ \varphi_{-n}^\# = a_n \text{ for all } n \geq 1\} \tag{19.25}$$

Remarks

1. This implies for any $\psi \in L^\infty$ that $\|H_\psi\| = \inf\{\|\varphi\|_\infty \mid \varphi \in L^\infty,\ \varphi - \psi \in H^\infty\}$
2. One can prove that $H(\{a_n\}_{n=1}^\infty)$ is bounded if and only if $\sum_{n=0}^\infty a_n z^n$ converges to an $H^2(\mathbb{D})$ function with boundary values in $BMO(\partial\mathbb{D})$ (what [328, Section 5.11] calls an HMO function).

Proof Since $H(a) = H_\varphi$ if $\varphi_{-n}^\# = a_n$, we have that for any such φ that

$$\|H(a)\| \leq \|\varphi\|_\infty \tag{19.26}$$

The idea of the other direction is to use the Hahn–Banach theorem to construct the required φ. The negative Fourier coefficients will determine the sup of $\int \varphi(e^{i\theta})f(e^{i\theta})\frac{d\theta}{2\pi}$ over norm-one f's in H_0^1 in terms of $H(a)$ because of the factorization of item (9) in the summary of Hardy space properties.

Suppose, for now, that

$$\sum_{n=1}^{\infty} |a_n| < \infty \tag{19.27}$$

so $\varphi_0(e^{i\theta}) \equiv \sum_{k=1}^{\infty} a_k e^{-ik\theta} \in L^\infty$. If $h_1(e^{i\theta}) = \sum_{n=1}^{\infty} \alpha_n e^{i(n-1)\theta}$ and $h_2(e^{i\theta}) = \sum_{m=1}^{\infty} \beta_m e^{im\theta}$ are both in H^2, then

$$\begin{aligned}\int \varphi_0(e^{i\theta})h_1(e^{i\theta})h_2(e^{i\theta})\tfrac{d\theta}{2\pi} &= \sum_{n,m=1}^{\infty} a_{n+m-1}\alpha_n\beta_m \\ &= \langle \alpha, H(a)\beta\rangle\end{aligned}$$

so

$$\left|\int \varphi_0(e^{i\theta})h_1(e^{i\theta})h_2(e^{i\theta})\tfrac{d\theta}{2\pi}\right| \le \|\alpha_n\|_{\ell^2}\|\beta_m\|_{\ell^2}\|H(a)\| \tag{19.28}$$

By item (9) in the summary of Hardy space properties, this implies that

$$\sup\{\left|\int \varphi_0(e^{i\theta})f(e^{i\theta})\tfrac{d\theta}{2\pi}\right| \mid f \in H_0^1, \|f\|_1 = 1\} \le \|H(a)\| \tag{19.29}$$

For general bounded sequences, a, define $a_n^{(r)} \equiv a_n r^n$ for $0 \le r < 1$ so that $a^{(r)}$ obeys (19.27). Then

$$r\langle \alpha, H(a^{(r)})\beta\rangle = \langle \alpha^{(r)}, H(a)\beta^{(r)}\rangle \Rightarrow \|H(a^{(r)})\| \le r^{-1}\|H(a)\|$$

Define H_{00}^1 to be the set of polynomials in $e^{i\theta}$, f with $f(0) = 0$ viewed as a dense subspace of H_0^1 and a subspace of L^1. For such f, define $L_a(f) = \sum_{n=1}^{\infty} a_n f_n^\#$ (finite sum) so $L_a(f) = \lim_{r\uparrow 1} L_{a^{(r)}}(f)$. Thus by (19.29),

$$|L_a(f)| \le \lim_{r\uparrow 1} \|H(a^{(r)})\|\|f\|_1 \le \|H(a)\|\|f\|_1 \tag{19.30}$$

By the Hahn–Banach Theorem and the known dual of L^1, there exists a function $\varphi \in L^\infty$ with $\|\varphi\|_\infty \le \|H(a)\|$ and so that for all $f \in H_{00}^1$, we have that $\int \varphi(e^{i\theta})f(e^{i\theta})\frac{d\theta}{2\pi} = L_a(f)$. Taking $f(e^{i\theta}) = e^{in\theta}$, for $n \ge 1$, we find that

$$\varphi_{-n}^\# = \langle e^{-in\theta}, \varphi\rangle = L_a(e^{in\theta}) = a_n \tag{19.31}$$

Because of (4.22), we have found the required φ. □

The next major cog in the machinery of this chapter concerns closed subspaces $V \subset H^2(\mathbb{D})$ with $zV \subset V$ (called invariant subspaces). It follows that V is left invariant under multiplication by polynomials and then, by a limiting argument, by multiplication by any function in H^∞. If $g \in H^\infty$, $V = \overline{gH^2}$, the closure of gH^2 is clearly an invariant subset. The following celebrated theorem says (more than) that they are only the only such spaces.

Theorem 19.2 (Beurling's Theorem) *Let $\{0\} \neq V \subset H^2$ be a subspace invariant under multiplication by z. Then there exists an inner function, Θ, so that $V = \Theta H^2$.*

Remark As we noted, for any $g \in H^\infty$, $V = \overline{gH^2}$ is an invariant subspace. This theorem says we don't need arbitrary g's but only inner ones (and the proof will show that the inner function is unique up to a multiplicative constant). This apparent puzzle can be understood by noting that it is known that any $g \in H^\infty$ can be written $g = \Theta h$ where Θ is inner and h, called an *outer function*, is of the form $h(z) = \exp\left(\int \frac{e^{i\theta}+z}{e^{i\theta}-z} q(e^{i\theta}) \frac{d\theta}{2\pi}\right)$ for some q which is bounded from above. By cutting off q, one can find $k_n \in H^\infty$ so $k_n h \to 1$ which shows that $\overline{hH^\infty} = H^2$ so $V = \Theta H^2$.

Proof Since multiplication by z is an isometry, zV is closed. It cannot be all of V because, if it was, $z^n V = V$ for all n so any $g \in V$ would vanish at 0 to infinite order and so be zero. But, by hypothesis, $V \neq \{0\}$.

Pick $\Theta \in V \cap (zV)^\perp$, $\|\Theta\|_2 = 1$. Because V is invariant, for all $n \geq 1$ we have that $z^n\Theta \in zV$ so $\langle z^n\Theta, \Theta\rangle = 0 = \langle \Theta, z^n\Theta\rangle$. Taking into account that $\|\Theta\|_2 = 1$, we have that for all $n \in \mathbb{Z}$ that $\int e^{-in\theta}(|\Theta(e^{i\theta})|^2 - 1) = 0$. Since Fourier series is injective on L^1 (by Fejér's theorem), we conclude that for a.e. θ, $|\Theta(e^{i\theta})| = 1$, i.e. Θ is an inner function.

Let $f \in V$. Then $\overline{\Theta} f \in L^2$ and for $n \geq 1$

$$\langle e^{-in\theta}, \overline{\Theta} f\rangle = \langle \Theta, z^n f\rangle = 0 \tag{19.32}$$

since $z^n f \in V$ also. Thus $h \equiv \overline{\Theta} f \in H^2$ so $f = \Theta\overline{\Theta} f = \Theta h \in \Theta H^2$. It follows that $V = \Theta H^2$.

□

Remark If also $\Psi \in V \cap (zV)^\perp$ with $\|\Psi\|_1 = 0$, as above Ψ is inner and we have for $n \geq 1$ that $\langle e^{-in\theta}, \overline{\Theta}\Psi\rangle = 0$ and $\langle e^{-in\theta}, \overline{\Psi}\Theta\rangle = 0$. This implies that Ψ is a multiple of Θ so $\dim(V \cap (zV)^\perp) = 1$.

To motivate the next issue we look at, we return to the Pick interpolation problem for Schur functions. We begin by looking for solutions in H^∞ of

$$f(z_j) = w_j \qquad j = 1, \ldots, n \tag{19.33}$$

Let f_0 be any H^∞ solution of these equations (it is easy to find such functions which are polynomials; the proof of Theorem 5.14 extends to complex x_j). Let B

be the Blaschke product vanishing exactly at $\{z_1, \ldots, z_n\}$. Any other H^∞ solution of (19.33) has the form $f_0 + Bg$ for some $g \in H^\infty$. Since $|B| = 1$ on $\partial\mathbb{D}$, we have that $\|f_0 + Bg\|_{H^\infty} = \|B^{-1} f_0 + g\|_{L^\infty}$. Thus, we want to know if $\min_{g \in H^\infty} \|B^{-1} f_0 + g\|_{L^\infty} \leq 1$ which by Nehari's theorem is equivalent to asking if $\|H_{B^{-1} f_0}\|_{\mathcal{L}(H^2, H^2_-)} \leq 1$. We thus see a connection between the Nehari problem and Pick's theorem involving the quotient space H^2/BH^2 which is, of course, isomorphic to $H^2 \cap (BH^2)^\perp$.

Returning to general inner functions, this motivates looking at $K_\Theta \equiv H^2 \cap (\Theta H^2)^\perp$. By Beurling's theorem, each K_Θ is an invariant subspaces for T_z^* (if we take the adjoint of the isometry on H^2; if we view T_z^* as a L^2 adjoint, then K_Θ is invariant for $P_+ T_z^*$) and every invariant subspace for T_z^* is a K_Θ. We'll let P_Θ be the projection (from either H^2 or L^2) onto K_Θ and let $T_f^\Theta \equiv P_\Theta M_f P_\Theta$, the Toeplitz operator.

For $\varphi \in H^\infty$ and $g \in H^2$, write $P_\Theta(\varphi g) = \varphi g - (1 - P_\Theta)(\varphi g)$. Note that $(1 - P_\Theta)(\varphi g) \in (K_\Theta)^\perp \cap H^2 = \Theta H^2$. Thus for any $\varphi \in H^\infty$ and $g \in H^2$, there is $h \in H^2$ so that

$$P_\Theta M_\varphi g = \varphi g + \Theta h \tag{19.34}$$

This suggests the operators T_φ^Θ are connected to Pick interpolation. We are heading towards proving that these operators are precisely all operators on K_Θ that commute with T_z^Θ with a norm relation. As a warmup we note that

Theorem 19.3

(a) *Let $A \in \mathcal{L}(L^2(\partial\mathbb{D}, \frac{d\theta}{2\pi}))$ so that*

$$A(e^{i\theta} f) = e^{i\theta}(Af) \tag{19.35}$$

for all $f \in L^2(\partial\mathbb{D}, \frac{d\theta}{2\pi})$. Then $A = M_h$ for some $h \in L^\infty(\partial\mathbb{D}, \frac{d\theta}{2\pi})$.

(b) *Let $A \in \mathcal{L}(H^2(\mathbb{D}))$ so that*

$$A(zf) = z(Af) \tag{19.36}$$

for all $f \in H^2(\mathbb{D})$. Then $A = M_h$ for some $h \in H^\infty(\mathbb{D})$.

Remarks

1. Part (a) is a special case of the result that if A is a normal operator, $\{B \mid BA = AB\}$ is the set of functions of A if and only if A has simple spectrum, i.e. A, A^* have a cyclic vector; see [329, Section 5.4].
2. While (b) appears as a warmup to Sarason's theorem, it is a special extension also to the case where $K = H^2$ as we'll discuss later.

Proof

(a) Let $h = A\mathbb{1}$. By (19.35) and $(e^{i\theta})^{-1} = e^{-i\theta}$ we see that $A(e^{in\theta}) = e^{in\theta}h$ for all $n \in \mathbb{Z}$. For any $r \in (0,1)$ and $e^{i\psi} \in \partial\mathbb{D}$ let

$$Q_{r,\psi}(e^{i\theta}) = \left(\frac{1-r^2}{1+r^2}\right)^{1/2} \sum_{n=-\infty}^{\infty} r^{|n|}e^{in(\theta-\psi)} \tag{19.37}$$

The sum converges in both L^2 and L^∞, so, since $h \in L^2$, we conclude that

$$(AQ_{r,\psi})(e^{i\theta}) = Q_{r,\psi}(e^{i\theta})h(e^{i\theta}) \tag{19.38}$$

The constant in front of the sum in (19.37) is picked so that $\int |Q_{r,\psi}(e^{i\theta})|^2 \frac{d\theta}{2\pi} = 1$ for all $r \in (0,1)$ and $e^{i\psi} \in \partial\mathbb{D}$. $Q_{r,\psi}(e^{i\theta})$ is a multiple of the Poisson kernel, so, uniformly in $|\theta - \psi| \geq \epsilon$ it goes to zero (like $(1-r^2)^{1/2}$) and thus $Q_{r,0}(e^{i\theta})^2$ is an approximate identity. This implies that $\lim_{r\uparrow 1} \int Q_{r,\psi}(e^{i\theta})^2 h(e^{i\theta})\frac{d\theta}{2\pi} = h(e^{i\psi})$ for a.e. ψ. On the other hand, since $\|Q_{r,\psi}\|_{L^2} = 1$, $|\langle Q_{r,\psi}, AQ_{r,\psi}\rangle| \leq \|A\|$. We conclude that $h \in L^\infty$ with $\|h\|_\infty \leq \|A\|$. It follows that $A = M_h$.

(b) Let U be multiplication by $e^{-i\theta}$ and $\mathcal{H}_n = U^n H^2$. Define A_n on $\mathcal{H}_n$ by $A_n(U^n g) = U^n(Ag)$ for all $g \in H^2$. Since U^n is a unitary map of H^2 onto $\mathcal{H}_n$, $\|A_n\| = \|A\|$. $AM_z = M_z A$ on H^2 implies that for $f \in H^2$ (so also $g = U^{-n} f \in H^2$), we have that $A_n f = A_n(U^n g) = U^n(Ag) = U^n AM_z^n f = U^n M_z^n Af = Af$, i.e. A_n is an extension of A. Similarly, A_{n+1} is an extension of A_n.

Thus A has an extension to $\cup\mathcal{H}_n$ with the same norm. Since $\cup\mathcal{H}_n$ is dense in L^2, A has an extension, A_∞, to L^2. Moreover, (19.35) holds for A_∞. By (a), $h \equiv A_\infty\mathbb{1} = A\mathbb{1}$ and $A_\infty = M_h$ on L^2. Since A maps H^2 to itself, $h \in H^2 \cap L^\infty = H^\infty$ and $A = A_\infty \restriction H^2$ is M_h as a map on H^2.

□

A second warmup for Sarason's theorem is

Proposition 19.4 *For any inner function,* Θ, *and* $f, g \in H^\infty$,

$$T_f^\Theta T_g^\Theta = T_{fg}^\Theta \tag{19.39}$$

In particular, $T_f^\Theta T_g^\Theta = T_g^\Theta T_f^\Theta$, *for all* $f, g \in H^\infty$.

Proof For any $h \in K_\Theta$, there is $q_1 \in H^2$ so that

$$T_g^\Theta h = gh + \Theta q_1$$

since $gh - T_g^\Theta h \in (K_\Theta)^\perp = (\Theta H^2)$. Similarly,

$$T_f^\Theta T_g^\Theta h = f T_g^\Theta h + \Theta q_2$$
$$= fgh + \Theta(fq_1 + q_2)$$

proving (19.39). □

Sarason's lifting theorem is a strong converse to this:

Theorem 19.5 (Sarason Lifting Theorem) *Let K to a closed subspace of $H^2(\mathbb{D})$ left invariant by $T_{\bar{z}}$. If $A \in \mathcal{L}(K)$ commutes with $P_K T_z P_K$, there exists $h \in H^\infty$ with $\|h\| = \|A\|_{\mathcal{L}(K)}$ so that $A = P_K M_h P_K$.*

We need some preliminaries before giving the proof. Since $T_{\bar{z}} = T_z^*$, $K^\perp$ is an invariant subspace for T_z—thus either $\{0\}$ or ΘH^2 so either $K = H^2$ or K_Θ. The first of these is Theorem 19.3 so we need only prove the K_Θ case. Given our discussion following (19.33), which says we want to look at the Nehari problem for $B^{-1} f_0$ and multiply that minimizer by B, it is reasonable to consider multiplication by $\overline{\Theta}$ (of course here, B is replaced by Θ and we use the fact that $\Theta^{-1} = \overline{\Theta}$ since Θ is an inner function). We systematically conflate Θ as function and as a multiplication operator on L^∞. Thus, given $A \in \mathcal{L}(K_\Theta)$ define as a map of $H^2 \to L^2$:

$$B_A = \overline{\Theta} A P_\Theta \tag{19.40}$$

Proposition 19.6

(a) *$\overline{\Theta} K_\Theta \subset H^2_-$, in particular,* $\text{Ran}(B_A) \subset H^2_-$.
(b) *If $C \in \mathcal{L}(H^2, H^2_-)$, then C is a Hankel operator if and only if $P_- M_z C = C M_z$ in the sense that for all Laurent polynomials, $p \in H^2$, $q \in H^2_-$, we have that*

$$\langle z^{-1} q, Cp \rangle = \langle q, Czp \rangle \tag{19.41}$$

(c) *B_A is a Hankel operator if and only if A commutes with T_z^Θ.*
(d) *If $B_A = H_t$ for some $t \in L^\infty$, then $\Theta t \in H^\infty$.*

Remark By Laurent polynomial, we mean a polynomial in z and z^{-1}. Such a polynomial is in H^2 if and only if it is an ordinary polynomial and is in H^2_- if and only if it is a polynomial in z^{-1} with vanishing constant term.

Proof

(a) If $f \in H^2$, $g \in K_\Theta$, then $\langle f, \overline{\Theta} g \rangle = \langle \Theta f, g \rangle = 0$ since $g \in (\Theta H^2)^\perp$. Therefore $\overline{\Theta} g \in (H^2)^\perp = H^2_-$.
(b) By Nehari's theorem, a bounded operator, $C \in \mathcal{L}(H^2, H^2_-)$ is a Hankel operator, H_t, associated to some $t \in L^\infty$ if and only if $a_{nm} \equiv \langle e^{-in\theta}, Ae^{im\theta} \rangle$, $m = 0, 1, \ldots, n = 1, 2, \ldots$ obeys $a_{n+1\; m-1} = a_{nm}$, $m \geq 1$. This is exactly (19.41) for $p(z) = z^{m-1}$, $q(z) = z^{-n}$.

(c) We begin by noting that since $\overline{\Theta}$ maps ΘH^2 bijectively and isometrically to H^2, $\overline{\Theta}$ is a unitary bijection of $H^2 = K_\Theta \oplus \Theta H^2$ to $\overline{\Theta} K_\Theta \oplus H^2 \equiv H^2_{\overline{\Theta}}$ preserving orthogonality. It follows that

$$P_- \restriction H^2_{\overline{\Theta}} = \overline{\Theta} P_\Theta \Theta \tag{19.42}$$

Since M_z maps H^2 to H^2 and commutes with multiplication by $\overline{\Theta}$, M_z maps $H^2_{\overline{\Theta}}$ to itself. And on $H^2_{\overline{\Theta}}$, $\overline{\Theta} P_\Theta M_z = P_- M_z P_\Theta$ which implies that

$$\overline{\Theta} P_\Theta M_z A P_\Theta = P_- M_z B_A \tag{19.43}$$

Next, note that $AT_z^\Theta = T_z^\Theta A \iff AP_\Theta M_z \restriction H^2 = P_\Theta M_z A P_\Theta \restriction H^2$ (because M_z maps ΘH^2 to itself) $\iff$ (multiplying by $\overline{\Theta}$)$\overline{\Theta} P_\Theta M_z A P_\Theta = \overline{\Theta} A P_\Theta M_z \restriction H^2$. In this last equality, the left side is $P_- M_z B_A$ by (19.43) while the right is $B_A M_z \restriction H^2$. By (b), this shows that $AT_z^\Theta = T_z^\Theta A$ is equivalent to B_A being a Hankel operator.

(d) $B_A = H_t$ applied to $\mathbb{1}$ says that $P_t = \overline{\Theta} A P_\Theta \mathbb{1}$. Thus $\Theta t = \Theta P_+ t + A P_\Theta \mathbb{1} \in H^2 \cap L^\infty = H^\infty$.

$\square$

Proof of Theorem 19.5 Clearly $\|B_A\| = \|A\|$ since $\overline{\Theta}$ is a unitary map of K_Θ onto $\mathrm{Ran}(B_A)$. If $AT_z^\Theta = T_z^\Theta A$, by (c) of the last Proposition, B_A is a Hankel operator. By Nehari's theorem (Theorem 19.1), there is $t \in L^\infty$ with $B_A = H_t$ and $\|t\|_\infty = \|A\|$.

Let $h = \Theta t$ which lies in H^∞ by (d) of the last Proposition and, since Θ is inner, has $\|h\| = \|t\| = \|A\|$.

If we show that $A = T_h^\Theta \restriction K_\Theta$, we have proven the theorem. Let $g \in K_\Theta$. By definition of B_A

$$\begin{aligned}
\overline{\Theta} A g &= B_A g \\
&= P_- M_t g \quad (\text{since } B_A = H_t) \\
&= \overline{\Theta} P_\Theta \Theta t g \quad (\text{by } (19.42)) \\
&= \overline{\Theta} P_\Theta M_h P_\Theta g
\end{aligned}$$

Multiplying by Θ and using $\Theta\overline{\Theta} = 1$, we see that $A = T_h^\Theta \restriction K_\Theta$. $\square$

We can now turn to using this machinery to prove Pick's theorem for Schur functions. We start by noting that solutions, $f \in H^2$, of

$$f(z_j) = w_j \qquad j = 1, 2, \ldots, n \tag{19.44}$$

are determined up to elements of $Z \equiv \{g \in H^2 \mid g(z_j) = 0\}$. Let $L_j(g) = g(z_j)$. L_j is a continuous functional on H^2 by a Cauchy estimate, so Z is closed. Indeed, since the L_j's are easily seen to be independent, Z has codimension n. It is clearly left invariant by multiplication by M_z—in fact $Z = BH^2$ where B is the product of Blaschke factors vanishing at each of the z_j's (without recourse to Beurling's theorem).

Since L_j is continuous, there are elements $s_j \in H^2$ with $L_j(g) = \langle s_j, g\rangle$. These functions, called *Szegő functions*, are explicitly given by

$$s_j(z) = \frac{1}{1 - \bar{z}_j z} \tag{19.45}$$

for, if $g(z) = \sum_{n=1}^{\infty} a_n z^n$, then $a_n = \langle z^n, g\rangle$. Writing out the Taylor series for s_j, which converges in H^2, this means that $\langle s_j, g\rangle = \sum_{n=1}^{\infty} z_j^n \langle z^n, g\rangle = \sum_{n=1}^{\infty} a_n z_j^n = g(z_j)$, i.e.

$$\langle s_j, g\rangle = g(z_j) \tag{19.46}$$

Since $g \in BH^2 \Rightarrow \langle s_j, g\rangle = 0$, we see that each $s_j \in (BH^2)^{\perp} = K_B \equiv K$. If $\sum_{j=1}^{n} \alpha_j s_j = 0$ in H^2, find $h_j \in H^2$ so that $\langle s_k, h_j\rangle = h_j(z_k) = \delta_{jk}$. Apply h_j to the sum to see that $\alpha_j = 0$ and conclude that $\{s_j\}_{j=1}^{n}$ are independent and so an algebraic basis of K. They are not orthogonal, indeed

$$\langle s_j, s_k\rangle = s_k(s_j) = \frac{1}{1 - \bar{z}_k z_j} \tag{19.47}$$

If $f \in H^{\infty}, h \in H^2$, then

$$\langle s_j, M_f h\rangle = f(z_j)h(z_j) = f(z_j)\langle s_j, h\rangle$$

It follows that $M_f^* s_j = \overline{f(z_j)} s_j$ (H^2 adjoint) so each M_f^* leaves K invariant (as we know already from the fact that M_f leave $K^{\perp}$ invariant) and

$$(T_f^B)^* s_j = \overline{f(z_j)} s_j$$

For any $\{w_j\}_{j=1}^{n}$, there exists $f \in H^{\infty}$ solving (19.44). Thus, if $X \in \mathcal{L}(K)$ is defined by

$$X s_j = \bar{w}_j s_j \tag{19.48}$$

then $X = (T_f^B)^*$ for some $f \in H^{\infty}$. In particular, by Proposition 19.4, X^* commutes with T_z^B.

Now use the Sarason's lifting theorem (Theorem 19.5). It says that there is $h \in H^{\infty}$ with $\|h\|_{\infty} = \|X^*\| = \|X\|$ and $X^* = T_h^B$. This means that $X = (T_h^B)^*$ so that h solves (19.44). This implies that Pick interpolation for Schur functions has a solution if $\|X\| \leq 1$. Conversely, if there is such a solution, then $\|X\| = \|T_h^B\| \leq$

$\|M_h\| = \|h\|_\infty \leq 1$. We conclude that Pick interpolation has a Schur function solution if and only if $\|X\| \leq 1$. This allows:

Proof of Theorem 16.4 Let $g = \sum_{j=1}^n g_j s_j \in K$ with $\{g_j\}_{j=1}^n \in \mathbb{C}^n$. By (19.47),

$$\|g\|^2 = \sum_{j=1}^n \bar{g}_j g_k \frac{1}{1 - \bar{z}_k z_j} \tag{19.49}$$

By (19.47) and (19.48)

$$\|Xg\|^2 = \sum_{j=1}^n \bar{g}_j g_k \frac{w_j \bar{w}_k}{1 - \bar{z}_k z_j} \tag{19.50}$$

so, if $M^{(\mathrm{S})}$ is given by (16.32), then

$$\|g\|^2 - \|Xg\|^2 = \sum_{j=1}^n M_{jk}^{(\mathrm{S})} \bar{g}_j g_k$$

Thus $M^{(\mathrm{S})} \geq 0 \iff \forall_{g \in K} \|Xg\|^2 \leq \|g\|^2 \iff \|X\| \leq 1$. □

That completes the discussion of using this method of proving Pick's theorem. Our proof of the hard half of Loewner's theorem in Chapter 33 will rely on the same ideas in a more general form. We'll need a generalization of Sarason's theorem that has essentially the same proof:

Theorem 19.7 *Let X be a bounded map of K to $H^2(\mathbb{D})$ where $K \subset H^2(\mathbb{D})$ is invariant under $T_{\bar{z}}$. Suppose that $XT_{\bar{z}}^K = T_{\bar{z}}X$. Then there exists $h \in H^\infty$ with $\|h\|_\infty = \|X\|$ so that $T_h^* \restriction K = X$.*

Remarks

1. This does generalize Sarason's theorem, for if $A \in \mathcal{L}(K)$ commutes with T_z^K, then $X = (AP_K)^*$ obeys the hypotheses of this theorem.
2. If $K = H^2$, this is just Theorem 19.3. It is only that special case we need in Chapter 33.

Proof Since the proof is essentially the same as Theorem 19.5, we only sketch it. Note that X^* maps H^2 to K so $B \equiv \overline{\Theta} X^*$ maps H^2 to $\overline{\Theta} K \subset H^2_-$. By a calculation identical to the one in the proof of Proposition 19.6c, $X^* T_z = T_z^K X^*$ translate to B being a Hankel operator. Thus, Nehari's theorem implies the existence of $t \in L^\infty$ with $H_t = B$ and $\|t\|_\infty = \|B\| = \|X\|$. By Proposition 19.6(d), $h \equiv \Theta t \in H^\infty$ and one shows that $T_h^* \restriction K = X$ as in the Proof of Theorem 19.5. □

We also need some other preliminary notation. V will denote a vector space over $\mathbb{C}$ with no specified topology, V' its algebraic dual and $\mathcal{L}(V)$ the set of *all* linear operators on V (since V has no topology, we abuse notation and use the same symbol we use for bounded operators when V is a Banach space).

Theorem 19.8 (Rosenblum–Rovnyak Theorem) *Let $b, c \in V$ and $A \in \mathcal{L}(V)$. Let $\mathcal{D} \subset V'$ be a linear subspace of V' so that $A'[\mathcal{D}] \subset \mathcal{D}$ and so that for all $\ell \in \mathcal{D}$*

$$\sum_{j=0}^{\infty} |\ell(A^j c)|^2 < \infty \tag{19.51}$$

Then the following are equivalent:

(a) *For all $\ell \in \mathcal{D}$*

$$\sum_{j=0}^{\infty} |\ell(A^j b)|^2 \leq \sum_{j=0}^{\infty} |\ell(A^j c)|^2 \tag{19.52}$$

(b) *There exists $g = \sum_{j=0}^{\infty} g_j z^j \in H^\infty(\mathbb{D})$ with $\|g\|_\infty \leq 1$ so that for $\ell \in \mathcal{D}$*

$$\ell(b) = \sum_{j=0}^{\infty} g_j \ell(A^j c) \tag{19.53}$$

Remarks

1. This is almost magical. We seem to have conjured $\mathbb{D}$ out of thin air! In some sense, the magic comes from the fact that (19.51) says that $\ell(A^j c)$ are the Taylor coefficients of a function in $H^2(\mathbb{D})$.
2. If g exists with $g \in H^2$, then $\{g_j\}_{j=0}^{\infty} \in \ell^2$ and, since (19.51) holds, the sum in (19.53) is convergent.
3. This is a kind of Pick's theorem on steroids. Equation (19.53) says that $b = g(A)c$ (formally, since we aren't claiming that $\sum_{j=0}^{\infty} g_j A^j$ converges other than in some weak sense in $\mathcal{D}'$. It is an interesting exercise to prove Pick's theorem for Schur functions (or, by changing V, for Herglotz functions) as a special case of this result.

Proof <u>*(b)* $\Rightarrow$ *(a)*</u> Let Y be the operator $M_{\tilde{g}}^*$ on $H^2(\mathbb{D})$ where $\tilde{g}(z) = \sum_{j=0}^{\infty} \bar{g}_j z^j$. By (19.53) and the invariance of $\mathcal{D}$ under A', we have that:

$$\ell(A^k b) = ((A')^k \ell)(b) = \sum_{j=0}^{\infty} g_j \left((A')^k \ell\right)(A^j c)$$

$$= \sum_{j=0}^{\infty} g_j \ell(A^{j+k} c) \tag{19.54}$$

On the other hand in the H^2 inner product

$$\begin{aligned}\left\langle z^k, Y(\sum_{j=0}^{\infty} \ell(A^j c) z^j)\right\rangle &= \left\langle Y^*(z^k), \sum_{j=0}^{\infty} \ell(A^j c) z^j\right\rangle \\ &= \left\langle \sum_{n=0}^{\infty} \bar{g}_n z^{n+k}, \sum_{j=0}^{\infty} \ell(A^j c) z^j\right\rangle \\ &= \sum_{n=0}^{\infty} g_n \left\langle z^{n+k}, \sum_{j=0}^{\infty} \ell(A^j c) z^j\right\rangle = \sum_{n=0}^{\infty} g_n \ell(A^{n+k} c) \\ &= \ell(A^k b) \qquad \text{(by (19.54))}\end{aligned}$$

Thus, since $\{z^k\}_{k=0}^{\infty}$ is an orthonormal basis

$$\begin{aligned}Y\left(\sum_{j=0}^{\infty} \ell(A^j c) z^j\right) &= \sum_{k=0}^{\infty} z^k \left\langle z^k, \sum_{j=0}^{\infty} \ell(A^j c) z^j)\right\rangle \\ &= \sum_{k=0}^{\infty} \ell(A^k b) z^k\end{aligned}$$

Since $\|Y\| \le 1$, we conclude that

$$\begin{aligned}\sum_{j=0}^{\infty} |\ell(A^j b)|^2 &= \left\| \sum_{k=0}^{\infty} \ell(A^k b) z^k \right\|^2 = \left\| Y(\sum_{j=0}^{\infty} \ell(A^j c) z^j) \right\|^2 \\ &\le \|Y\|^2 \left\| \sum_{j=0}^{\infty} \ell(A^j c) z^j \right\|^2 \le \sum_{j=0}^{\infty} |\ell(A^j c)|^2\end{aligned}$$

proving (19.52)

<u>$(a) \Rightarrow (b)$</u> For $\ell \in \mathfrak{D}$, define

$$f_\ell(z) = \sum_{j=0}^{\infty} \ell(A^j c) z^j \qquad h_\ell(z) = \sum_{j=0}^{\infty} \ell(A^j b) z^j \tag{19.55}$$

which, by (19.51) and the hypotheses (19.52), defines functions in H^2. Note that

$$(T_{\bar{z}} f_\ell)(z) = \sum_{j=0}^{\infty} \ell(A^{j+1} c) z^j = f_{A'\ell}(z) \tag{19.56}$$

and similarly for h_ℓ. Let K be the closure in $H^2(\mathbb{D})$ of $\{f_\ell\}_{\ell \in \mathfrak{D}}$. By (19.56), K is left invariant by $T_{\bar{z}}$. Define $X : K \to H^2(\mathbb{D})$ initially by $X f_\ell = h_\ell$. Since $\ell \mapsto f_\ell$ and $\ell \mapsto h_\ell$ are linear and, by (19.52), $f_\ell = 0 \Rightarrow h_\ell = 0$, X is well-defined and linear. By (19.52) again, $\|h_\ell\| \le \|f_\ell\|$, so X extends to the closure and $\|X\| \le 1$.

Equation (19.56) and its analog for h_ℓ shows that $T_{\bar{z}} X = X T_{\bar{z}}^K$. Thus, by Theorem 19.7, there is $q(z) = \sum_{j=0}^\infty \bar{g}_j z^j$ so that $T_q^* \restriction K = X$ and $\|q\|_\infty \le 1$. Therefore $T_q^* f_\ell = h_\ell$. By (19.56)

$$(T_q^* f_\ell)(z) = \sum_{j=0}^\infty g_j \left((T_{\bar{z}})^j f_\ell \right)(z) = \sum_{j=0}^\infty g_j f_{(A')^j \ell}(z)$$

Thus,

$$\ell(b) = h_\ell(0) = (X f_\ell)(0) = \sum_{j=0}^\infty g_j f_{(A')^j \ell}(0)$$

$$= \sum_{j=0}^\infty g_j \left((A')^j \ell \right)(c) = \sum_{j=0}^\infty g_j \ell(A^j c)$$

proving (19.53). □

This completes what we need for the rest of the book, but we end the chapter with an aside showing how to use Beurling's theorem to provide proofs of the following pair of theorems:

Theorem 19.9 (Titchmarsh Convolution Theorem) *Let $f, g \in L^2(0, \infty)$. Let $\iota(h) \equiv \inf \operatorname{supp}(h)$. Then $\iota(f) = \iota(g) = 0 \Rightarrow \iota(f * g) = 0$.*

Remarks

1. That $\iota(f * g) \ge 0$ is trivial. At first sight it seems obvious that equality must hold until one begins to try to eliminate possible cancelations in the integral defining the convolution. It is a surprising subtle result and one whose proofs are all more than a few lines.
2. This easily implies that if $f, g \in L^2(\mathbb{R})$ and if $\iota(f) > -\infty$, $\iota(g) > -\infty$, then $\iota(f * g) = \iota(f) + \iota(g)$. We'll extend this to ν-dimensions at the end of the chapter (Lions' theorem—Theorem 19.14).
3. One easily goes from this case to the case where f and g are L^p-functions or even distributions of compact support.
4. There are many other equivalent forms of this result; see the Notes.

Theorem 19.10 (Agmon Unicellularity Theorem) *Define $V : L^2([0, 1], dx) \to L^2([0, 1], dx)$ by*

$$Vf(x) = \int_0^x f(y)dy \tag{19.57}$$

The only closed invariant subspaces for V are $L^2([a, 1], dx)$; $0 \le a \le 1$.

Remarks

1. V is called the *Volterra operator*. Sometimes the family of its invariant subspaces is called the *Volterra nest*. Note that this family is linearly ordered under inclusion—the structure of invariant subspaces of a self-adjoint operator under inclusion is very different.
2. $L^2([a, 1], dx)$ for $a = 1$ is intended to mean $\{0\}$.
3. It is easy to go from this to characterizing the closed invariant subspaces in L^p or in $C([0, 1])$.

We'll prove Theorem 19.10 first and use it to prove Theorem 19.9. (One go the other way; see the Notes.) Since they deal with invariant subspaces, Beurling's theorem and Theorem 19.10 would seem to be related but there are several issues that might appear to bring this into question. V is not an isometry while T_z is an isometry. We'll see, though, that $\frac{1-V}{1+V}$ is associated to T_z though. Second, there is no H^2 in sight! But inverse Fourier transform will map $L^2([0, \infty))$ unitarily to $H^2(\mathbb{C}_+)$. Finally, the invariant subspaces in Beurling's theorem are far from linearly ordered under inclusion. This will be resolved by noting that $L^2([0, 1])$ doesn't map onto all of $H^2(\mathbb{C}_+)$, so the following will be relevant:

Proposition 19.11 *The invariant subspaces for $T_{\bar{z}}^{\Theta}$ are those K_{Θ_1} where there exists a third inner function Θ_2 so that $\Theta = \Theta_1\Theta_2$.*

Remarks

1. If this happens, we say that Θ_1 *divides* Θ.
2. Of course, $K_{\Theta_1} = K_{\Theta_3}$ if and only if $\Theta_1 = \zeta_0\Theta_3$ for some $\zeta_0 \in \partial\mathbb{D}$, so to study the set of invariant subspaces, we need to consider divisors of Θ up to overall phase. For example, if $\Theta(0) > 0$, we can restrict to those Θ_1 with $\Theta_1(0) > 0$ and thereby count each invariant subspace once.

Proof Any invariant subspace for $T_{\bar{z}}^{\Theta}$ is also invariant for $T_{\bar{z}}$, so a K_{Θ_1}. $K_{\Theta_1} \subset K_\Theta \iff K_\Theta^\perp = \Theta H^2 \subset K_{\Theta_1}^\perp = \Theta_1 H^2 \iff \Theta \in \Theta_1 H^2 \iff \Theta = \Theta_1\Theta_2$ for some $\Theta_2 \in H^2$. But then since $|\Theta_1(e^{i\theta})| = |\Theta(e^{i\theta})| = 1$ for a.e. θ we see that Θ_2 is inner. □

We'll use this in proving Theorem 19.10 via

Proposition 19.12 *Let $\Theta(z) = e^{iz}$ as a function in $H^\infty(\mathbb{C}_+)$. The inner functions $\Theta_1 \in H^\infty(\mathbb{C}_+)$ so that $\Theta = \Theta_1\Theta_2$ for some inner function Θ_2 and with $\Theta_1(i) > 0$ are precisely the functions $\Theta_1(z) = e^{iaz}$ for some $a \in [0, 1]$.*

Remarks

1. One can also prove this by looking at a representation for singular inner functions in terms of singular measures, ν_s on $\mathbb{R} \cup \{\infty\}$. In this case ν_s is a unit point mass at infinity and the Proposition follows from the fact that the only smaller measures are multiples of this point mass. We give the proof here since one can use "bare hands".
2. This implies the invariant subspaces for $T_{\bar{z}}^{\Theta}$ with this Θ are linearly ordered!

Proof Since Θ is non-vanishing on $\mathbb{C}_+$, so are Θ_1, Θ_2. We can therefore define $f_j(z) = -i \log \Theta_j(z)$, $j = 1, 2$ picking the branch with $\operatorname{Re} f_j(i) = 0$. $\Theta = \Theta_1\Theta_2 \Rightarrow f_1 + f_2 = z$ and $|\Theta_j(z)| \leq 1$ on $\mathbb{C}_+ \Rightarrow \operatorname{Im} f_j \geq 0$. Thus

$$0 \leq \operatorname{Im} f_1(x + iy) \leq y \tag{19.58}$$

It follows that for all $x \in \mathbb{R}$, $\lim_{y \downarrow 0} \operatorname{Im} f_1(x + iy) = 0$, so by the strong reflection principle (see [326, Section5.5]), f_1 extends to an entire function with $f(\bar{z}) = \overline{f(z)}$. This implies that on all of $\mathbb{C}$, $|\operatorname{Im} f_1(x + iy)| \leq |y| \leq |z|$. A Poisson formula then implies that $|f_1(z)| \leq c_1 + |z|$. This in term means that $f_1(z) = c + az$. f_1 real on $\mathbb{R}$ means c and a are real. $\operatorname{Re} f_j(i) = 0$ then implies that $c = 0$ and 19.58 implies $0 \leq a \leq 1$ so $\Theta_1(z) = e^{ia\theta}$ as required. □

We next turn to analyzing V. We need some notation. $K \equiv L^2([0, 1], dx)$ thought of as a subset of either $L^2(\mathbb{R})$ or $L^2([0, \infty)) \equiv L^2_+$. P_K is the orthogonal projection onto K. $C = \frac{1-V}{1+V}$ is the Cayley transform of V. It is a contraction; see the Notes.

Proposition 19.13 *V has the following properties:*

(a) *For $j = 0, 1, \ldots$, let $g_j(x) = \frac{x^j}{j!}\chi_{[0,1]}(x)$ as a function on $[0, 1]$. Then, for $j \geq 1$ and $f \in K$, we have that*

$$V^j f = g_{j-1} * f \tag{19.59}$$

(or $P_K(g_{j-1} * f)$ if we think of $f \in L^2_+$ and convolution on $[0, \infty)$ rather than $[0, 1]$).

(b) *A subspace $\mathcal{K} \subset K$ is invariant for V if and only if it is invariant for C.*

(c) *Let $q(x) = e^{-x}\chi_{[0,1]}(x)$ and $Qf = q * f$. Then for $f \in K$, we have that $Cf = f - 2Qf$.*

(d) *Let $\tilde{q}(x) = e^{-x}\chi_{[0,\infty)}(x)$ and $\widetilde{Q}f = \tilde{q} * f$ on L^2 or L^2_+. Then for $f \in K$*

$$Cf = P_K(1 - 2\widetilde{Q})f \tag{19.60}$$

Proof

(a) For two functions in K, $g * f(x) = \int_0^x g(x - y)f(y)dy$. This shows first that $Vf = g_0 * f$ proving (19.59) for $j = 1$. It also proves that $g_0 * g_j = g_{j+1}$

by evaluating the integral. This proves by induction that $V^j f = V(V^{j-1}f) = g_0 * (g_{j-2} * f) = (g_0 * g_{j-2}) * f = g_{j-1} * f$ completing the proof of (19.59).

(b) By (a), $|V^j g(x)| \le [(j-1)!]^{-1} \int_0^1 |g(y)|dy \le [(j-1)!]^{-1}\|g\|_2$ by the Schwarz inequality. Thus $\|V^j\| \le [(j-1)!]^{-1}$. Therefore the series expansion

$$C = \mathbb{1} + 2\sum_{j=1}^{\infty}(-V)^j \tag{19.61}$$

converges in norm implying that any closed invariant subspace for V is invariant for C.

On the other hand, if $D = C - \mathbb{1}$, then $D = V(-2\sum_{j=1}^{\infty}(-V)^{j-1}) \equiv VE$ where E is a bounded operator commuting with V. This means that $\|D^n\| \le \|V^n\|\|E\|^n$ implying that $\lim_{n\to\infty}\|D^n\|^{1/n} = 0$. Some algebra shows that $V = -D(D+2)^{-1} = -\frac{D}{2}\sum_{j=0}^{\infty}\frac{-D}{2}^j$ which converges uniformly. Therefore any closed invariant subspace for C (and so for D) is invariant for V.

(c) By (a), $\sum_{j=1}^{\infty}(-V)^j$ is convolution with $\sum_{j=0}^{\infty}(-1)^j\frac{x^j}{j!}\chi_{[0,1]}(x)$, which is $-q$ where the sum and convolution integral can be interchanged by an easy uniform estimate. The formula (19.61) completes the proof.

(d) For $0 \le x \le 1$, $\int_0^1 \tilde{q}(x-y)f(y)dy = \int_0^1 q(x-y)f(y)dy$.

□

Proof of Theorem 19.10 (Agmon Unicellularity Theorem) Use $\check{}$ to denote the inverse Fourier transform given by (19.20). It is a unitary map of $L^2(\mathbb{R})$ to itself and of L^2_+ to $H^2(\mathbb{C}_+)$. If $\tau_a : L^2(\mathbb{R}) \to L^2(\mathbb{R})$ by $\tau_a f(x) = f(x-a)$, then $\widecheck{\tau_a f}(z) = e^{iza}\check{f}(z)$. That means if $\Theta_a(z) = e^{iza}$, we have that $\check{}$ maps $L^2([a,\infty))$ to $\Theta_a H^2(\mathbb{C}_+)$ and so $L^2([0,a])$ to $K_a \equiv (\Theta_a H^2(\mathbb{C}_+))^\perp$. Let P_a be the projection onto K_a from either $H^2(\mathbb{C}_+)$ or $L^2(\mathbb{R})$.

Define $\check{C}$ on K by $\check{C}\check{h} = \widecheck{Ch}$. Inverse Fourier transform turns convolution into multiplication. Thus, since

$$\int_0^\infty e^{-k}e^{ikx}dk = (1-ix)^{-1}$$

we have that $(\widecheck{\tilde{Q}f})(x) = (1-ix)^{-1}\check{f}(x)$. Noticing that $1 - 2(1-ix)^{-1} = (x-i)(x+i)^{-1} = \zeta(x)$, (19.60) becomes

$$\check{C}g = P_1 M_\zeta P_1 g \tag{19.62}$$

We note that K_1 is invariant under $T_\zeta^{K_1}$ so to apply Proposition 19.11, we need to look at $\check{C}^*$, not $\check{C}$. Translating the Proposition to $H^2(\mathbb{C}_+)$, we see that the invariant subspaces for $\check{C}^*$ are $K_{\tilde{\Theta}}$ for all inner functions $\tilde{\Theta}$ dividing Θ_1. By Proposition 19.12, these are precisely the spaces K_a. Thus the invariant subspaces for C^* are precisely the $L^2([0,a]), 0 \le a \le 1$. In $L^2([0,1]), L^2([0,a]^\perp = L^2([a,1])$ so these are precisely the invariant subspaces of C and so of V. □

Proof of Theorem 19.9 (Titchmarsh Convolution Theorem) Let $\iota(f) = 0$. Let M be the closure of $\{\alpha f + f * h \mid \alpha \in \mathbb{C}, h \in L^2([0,1])\}$. Since $Vr = r * \chi_{[0,1]}$, M is an invariant subspace for V. By Theorem 19.10, $M = L^2([a,1])$ for some $a \ge 0$. Since $\iota(f) = 0$ and $f \in M$, $a = 0$. Thus, f is a cyclic vector for V.

Now suppose $\iota(f) = \iota(g) = 0$. Suppose that $a = \iota(f * g) > 0$. Then for any $h \in L^2([0,1])$, $(f * h) * g = h * (f * g) \in L^2([a,1])$, so by the first paragraph, $h * g \in L^2([a,1])$ for all $h \in L^2([0,1])$.

Notice that $h * g$ is continuous so, $h * g(x) = 0$ for a.e. $x \in [0,a]$ implies that $h * g(a) = 0$ for all h. Define

$$h(x) = \begin{cases} \overline{g(a-x)} & 0 < x \le a \\ 0 & x > a \end{cases}$$

Then

$$h * g(a) = \int_0^a h(a-y)g(y)dy = \int_0^a |g(y)|^2 dy$$

so $h*g(a) = 0 \Rightarrow \iota(g) \ge a$ which means that $\iota(f) = \iota(g) = 0 \Rightarrow \iota(f*g) = 0$. □

Finally, we state and sketch the proof of a ν-dimensional version of Titchmarsh's convolution theorem.

Theorem 19.14 (Lions' Theorem) *Let T, S be distributions of compact support. Let $\gamma(T)$ be the convex hull of the support of T. Then*

$$\gamma(T * S) = \gamma(T) + \gamma(S) \tag{19.63}$$

Remark In general, convolution is not defined for arbitrary pairs of tempered distributions. An easy way to see this is to note that, in general, one cannot take the products of distributions (see [328, Section 4.2]) and Fourier transforms take convolution to products. But the Fourier transform of a distribution of compact support is a polynomially bounded function, and one can define the products of such functions as a distribution. This allows one to define the convolution of distributions of compact support.

Proof It is fairly easy to see that $\gamma(T * S) \subseteq \gamma(T) + \gamma(S)$. Since this is an aside on an aside, we only sketch the proof of the other implication:

(1) We extend the Titchmarsh convolution theorem to distributions of compact support. Note first that $\iota(\frac{dT}{dx}) = \iota(T)$ since $\frac{dT}{dx} = 0$ on $[0,a]$ would imply that T is a constant on $(-\infty, a)$ and that constant would have to be zero if supp(T) is bounded below. If T is supported on $[0,\infty)$, one can write $T = \frac{d^k f}{dx^k}$ for

some k and some continuous function f on $[0,\infty)$. Then $V^k T = f$. Similarly, $V^\ell S = g$. If $\mathrm{supp}(f) = [a,\infty)$, then its derivative including T is supported there. Thus $\iota(T) = \iota(S) = 0 \Rightarrow \iota(f) = \iota(g) = 0$. By the Titchmarsh theorem, $\iota(f * g) = 0$. This implies that $\iota(T * S) = \iota(\frac{d^{k+\ell}(f*g)}{dx^{k+\ell}}) = 0$.

(2) It is a basic geometric fact that if K is a compact subset in $\mathbb{R}^\nu$, then it is the intersection of all closed half spaces containing K and with a point of K in the boundary of that half space. To see this pick $x \notin K$ and pick $y \in K$ minimizing $\mathrm{dist}(x, y)$. The closed half space perpendicular to $x - y$ containing y in its boundary but not containing x is a half space of the required type proving the claim. Using this and coordinate change, and that $\gamma(T * S) \subseteq \gamma(T) + \gamma(S)$, it suffices to prove if $\mathrm{supp}(T), \mathrm{supp}(S) \subset \mathbb{R}^\nu_+ \equiv \{\boldsymbol{x} \mid,\ x_1 \geq 0\}$ with $0 \in \mathrm{supp}(T) \cap \mathrm{supp}(S)$, then $\mathrm{supp}(T) \cap \mathbb{R}^\nu_+ \neq \emptyset$.

(3) Now, we consider the partial Fourier transform

$$\mathcal{P}T(x_1, k_\perp) = \int_{x_\perp} T(x_1, x_\perp) \exp(ix_\perp \cdot k_\perp)\, d^{\nu-1}x_\perp$$

defined on functions initially and then extended to distributions. If T is a distribution of compact support, $k_\perp \mapsto \mathcal{P}T(x, k_\perp)$ is a distribution valued analytic map. By this analyticity, if $0 \in \mathrm{supp}(T), 0 \in \mathrm{supp}(S)$, then for a.e. $k_\perp$, $0 \in \mathrm{supp}_{x_1}(\mathcal{P}T(x_1, k_\perp)), 0 \in \mathrm{supp}_{x_1}(\mathcal{P}S(x_1, k_\perp))$. Moreover for $k_\perp$ fixed $\mathcal{P}(T * S)(\cdot, k_\perp)$ is the one-dimensional convolution of $\mathcal{P}T(\cdot, k_\perp)$ and $\mathcal{P}S(\cdot, k_\perp)$. By step (1), $0 \in \mathrm{supp}_{x_1}(\mathcal{P}(T * S)(x_1, k_\perp))$ for a.e. $k_\perp$ which implies that $\mathrm{supp}(T * S) \cap R^\nu_+ \neq \emptyset$. By step (2), this proves the result.

□

Notes and Historical Remarks Central to this chapter are two papers of Sarason. First, the seminal [294], which has Theorem 19.5 and its application to prove Pick's theorem. Second, [293], which deduced the Agmon unicellularity theorem from Beurling's theorem.

For much more on classical $H^p(\mathbb{D})$ spaces, see the books of Duren [86] and Koosis [185] or the chapters in Garnett [119], Katznelson [179], Nikolski [244], Rudin [290], or Simon [328, Chapter 5]. For $H^p(\mathbb{C}_+)$, see Koosis [185] or Simon [328, Section 5.9].

Hankel matrices originally arose in the study of the Stieltjes moment problem. For books on the subject, see Nikolski [244] and Peller [266]. Toeplitz matrices arose in work on Carathédory functions (analytic functions on $\mathbb{D}$ with positive real part) and the closely related trigonometric problem. For books on the subject, see Böttcher–Silbermann [44] and Nikolski [244].

The terms Hankel matrix and Toeplitz matrix are universal for the one-sided infinite matrices we define. In the older literature, the phrases Hankel operator and Toeplitz operator are used for the operators on $\ell^2(\mathbb{Z}_+)$ defined by the matrices using the canonical basis for ℓ^2. Many modern approaches talk about the operators, T_f and H_f associated to $f \in L^\infty$, its symbol, as a representation of the Toeplitz and

Hankel operators. I prefer to reserve the phrases Toeplitz and Hankel operators for the maps from H^2 to itself and from H^2 to H^2_- and use the names Hankel and Toeplitz matrices for the operators on $\ell^2(\mathbb{Z}_+)$. One does need to bear in mind that a Hankel matrix defines an operator from a space to itself but we view H_f as a map from one space to another. There is of course a natural map of H^2_- to H^2 by taking e_{-n-1} to e_n for $n = 0, 1, \ldots$ and by combining H_f with this map, one gets a map from H^2 to itself. The advantage of using these maps instead of H_f is that one can take products. This is essential for beautiful formulae of Widom and Borodin–Okounkov as discussed, for example, in [323, Section 6.2].

We also note that while we think of H_f as a map of H^2 to H^2_-, the definition as $P_- M_f P_+$ would just as well allow us to think of it as a map of L^2 to itself which of course has the same norm as it's restriction to a map from H^2 to H^2_-.

Nehari's theorem is from Nehari [236]. There are two classes of proofs of this result. One used by Nehari and by this book relies on the Hahn–Banach theorem. Another type adds Fourier coefficients one at a time—an idea going back to Adayam, Arov, and Krein [1] who seem to have found their result without knowing of Nehari which they mention only in a note added in proof. Of course, the usual proof of the Hahn–Banach theorem is doing one dimension at a time (see [325, Section 5.5]) so these approaches aren't totally unrelated. One advantage of the AAK approach is that it allows a description of all solutions of the Nehari problem.

Beurling's theorem is from Beurling [30] with important extensions of Lax [197] to the vector-valued case. The idea of using orthogonal complements to find the inner function is due to Halmos [129] whose paper opened up new vista in the general study of shift operators on Hilbert spaces. Our proof here is a variant of Halmos.

The idea of using Nehari's theorem to prove Sarason's theorem is due to Douglas Clark (in unpublished Lecture Notes) and appeared first in Nikolski [243] who found it independently. Sarason [297] has a history of his discovery of his commutant lifting theorem and he (in [295, 296] and Shapiro [314] put the theorem in historical context.

Commutant lifting theorems have been extended to a much more general context. T_z on $H^2(\mathbb{D})$ is an isometry, but not unitary. It is however the restriction of the unitary operator M_z on $L^2(\partial\mathbb{D})$ and one can prove it is the unique such restriction (up to unitary equivalence) if one demands the unitary extension, $\widetilde{M}_z$, has $\bigcup_{n=0}^{\infty} \widetilde{M}_z^{-n}[H^2]$ dense in the extended space. More generally, Sz.-Nagy [339] proved in 1953 that any contraction has a *minimal dilation*, i.e. if $A : \mathcal{H} \to \mathcal{H}$ is a contraction on a Hilbert space, $\mathcal{H}$, there is a unique (up to unitary equivalence) unitary operator, $B : \mathcal{K} \to \mathcal{K}$ where $\mathcal{H} \subset \mathcal{K}$ and if P is the orthogonal projection from $\mathcal{K}$ to $\mathcal{H}$, then for all integers $n \geq 0$, $A^n = PB^nP$ and so that $\bigcup_{n=0}^{\infty} B^{-n}[\mathcal{H}]$ is dense $\mathcal{K}$. Ando [13] extended this to pairs of commuting contractions.

Motivated by Sarason, AAK and Ando, Sz.-Nagy–Foiaş [340] proved under certain conditions on A, one could also lift C's commuting with A to the minimal dilation space without increasing the norm. This commutant lifting theorem has become part of a huge industry, see [152, 244, 245, 258, 341]. In particular,

Foiaş–Frazho have a book [101] on using commutant lifting to prove interpolation theorems.

We will say more about the Rosenblum–Rovnyak work in the Notes to Chapter 33. We note here that Theorem 19.8 is from [288, 289].

The remainder of these Notes deals with issues connected with the Titchmarsh convolution theorem (Theorem 19.9) and Agmon unicellularity theorem (Theorem 19.10).

In 1938, Gel'fand [120], at the time a not well-known student, raised the question of for which $f \in L^2([0,1])$, the linear combinations of $\{V^n f\}_{n=0}^\infty$ were dense. This was first settled by Agmon [4] in 1949 who used complex variable techniques to prove that this was true if and only if $\iota(f) = 0$. We use the name Agmon unicellularity theorem after this work since his result is easily seen to be equivalent to unicellularity as follows:

Unicellularity ⇒ Cyclicity Let M be the closure of the span of $\{V^n f\}_{n=0}^\infty$. M is clearly invariant for V, so by unicellularity, $M = L^2([a,1])$ for some $a \geq 0$. Since $f \notin L^2([a,1])$ if $a \neq 0$, we must have that $M = L^2([0,1])$. □

Cyclicity ⇒ Unicellularity It is easy to see by translation and scaling that the result on density when $\iota(f) = 0$ shows that the closed span of $\{V^n f\}_{n=0}^\infty$ is $L^2([\iota(f),1])$. If M is an invariant subspace for V, this shows that M contains $\overline{\bigcup_{f\in M} L^2([\iota(f),1])}$ which is easily seen to be $L^2([a,1])$ with $a = \inf_{f\in M}\iota(f)$. But this space clearly contains M and so is M. □

We proved that Agmon Unicellularity ⇒ Titchmarsh Convolution in a few lines; one can go in the other direction so the two results are "equivalent":

Proof That Titchmarsh Convolution ⇒ Agmon Unicellularity We begin by noting that, by a simple translation argument, the Titchmarsh convolution theorem shows that $\iota(f*g) = \min[\iota(f)+\iota(g),1]$ so $[\iota(f) = 0\,\&\, f*g = 0] \Rightarrow g = 0$. Now suppose $\iota(f) = 0$. As we've seen, $V^n f = g_{n-1} * f$ where $g_m = x^m/m!$ (Proposition 19.13(i)) so $\bar{h} \perp \{V^n f\}_{n=1}^\infty$ implies that $\langle h, p*f\rangle = 0$ for all polynomials p. Define $\tilde{q}(x) = q(1-x)$. Then any easy calculation shows that $\langle h, p*f\rangle = \langle \bar{\tilde{p}}, \bar{\tilde{h}} * f\rangle$ for all $p, h \in L^1 \cap L^\infty$, $f \in L^2$. Thus $\bar{h} \perp \{V^n f\}_{n=1}^\infty \Rightarrow \bar{\tilde{h}} * f = 0$. By the consequence of the Titchmarsh convolution theorem above, $\bar{\tilde{h}} = 0$ so $h = 0$. Thus the span of $\{V^n f\}_{n=1}^\infty$ is dense. □

The early proofs of Titchmarsh's theorem use complex variables—namely Titchmarsh's original proof [343] and of Crum [65] and Dufresnoy [85]. Real variable proofs are due to Mikusinski [221–224], Yosida [368], and Doss [84]. Doss' proof is elementary in that it only uses Fubini's theorem and the Parseval formula but none of the proofs are straight-forward.

There are many equivalent versions of Titchmarsh's theorem. It holds for L^p spaces and for distributions. As we've seen, it implies and in turn is equivalent to $[\iota(f) = 0\,\&\, f*g = 0] \Rightarrow g = 0$. It is also equivalent to $\iota(f*g) = \min[\iota(f)+\iota(g),1]$. Lions proved his extension to higher dimensions in [206–208].

The unicellularity question goes back, as we mentioned, to Gel'fand's question. There were several papers in the 1950s that seemed unaware of Agmon's work. In 1957, Donoghue [77] noted that Titchmarsh's theorem implies that the only invariant subspace for V is $L^2([a, 1])$ (see above) and in that same year, Brodski [47] and Sakhnovich [292] used results in the theory of non-self-adjoint operators to identify invariant subspaces of the Volterra operator. That unicellularity implies Titchmarsh's theorem was noted by Kalisch [173].

We conclude with a few additional observations about the Volterra operator. It is easy to check that $V^* f(x) = \int_x^1 f(y)\, dy$ by thinking of V as an integral operator with kernel which is 1 (resp. 0) below (resp. above) the diagonal and looking at the transpose kernel. Thus $(V + V^*) f(x) = \int_0^1 f(y) dy$ for all x. Therefore $\langle f, (V + V^*) f \rangle = \left[\int_0^1 f(y) dy \right]^2$ so V is a dissipative operator which explains why its Cayley transform is a contraction. For more along these lines see Nikolski [245].

Chapter 20
A Proof of Loewner's Theorem as a Degenerate Limit of Pick's Theorem

In this chapter, we will give our first proof of the hard part of Loewner's theorem. Despite the fact that the approach is very natural, this proof seems to be new (although, as I will explain in the Notes, it can be regarded as a kind of variant of Korányi's proof of Loewner's theorem and also Loewner's). Since L is a degenerate limit of the Pick matrix, it is natural to try to move points and values on $\mathbb{R}$ into $\mathbb{C}_+$ in such a way that the Pick matrix is positive, and take a limit of the resulting Pick functions as the points and values are moved back to $\mathbb{R}$. Explicitly, given f on $[-1, 1]$, we will consider the $2 \cdot 2^n + 1$ points $j/2^n$, $j = -2^n, \ldots, 0, \ldots, 2^n$ and find a Pick function with

$$f_n\left(\frac{j}{2^n} + \varepsilon_n \frac{i}{2^n}\right) = f\left(\frac{j}{2^n}\right) + \frac{i\varepsilon_n}{2^n} f'\left(\frac{j}{2^n}\right) \tag{20.1}$$

To do this, we will need to prove that the Pick matrix for f_n is positive for ε_n sufficiently small. This will be true only because we will be able to assume that the Loewner matrix is strictly positive. Thus, we will need the following technical fact, which we will use many times later in this monograph:

Theorem 20.1 *Let* $\widetilde{\mathcal{M}}_\infty(-1, 1)$ *be the functions* $f \in \mathcal{M}_\infty(-1, 1)$ *for which all* $L_n(x_1, \ldots, x_n; f)$ *are strictly positive and which are* C^∞ *in a neighborhood of* $[-1, 1]$. *Then to prove for all* $f \in \mathcal{M}_\infty(-1, 1)$,

$$f(x) = f(0) + \int_{-1}^{1} \frac{x}{1 + \lambda x}\, d\nu(\lambda) \tag{20.2}$$

holds for a measure ν *on* $[-1, 1]$, *it suffices to prove this for all* $f \in \widetilde{\mathcal{M}}_\infty(-1, 1)$.

Proof Let $g_0(x) = \log(\frac{3-x}{2-x})$ as in (5.74). Given $f \in \mathcal{M}_\infty(-1, 1)$, let

$$f_{\varepsilon,\delta}(x) = f(x(1 - \delta)) + \varepsilon g_0(x) \tag{20.3}$$

B. Simon, *Loewner's Theorem on Monotone Matrix Functions*, Grundlehren der mathematischen Wissenschaften 354, https://doi.org/10.1007/978-3-030-22422-6_20

which lies in $\widetilde{\mathcal{M}}_\infty(-1,1)$ since an argument similar to Lemma 5.17 shows $L_n(x_1,\dots,x_n;g_0) > 0$, so $L_n(x_1,\dots,x_n;f+\varepsilon g_0) = L_n(x_1,\dots,x_n;f) + \varepsilon L_n(x_1,\dots,x_n;g_0) > 0$. Thus, by hypothesis,

$$f_{\varepsilon,\delta}(x) = f_{\varepsilon,\delta}(0) + \int \frac{x}{1+\lambda x}\, d\nu_{\varepsilon,\delta}(\lambda) \tag{20.4}$$

Since

$$\int d\nu_{\varepsilon,\delta}(x) = f'_{\varepsilon,\delta}(0) = (1-\delta) f'(0) + \varepsilon g_0'(0) \tag{20.5}$$

we see, by (20.4) that $d\nu_{\varepsilon,\delta}$ has a weak-* limit $d\nu$ for which (20.2) holds. □

Henceforth, we will suppose $f \in \widetilde{\mathcal{M}}_\infty(-1,1)$. We begin by showing Pick's problem can be solved for ε_n small:

Lemma 20.2 *For $m = 2\cdot 2^n + 1$, let $\{z_\ell\}_{\ell=1}^m$ be the points $\frac{j}{2^n} + \frac{\varepsilon_n i}{2^n}$ where $i = \sqrt{-1}$ and $j = \ell - 1 - 2^n$. Let w_m be given by the right side of (20.1). Then for all $\varepsilon_n > 0$ sufficiently small, the Pick matrix associated to these data is strictly positive.*

Proof We will label points by $j = -2^n, -2^n+1, \dots, 2^n$ rather than ℓ. The Pick matrix has the form

$$M_{jk} = \begin{cases} f'\left(\frac{j}{2^n}\right), & j = k \\ \dfrac{f(\frac{j}{2^n}) - f(\frac{k}{2^n}) + i\varepsilon_n[f'(\frac{j}{2^n}) + f'(\frac{k}{2^n})]}{\frac{j-k}{2^n} + \frac{2i\varepsilon_n}{2^n}}, & j \neq k \end{cases} \tag{20.6}$$

M is Hermitian, and as $\varepsilon_n \downarrow 0$, M approaches the Loewner matrix for $f\colon L_m(\{\frac{j}{2^n}\}; f)$. Since L is assumed strictly positive, M is strictly positive for ε small. □

Henceforth, we pick, once and for all, ε_n so

$$0 \le \varepsilon_n \le 1 \tag{20.7}$$

and so that the Pick matrix of Lemma 20.2 is strictly positive. Thus, by Theorem 17.2, there exist measures $d\mu_n$ on $\mathbb{R}$ (indeed, point measures supported at $m = 2\cdot 2^n + 1$ points) so that

$$f_n(z) = \int \frac{d\mu_n(x)}{x - z} \tag{20.8}$$

obeys

$$f_n\left(\frac{j}{2^n} + \varepsilon_n \frac{i}{2^n}\right) = f\left(\frac{j}{2^n}\right) + i\frac{\varepsilon_n}{2^n} f'\left(\frac{j}{2^n}\right) \tag{20.9}$$

A common element of many proofs of Loewner's theorem is that we get a measure like (20.8) where to shift to $[-1, 1]$ using the construction in Theorem 3.12 and take limits, we need to show that $d\mu_n([-1, 1])$ is small. In some proofs (but not this one!), one shows it is zero. So the following is important:

Proposition 20.3

(i) *We have that, for all* $j = -2^n, \ldots, 2^n$

$$\int \frac{d\mu_n(x)}{|x - z_j|^2} \leq \|f'\|_\infty \tag{20.10}$$

where

$$z_j = \frac{j}{2^n} + \frac{i\varepsilon_n}{2^n} \tag{20.11}$$

(ii)

$$\mu_n\left(\left[\frac{j}{2^n} - \frac{1}{2^{n+1}}, \frac{j}{2^n} + \frac{1}{2^{n+1}}\right]\right) \leq \frac{5\|f'\|_\infty}{4} 2^{-2n} \tag{20.12}$$

(iii)

$$\mu_n([-1, 1]) \leq \tfrac{5}{4} \|f'\|_\infty 2^{-n}[2 + 2^{-n}] \tag{20.13}$$

Proof

(i) By (20.8),

$$\operatorname{Im} f_n(z) = \operatorname{Im} z \int \frac{d\mu_n}{|x - z|^2} \tag{20.14}$$

By (20.9),

$$\frac{\operatorname{Im} f_n(z_j)}{\operatorname{Im} z_j} = f'\left(\frac{j}{2^n}\right) \tag{20.15}$$

(20.14) and (20.15) imply (20.10).

(ii) If $x \in I_j^{(n)} \equiv [\frac{j}{2^n} - \frac{1}{2^{n+1}}, \frac{j}{2^n} + \frac{1}{2^{n+1}}]$, then

$$|x - z_j|^2 \leq \left[\frac{1}{2^{n+1}}\right]^2 + \frac{|\varepsilon_n|^2}{(2^n)^2} \leq \frac{5}{4} 2^{-2n} \tag{20.16}$$

by (20.7). Thus, on $I_j^{(n)}$, $|x - z_j|^{-2} \geq \frac{4}{5} 2^{2n}$, so

$$\mu_n(I_j^{(n)}) \leq \frac{5}{4} 2^{-2n} \int \frac{d\mu_n(x)}{|x - z_j|^2} \tag{20.17}$$

so (20.10) implies (20.12).

(iii) Since $[-1, 1] \subset \cup_{j=-2^n}^{2^n} I_j^{(n)}$ and $(2 \cdot 2^n + 1)(2^{-2n}) = 2^{-n}(2 + 2^{-n})$, (20.12) implies (20.13).

□

Now define a measure $\tilde{\mu}_n$ by

$$\tilde{\mu}_n(A) = \mu_n(A \backslash [-1, 1]) \tag{20.18}$$

and

$$\tilde{f}_n(z) = \int \frac{d\tilde{\mu}_n(x)}{x - z} \tag{20.19}$$

Proposition 20.4 *If z_j is given by* (20.11), *then*

$$|f_n(z_j) - \tilde{f}_n(z_j)| \leq \|f'\|_\infty [\tfrac{5}{4} 2^{-n}(2 + 2^{-n})]^{1/2} \tag{20.20}$$

and

$$\operatorname{Im} \tilde{f}_n\left(\frac{i\varepsilon_n}{2^n}\right) \leq f'(0) \frac{\varepsilon_n}{2^n} \tag{20.21}$$

Proof Clearly,

$$\begin{aligned} |f_n(z_j) - \tilde{f}_n(z_j)| &\leq \int_{[-1,1]} \frac{d\mu_n}{|x - z_j|} \\ &\leq \left(\int \frac{d\mu_n}{|x - z_j|^2}\right)^{1/2} \mu_n([-1, 1])^{1/2} \end{aligned} \tag{20.22}$$

by the Schwarz inequality. Equation (20.20) follows immediately from (20.10) and (20.13).

Since $d\tilde{\mu}_n \leq d\mu_n$ and $\operatorname{Im} \int \frac{d\eta(x)}{x-z} = \operatorname{Im} z \int |x - z|^{-2} \, d\eta$, we see

$$\operatorname{Im} \tilde{f}_n(z) \leq \operatorname{Im} f_n(z)$$

for all $z \in \mathbb{C}_+$, so (20.21) follows from (20.9). □

Given Theorem 20.1, the following proves the hard part of Loewner's theorem:

Theorem 20.5 *Let $f \in \widetilde{\mathfrak{M}}_\infty(-1, 1)$. Then there is a measure $d\nu$ on $[-1, 1]$ so that for all $x \in (-1, 1)$,*

$$f(x) = f(0) + \int \frac{x}{1+\lambda x}\, d\nu(\lambda) \tag{20.23}$$

Proof $\tilde{f}_n(z)$ is a Herglotz function with an analytic continuation to $(\mathbb{C}\backslash\mathbb{R}) \cup (-1, 1)$ since $\tilde{\mu}_n((-1, 1)) = 0$. Thus, by Theorem 3.8,

$$\tilde{f}_n(z) = \tilde{f}_n(0) + \int \frac{z}{1+\lambda z}\, d\nu_n(\lambda) \tag{20.24}$$

for some measure $d\nu_n$ on $[-1, 1]$ and all $z \in (\mathbb{C}\backslash\mathbb{R}) \cup (-1, 1)$. In particular, since

$$\operatorname{Im}\left(\frac{z}{1+\lambda z}\right) = \frac{\operatorname{Im} z}{|1+\lambda z|^2}$$

we see

$$\operatorname{Im} \tilde{f}_n\left(\frac{i\varepsilon_n}{2^n}\right) = \frac{\varepsilon_n}{2^n} \int \frac{d\nu_n(\lambda)}{|1+\frac{i\lambda\varepsilon_n}{2^n}|^2} \tag{20.25}$$

Since $\varepsilon_n < 1$, for $\lambda \in [-1, 1]$, $|1+i\lambda\varepsilon_n/2^n|^2 \leq 4/5$ for all n, so (20.25) and (20.21) imply

$$\int d\nu_n(\lambda) \leq \tfrac{4}{5} f'(0) \tag{20.26}$$

so the $d\nu_n$'s are uniformly bounded.

We compute

$$\begin{aligned}
\left|\frac{x+i\varepsilon}{1+\lambda(x+i\varepsilon)} - \frac{x}{1+\lambda x}\right| &\leq \left|\int_0^\varepsilon \frac{d}{dy} \frac{x+iy}{1+\lambda(x+iy)}\, dy\right| \\
&\leq \int_0^\varepsilon \frac{1}{1+\lambda(x+iy)^2} \\
&\leq \frac{\varepsilon}{(1-\lambda|x|)^2}
\end{aligned} \tag{20.27}$$

so, by (20.24) we see that

$$|\tilde{f}_n(x) - \tilde{f}_n(x+i\varepsilon)| \leq \frac{4\varepsilon f'(0)}{(1-|x|)^2} \tag{20.28}$$

Thus, for any $\delta > 0$:

$$\lim_{n\to\infty} \sup_{|j|\le(1-\delta)2^n} \left| \tilde{f}_n\left(\frac{j}{2^n}\right) - \tilde{f}\left(z_j\right) \right| = 0 \tag{20.29}$$

By (20.9) and (20.20), we have that

$$\left| \tilde{f}_n(z_j) - f\left(\frac{j}{2^n}\right) \right| \le \|f'\|_\infty \left[\tfrac{5}{4} 2^{-n}(2+2^{-n})\right]^{1/2} + 2^{-n}\|f'\|_\infty \tag{20.30}$$

and thus

$$\lim_{n\to\infty} \sup_{|j|\le(1-\delta)2^n} \left| \tilde{f}_n(z_j) - f\left(\frac{j}{2^n}\right) \right| = 0 \tag{20.31}$$

Let $\mathfrak{D}$ be the set of dyadic rationals in $(-1, 1)$. Combining (20.29) and (20.31), we conclude that

$$\lim_{n\to\infty} \sup_{y\in\mathfrak{D}\ |y|<1-\delta} \left| \tilde{f}_n(y) - f(y) \right| = 0 \tag{20.32}$$

In particular, $\int g(\lambda)\, d\nu_n(\lambda)$ converges for any $g(\lambda) = y(1+\lambda y)^{-1}$ with $y \in \mathfrak{D}$. By Lemma 1.4, these are total, so by (20.27) $d\nu_n$ has a weak-*–limit, $d\nu$, and (20.23) holds for $x = j/2^n$, and so for all x. □

Notes and Historical Remarks Given Pick's theorem, the proof is exceedingly natural, but it does not seem to have been noted before. However, if you note the Sz.-Nagy–Korányi proof of Pick's theorem and Korányi's estimates of the contribution of his measure to $[-1, 1]$ (close to the estimates in Proposition 20.3), you see that our proof is closely related to Korányi's work [186]. The closeness is especially manifest in our presentation of Korányi's proof: where he does things directly for all $x \in (-1, 1)$ and then estimates the measure, our version of his proof uses finite approximations and our version of his estimates is closer to Proposition 20.3 than his.

While we used Theorem 17.2, we never used the finiteness of the support of $d\mu_n$. Moreover, while it was notationally useful to have $A = 0$, an Az term can easily be accommodated. Thus, we could have used any proof of Pick's theorem, which somewhat decouples this argument from Korányi's proof.

There is also a relation of this proof to Loewner's proof. By refined versions of Pick's theorem, the approximations f_n are rational functions and we get Herglotz approximations by rational functions from positivity of Loewner matrices as does Loewner. But we can use any Herglotz function, not just the rational ones and the proof of the key fact that the approximations do not (in our case, at least in the limit) have poles in the interval is very different so the connection isn't very close.

Chapter 21
Rational Approximation and Orthogonal Polynomials

A key element of many proofs of Loewner's theorem is the approximation of f by rational Herglotz functions. This is true of the proof in the last chapter. But in those cases, the rationality of the approximation is not essential. Indeed, Korányi's original proof does not have rational approximations, and it is only our version that does—and, as noted, the proof of the last chapter works even if one takes nonrational solutions of the Pick problem. That rationality is not key is seen in part by the fact that the approximations are not exact at any points in the proof of Chapter 20.

In contrast, Loewner's proof (Chapter 25) and the Wigner–von Neumann proof (Chapters 31 and 32) depend on rational interpolation, that is, rational functions that are equal to f (or have equal derivatives) at specific points. So this is the first of seven chapters on polynomial and rational interpolation. In this chapter and the next, we discuss interpolation of more and more derivatives. We will see close connections to the Dobsch matrix.

This chapter and the next three chapters are preparation for Loewner's proof in Chapter 25. This chapter deals with some aspects of rational approximation that are not closely connected to Loewner's proof but provide important background and connection to the proofs of Bendat–Sherman (Chapter 26) and Wigner–von Neumann (Chapter 31). The last two chapters on rational approximation are the ones on continued fraction approximation (Chapters 31 and 32).

Given a formal power series, $\sum_{n=0}^{\infty} \gamma_n z^n$ about $z = 0$ (or, alternatively, a C^∞ function with $\gamma_n = f^{(n)}(0)/n!$), and integers N, M, the *Padé approximant, $f^{[N,M]}$, of type* $[N, M]$ is the rational function $X^{[N,M]}/Y^{[N,M]}$ where

$$\deg(X) \leq N \quad \text{and} \quad \deg(Y) \leq M \tag{21.1}$$

so that

$$f^{[N,M]}(z) - \sum_{j=0}^{N+M} \gamma_j z^j = O(z^{N+M+1}) \tag{21.2}$$

B. Simon, *Loewner's Theorem on Monotone Matrix Functions*, Grundlehren der mathematischen Wissenschaften 354, https://doi.org/10.1007/978-3-030-22422-6_21

If $M = 0$, this is just the Taylor approximation. For $M \neq 0$, these go beyond the Taylor approximation and can converge in cases where the Taylor series diverges; see the Notes.

The parameter count in (21.2) is right: X has $N+1$ free parameters, Y has $M+1$, but the ratio has one less, so $N+M+1$ free parameters, as does $\{\gamma_j\}_{j=0}^{N+M}$. Basically, given $\{\gamma_j\}_{j=0}^{N+M}$, (21.2) can be written as $N+M+1$ homogeneous linear equations in the $N+M+2$ coefficients of X and Y, which generically (depending on non-vanishing of certain determinants) have a straight line of solutions. So $f^{[N,M]}$ exists and is uniquely determined by (21.2) (again for generic $\{\gamma_j\}_{j=0}^{N+M}$).

We will concentrate on the cases $N = M$, $M \pm 1$, which are most relevant to our discussion here since rational Herglotz functions

$$Az + B + \sum_{j=1}^{k} \frac{\alpha_j}{x_j - z} \tag{21.3}$$

with all $\alpha_j > 0$ are, depending on A and B, rationals of degree $[k-1, k]$ ($A = B = 0$), $[k, k]$ ($A = 0$, $B \neq 0$), or $[k+1, k]$ ($A \neq 0$). It also turns out, going back to fundamental work of Jacobi, Chebyshev, Markov, and especially Stieltjes, that it is known these Padé approximants are intimately connected to the theory of orthogonal polynomials on the real line (OPRL). So we start with that subject.

Let $d\mu$ be a probability measure and consider first the function

$$f(z) = f(0) + \int \frac{z\, d\mu(\lambda)}{1 + z\lambda} \tag{21.4}$$

To make links to the more usual theory, consider the *m-function* defined by

$$m(z) = \int \frac{d\mu(\lambda)}{\lambda - z} \tag{21.5}$$

m and f are related by

$$f(z) = f(0) + m(-z^{-1}) \tag{21.6}$$

In particular, asymptotics of f near $z = 0$ are related to asymptotics of m near $z = \infty$.

If the measure $d\mu$ has moments

$$c_n = \int \lambda^n \, d\mu(\lambda) \tag{21.7}$$

and $d\mu$ has compact support, then

$$\begin{aligned} m(z) &= -z^{-1} \int \frac{d\mu(\lambda)}{1 - \lambda z^{-1}} \\ &= -z^{-1} \sum_{n=0}^{\infty} c_n z^{-n} \end{aligned} \tag{21.8}$$

where the series converges for $|z^{-1}|$ small. If $d\mu$ does not have compact support, then (21.8) does not converge but does describe an asymptotic series in the region

$$\{z \mid \varepsilon < |\arg z| < \pi - \varepsilon\} \tag{21.9}$$

as $|z| \to \infty$.

By this convergent (or asymptotic) series, we conclude that

Proposition 21.1 *If μ and μ_1 obey*

$$c_n(d\mu) = c_n(d\mu_1) \qquad 0 \le n \le N \tag{21.10}$$

then

$$m(z) = m_1(z) + O(z^{-2-N}) \tag{21.11}$$

If μ and μ_1 both have compact support, (21.11) holds as $|z| \to \infty$ uniformly in direction. If not, it holds as $|z| \to \infty$ uniformly in each region of the form (21.9).

Note that if $d\mu_1$ has finite support, then $m_1(z)$ is a rational function and (21.11) is a statement of rational approximation or even rational interpolation. Thus, we are interested in finding μ_1's where (21.10) holds for as many n as possible (say, if we fix the number of points in supp($d\mu_1$)). For $d\mu = dx \restriction [-1, 1]$, this problem was solved in 1814 by Gauss using Legendre polynomials, so the solution is sometimes called Gaussian quadrature. The names of Jacobi and/or Christoffel are sometimes added for extensions. In particular, Christoffel handled general absolutely continuous $d\mu$'s in 1877.

The key is the family of orthogonal polynomials associated to $d\mu$. A measure, $d\mu$, is called *nontrivial* if its support is not a finite set of points. If the moments are all finite (i.e., $\int |x|^n \, d\mu(x) < \infty$ for all n), then in $L^2(d\mu)$, $\{x^j\}_{j=0}^{\infty}$ are linearly independent (for if $P(x) = \sum_{j=0}^{n} \alpha_j x^j \equiv 0$ in L^2, then $P(x) = 0$ on all of support μ, which is impossible if supp($d\mu$) is infinite). Thus, by using Gram–Schmidt, we can define *monic orthogonal polynomials*, $P_n(x)$, by

$$\begin{aligned} P_n(x) &= x^n + \text{ lower order} \\ \int P_n(x) x^j \, d\mu(x) &= 0 \text{ for } j = 0, 1, \ldots, n-1 \end{aligned} \tag{21.12}$$

and the *orthonormal polynomials*,

$$p_n(x) = \frac{P_n(x)}{\|P_n\|} \tag{21.13}$$

where $\|\cdot\|$ is $(\int |\cdot|^2 \, d\mu(x))^{1/2}$, the $L^2(\mathbb{R}, d\mu)$ norm.

For later purposes, we note that (21.12) says $\langle P_n, P_n - x^n \rangle = 0$, so

$$\langle x^n, P_n \rangle = \|P_n\|^2 \tag{21.14}$$

Theorem 21.2 (Heine Formula) *We have*

$$P_n(x) = \begin{vmatrix} c_0 & c_1 & \dots & c_n \\ c_1 & c_2 & \dots & c_{n+1} \\ \vdots & \vdots & & \vdots \\ 1 & x & \dots & x^n \end{vmatrix} \Bigg/ h_n \tag{21.15}$$

where

$$h_n = \begin{vmatrix} c_0 & \dots & c_{n-1} \\ \vdots & & \vdots \\ c_{n-1} & \dots & c_{2n-2} \end{vmatrix} \tag{21.16}$$

Moreover,

$$p_n(x) = \begin{vmatrix} c_0 & c_1 & \dots & c_n \\ c_1 & c_2 & \dots & c_{n+1} \\ \vdots & \vdots & & \vdots \\ 1 & x & \dots & x^n \end{vmatrix} \Bigg/ \sqrt{h_n h_{n+1}} \tag{21.17}$$

Remarks

1. Here $|\cdot|$ is a determinant.
2. This needs that $h_n \neq 0$. In fact, the matrix $\mathcal{H}_n$ whose determinant is h_n is strictly positive since, by a variant of the calculation we have done before,

$$\langle \zeta, \mathcal{H}_n \zeta \rangle = \int \left| \sum_{j=1}^{n} \zeta_j x^{j-1} \right|^2 d\mu(x) \tag{21.18}$$

is strictly positive by nontriviality. Thus, $h_n > 0$,

Proof Let $\pi_n(x)$ be the determinant in the numerator of (21.15). Then

$$\int x^j \pi_n(x) \, d\mu(x) = \delta_{jn} h_{n+1} \qquad 0 \leq j \leq n \tag{21.19}$$

for integrating turns the last row to $c_j \dots c_{n+j}$. Thus, (21.19) is obvious since, for $0 \le j \le n-1$, two rows in the determinant are equal.

Since $\pi_n(x) = h_n x^n +$ lower order, (21.15) is immediate. By (21.19) for $j = n$ and (21.14),

$$\| P_n \|^2 = \frac{h_{n+1}}{h_n} \tag{21.20}$$

which implies (21.17). □

A key feature of the OPRL is that there is a second-order difference equation in n that they obey:

Theorem 21.3 *Given a nontrivial probability measure $d\mu$ on $\mathbb{R}$, there exist positive constants $\{a_n\}_{n=1}^{\infty}$ and real constants $\{b_n\}_{n=1}^{\infty}$ so that for $n = 0, 1, 2, \dots,$*

$$P_{n+1}(x) = (x - b_{n+1}) P_n(x) - a_n^2 P_{n-1}(x) \tag{21.21}$$

where $P_{-1} \equiv 0$ (so a_0 is not needed). Moreover, for $n \ge 1$

$$\| P_n \| = a_n \| P_{n-1} \| \tag{21.22}$$

and

$$\| P_n \| = a_n \dots a_1 \tag{21.23}$$

so

$$p_n = \frac{P_n}{a_n \dots a_1} \tag{21.24}$$

and

$$x p_n = a_{n+1} p_{n+1}(x) + b_{n+1} p_n(x) + a_n p_{n-1}(x) \tag{21.25}$$

Remarks

1. The a's and b's are called *Jacobi parameters* and the matrix

$$J = \begin{pmatrix} b_1 & a_1 & 0 & & \\ a_1 & b_2 & a_2 & \ddots & \\ 0 & a_2 & b_3 & \ddots & \ddots \\ & & \ddots & \ddots & \ddots \end{pmatrix} \tag{21.26}$$

is called the *Jacobi matrix* associated to $d\mu$.

2. Equation (21.25) says that the matrix J of (21.26) has

$$\langle p_n, xp_m \rangle = J_{nm} \tag{21.27}$$

3. Many authors start labeling a and b at $n = 0$. Others interchange the symbols a and b or use different parameters all together.

Proof Clearly, since xP_n has degree $n+1$ and $\{P_j\}_{j=0}^{n+1}$ are a basis,

$$xP_n(x) = \sum_{j=0}^{n+1} \gamma_{n,j} P_j(x) \tag{21.28}$$

Since P_n and P_{n+1} are monic, $\gamma_{n,n+1} = 1$, and we set $\gamma_{n,n} \equiv b_{n+1}$.
If $j \leq n-2$,

$$\gamma_{n,j} \|P_j\|^2 = \langle P_j, xP_n \rangle = \langle xP_j, P_n \rangle = 0 \tag{21.29}$$

since $\deg(xP_j) < n$. For $j = n-1$,

$$\begin{aligned}\gamma_{n,n-1} \|P_{n-1}\|^2 &= \langle xP_{n-1}, P_n \rangle \\ &= \langle x^n, P_n \rangle = \|P_n\|^2\end{aligned}$$

so, with a_n given by (21.22), we have $\gamma_{n,n-1} = a_n^2$. This proves (21.21).

Equation (21.23) follows from (21.22) and $\|P_0\| = 1$ ($d\mu$ is a probability measure). Equation (21.23) implies (21.24) and (21.24)/(21.21) imply (21.25). □

We can write the a_n's in terms of the c_j's—indeed, in terms of the determinants h_n of (21.16). There are more complicated formulae for the b_n which are less important to us.

Theorem 21.4 *We have that*

$$a_n^2 = \frac{h_{n+1} h_{n-1}}{h_n^2} \tag{21.30}$$

Proof By (21.22),

$$a_n^2 = \frac{\|P_n\|^2}{\|P_{n-1}\|^2}$$

which, by (21.20), implies (21.30). □

Theorem 21.5 *All the zeros of P_n lie on $\mathbb{R}$, indeed, within $[\alpha, \beta]$, where $\alpha = \inf(\text{supp}(d\mu))$ and $\beta = \sup(\text{supp}(d\mu))$. These zeros are simple. Moreover, the zeros of P_n and P_{n-1} interlace.*

Proof Suppose $x_0 > \beta$ is a zero of P_n. Then $P_n(x)^2/(x_0 - x) \geq 0$ on $[\alpha, \beta]$. So since $d\mu$ is nontrivial,

$$\int \left[\frac{P_n(x)^2}{x_0 - x}\right] d\mu(x) > 0 \tag{21.31}$$

Since $P_n(x_0) = 0$, $P_n(x)/(x_0 - x) = Q_{n-1}(x)$ is a polynomial of degree $n-1$, so orthogonal to P_n. Thus, the integral in (21.31) is zero. This contradiction proves P_n has no zeros in (β, ∞). Similarly, it has no zeros in $(-\infty, \alpha)$.

For $n \geq 1$, let S_n be the statement: S_n: P_n has all real zeros, all simple, and the zeros of P_{n-1} lie one between each zero of P_n.

S_1 says that P_1 has a real zero, which is immediate, given that it is a real polynomial.

Suppose S_n holds. Let $x_1^{(n)} < x_2^{(n)} < \cdots < x_n^{(n)}$ be the zeros of P_n. Because P_{n-1} is monic, it is positive near infinity and so, given its zeros, we have

$$(-1)^j P_{n-1}(x_{n-j}^{(n)}) > 0 \qquad j = 0, 1, \ldots, n-1 \tag{21.32}$$

By (21.21), at a zero of P_n, P_{n-1}, and P_{n+1} have opposite signs, so

$$(-1)^{j+1} P_{n+1}(x_{n-j}^{(n)}) > 0 \qquad j = 0, \ldots, n-1$$

Since P_{n+1} is positive near ∞ and $(-1)^{n+1} P_n(x)$ is positive near $-\infty$, we conclude that P_{n+1} has an odd number of zeros in each of the $n+1$ intervals $(-\infty, x_1^{(n)})$, $(x_1^{(n)}, x_2^{(n)})$, $\ldots$, $(x_n^{(n)}, \infty)$. Since there are only $n+1$ zeros, there is precisely one zero in each interval. Thus, the zeros are all real and simple, and S_{n+1} holds. □

Remark The intertwining property only depended on the fact that P_n has a positive leading coefficient and P_{n+1} and P_{n-1} have opposite signs at each zero of P_n. Thus, if Q_n are polynomials with

$$Q_{n+1}(x) = (\sigma_n x + \tau_n) Q_n(x) + \rho_n Q_{n-1}(x) \tag{21.33}$$

and

$$\sigma_n > 0 \qquad \rho_n < 0 \tag{21.34}$$

with $Q_0(x) = c > 0$, then the zeros interlace.

The *truncated Jacobi matrix* $J_{n,F}$ is the $n \times n$ matrix

$$J_{n,F} = \begin{pmatrix} b_1 & a_1 & \dots & & \\ a_1 & b_2 & \ddots & \ddots & \\ & \ddots & \ddots & a_{n-1} \\ & & a_{n-1} & b_n \end{pmatrix} \tag{21.35}$$

This matrix is critical because its resolvent will yield the rational approximation to m.

Theorem 21.6 *We have*

$$P_n(x) = \det(x\mathbf{1} - J_{n;F}) \tag{21.36}$$

Proof Let $v(z)$ be the vector in $\mathbb{C}^n$ with

$$v_j(z) = p_{j-1}(z) \qquad j = 1, 2, \dots, n \tag{21.37}$$

Then, by (21.25),

$$(J_{n;F} - z\mathbf{1})v(z) = -\delta_n a_n p_n(z) \tag{21.38}$$

where δ_n is the vector $(0 \dots 0\, 1)^t$.

It follows that if $p_n(z_0) = 0$, then z_0 is an eigenvalue of $J_{n;F}$. Since $J_{n;F}$ has at most n eigenvalues and p_n has exactly n zeros, $J_{n,F}$ has n distinct eigenvalues, each simple, and they are the zeros, $x_1 < \cdots < x_n$ of $p_n(x)$. Thus,

$$\begin{aligned} \det(x\mathbf{1} - J_{n;F}) &= \prod_{j=1}^{n} (x - x_j) \\ &= P_n(x) \end{aligned}$$

since P is monic and has zeros precisely at $\{x_j\}_{j=1}^n$. □

Remark One can show directly that the eigenvalues of J are simple, essentially because the eigenvectors are defined by v_1, and so make this proof self-contained. Thus, (21.36) can prove Theorem 21.5.

The proof shows that not only are the eigenvalues of $J_{n;F}$ the zeros of $P_n(x)$, but the normalized eigenvectors are given by

$$\tilde{v}_j(x) = \frac{p_{j-1}(x)}{K_{n-1}(x,x)^{1/2}} \qquad j = 1, \ldots, n \tag{21.39}$$

where

$$K_{n-1}(x,x) = \sum_{j=0}^{n-1} p_j(x)^2 \tag{21.40}$$

the diagonal *Christoffel–Darboux kernel*. We conclude:

Theorem 21.7 *The spectral measure for $J_{n;F}$ and vector $\delta_1 = (1\,0\ldots0)^t$ is*

$$d\mu_{n,F} = \sum_{j=1}^{n} \lambda_{n-1}(x_j)\delta_{x_j} \tag{21.41}$$

where $x_1 < \cdots < x_n$ are the zeros of P_n and

$$\lambda_{n-1}(x_j) = K_{n-1}(x_j, x_j)^{-1} \tag{21.42}$$

(called Christoffel *or* Cotes numbers*). Moreover,*

$$m_{n;F}(z) = \int \frac{d\mu_{n;F}(x)}{x - z} = -\frac{Q_{n-1}(z)}{P_n(z)} \tag{21.43}$$

where Q_{n-1} is the $(n-1)$st orthogonal polynomials for the Jacobi matrix with Jacobi parameters $\{a_{n+1}, b_{n+1}\}_{n=1}^{\infty}$.

Proof By the definition of spectral measure [329, Chapter 5], the eigenvalues and eigenfunctions (note $p_0(x) = 1$, so $\tilde{v}_1 = 1/K_{n-1}^{1/2}$) determine $d\mu_{n;F}$. If we note that, by Theorem 21.6, $Q_{n-1}(z) = \det(z - J_{n-1;F}^{(1)})$ with

$$J_{n-1,F}^{(1)} = \begin{pmatrix} b_2 & a_2 & \ldots & \\ a_2 & b_3 & a_3 & \ddots \\ & \ddots & \ddots & a_{n-1} \\ & & a_{n-1} & b_n \end{pmatrix} \tag{21.44}$$

then (21.43) is just Cramer's rule. $\square$

Theorem 21.8 *For* $k = 0, 1, 2, \dots, 2n - 1$,

$$\int x^k \, d\mu(x) = \int x^k \, d\mu_{n;F}(x) \tag{21.45}$$

Remark This is known as *Gauss–Jacobi quadrature.*

Proof J is the matrix of multiplication by x in the orthonormal basis $\{p_j\}_{j=0}^{\infty}$, so

$$c_k \equiv \int x^k \, d\mu = \langle \delta_1, J^k \delta_1 \rangle \tag{21.46}$$

Now $J^k \delta_1$ clearly depends only on $\{a_\ell, b_\ell\}_{\ell=1}^{k}$, so c_k, $k = 1, \dots, 2n - 2$, is a function of only $\{a_\ell, b_\ell\}_{\ell=1}^{n-1}$ (since $c_{k+\ell} = \langle J^k \delta_1, J^\ell \delta_1 \rangle$). Moreover, since $J^{n-1}\delta_1$ is an n-component vector $c_{2n-1} = \langle J^{n-1}\delta_1, J(J^{n-1}\delta_1)\rangle$ only depends on $\{a_\ell, b_\ell\}_{\ell=0}^{n-1} \cup \{b_n\}$.

By definition of $d\mu_{n;F}$,

$$\int x^k \, d\mu_{n;F} = \langle \delta_1, J_{n;F}^k \delta_1 \rangle \tag{21.47}$$

which, by the same analysis, is the same function of the a's and b's, that is, (21.45) holds. □

By Proposition 21.1, this result and (21.43) prove:

Theorem 21.9 *Let* $f(z)$ *be given by* (21.4). *Then*

$$f(z) = f(0) - \frac{Q_{n-1}(-z^{-1})}{P_n(-z^{-1})} + O(z^{2n+1}) \tag{21.48}$$

where P_n *is an orthogonal polynomial given by* (21.15) *and* Q_{n-1} *is a polynomial of degree* $n - 1$. *Equivalently,*

$$m(z) = -\frac{Q_{n-1}(z)}{P_n(z)} + O(z^{-2n-1}) \tag{21.49}$$

Finally, we want to discuss the continued fraction expansion of $m(z)$ about $z = \infty$ and its relation to Jacobi parameters and OPRL. We start with

Theorem 21.10 *For* $d\mu$ *of compact support,*

$$m(z) = \lim_{n\to\infty} -\frac{Q_{n-1}(z)}{P_n(z)} \tag{21.50}$$

uniformly on compact subsets of $\mathbb{C}\backslash[\alpha, \beta]$ *with* α, β *given by Theorem 21.5.*

Proof Since P_n has its zeros in $[\alpha, \beta]$, for all n, $d\mu_{n;F}$ is supported on $[\alpha, \beta]$. By (21.45) and the density of polynomials in $\mathbb{C}([\alpha, \beta])$, $d\mu_{n;F} \to d\mu$ weakly so $m_{n;F}(z) \to m(z)$ for each $z \in \mathbb{C}\backslash[n, \beta]$. Uniformity follows from boundedness, analyticity, and pointwise convergence. Equation (21.43) completes the proof. □

Theorem 21.11 *Let $d\mu$ have compact support. Let J_1 be the Jacobi matrix obtained by deleting the top row and leftmost column of J (i.e., J_1 has Jacobi parameters $\{a_{n+1}, b_{n+1}\}_{n=1}^{\infty}$) and let m_1 be its m-function ($\langle\delta_1, (J_1 - z)^{-1}\delta_1\rangle$). Then*

$$m(z) = \frac{1}{-z + b_1 - a_1^2 m_1(z)} \tag{21.51}$$

Proof Let J_ℓ be the Jacobi matrix with Jacobi parameters $\{a_{n+\ell}, b_{n+\ell}\}_{n=1}^{\infty}$. By expanding in minors in the top row, we see

$$\det(J_{n;F} - z) = (-z + b_1)\det(J_{1,n-1;F} - z) - a_1^2 \det(J_{2,n-2;F} - z) \tag{21.52}$$

Dividing by $\det(J_{1,n-1;F} - z)$ and using Cramer's rule, we see that

$$m_{n;F}(z)^{-1} = -z + b_1 - a_1^2 m_{1,n-1;F}(z)$$

(21.51) follows by taking $n \to \infty$ using Theorem 21.10. □

Above we saw how to go from the measure to the OPRL to the continued fractions. There is useful piece of algebra that goes in the other direction. We want to consider continued fractions of the form

$$f = \cfrac{B_1}{A_1 + \cfrac{B_2}{A_2 + \cfrac{B_3}{A_3 + \cdots}}} \tag{21.53}$$

with f_n defined by replacing B_{n+1} by zero. Continued fractions are shorthand for iterated fractional linear transformations. Define

$$\tau_j(w) = \frac{B_j}{A_j + w} \tag{21.54}$$

$$f_n(w) = \tau_1(\tau_2(\ldots\tau_n(w))\ldots) \tag{21.55}$$

so

$$f_n = f_n(0) \tag{21.56}$$

Then:

Theorem 21.12 (Euler–Wallis Formulae) *Define* $\widetilde{P}_n$, $\widetilde{Q}_n$ *inductively by* $n = 1, 2, \ldots,$

$$\widetilde{P}_n = A_n \widetilde{P}_{n-1} + B_n \widetilde{P}_{n-2} \tag{21.57}$$

$$\widetilde{Q}_n = A_n \widetilde{Q}_{n-1} + B_n \widetilde{Q}_{n-2} \tag{21.58}$$

$$\widetilde{P}_0 = \widetilde{Q}_{-1} = 1 \qquad \widetilde{P}_{-1} = \widetilde{Q}_0 = 0 \tag{21.59}$$

Then

$$f_n(w) = \frac{\widetilde{Q}_n + w\widetilde{Q}_{n-1}}{\widetilde{P}_n + w\widetilde{P}_{n-1}} \qquad f_n = \frac{\widetilde{Q}_n}{\widetilde{P}_n} \tag{21.60}$$

Proof Let W_j be the matrix

$$W_j = \begin{pmatrix} 0 & B_j \\ 1 & A_j \end{pmatrix} \tag{21.61}$$

so

$$W_j \begin{pmatrix} w \\ 1 \end{pmatrix} = \begin{pmatrix} B_j \\ w + A_j \end{pmatrix} = (w + A_j) \begin{pmatrix} \tau_j(w) \\ 1 \end{pmatrix} \tag{21.62}$$

Then composition of τ's is matrix multiplication, that is, if

$$T_n = W_1 \ldots W_n \tag{21.63}$$

then

$$T_n \begin{pmatrix} w \\ 1 \end{pmatrix} = s_n(w) \begin{pmatrix} f_n(w) \\ 1 \end{pmatrix} \tag{21.64}$$

The form of s_n doesn't matter since it drops out of ratios but it is in fact $\prod_{j=1}^{n}(w_j + A_j)$ where $w_n = w$ and $w_{j-1} = \tau_j(w_j)$. It is a straightforward induction to prove that

$$T_n = \begin{pmatrix} \widetilde{Q}_{n-1} & \widetilde{Q}_n \\ \widetilde{P}_{n-1} & \widetilde{P}_n \end{pmatrix} \tag{21.65}$$

where $\widetilde{P}_n$, $\widetilde{Q}_n$ obey (21.57)/(21.58)/(21.59).

The formula for $f_n(w)$ in (21.60) is thus (21.64), and we get the formula for f_n by setting $w = 0$. □

Example 21.13 The continued fraction obtained by iterating (21.51) has the form (21.53) if we pick

$$A_n = -z + b_n \qquad B_n = -a_{n-1}^2 \tag{21.66}$$

where $B_1 = 1$. The Euler–Wallis formula for $\widetilde{P}_n$ is thus

$$\widetilde{P}_{n+1} = (-z + b_{n+1})\widetilde{P}_n - a_n^2 \widetilde{P}_{n-1} \tag{21.67}$$

which shows that

$$\widetilde{P}_n(z) = (-1)^n P_n(z)$$

with P_n the OPRL. We get a similar formula for $\widetilde{Q}$ and (21.60) for f_n is just (21.43). □

This is the ideal language for one-point rational approximation when there is positivity, and we will return to it in Chapter 31. Since there are algebraic expressions in the c_j's behind this, the formulae apply also if the determinants in h_n are all non-zero. In terms of the Taylor coefficients of f's, these determinants are of the form

$$\Delta_n(f) = \det\left(\left[\frac{f^{(i+j-1)}(0)}{(i+j-1)!}\right]_{1\le i,j\le n}\right) \tag{21.68}$$

that is, the determinants of the Dobsch matrix (see (5.38)).

Notes and Historical Remarks The history of OPRL and continued fractions are intertwined, with continued fractions going back to the Middle Ages with important contributions by Wallis and Euler in the seventeenth and eighteenth centuries. Indeed, Theorem 21.12 goes back to that period. For a history of continued fractions, see Brezinski [46]. For more on the theory of continued fractions, see Simon [326, Section 7.5].

Heine's formula appeared in Heine [150]; it has been rediscovered often. It is a special case of explicit determinantal formulae that solve the Gram–Schmidt problem in terms of inner products.

Jacobi realized that the resolvent of a finite Jacobi matrix had a continued fraction generated by (21.51). Such continued fractions are often called J-functions; see, for example, Wall [359].

Orthogonal polynomials started with the concrete examples of Legendre and Jacobi. The general theory was developed by Chebyshev, Markov, and Christoffel. A watershed paper was that of Stieltjes [335] which discussed the moment problem in the context of OPRL and continued fractions. In particular, Stieltjes showed that in certain cases the rational approximations have a much larger region of convergence than the Taylor series (see [329, Section 7.7]).

Chapter 22
Divided Differences and Polynomial Approximation

In this chapter, as warmup for rational approximation, we discuss polynomial approximation. Since rational approximation has $f = P/Q \Leftrightarrow Qf = P$, finding P and Q can be viewed as a kind of polynomial approximation. We want to begin by reminding the reader of the well-known fact that a polynomial, P, of degree n is determined uniquely by its values at any $n + 1$ distinct points, $x_1, \dots, x_{n+1}$. As a preliminary, we define the Vandermonde determinant.

$$V_{n+1}(x_1, \dots, x_{n+1}) = \det(x_i^{n+1-j})_{i,j=1,\dots,n+1} \tag{22.1}$$

that is

$$\begin{vmatrix} x_1^n & x_1^{n-1} & \dots & x_1 & 1 \\ x_2^n & x_2^{n-1} & \dots & x_2 & 1 \\ x_3^n & x_3^{n-1} & \dots & x_3 & 1 \\ \vdots & \vdots & \vdots & \ddots & \vdots \\ x_{n+1}^n & x_{n+1}^{n-1} & \dots & x_{n+1} & 1 \end{vmatrix}$$

Replacing x_1 by x, we get a polynomial of degree n in x whose highest order term is $x^n V_n(x_2, \dots, x_{n+1})$. By the antisymmetry of determinants, this polynomial has zeros when x is $x_2, \dots, x_{n+1}$ so

$$V_{n+1}(x_1, \dots, x_{n+1}) = V_n(x_2, \dots, x_{n+1}) \prod_{j=2}^{n+1} (x_1 - x_j)$$

Since $V_2(x_1, x_2) = x_1 - x_2$, by induction we get

B. Simon, *Loewner's Theorem on Monotone Matrix Functions*, Grundlehren der mathematischen Wissenschaften 354, https://doi.org/10.1007/978-3-030-22422-6_22

Proposition 22.1

$$V_{n+1}(x_1, \dots, x_{n+1}) = \prod_{1 \le i < j \le n+1} (x_i - x_j) \tag{22.2}$$

As an immediate consequence

Theorem 22.2 *Let $\{x_j\}_{j=1}^{n+1} \in \mathbb{C}^{n+1}$ be $n+1$ distinct points. Let $\boldsymbol{\alpha} \equiv \{\alpha_j\}_{j=1}^{n+1} \in \mathbb{C}^{n+1}$. Then there is a unique polynomial, P, of degree at most n so that*

$$P(x_j) = \alpha_j \qquad j = 1, \dots, n+1 \tag{22.3}$$

P is a linear function of $\boldsymbol{\alpha}$.

Proof Write

$$P(x) = \sum_{j=1}^{n+1} \rho_j x^{n+1-j} \tag{22.4}$$

This says that $V\boldsymbol{\rho} = \boldsymbol{\alpha}$ where V is the matrix with $V_{i,j} = x_i^{n+1-j}$. Since $\det(V) \neq 0$, V is invertible, so $\boldsymbol{\rho} = V^{-1}\boldsymbol{\alpha}$ is the unique solution of (22.4) and is linear in $\boldsymbol{\alpha}$. □

If instead of (22.4), we have only $x_1, x_2, \dots, x_n$ and we replace $P(x_{n+1}) = \alpha_{n+1}$ by $P'(x_n) = \alpha_{n+1}$, then the last row of V is replaced by $\widetilde{V}_{n+1,j} = (n+1-j)x_n^{n-j}$. It is easy to see that

$$\begin{aligned} \det(\widetilde{V}) &= \frac{\partial}{\partial x_{n+1}} V(x_1, \dots, x_{n+1}) \bigg|_{x_{n+1} = x_n} \\ &= \prod_{1 \le i < j \le n} (x_i - x_j) \prod_{i=1}^{n-1} (x_i - x_n) \end{aligned} \tag{22.5}$$

is still non-zero. Doing a similar calculation for higher derivatives, one sees that (22.3) can be replaced by

$$P^{(\ell)}(x_j) = \alpha_{j,\ell} \qquad j = 1, \dots, m \quad \ell = 0, \dots, L_j - 1 \tag{22.6}$$

where $P^{(\ell)}(x)$ is the ℓth derivative ($P^{(0)} \equiv P$). Here $m \le n+1$ and $\sum_{j=1}^m L_j = n+1$ so that (22.6) has $n+1$ conditions. By generalizing (22.5), these linear equations have a non-zero determinant so that (22.6) has a unique solution which is linear in $\{\alpha_{j,\ell}\}$.

$\{x^\ell\}_{\ell=0}^n$ is a basis for $\mathcal{P}_n$, the polynomials of degree at most n. The ρ_j are involved functions of $\boldsymbol{\alpha}$ so it is often convenient to pick a different basis for $\mathcal{P}_n$

and so a different set of coordinates. In some ways, the most natural basis are the *Lagrange polynomials*

$$L_j^{(n)}(x, \{x_j\}_{j=1}^{n+1}) = \frac{\prod_{k \neq j}(x - x_k)}{\prod_{k \neq j}(x_j - x_k)} \tag{22.7}$$

which has

$$L_j^{(n)}(x_k) = \delta_{jk} \tag{22.8}$$

so, as we saw in Theorem 5.14, (22.3) is solved by (*Lagrange interpolation*)

$$P(x) = \sum_{j=1}^{n+1} \alpha_j L_j^{(n)}(x) \tag{22.9}$$

Note that the $L^{(n)}$ are singular as $x_j - x_k \to 0$.

For our purposes, a useful basis comes from the *Newton polynomials*, $j = 0, \dots .n \quad (N_0 \equiv \mathbf{1})$

$$N_j(x, \{x_k\}_{k=1}^{j}) = (x - x_1)(x - x_2) \dots (x - x_j) \tag{22.10}$$

N_j is a monic polynomial of degree j. It follows by an easy induction that $\{N_j\}_{j=0}^{n}$ is a basis for $\mathcal{P}_n$ for any choice of $\{x_j\}_{j=1}^{n}$. This remains true even if some of the x_j's are equal (although for now, we'll assume them unequal).

Our main goal in this chapter is to find the coefficients of the expansion of the solution of (22.3) in a Newton polynomial basis. As a warmup consider $n = 1$ so $P(x) = a + b(x - x_1)$. Clearly, $a = P(x_1)$. Plugging, $x = x_2$, into the expansion, we see that $b = (P(x_2) - P(x_1))/(x_2 - x_1)$, the first order divided difference. What we are heading for is a proof that in general, the coefficients of the expansion are given by higher order divided differences.

Lemma 22.3

(a)

$$[x_1, x; N_j(\cdot, \{x_k\}_{k=1}^{j})] = N_{j-1}(x, \{x_k\}_{k=2}^{j}) \tag{22.11}$$

(b) *If* $\ell \leq j$

$$[x_1, \dots, x_\ell, x; N_j(\cdot, \{x_k\}_{k=1}^{j})] = N_{j-\ell}(x, \{x_k\}_{k=\ell+1}^{j}) \tag{22.12}$$

(c) *Fix* x_1. *If* P *is any polynomial of degree* d, *then* $[x_1, x; P]$ *is a polynomial in* x *of degree* $d - 1$ *and the coefficients of the leading terms are the same.*

(d) *Fix* $x_1, \dots, x_\ell$. *If* P *is any polynomial of degree* d, *and* $\ell > d + 1$, *then* $[x_1, \dots, x_\ell; P] = 0$.

Proof

(a) $N_j(x_1, \{x_k\}_{k=1}^j) = 0$ so

$$(N_j(x, \{x_k\}_{k=1}^j) - N_j(x_1, \{x_k\}_{k=1}^j))/(x - x_1) = N_{j-1}(x, \{x_k\}_{k=2}^j).$$

(b) Immediate from (a) and induction.
(c) We have that

$$x^d = (x_1 + (x - x_1))^d = \sum_{j=0}^{d} \binom{d}{j} (x - x_1)^j x_1^{d-j}$$

so

$$[x_1, x; x^d] = \sum_{j=1}^{d} \binom{d}{j} (x - x_1)^{j-1} x_1^{d-j} \tag{22.13}$$

is a polynomial in x of degree $d - 1$.
(d) By induction and (c) $[x_1, \dots, x_{d+1}; P]$ as a function of x_{d+1} is constant, so $[x_1, \dots, x_{d+2}; P] = 0$.

□

Corollary 22.4 $[x_1, x_2, \dots, x_{n+1}; x^n] = 1$

Proof By the proof of (d) above, $[x_1, x_2, \dots, x_{n+1}; x^n]$ is a constant independent of the x_j, so we need only compute when all $x_j = 0$. $[0, x; x^n] = \frac{x^n - 0}{x - 0} = x^{n-1}$ so long as $n \geq 1$. By induction $[x_1 = 0, x_2 = 0, \dots, x_n = 0, x; x^n] = 1$. □

Proposition 22.5 *Fix* $x_1, \dots, x_n$. *Then for any* $j = 1, \dots, n$ *and* $\ell = 0, \dots, n$

$$[x_1, \dots, x_j; N_\ell(\cdot, \{x_k\}_{k=1}^\ell)] = \delta_{j(\ell+1)} \tag{22.14}$$

Proof For $j \leq \ell$, this follows from (22.12) and

$$N_j(x_j, \{x_k\}_{k=j}^\ell) = 0$$

For $j = \ell + 1$, we get (22.14) from $N_0 \equiv 1$. For $j > \ell + 1$, we use (d) of the Lemma. □

Corollary 22.6 *Let* $x_1, \dots, x_n$ *be* n *(not necessarily distinct) points in* $\mathbb{C}$. *Then* $\{N_{\ell-1}(\cdot, \{x_k\}_{k=1}^{\ell-1})\}_{\ell=1}^{n+1}$ *are linearly independent functions and are a basis for* $\mathcal{P}_n$, *the polynomials of degree at most* n

Proof Since the dimensions are the same, it suffices to prove independence. So suppose that $f(x) = \sum_{\ell=1}^{n+1} \gamma_\ell N_{\ell-1}(x, \{x_k\}_{k=1}^{\ell-1})$ is the zero polynomial. By (22.14), $\gamma_j = [x_1, \ldots, x_j; f] = 0$ for $j = 1, \ldots, n$. That leaves $\gamma_{n+1} N_n(\cdot, \{x_k\}_{k=1}^n) = 0 \Rightarrow \gamma_{n+1} = 0$ since $N_n(\cdot, \{x_k\}_{k=1}^n)$ is a monic polynomial.

Alternatively, one can note that any sequence of monic polynomials of sequentially increasing degrees is a basis for $\mathcal{P}_n$. □

Lemma 22.7 *Let $x_1, \ldots, x_n$ be n (not necessarily distinct) points in $\mathbb{C}$. Let f, g be two suitably smooth functions. If*

$$[x_1, \ldots, x_j; f] = [x_1, \ldots, x_j; g]; \qquad j = 1, \ldots, n \tag{22.15}$$

then

$$f(x_j) = g(x_j); \qquad j = 1, \ldots, n \tag{22.16}$$

Remarks

1. Equation (22.16) is stated for $x_1, \ldots, x_n$ distinct in which case "suitably smooth" just means that values are given at each point. In general, if $x_1, \ldots, x_n$ are y_1 k_1 times, ... y_ℓ k_ℓ times with $\sum_{j=1}^{\ell} k_j = n$, then (22.16) is replaced by

$$f^{(m)}(y_j) = g^{(m)}(y_j); \qquad m = 0, \ldots, k_j - 1;\ j = 1, \ldots, \ell \tag{22.17}$$

and "suitably smooth" means that f and g are $C^{(k_j-1)}$ near y_j; in fact, all one needs is C^{k_j-1} germs.

2. We will use this lemma to prove Theorem 22.8 below but we note that if one has another proof of the theorem, the theorem immediately implies that lemma.

Proof Suppose all the x's are distinct. $j = 1$ is precisely the statement that $f(x_1) = g(x_1)$. Suppose we have proven the result for all sets of ℓ points and all f's and we are given an f and a set of $\ell + 1$ x's for which (22.15) holds. Let $\tilde{f}(x) = [x_1, \ldots, x_{\ell-1}; f]$, $\tilde{g}(x) = [x_1, \ldots, x_{\ell-1}; g]$. Then (22.15) for $j = \ell + 1$ says that

$$[x_\ell, x_{\ell+1}; \tilde{f}] = [x_\ell, x_{\ell+1}; \tilde{g}] \tag{22.18}$$

By (22.15) for $j = \ell$, we have that $\tilde{f}(x_\ell) = \tilde{g}(x_\ell)$ which together with (22.18) implies that $\tilde{f}(x_{\ell+1}) = \tilde{g}(x_{\ell+1})$. Thus (22.15) holds for $j = 1, \ldots, \ell$ for the set $\{x_1, \ldots, x_{\ell-1}, x_{\ell+1}\}$ so by the induction hypothesis $f(x_{\ell+1}) = g(x_{\ell+1})$. This proves the induction hypothesis for $\ell + 1$ point sets. The proof for sets with coincident sets is similar. □

Theorem 22.8 (Hermite Interpolation) *Fix n. Let f be a suitably smooth function and let $x_1, \ldots, x_{n+1}$ be $n + 1$ (not necessarily distinct) points in $\mathbb{C}$. Define the degree n polynomial*

$$P(x; x_1, \ldots, x_n; f) = P(x) \equiv \sum_{j=1}^{n+1} [x_1, \ldots, x_j; f]\, N_{j-1}(x, \{x_k\}_{k=1}^{j-1}) \tag{22.19}$$

Then

$$P(x_j) = f(x_j); \qquad j = 1, \ldots, n+1 \tag{22.20}$$

Remarks

1. For the meaning of "suitably smooth" and of (22.20) when there are coincident points, see the first remark after Lemma 22.7.
2. It may be surprising that (22.20) holds for $j = n+1$ since the N's only depend on $\{x_k\}_{k=1}^n$. But consider the case where f vanishes at x_j; $j = 1, \ldots, n$. Then $[x_1, \ldots, x_{n+1}; f] = f(x_{n+1})/\prod_{j=1}^n (x_{n+1} - x_j)$ while $N_n(x_{n+1}, \{x_j\}_{j=1}^n) = \prod_{j=1}^n (x_{n+1} - x_j)$.
3. Equation (22.19) is called *Hermite interpolation*. If f is a polynomial of degree at most n, then $f - P$ vanishes at $n+1$ points and so is identically zero. Thus (22.19) gives an expansion for f that holds for all x and is called the Hermite expansion.
4. Notice the big difference from Lagrange interpolation. That is singular as $x_j - x_k \to 0$. In distinction, (22.19) has a simple limit as some of the x's become coincident.

Proof By (22.14), we have that

$$[x_1, \ldots, x_j; P] = [x_1, \ldots, x_j; f] \tag{22.21}$$

for $j = 1, \ldots, n+1$, so by Lemma 22.7, (22.20) holds. □

Theorem 22.9 *Let $x_1, \ldots, x_n$ be n (not necessarily distinct) points in $\mathbb{R}$ with $x_1 \leq x_2 \leq \cdots \leq x_n$. Let f be a C^{n+1} function. Then there exists $\{y_j\}_{j=1}^{n-1}$ with $x_j \leq y_j \leq x_{j+1}$ so that*

$$\frac{d}{dx} P(x; x_1, \ldots, x_n; f) = P\left(x; y_1, \ldots, y_{n-1}; \frac{df}{dx}\right) \tag{22.22}$$

Remark If the x's are distinct, one can show that the y's strictly interlace the x's (i.e. $x_j < y_j < x_{j+1}$) and f need only be C^1.

Proof Let

$$Q(x) \equiv P(x; x_1, \ldots, x_n; f) \tag{22.23}$$

Then $f - Q$ vanishes at $x_1, \ldots, x_n$ so by Rolle's theorem, there exists y_j with $x_j \leq y_j \leq x_{j+1}$ so that $\frac{d}{dx}[f - Q](y_j) = 0$ (if some x's are coincident, one either needs to take suitable limits or think about the fact that if, for example, $x_{j-1} < x_j = x_{j+1} < x_{j+2}$, then $Q'(x_j) = f(x_j)$). Thus $\frac{dQ}{dx}$ is a polynomial of degree at most $n-1$ which agrees with $\frac{df}{dx}$ at $y_1, \ldots, y_{n-1}$ and so equal to the right side of (22.22) by the uniqueness of interpolating polynomials. □

Corollary 22.10 *Under the hypotheses and notation of Theorem 22.9, we have that*

$$[y_1, \ldots, y_{n-1}; f'] = (n-1)[x_1, \ldots, x_n; f] \tag{22.24}$$

Proof If Q is given by (22.23), then, by (22.19) or (5.62)

$$Q(x) = [x_1, \ldots, x_n; f]x^{n-1} + \text{ lower order} \tag{22.25}$$

Similarly,

$$\text{RHS of (22.22)} = [y_1, \ldots, y_{n-1}; f']x^{n-2} + \text{ lower order} \tag{22.26}$$

Thus (22.22) implies (22.24) by identifying x^{n-2} terms on both sides. □

Theorem 22.11 (≡ Theorem 5.15; Mean Value Theorem for Divided Differences) *Let f be a C^{n-1} function on $(a, b) \subset \mathbb{R}$. Let*

$$a < x_1 \leq x_2 \leq \cdots \leq x_n < b \tag{22.27}$$

Then there exists $w \in [x_1, x_n]$ so that

$$[x_1, \ldots, x_n; f] = \frac{f^{(n-1)}(w)}{(n-1)!} \tag{22.28}$$

Remark

1. The case $n = 2$ is exactly the usual mean value theorem for derivatives.
2. This was proven in a different way using the Genocchi–Hermite formula in Chapter 5 (Theorem 5.15).

Proof Follows by iterating (22.24). □

There is one final fact about divided differences we want to include in this chapter—we will use Hermite interpolation to prove

Theorem 22.12 (Leibniz Rule for Divided Differences) *For any two C^{n-1} functions, f and g, and any $x_1, \ldots, x_n \in \mathbb{C}$*

$$[x_1, \ldots, x_n; fg] = \sum_{j=1}^{n} [x_1, \ldots, x_j; f][x_j, \ldots, x_n; g] \tag{22.29}$$

Remarks

1. This is algebraic so it holds for any function if $x_1, \dots, x_n$ are distinct. The C^{n-1} condition is only needed for the symbols to be defined if the x's can be coincident.
2. Leibniz rule for derivatives says that

$$(fg)^{(k)}(x) = \sum_{j=1}^{k} \binom{k}{j} f^{(j)}(x) g^{(k-j)}(x) \tag{22.30}$$

(where $f^{(0)}(x) = f(x)$) so it may be surprising that (22.29) has no binomial coefficient. However we recall that $[x_1, \dots, x_{n+1}; f]|_{x_1=\dots=x_{n+1}=x} = f^{(n)}(x)/n!$. Because of these factors of $n!$, (22.29) evaluated at $n = k + 1$ and $x_1 = \dots = x_{n+1} = x$ is just (22.30).
3. There is also an analog of the chain rules; see the notes.

We need a lemma:

Lemma 22.13 *Fix g, an arbitrary function and $\{x_j\}_{j=1}^n$ distinct points in $\mathbb{C}$. Then for $k = 1, \dots, n-1$*

$$[x_1, \dots, x_n; N_k(\cdot, \{x_j\}_{j=1}^k)g] = [x_{k+1}, \dots, x_n; g] \tag{22.31}$$

Proof If $h(x) = N_1(x, x_1)g(x)$, then $h(x_1) = 0$ so

$$[x_1, x; h] = \frac{h(x)}{x - x_1} = g(x) = [x, g]$$

Taking divided differences of this equality, we see that

$$[x_1, \dots, x_{n-1}, x; h] = [x_2, \dots, x_{n-1}, x; g]$$

Setting $x = x_n$, we get (22.31) for $k = 1$. In particular, replacing g by $N_{k-1}(\cdot, \{x_j\}_{j=2}^k)g$ in this result shows the left side of (22.31) is $[x_2, \dots, x_n; N_{k-1}(\cdot, \{x_j\}_{j=2}^k)g]$. Iteration yields (22.31) for general k. □

Proof of Theorem 22.12 Let $x_1, \dots, x_n$ be distinct points. Define

$$P(x) = \sum_{j=1}^{n} [x_1, \dots, x_j; f] N_{j-1}(x, \{x_k\}_{k=1}^{j-1}) \tag{22.32}$$

Then, by Theorem 22.8, P is the unique polynomial of degree $n-1$ with $f(x_j) = P(x_j)$. Since $[x_1, \dots, x_n; fg]$ depends only on $\{fg(x_j)\}_{j=1}^n$, we have that

$$\begin{aligned}[x_1, \dots, x_n; fg] &= [x_1, \dots, x_n; Pg] \\ &= \sum_{j=1}^{n} [x_1, \dots, x_j; f][x_1, \dots, x_n; N_{j-1}(x, \{x_k\}_{k=1}^{j-1})g]\end{aligned} \tag{22.33}$$

$$= \sum_{j=1}^{n} [x_1, \dots, x_j; f][x_j, \dots, x_n; g] \tag{22.34}$$

(22.33) comes from (22.32) using the fact that $[x_1, \dots, x_n; f]$ is a number, i.e. x-independent. Equation (22.34) comes from (22.29).

For general, perhaps coincident, points, take limits from sets of distinct points. □

Notes and Historical Remarks Lagrange interpolation goes back to Lagrange [195] in 1793 and Hermite interpolation to Hermite [154] in 1878. Hermite's big contribution is the realization that the formula works for coincident points. Newton polynomials and Hermite interpolation for non-coincident points go back to Newton [242] including the idea of divided differences. Eraser [91] discusses Newton's work on interpolation in detail.

For a modern exposition of divided differences, see de Boor [72] whose exposition we follow in some places. The mean value theorem for divided differences, Theorem 22.11, goes back to Schwarz [311] in 1881 while Theorem 22.9 is due to E. Hopf [165]. There is a formula that relates divided differences to derivatives that illuminates this mean value theorem. Let f be a C^{n-1} function on (a, b). Taylor's theorem with remainder can be written for $x, c \in (a, b)$ with $x > c$ and $n \geq 2$:

$$f(x) = P_n(x) + \int_c^b \frac{(x-t)_+^{n-2}}{(n-2)!} f^{(n-1)}(t)\, dt; \qquad P_n(x) = \sum_{j=1}^{n-2} \frac{(x-c)^j}{j!} f^j(c) \tag{22.35}$$

Since P_n is a polynomial of degree at most $n-2$, by Lemma 22.3, we see that $[x_1, \dots, x_n; P_n] = 0$ so

$$[x_1, \dots, x_n; f] = \int_{-\infty}^{\infty} K_n(t; x_1, \dots, x_n) \frac{f^{(n-1)}(t)}{(n-1)!}\, dt \tag{22.36}$$

where

$$K_n(t; x_1, \dots, x_n) = [x_1, \dots, x_n, g_{t,n}(\cdot)]; \qquad g_{t,n}(x) = (n-1)(x-t)_+^{n-2} \tag{22.37}$$

Since $(x-t)_+^{n-2}$ is only C^{n-3}, we need that no x has $x_j = x$ for more than $n-2$ values of j. When the x_j are all distinct, we have that

$$K_n(t; x_1, \ldots, x_n) = \sum_{j=1}^{n} \frac{(x_j - t)_+^{n-1}}{\prod_{k \neq j}(x_k - x_j)} \tag{22.38}$$

K_n is called the Peano kernel after Peano [260, 261] (see also Sard [298]) although what we call K_n is sometimes called K_{n-1} and is sometimes normalized differently with the two differing by an n dependent constant. K_n has the following properties:

(a) $K_n(t)$ is continuous in t for each set of distinct $\{x_j\}_{j=1}^n$ since $t \mapsto g_{t,n}$ is continuous in t (uniformly in x).
(b) $K_n(t) = 0$ if $t < \min(x_j)$ or if $t > \max(x_j)$ so the integral is only from $\min(x_j)$ to $\max(x_j)$. For $t > \max(x_j)$, this is because for such t, one has that $g_{t,n}(x_j) = 0$ for all j and, for $t < \min(x_j)$, one has that for x in a neighborhood of all the x_j's, $g_{t,n}(x)$ agrees with the degree $n-2$ polynomial $(n-1)(x-t)^{n-2}$, so, by Lemma 22.3(d), the nth divided difference is 0.
(c) We have that

$$\int_{-\infty}^{\infty} K_n(t)\, dt = 1 \tag{22.39}$$

for we can take $f(x) = x^{n-1}$ in which case $f^{(n-1)}(t) = 1$ for all t while, by Corollary 22.4, $[x_1, \ldots, x_n; f] \equiv 1$.
(d) $K_n(t) \geq 0$ for all t. We will not provide a proof here but refer the reader to either deBoor [73] who uses induction and a recursion relation or Heinävaara[147] who uses induction and an integral relation (if one wants to use the mean value theorem for divided differences, it is easy to prove positivity by picking f carefully but see (e) below).
(e) Equation (22.39) and $K_n(t) \geq 0 \Rightarrow \frac{\min_t f^{(n-1)}(t)}{(n-1)!} \leq [x_1, \ldots, x_n; f] \leq \frac{\max_t f^{(n-1)}(t)}{(n-1)!}$, so by continuity of $f^{(n-1)}$ there is t_0 with $[x_1, \ldots, x_n; f] = \frac{f^{(n-1)}(t_0)}{(n-1)!}$, providing another proof of the mean value theorem for divided differences.

Modern treatments of the Peano kernel involve the theory of splines [73]. A spline is a piecewise polynomial function with some kind of smoothness condition at joining points. Notice that (22.38) is piecewise polynomial and globally C^{n-2}. The Peano kernel is a Curry–Schoenberg spline named after [66].

Another way of understanding the Peano kernel is to note that by the Genocchi–Hermite formula (Theorem 5.15), $K_n(s)$ is just the area of the slice of the simplex $\{(t_1, \ldots, t_n) \in \Delta_{n-1} \mid x(t_j, x_j) = s\}$. This immediately implies positivity of K_n.

The Leibniz rule for divided differences goes back to Popoviciu [274] in 1933 although it is often attributed to Steffensen [333] whose 1937 paper was more widely available. One compact way to write the Leibniz rule is to fix $x_1, x_2, \ldots, x_n$ and define the $n \times n$ matrix, T_f, by

$$\begin{pmatrix} [x_1; f] & [x_1, x_2; f] & [x_1, x_2, x_3; f] & \dots & [x_1, \dots, x_n; f] \\ 0 & [x_2; f] & [x_2, x_3; f] & \dots & [x_2, \dots, x_n; f] \\ \vdots & \vdots & \vdots & \ddots & \vdots \\ 0 & 0 & 0 & \dots & [x_n; f] \end{pmatrix} \tag{22.40}$$

The Leibniz rule says that $T_{fg} = T_f T_g$.

The analog of first derivatives are second (i.e. the simplest) divided differences. It is easy to prove the analog of the chain rule:

$$[x_1, x_2, f \circ g] = [g(x_1), g(x_2); f][x_1, x_2; g] \tag{22.41}$$

There are formulae for higher derivatives of $f \circ g$ that go under the name *Faa di Bruno's formula*. One can find analogs for divided differences in Floater–Lyche [100].

Chapter 23
Divided Differences and Multipoint Rational Interpolation

This chapter is the most involved and most central of the chapters on rational approximation. In it, we study approximations by rational functions, $f = p/q$ where p and q are polynomials. If p and q have no common factors (equivalently, no common complex zeros), the *degree* of f, $\deg(f)$, is $\max(\deg(p), \deg(q))$. This is the same as the topological degree of f as a map of $\widehat{\mathbb{C}}$, the Riemann sphere, to $\widehat{\mathbb{C}}$ (see [326, Sections 7.1, 7.2, and 10.1]). Counting multiplicity, f takes each value in $\widehat{\mathbb{C}}$ exactly $\deg(f)$ times.

We are interested in the interpolation

$$f(z_j) = \alpha_j \tag{23.1}$$

where $\{z_j\}_{j=1}^N$ are N distinct points in $\mathbb{C}$ and $\{\alpha_j\}_{j=1}^N$ are N values in $\mathbb{C}$ and f is a rational function. For most of this chapter, and, in particular until further notice, we take

$$N = 2n \quad \text{and} \quad \deg(f) \leq n \tag{23.2}$$

One can consider a more general interpolation problem than (23.1). Namely, pick $\{x_j\}_{j=1}^{\ell}$; $\ell \leq 2n$ and $k_1, \ldots, k_\ell$ in $\{1,2,\ldots\}$ so that $\sum_{j=1}^{\ell} k_j = 2n$. The interpolation data is $\{\alpha_{j,q}\}_{0 \leq q \leq k_j - 1; 1 \leq j \leq \ell}$ and looks for solutions of (here $f^{(0)} \equiv f$)

$$f^{(q)}(z_j) = \alpha_{j,q} \qquad 0 \leq q \leq k_j - 1; 1 \leq j \leq \ell \tag{23.3}$$

Equivalently, one can speak of $2n$ points but drop the condition that they are distinct, repeating a point k_j times. There are few changes in the arguments, so for simplicity, we only consider the case of distinct points—the reader can consult Donoghue [80, 81] for the more general case. In the next two chapters, we only need the notationally simpler problem we treat in this chapter. Chapters 31 and 32 will involve rational

B. Simon, *Loewner's Theorem on Monotone Matrix Functions*, Grundlehren der mathematischen Wissenschaften 354, https://doi.org/10.1007/978-3-030-22422-6_23

approximation with derivatives although the methods there will be different from those of this chapter.

Since p and q each have $n+1$ free parameters, and the ratio is unchanged upon multiplication by a constant, the set of rational functions of degree at most n is a variety of dimension $2n+1$, so, in general, we expect (23.1) to have a one parameter set of solutions.

It is natural to replace (23.1) by

$$p(z_j) - \alpha_j q(z_j) = 0 \tag{23.4}$$

which is a set of $2n$ linear equations on $\mathcal{P}_n \times \mathcal{P}_n \equiv \mathcal{P}_n^2$, the space of pairs of polynomials (p, q) of degree at most n. A solution of (23.4) is called a *linear solution* of the interpolation problem. Since $\dim(\mathcal{P}_n^2) = 2n+2$, the set of solutions is a subspace of dimension at least 2.

Example 23.1 Let

$$f(1) = 0 \quad f(2) = \frac{1}{2} \quad f(3) = \frac{2}{3} \quad f(4) = \frac{1}{2} \tag{23.5}$$

The first three values were picked so they are matched by $g(x) = (x-1)/x$, but notice that $g(4) = 3/4 \neq 1/2$, so g does not solve the original interpolation problem. However $p(x) = (x-1)(x-4) \quad q(x) = x(x-4)$ does solve the linear equations since both sides are zero at $x = 4$ and the $x - 4$ factors cancel at the other three points. This shows that while (23.1) implies (23.4), the converse may not be true if $p(z_j) = q(z_j) = 0$ for some z_j. Such a solution is called an *exceptional linear solution*. For the current example, we note that (23.1) has solutions (which, of course, solve (23.4) also), for example,

$$p(x) = x - 1 \qquad q(x) = x^2 - 4x + 6$$

□

We do note however the easy to prove

Proposition 23.2 *Let* $(p, q) \in \mathcal{P}_n^2$ *solve* (23.4). *Suppose that no common zero of* p *and* q *is a* z_j. *If* $p = p_0 h$, $q = q_0 h$ *where* h *is the product of the common zeros, then* $f = p_0/q_0$ *is a solution of* (23.1) *and* (p_0, q_0) *also solves* (23.4).

Example 23.3 Let

$$f(1) = 0 \quad f(2) = 0 \quad f(3) = 0 \quad f(4) = 1 \tag{23.6}$$

Then (23.4) implies $p(1) = p(2) = p(3) = 0$. Since $\deg p \leq 2$, we have that $p(z) \equiv 0$. In order for (23.4) to hold at $z_4 = 4$, we must have that $q(4) = 0$. Thus, the two-dimensional subspace of solutions of (23.4) has

$$p(z) = 0; \qquad q(z) = (z-4)(\alpha z + \beta), \quad \alpha, \beta \in \mathbb{C}$$

For all such solutions, $p/q \equiv 0$, so $f(4) \neq 1$. That is (23.6) has no solutions! Note that there is nothing special about the value zero in (23.6). If $\deg(f) = n$, $n > 0$, a single value is taken at most n times so if f has $\deg(f) \leq 2$ and takes the same value 3 times, then f is constant. □

We want to break up the set of interpolation problems (23.1) into two classes.

Definition We say the interpolation problem (23.1) is of *type 2* if (23.4) has a non-zero solution with $\deg(p) \leq n-1$ and $\deg(q) \leq n-1$. If it is not of type 2, we say it is of *type 1*.

Our main goal in this chapter is to develop the type 1 theory which will suffice for the applications in the next two chapters. In the type 2 case, one can show that either there a unique solution, f, of (23.1) with degree strictly smaller than n (in that case if $f = p_0/q_0$ where p_0 and q_0 have no common zeros, all solutions of (23.4) have the form $p = p_0 r$, $q = q_0 r$ where $\deg(r) \leq n - \deg(f)$) or else, as in Example 23.3 (where there is a degree 1 linear solution), (23.1) has no solutions.

While we will not extensively develop the type 2 case, we do want to note the simple uniqueness result. If $(p, q), (\tilde{p}, \tilde{q})$ both solve the linear problem and all polynomials have degree at most $n-1$, then $p\tilde{q} - \tilde{p}q$ vanishes at $2n$ points but is of degree at most $2n-2$ so identically zero. Thus after canceling common factors, $\frac{p}{q} = \frac{\tilde{p}}{\tilde{q}}$ showing there is at most one solution of lower degree. In fact, this argument only requires (p, q) to have degree $n-1$ so one sees that if (p_0, q_0) is a solution with common factor removed and $\deg(p_0, q_0) = r \leq n-1$, then all solutions of the linear problem have the form $p = p_0 h$, $q = q_0 h$, $\deg(h) \leq n-r$.

We focus first on obtaining computable conditions for a problem to be of type 1. We begin our analysis by noting that

Proposition 23.4 *For any type 1 problem, the dimension of S, the set of linear solutions, is exactly 2.*

Proof We already noted that $\dim(S) \geq 2$. If $\dim(S) = k > 2$, the leading coefficients of p and q are linear functions which vanish on sets of codimension at most 1. It follows that these two subspaces have an intersection of codimension at most 2, hence not just 0. It follows that the problem is type 2. □

Let (p_0, q_0) and (p_∞, q_∞) be two linear solutions. We form the determinant

$$\delta_{p_0, q_0; p_\infty, q_\infty}(z) = \begin{vmatrix} p_0(z) & q_0(z) \\ p_\infty(z) & q_\infty(z) \end{vmatrix} \tag{23.7}$$

We will often use δ without subscripts when $(p_0, q_0), (p_\infty, q_\infty)$ are fixed.

Proposition 23.5 *For any $(p_0, q_0), (p_\infty, q_\infty)$, we have that*

$$\delta(z) = C \prod_{j=1}^{2n} (z - z_j) \tag{23.8}$$

where C is a constant depending only on $(p_0, q_0), (p_\infty, q_\infty)$.

Proof Suppose that the α_j of (23.1) has $\alpha_j \neq 0$. Since $p_0(z_j) = \alpha_j q_0(z_j)$ and $p_\infty(z_j) = \alpha_j q_\infty(z_j)$, we see that

$$\alpha_j (q_0(z_j) p_\infty(z_j) - p_0(z_j) q_\infty(z_j)) = 0$$

so $\delta(z_j) = 0$. If $\alpha_j = 0$, then $p_0(z_j) = p_\infty(z_j) = 0$, so again $\delta(z_j) = 0$. Since there are $2n$ z_j's and $\deg(\delta) \leq 2n$, we get (23.8). □

Definition If the constant C in (23.8) is 1, we say that the pair $(p_0, q_0), (p_\infty, q_\infty)$ is *normalized.*

Theorem 23.6 *If* (23.1) *is of type 2, then* $\delta(z) = 0$ *for any pair of solutions. If* (23.1) *is of type 1, and* $(p_0, q_0), (p_\infty, q_\infty)$ *are linearly independent linear solutions, then the constant C of* (26.7) *is non-zero.*

Remark The proof shows once again that if the problem is of type 2, there is at most one rational function, f, solving (23.1).

Proof Suppose (23.1) is of type 2. Let $(\tilde{p}, \tilde{q})$ be a non-zero solution with $\deg(\tilde{p}) \leq n - 1$, $\deg(\tilde{q}) \leq n - 1$. For any other solution (p_0, q_0), the associated δ is a polynomial of degree at most $2n - 1$. Since it has $2n$ zeros, $\delta(z) = 0$ for all z. That means at any z in $\mathbb{C}$, except perhaps for the zeros of q_0 and $\tilde{q}$, we have that

$$\frac{p_0}{q_0} = \frac{\tilde{p}}{\tilde{q}} \tag{23.9}$$

If (p_∞, q_∞) is another linear solution, then at all z in $\mathbb{C}$ except perhaps for zeros of q_0, q_∞, and $\tilde{q}$, we have that

$$\frac{p_0(z)}{q_0(z)} = \frac{p_\infty(z)}{q_\infty(z)} \Rightarrow \delta_{p_0, q_0; p_\infty, q_\infty}(z) = 0 \tag{23.10}$$

Conversely, suppose (23.1) is of type 1. Pick $(p_0, q_0), (p_\infty, q_\infty)$ two linearly independent solutions. We claim that we can suppose that $\deg(q_0) \leq n - 1$. For if, $\deg(q_0) = n$, for some $t \in \mathbb{C}$, $\deg(tq_0 + q_\infty) = n - 1$. Replacing (p_∞, q_∞) by $(tp_0 + p_\infty, tq_0 + q_\infty)$ and then interchanging the pairs, we get the same δ up to a sign change.

Since we are type 1, $\deg(p_0) = n$. We claim that also $\deg(q_\infty) = n$. For, if not, $\deg(tq_0 + q_\infty) \leq n - 1$ for all $t \in \mathbb{C}$. Since $\deg(p_0) = n$ for some $t \in \mathbb{C}$, $\deg(tp_0 + p_\infty) \leq n - 1$ and the problem would be type 2.

Thus $\deg(q_\infty) \leq n - 1$, $\deg(p_0) \leq n$, while $\deg(q_0) = \deg(p_\infty) = n$. It follows that the coefficient of z^{2n} in $\delta(z)$ is non-zero. Therefore in (23.8), $C \neq 0$. □

Next, we turn to Hermite interpolation to shed some light on the type 1 question and on the structure of the solutions of (23.1). Initially, we consider $f = p/q$ where $\deg(p)$ and $\deg(q)$ are at most $2n$ (rather than n as we discussed above). We claim for any $q \in \mathcal{P}_{2n}$, there is a unique $p \in \mathcal{P}_{2n}$ solving (23.4) because (23.4) determines $\{p(z_j)\}_{j=1}^{2n}$ and thus p uniquely by Theorem 22.2.

We expand p and q using (22.19)

$$p(z) = \sum_{j=1}^{2n} a_j N_{j-1}(z) \qquad q(z) = \sum_{j=1}^{2n} b_j N_{j-1}(z) \tag{23.11}$$

where $N_{j-1}(z)$ is shorthand for $N_{j-1}(z, \{z_\ell\}_{\ell=1}^{j-1})$ for $j \geq 2$ and $N_0 = \mathbf{1}$. By Theorem 22.8, we have that

$$a_j = [x_1, \dots, x_j; p] \qquad b_j = [x_1, \dots, x_j; q] \tag{23.12}$$

By Theorem 22.12, the equation $p = qf$ says that:

$$a_j = \sum_{k=1}^{j} b_k [x_k, \dots, x_j; f] \tag{23.13}$$

Thus, given $q \in \mathcal{P}_{2n}$, the p solving $p = qf$ can be computed by expanding q via (23.11), finding $\{a_j\}_{j=1}^{2n}$ by (23.13) and using (23.11) again to find p.

In order that $\deg(p) \leq n$, $\deg(q) \leq n$, we demand that

$$a_j = b_j = 0 \quad \text{for} \quad j = n+2, \dots, 2n \tag{23.14}$$

Thus we have $n+1$ b_j's that we are free to vary and the condition on the a_j's sets up $n-1$ linear conditions on these b's. We see once more that the set of linear solutions is of dimension at least 2 since any linear map $S : \mathbb{C}^{n+1} \to \mathbb{C}^{n-1}$ has a kernel of dimension at least 2.

We can also understand why type 2 is special. For such a problem to have a lower degree solution, we demand

$$a_j = b_j = 0 \quad \text{for} \quad j = n+1, \dots, 2n \tag{23.15}$$

instead of (23.14). The associated linear map, T, defined explicitly in (23.16) below, takes $\mathbb{C}^n$ to $\mathbb{C}^n$ so that "normally," there is zero kernel, i.e., no solution of lower degree. We conclude that there is a type 2 solution if and only if $\det(T) = 0$. We have thus proven:

Theorem 23.7 *Given a problem* (23.1), *let T be the $n \times n$ matrix*

$$T_{ij} = [x_i, \dots, x_{n+j}; f] \qquad 1 \leq i, j \leq n \tag{23.16}$$

(which only depends on $\{\alpha_j\}_{j=1}^{2n}$*). Then the problem is type 1 if and only if*

$$\det(T) \neq 0 \tag{23.17}$$

Let $A_{ij}^{(e)} = T_{n+1-i\,j}$, i.e.,

$$A_{ij}^{(e)} = [x_n, x_{n-1}, \dots, x_{n+1-i}, x_{n+1}, \dots, x_{n+j}; f] \tag{23.18}$$

which permutes the rows, so $\det(A^{(e)}) = \pm\det(T)$.

This suggests we relabel $(x_1, \dots, x_{2n})$ so

$$\lambda_1 = x_n, \lambda_2 = x_{n-1}, \dots, \lambda_n = x_1; \mu_1 = x_{n+1}, \dots, \mu_n = x_{2n} \tag{23.19}$$

so

$$A_{ij}^{(e)} = [\lambda_1, \dots, \lambda_i, \mu_1, \dots, \mu_j; f] \tag{23.20}$$

This is related to the matrix, A, of (5.72) as is the matrix, $L^{(e)}$, of (5.100) to the Loewner matrix, L, of (5.1). Recall that

$$L^{(e)}(\lambda_1, \dots, \lambda_n; \mu_1, \dots, \mu_n)_{ij} = [\lambda_i, \mu_j; f] \tag{23.21}$$

The same argument that proves (5.78) proves that

$$\det(L^{(e)}(\lambda_1, \dots, \lambda_n; \mu_1, \dots, \mu_n)) = \prod_{i<j}(\lambda_i - \lambda_j)(\mu_i - \mu_j)\det(A^{(e)}(\lambda_1, \dots, \lambda_n; \mu_1, \dots, \mu_n)) \tag{23.22}$$

So long as we are dealing with distinct points (or, if one picks, μ, λ carefully so long as there are no points of multiplicity 3 or more), this shift from $A^{(e)}$ to $L^{(e)}$ is fine, but if one considers the more general problem with coincident points, one must stick with $A^{(e)}$.

We thus have:

Restatement of Theorem 23.7 If the distinct points of a problem (23.1) *are* $(\lambda_1, \dots, \lambda_n; \mu_1, \dots, \mu_n)$*, then the problem is of type 1 if and only if* $\det(L^{(e)}(\lambda_1, \dots, \lambda_n; \mu_1, \dots, \mu_n)) \neq 0$. □

Fix a type one problem with $2n$ points and some basis, (p_0, q_0), (p_∞, q_∞) for the linear solutions. For $t \in \widehat{\mathbb{C}}$, the Riemann sphere, $\mathbb{C} \cup \infty$, we define the rational functions:

$$f_t(z) = \frac{p_t(z)}{q_t(z)}, \qquad p_t = p_0 + tp_\infty, \qquad q_t = q_0 + tq_\infty \tag{23.23}$$

for $t \in \mathbb{C}$ and $f_\infty(z) = p_\infty(z)/q_\infty(z)(= \lim_{t\to\infty; t\neq\infty} f_t(z))$. Among $t \in \widehat{\mathbb{C}}$, there are at most $2n$ values where some $q_t(z_j) = 0$ (for some $j \in \{1, \dots, 2n\}$). It may be fewer since some of the values $t_j = -q_0(z_j)/q_\infty(z_j)$ may be equal. These are called *exceptional values* and the corresponding f_{t_j} *exceptional solutions* where we remove any common factors of p_t and q_t of which there is at least one since $q_t(z_j) = 0 \Rightarrow p_t(z_j) = \alpha_j q_t(z_j) = 0$. Thus, these functions, f_{t_j}, are of lower degree, so, since the problem is type 1, they cannot obey (23.1) at least one of the exceptional points.

Naively, one might think that $f_t(z)$ is continuous in t so one could argue by continuity that $f_{t_0}(z_j) = \alpha_j$ at an exceptional value, t_0 where $q_{t_0}(z_j) = p_{t_0}(z_j) = 0$ by taking t non-exceptional to t_0. In fact, $f_t(z)$ is continuous in t except at those pairs (t_0, z_j) where $q_{t_0}(z_j) = 0$. To understand why, suppose for simplicity that $q_{t_0}(z_k) \neq 0$ for all $k \neq j$ and that $q'_{t_0}(z_j) \neq 0$.

We can write f_{t_0} as a rational function of degree $n-1$ which, in general, has $n-1$ poles. For t near t_0, f_t has a piece with poles and residues near those of f_{t_0} plus an additional pole near z_j. Thus

$$f_t(z) = h_t(z) + \frac{r_t}{z - z_t} \tag{23.24}$$

As $t \to t_0$, $h_t(z) \to f_{t_0}(z)$ for all z including z_j but $r_t \to 0$ and $z_t \to z_j$. Suppose $r_t/t \to c \neq 0$ and $(z_t - z_j)/t \to d \neq 0$. Then

$$f_t(z_j) - h_t(z_j) = \frac{r_t}{z_j - z_t} \to \frac{c}{d} \neq 0 \tag{23.25}$$

so since $f_t(z_j) = \alpha_j$ for t near but distinct from t_0, we see that $f_{t_0}(z_j) = \alpha_j - c/d$ and we understand the lack of continuity. We will discuss this further in Theorem 23.13.

We summarize in:

Theorem 23.8 *Let* (23.1) *be a type 1 problem. Then there is a nonempty set,* $\mathcal{E}$, *of exceptional values with at most* $2n$ *points so that the set of solutions of* (23.1) *is* $\{f_t\}_{t\in\widehat{\mathbb{C}}\setminus\mathcal{E}}$.

We can use this to solve and study the interpolation problem for $N = 2n + 1$ points:

Theorem 23.9 *Consider* (23.1) *for* $N = 2n + 1$. *Suppose the problem with* z_{2n+1} *dropped is type 1. Then there is a unique* $t \in \widehat{\mathbb{C}}$ *so that* $f_t(z_{2n+1}) = \alpha_{2n+1}$. *If* t *is not exceptional,* f_t *is a degree* n *solution of* (23.1). *If* t *is exceptional,* (23.1) *has no degree* n *solution for all* N *points. Even in the exceptional case,* (p_t, q_t) *is a linear solution for the* $2n + 1$*-point problem and is the unique such solution up to multiplication by a constant.*

Proof Since z_{2n+1} is distinct from the other z_j's, $\delta(z_{2n+1}) \neq 0$. Thus, $t \mapsto f_t(z_{2n+1})$ is a non-degenerate fractional linear transformation and thus a bijection of $\widehat{\mathbb{C}}$ implying the existence of a unique t solving (23.1) for $j = 2n + 1$. If t is not exceptional, we have the required solution. If there is a degree n solution of (23.1), it must be that f_t obeying (23.1) which implies that t is not exceptional. As for the final assertion, the additional linear equation on a two-dimensional space defines a one-dimensional space. □

Given $S = 1, \dots, 2n + 1$, let $S_j = S \setminus \{j\}$ for $j = 1, \dots, 2n + 1$.

Theorem 23.10 *Consider* $2n + 1$ *points* $\{z_j\}_{j=1}^{2n+1}$ *and a set of conditions* (23.1) *on those points. Suppose for each* $j \in \{1, \dots, 2n + 1\}$, *the problem for* $\{z_k\}_{k\in S_j}$ *is of type 1. Then there is a unique rational function,* f, *obeying* (23.1). *If* $f_t^{(j)}$ *are the solutions for the* S_j *problem, there is for each* j, *a* t_j *which is not exceptional for the* S_j *problem and with* $f_{t_j}^{(j)} = f$.

First Proof Let g be the ratio of the polynomials constructed for the S_1 linear problem that also obeys the condition at z_1. Suppose this solution is exceptional, i.e., for some $k \in S_1$, $p(z) = (z - z_k)\tilde{p}(z)$, $q(z) = (z - z_k)\tilde{q}(z)$. Then $(\tilde{p}, \tilde{q})$ is a solution of lower degree of the linear problem for S_k violating the type 1 hypothesis. It follows that $q(z_j) \neq 0$ for all $j \in S$ and g (which is an $f_{t_1}^{(1)}$) solves (23.1) for all $j \in S$. Since S_1 can be replaced by S_k, we get the $f_{t_k}^{(k)}$ result for each k. □

Second Proof Since δ for an S_j problem is non-vanishing at z_j (by the type 1 hypothesis), $f_{t_j}^{(j)}(z_j) = \alpha_j$. By the uniqueness of solutions of the linear problem, even in the exceptional case, for all j and k, $f_{t_j}^{(j)} = f_{t_k}^{(k)}$ so this common function obeys (23.1) at all z_j. □

This $2n + 1$ solution can sometimes work at one additional point.

Theorem 23.11 *Let* $\{z_j\}_{j=1}^{2n+2}$ *distinct points and values* $\{\alpha_j\}_{j=1}^{2n+2}$ *be given. For* $j \neq k$ *in* $\{1, 2, \dots, 2n + 2\}$, *let* $S_{jk} = \{1, 2, \dots, 2n + 2\} \setminus \{j\, k\}$. *Suppose that each* $2n$ *problem for* S_{jk} *is type 1. Then there is a rational function,* f, *of degree* n *solving* (23.1) *for* $j = 1, 2, \dots, 2n + 2$ *if and only if* $\det(L^{(e)}(z_1, \dots, z_{n+1}; z_{n+2}, \dots, z_{2n+2})) = 0$.

Proof If there is an f of degree n solving (23.1), by Proposition 23.14 below, the $(n + 1) \times (n + 1)$ matrix, $L^{(e)}(z_1, \dots, z_{n+1}; z_{n+2}, \dots, z_{2n+2})$ is of rank n so not invertible so its determinant is 0.

Conversely, suppose $\det(L^{(e)}) = 0$. Then the $2n + 2$ problem is of type 2 by Theorem 23.7, so there is a solution of the linear problem of the form $f_0 = p_0/q_0$ with $\deg(f_0) \leq n$. Fix j and pick any $k \neq j$. By Theorem 23.10 and the fact that each S_{jk} problem is type 1, there is a rational solution, $f = p/q$ of the S_k problem of degree $n - 1$ with $q(z_j) \neq 0$. It is the unique linear solution for the S_k problem so it must agree with (p_0, q_0). Since $j\,k$ are arbitrary, $q_0(z_j) \neq 0$ for all j and the linear solution is a solution for the full $2n + 2$ problem. □

Our final topic in this chapter concerns computing the residue of poles of a rational function interpolation $f = p/q$ at a simple zero of q. This will be a key component of our applications in the next two chapters where we want to prove certain such interpolations are Herglotz functions. For a rational function, $f = p/q$, of degree n with $\deg(q) = n$ is a Herglotz function if and only if f is real on $\mathbb{R}$ and all the zeros of q are simple, lie in $\mathbb{R}$, and the residues are all negative. In fact, we'll look at "residues" by which we mean $\lim_{z\to z_0; z\neq z_0}(z - z_0)f(z)$. If f has a simple pole at z_0, this is the usual residue but if f has a higher order pole at z_0, then this "residue" is ∞, not the standard residue of complex analysis (see [326, Section 3.8]).

We'll need an additional function associated to a pair, $(p_0, q_0), (p_\infty, q_\infty)$ of linearly independent solutions, their *Wronskian*,

$$\begin{aligned} W(z) &= q_0'(z)q_\infty(z) - q_0(z)q_\infty'(z) \\ &= \begin{vmatrix} q_0'(z) & q_0(z) \\ q_\infty'(z) & q_\infty(z) \end{vmatrix} \end{aligned} \tag{23.26}$$

W is a polynomial of degree at most $2n - 1$. It is not the zero polynomial, for if it were, at all z which are not zeros of either q_0 or q_∞, one has $\frac{d}{dz}\log\left(\frac{q_0}{q_\infty}\right) = 0$ which would imply that $q_0 = cq_\infty$. This implies that $p_0 - cp_\infty$ vanishes at the $\{z_j\}_{j=1}^{2n}$ so $p_0 = cp_\infty$ also violating the assumption that $(p_0, q_0), (p_\infty, q_\infty)$ are linearly independent.

Theorem 23.12 *Suppose $q_t(z_0) = 0$ with $z_0 \notin \{z_j\}_{j=1}^{2n}$. Then the "residue" of f_t at z_0 is given by*

$$\lim_{\substack{z\to z_0 \\ z\neq z_0}} (z - z_0)f_t(z) = \frac{\delta(z_0)}{W(z_0)} \tag{23.27}$$

Proof The "residue" is infinite if and only if $q_t'(z_0) = 0$. Suppose $t \neq \infty$. Since $q_t(z_0) = 0$ and

$$W(z) = \begin{vmatrix} q_t'(z) & q_t(z) \\ q_\infty'(z) & q_\infty(z) \end{vmatrix}$$

$W(z_0) = q_t'(z_0)q_\infty(z_0)$. Since the problem is not type 2, $q_\infty(z_0) \neq 0$ so $W(z_0) = 0$ if and only if the "residue" is infinite. This proves (23.27) in this infinite residue case. If $t = \infty$, $W(z_0) = -q_\infty'(z_0)q_0(z_0)$ and then same argument works.

Thus we can suppose the "residue" is finite, i.e., z_0 is a simple zero of q_t. Suppose $t \neq \infty$. Then $q_t(z_0) = 0 \Rightarrow t = -q_0(z_0)/q_\infty(z_0)$. Since $q_\infty(z_0) \neq 0$, we can plug this value into the definition of f_t and multiply numerator and denominator by $q_\infty(z_0)$ to get for $z \neq z_0$

$$(z - z_0)f_t(z) = \frac{p_0(z)q_\infty(z_0) - q_0(z_0)p_\infty(z)}{\left[\left(\frac{q_0(z)-q_0(z_0)}{z-z_0}\right)q_\infty(z_0) - \left(\frac{q_\infty(z)-q_\infty(z_0)}{z-z_0}\right)q_0(z_0)\right]} \tag{23.28}$$

The numerator converges to $\delta(z_0)$ and the denominator to $W(z_0)$ proving (23.27). If $q_\infty(z_0) = 0$, (23.28) still holds and we get (23.27). □

Remark This result (or alternatively, some linear algebra) shows that $W(x)$ is the same for all normalized pairs.

With Wronskian formulae at our disposal, we can return to the discussion prior to Theorem 23.8:

Theorem 23.13 *Suppose we have a type 1 problem for points* $\{z_j\}_{j=1}^{2n}$ *and* $q_{t_0}(z_j) = 0$, $q'_{t_0}(z_j) \neq 0$. *Let* $\tilde{f}_{t_0}(z)$ *be* p_{t_0}/q_{t_0} *with the common factors (and, in particular* $(z - z_j)$*) removed. Then*

$$\tilde{f}_{t_0}(z_j) = \alpha_j + W(z_j)^{-1} \prod_{k=1, k\neq j}^{2n} (z_j - z_k) \tag{23.29}$$

Proof As in the discussion in (23.24)/(23.25)

$$\alpha_j - h_t(z_j) = \frac{\delta(z_t)}{W(z_t)(z_j - z_t)} \tag{23.30}$$

where t is near but not equal to t_0, z_t is the pole of f_t near z_j, and h_t is f_t with the pole at z_t dropped so $h_t(z_j) \to \tilde{f}_{t_0}(z_j)$ as $t \to t_0$. Note that $\delta(z_t)/(z_j - z_t) = \prod_{k=1}^{2n}(z_t - z_k)/(z_j - z_t) = -\prod_{k=1, k\neq j}^{2n}(z_t - z_k)$. This and (23.30) yield (23.29). □

We next turn to an expression for W in terms of extended Loewner matrices. We need several preliminaries.

Proposition 23.14 *Let f be a rational function of degree d and* $\lambda_1, \dots, \lambda_n, \mu_1, \dots, \mu_n \in \mathbb{C}$. *Then the extended Loewner matrix* $[\lambda_i, \mu_j; f]_{i,j=1,\dots,n}$ *is a matrix of rank at most d.*

Remark In fact, using ideas from Chapter 24, one can show that the rank is $\min(d, n)$

Proof By a partial fraction expansion,

$$f(z) = P_0(z) + \sum_{k=1}^{\ell} P_k((z - z_k)^{-1}) \tag{23.31}$$

where P_0, $\{P_j\}_{j=1}^{\ell}$ are polynomials with

$$\deg(P_0) + \sum_{k=1}^{\ell} \deg(P_k) = d \tag{23.32}$$

so it is sufficient to show that

$$\operatorname{rank}([\lambda_i, \mu_j; P_j]) \leq \deg(P_j) \quad \text{for} \quad j = 0, 1, \ldots, \ell \tag{23.33}$$

Let $D_\mu(f)$ be the diagonal matrix with entries $f(\mu_j)$ and similarly for $D_\lambda(f)$. Then

$$[\lambda_i, \mu_j; fg] = D_\lambda(f)[\lambda_i, \mu_j; g] + [\lambda_i, \mu_j; f]D_\mu(g) \tag{23.34}$$

so, if $R(f) = \operatorname{rank}([\lambda_i, \mu_j; f])$

$$R(fg) \leq R(f) + R(g) \tag{23.35}$$

Since any polynomial, Q of degree q is a product of q linear factors and $R(f + c\mathbf{1}) = R(f)$ for any constant c, it suffices to prove that

$$R(z) = 1, \qquad R((z - z_k)^{-1}) = 1 \tag{23.36}$$

These are immediate since

$$L^{(e)}(z)_{ij} = 1, \qquad L^{(e)}((z - z_k)^{-1})_{ij} = -(\lambda_i - z_k)^{-1}(\mu_j - z_k)^{-1} \tag{23.37}$$

(since $(\lambda - \mu)^{-1}[(\lambda - z)^{-1} - (\mu - z)^{-1}] = -(\lambda - z)^{-1}(\mu - z)^{-1}$) showing that each is rank 1. □

Proposition 23.15 *Let f be a rational function of degree d and z_0 a simple pole of f with residue r. Then*

$$f(z) = h(z) + \frac{r}{z - z_0} \tag{23.38}$$

where $\deg(h) = d - 1$.

Proof By adding fractions, it is immediate that $\deg(f) \leq \deg(h) + 1$ so $\deg(h) \geq d - 1$.

On the other hand, if $f = p/q$ where p, q have no common factors, note that $q(z_0) = 0$ so $q(z) = (z - z_0)q_1(z)$, $p(z) = (z - z_0)p_1(z) + r_1$. Then

$$\begin{aligned} f(z) &= \frac{p_1(z)}{q_1(z)} + \frac{r_1}{(z - z_0)q_1(z)} \\ &= \frac{p_1(z)}{q_1(z)} + \frac{r_1}{(z - z_0)q_1(z_0)} + r_1 \frac{q_1(z) - q_1(z_0)}{(z - z_0)q_1(z)q_1(z_0)} \\ &= \frac{p_1(z) + p_2(z)}{q_1(z)} + \frac{r_1}{q_1(z_0)} \frac{1}{z - z_0} \end{aligned}$$

where p_2 is the polynomial $p_2(z) = \frac{r_1}{q_1(z_0)} \frac{q_1(z)-q_1(z_0)}{z-z_0}$. Clearly $\deg(p_1) \leq d - 1$, $\deg(p_2) \leq d-1$, $\deg(q_1) \leq d-1$ so in (23.23), $\deg(h) \leq d-1$. □

Recall Cramer's rule $\det(A)A^{-1} = C(A)$, the cofactor matrix (whose (ij) element is $(-1)^{i+j}$ times the ij minor; see [329, Section 1.3]).

Proposition 23.16 *Let A be an invertible finite matrix and B a rank one matrix. Then*

$$\det(A+B) = \det(A)[1 + \mathrm{Tr}(A^{-1}B)] \tag{23.39}$$

In all cases (even A non-invertible):

$$\det(A+B) = \det(A) + \mathrm{Tr}(C(A)B) \tag{23.40}$$

Proof In terms of exterior algebra (see [329, Section 1.3]), for any $n \times n$ matrix, Q,

$$\det(\mathbf{1} + Q) = \sum_{j=0}^{n} \mathrm{Tr}_{\wedge^j(\mathbb{C})}(\wedge^j(Q)) \tag{23.41}$$

If Q is rank 1, we have that $\wedge^j(Q) = 0$ for $j \geq 2$. Since $A^{-1}B$ is rank 1:

$$\begin{aligned} \det(A+B) &= \det(A(\mathbf{1} + A^{-1}B)) \\ &= \det(A) \sum_{j=0}^{n} \mathrm{Tr}_{\wedge^j(\mathbb{C})}(\wedge^j(A^{-1}B)) \\ &= \det(A)[1 + \mathrm{Tr}(A^{-1}B)] \end{aligned}$$

By Cramer's rule, (23.40) holds if A is invertible. If A is not invertible, (23.40) holds for $A + \epsilon\mathbf{1}$, so taking ϵ to 0, it holds for A. □

Theorem 23.17 *Consider a type 1 problem of the form* (23.1)*. Suppose that* $z_0 \notin \{z_j\}_{j=1}^{2n}$*. Let* $f = p/q$ *solve* (23.1) *and suppose* $q(z_0) = 0$*. Let* λ, μ *be given by* (23.19) *and* $L_{ij}^{(e)}(f) = [\lambda_i, \mu_j; f]$*. Suppose that* $L^{(e)}(f)$ *is invertible. Let W be the Wronskian for a normalized basis. Then*

$$W(z_0) = -\delta(z_0) \sum_{i,j} [(L^{(e)})^{-1}]_{ij} (z_0 - \mu_i)^{-1} (z_0 - \lambda_j)^{-1} \tag{23.42}$$

Proof Suppose first that the zero is simple. By Proposition 23.15 and Theorem 23.12,

$$h(z) = f(z) - \frac{\delta(z_0)}{W(z_0)} \frac{1}{z - z_0} \tag{23.43}$$

has h with degree at most $n-1$. By Proposition 23.14, $\det(L^{(e)}(h)) = 0$, so

$$\det\left(L^{(e)}(f) - \frac{\delta(z_0)}{W(z_0)} L^{(e)}\left(\frac{1}{z-z_0}\right)\right) = 0$$

As noted in (23.37), $L^{(e)}(1/(z-z_0))$ is rank 1, so by Proposition 23.16 and $\det(L^{(e)}(f) \neq 0$,

$$W(z_0) = \delta(z_0)\mathrm{Tr}\left(L^{(e)}(f)^{-1} L^{(e)}\left(\frac{1}{z-z_0}\right)\right) \tag{23.44}$$

By (23.37)

$$\mathrm{Tr}\left(L^{(e)}(f)^{-1} L^{(e)}\left(\frac{1}{z-z_0}\right)\right) = -\sum_{i,j=1}^{n} [(L^{(e)})^{-1}]_{ij}(z_0-\mu_i)^{-1}(z_0-\lambda_j)^{-1}$$

proving (23.42).

Suppose $(p,q) = (p_{t_0}, q_{t_0})$. If $W(z_0) = 0$, it is an isolated zero so for t near but not equal to t_0, the corresponding zero is simple and (23.42) holds for the nearby z's. Since both sides of (23.42) are continuous in z, we get the result at the zero of W. □

For our last result, we shift back to $A^{(e)}$ rather than the simpler $L^{(e)}$ because we want to use the Hermite interpolation (23.11)–(23.13). Recall that $\lambda_1, \ldots, \lambda_n$ are $x_n, \ldots, x_1$ so the Newton polynomials are the N below and the $M's$ are analogous:

$$N_i(z) = \prod_{j=1}^{i}(z-\lambda_{n+1-j}) \quad M_i(z) = \prod_{j=1}^{i}(z-\mu_{n+1-j}); \quad i = 0, \ldots, n \tag{23.45}$$

Theorem 23.18 *Let*

$$f(z) = h(z) + \frac{r}{z-z_0} \tag{23.46}$$

$h = \sigma/\tau$, where σ, τ are polynomials with no common factor, with $\deg(\tau) = n-1$ $\deg(\sigma) \leq n-1$ *and* $\tau(z_0) \neq 0$. *Then* $\mathrm{rank}(A^{(e)}(h)) = n-1$ *and*

$$\det(A^{(e)}(f)) = -r\chi(z_0) \tag{23.47}$$

with $\chi(z_0) = \chi(x,y)|_{x=y=z_0}$ *and*

$$\chi(x,y) = \sum_{i,j=1,\ldots,n} C(A^{(e)}(h))_{ij} \frac{N_{n-j}(x)}{N_n(x)} \frac{M_{n-j}(y)}{M_n(y)} \tag{23.48}$$

Moreover, for some constant, C, depending only on h and not on r or z_0

$$\delta(z)\chi(z) = C\tau(z)^2 \tag{23.49}$$

Proof We write $A^{(e)}(f) = A^{(e)}(h) + A^{(e)}(g)$ where $g(z) = r/(z - z_0)$. Since $\det(A^{(e)}(h)) = 0$, (23.40) yields (23.47) if we note that, by direct calculation

$$A^{(e)}(g)_{ij} = -\prod_{k=1}^{i} \frac{1}{z - \lambda_k} \prod_{k=1}^{j} \frac{1}{z - \mu_k}$$

and we use $\frac{N_{n-i}(z)}{N_n(z)} = \prod_{k=1}^{i}(z - \lambda_k)^{-1}$ and similarly for M.

Since σ and τ are relatively prime and $\deg(\tau) = n-1$, we know that $\ker(A^{(e)}(h))$ is one-dimensional spanned by $(b_1, \dots, b_n)$ where $\sum_{j=1}^{n} b_j M_{n-j}(y) = \tau(y)$. Moreover

$$\det(A^{(e)}(h)) = 0 \Rightarrow \sum_{j=1}^{n} a_{ij} C(A)_{jk} = 0$$

for all i and k.

It follows that $C(A)$ maps $\mathbb{C}^n$ to $\ker(A)$ so that for any x

$$M_n(y)N_n(x)\chi(x, y) = c_1(x)\tau(y) \tag{23.50}$$

(Note that both sides are, for fixed x, polynomials in y of degree $n - 1$.) Interchanging μ's and λ's and x and y, we see that

$$M_n(y)N_n(x)\chi(x, y) = c_2(y)\tau(x) \tag{23.51}$$

It follows that

$$M_n(y)N_n(x)\chi(x, y) = C\tau(x)\tau(y) \tag{23.52}$$

Setting $x = y$ and noting that $M_n(x)N_n(x) = \delta(x)$, we get (23.49). □

Notes and Historical Remarks Our presentation in this and the next two chapters is heavily impacted by the write-up in Donoghue [81] who, in turn, was following the work of Loewner [209]. That said, there are some important differences in presentation. For example, Theorem 23.10 is new and allows replacement of some involved argument of Donoghue when we get to Chapter 25. Donoghue uses the fact that

$$
-\det(A)\sum_{i,j=1}^{n}\alpha_i\beta_j(A^{-1})_{ij} = \begin{vmatrix} 0 & \beta_1 & \beta_2 & \dots & \beta_n \\ \alpha_1 & a_{11} & a_{12} & \dots & a_{1n} \\ \alpha_2 & a_{21} & a_{22} & \dots & a_{2n} \\ \vdots & \vdots & \vdots & \ddots & \vdots \\ \alpha_n & a_{n1} & a_{n2} & \dots & a_{nn} \end{vmatrix}
$$

and we never need this—we use the inner product which we believe is simpler.

Both Loewner and Donoghue use the term "Cauchy Interpolation" for the method of this chapter but a Google search shows this term, unlike "Hermite interpolation", is hardly used in literature so we haven't used it.

Chapter 24
Pick Interpolation, V: Rational Interpolation Proof

In this chapter, we'll use the theory of rational interpolation of the previous chapter to prove the harder direction of Pick's theorem for Herglotz functions, Theorem 16.1. Extended Loewner matrices were defined for points in (a, b) in (5.100). When G is rational, that expression can be used to define $L^{(e)}(G)$ for any points which are not poles of G. Note that if G obeys (16.8) and I is a set with n points, then the matrix, M, of (16.10) is an extended Loewner matrix, namely

$$M = L^{(e)}(z_1, \dots, z_n; \bar{z}_1, \dots, \bar{z}_n; G) \tag{24.1}$$

Since $L^{(e)}(G) \geq 0$ when G is z or $(z_0 - z)^{-1}$ (see (23.37)) and the Herglotz representation says that any Herglotz G is a constant (which doesn't change $L^{(e)}$) plus an integral/sum of these two simplest cases, we see that G Herglotz $\Rightarrow M \geq 0$ proving the easy half of Theorem 16.1 for I finite. We can then handle arbitrary I's by taking limits. Before leaving this easy direction, we want to study when M is strictly positive.

Lemma 24.1 *Fix $\boldsymbol{\mu} = (\mu_1, \dots, \mu_n)$ in $\mathbb{C}^n$. Let $v_n(z, \boldsymbol{\mu}) \in \mathbb{C}^n$ be given by*

$$[v_n(z; \boldsymbol{\mu})]_j = (z - \mu_j)^{-1} \tag{24.2}$$

Then for any distinct $\{z_j\}_{j=1}^n \in \mathbb{C}$, the set of vectors $\{v_n(z_j; \boldsymbol{\mu})\}_{j=1}^n$ are linearly independent.

Remark We can include $z_j = \infty$ if we take $[v_n(\infty; \boldsymbol{\mu})]_j = \mu_j$. We then take a linear term in (24.3).

Proof Suppose that $\sum_{j=1}^n \alpha_j v_n(z_j; \boldsymbol{\mu}) = 0$. Let

$$f(w) = \sum_{j=1}^n \alpha_j (z_j - w)^{-1} = \frac{p(w)}{q(w)} \tag{24.3}$$

B. Simon, *Loewner's Theorem on Monotone Matrix Functions*, Grundlehren der mathematischen Wissenschaften 354, https://doi.org/10.1007/978-3-030-22422-6_24

where $\deg(q) = n, \deg(p) \leq n - 1$. By assumption, $f(\mu_j) = 0$ which implies that $p(\mu_j) = 0;\ j = 1, \ldots, n$. Since p has degree at most $n - 1$ and vanishes at n points, $p \equiv 0$, i.e., $f \equiv 0$. Thus $\alpha_j = \lim_{w \to z_j; w \neq z_j} (z_j - w) f(w) = 0$ proving independence. □

Proposition 24.2 *Let G be a Herglotz function written in the form* (3.33) *with μ a measure on $\mathbb{R} \cup \{\infty\}$. Then $L^{(e)}$ is strictly positive if and only if μ has at least n points in its support.*

Proof If μ has $k < n$ points in its support, then $L^{(e)}(G)$ has rank k, so a non-zero kernel, and thus is not strictly positive.

(23.37) and (3.39) show that (note that $\frac{1+xz}{x-z} = x + \frac{1+x^2}{x-z}$ for $x \in \mathbb{R}$ and is z for $x = \infty$)

$$L^{(e)}(G) = \mu(\{\infty\}) P_\infty(z) + \int (1 + x^2) P_x(z) d\mu(x) \tag{24.4}$$

where $[P_x(z)]_{ij} = (x - z_i)^{-1}(x - \bar{z}_j)^{-1}$ and $[P_\infty(z)]_{ij} = z_i \bar{z}_j$. It follows that if $\langle \varphi, L^{(e)} \varphi \rangle = 0$, then $\langle \varphi, v_n(x; \{z_j\}_{j=1}^n) \rangle = 0$ for a.e. x. If μ has at least n points in its support, this holds for a spanning set by Lemma 24.1 and its remark if ∞ is one of the n points. Thus $L^{(e)}$ is strictly positive. □

We now turn to the harder half of the theorem. We are heading towards:

Theorem 24.3 *Given $\{z_j, w_j\}_{j=1}^n$ in $\mathbb{C}_+$, form the matrix $\{M_{ij}\}_{1 \leq i,j \leq n}$ of* (16.10). *If M is strictly positive, there is a rational Herglotz function, f, of degree n with $f(z_j) = w_j$. If M is positive but not strictly positive, there is a rational Herglotz function, f of degree at most $n - 1$ so that $f(z_j) = w_j$.*

We remind the reader that the term "rational Herglotz function" includes the condition that f is real on $\mathbb{R}$ except for possible poles. We begin by noting that it suffices to prove the result for strictly positive M. For suppose we have that result and that we are given M that is positive, but not strictly positive. Let g be a rational Herglotz function of degree n (e.g., $g(z) = \sum_{j=1}^n (j - z)^{-1}$) and let N be the extended Loewner matrix for the data $(z_j, g(z_j))$. Then N is strictly positive by Proposition 24.2, so, for each $m = 1, 2, \ldots$, we have that $M + m^{-1}N$ is strictly positive and we can apply the supposed result in that case. Thus we can find rational, degree n, Herglotz functions, f_m, with $f_m(z_j) = w_j + m^{-1} g(z_j)$.

By the compactness results on Herglotz functions (see Remark 4 after Theorem 3.9), we can find a limit point, f_∞, of $\{f_m\}_{m=1}^\infty$. As a limit of rational Herglotz functions of degree n, f_∞ is a rational Herglotz function of degree at most n which clearly obeys $f_\infty(z_j) = w_j$. If $\deg(f_\infty)$ were n, M would be strictly positive by Proposition 24.2. Thus $\deg(f_\infty) \leq n - 1$ as claimed.

Thus we are reduced to the case M strictly positive which we henceforth assume. We'll construct f by solving the $2n$ point, degree n, rational interpolation problem with

$$f(z_j) = w_j \qquad f(\bar{z}_j) = \bar{w}_j \tag{24.5}$$

If f solves (24.5), so does $z \mapsto \overline{f(\bar{z})}$ and thus so does $z \mapsto \frac{1}{2}\left(f(z) + \overline{f(\bar{z})}\right)$. It follows that there is a two real dimensional space of solutions of the linear problem with $\overline{p(\bar{z})} = p(z)$, $\overline{q(\bar{z})} = q(z)$. Thus we fix once and for all a normalized pair of real solutions (p_0, q_0), (p_∞, q_∞). We'll prove below that for all real t, $f_t(z)$ is a rational Herglotz function, so for all non-exceptional t, we get a rational Herglotz function obeying (24.5).

We'll only look at the case where $\deg(q_t) = n$. For there is a single t_1 for which $\deg(q_{t_1}) \leq n - 1$ and if f_t is a rational Herglotz function for $t \neq t_1$, this is true for t_1 by taking limits.

Proposition 24.4 *Suppose $L^{(e)}$ is strictly positive, Then for all real x, $W(x) < 0$.*

Proof By (23.42)

$$W(x) = -\delta(x)\langle\varphi, (L^{(e)})^{-1}\varphi\rangle \tag{24.6}$$

where $\varphi_j = (x - z_j)^{-1}$ (since $\lambda_j = z_j$, $\mu_j = \bar{z}_j$ means $(x - \mu_j)^{-1} = \overline{\varphi}_j$). Since we are defining δ in a normalized basis

$$\delta(x) = \prod_{j=1}^{n}(x - z_j)(x - \bar{z}_j) = \prod_{j=1}^{n} |x - z_j|^2 > 0 \tag{24.7}$$

for x real. Since $L^{(e)}$ is strictly positive, the inner product is strictly positive. Thus (24.6) implies that $W(x) < 0$. □

Proposition 24.5 *Suppose $L^{(e)}$ is strictly positive. For all real t, q_t has all its zeros on the real axis.*

Proof Consider

$$g(z) = -\sum_{j=0}^{n-1} \frac{1}{z - j} = \frac{P(z)}{Q(z)} \tag{24.8}$$

where

$$Q(z) = \left[\prod_{j=0}^{n-1}(j - z)\right] / (n - 1)! \tag{24.9}$$

which obeys

$$Q(0) = 0 \qquad Q'(0) = -1 \tag{24.10}$$

Consider the interpolation problems ($0 \le \theta \le 1$)

$$f(z_j) = \theta w_j + (1-\theta)g(z_j); \qquad f(\bar{z}_j) = \theta \bar{w}_j + (1-\theta)g(\bar{z}_j) \tag{24.11}$$

The extended Loewner matrix at $[z_1, \dots, z_n; \bar{z}_1, \dots, \bar{z}_n]$ is

$$\theta M + (1-\theta)L^{(e)}(g) \tag{24.12}$$

where M is the Loewner matrix for the original problem. These are all strictly positive.

There is a unique $Q^{(\theta)}$ associated to the linearized problem for (24.11) obeying (24.10). It is easy to see that $Q^{(\theta)}$ varies continuously in θ, so its zeros move continuously. At $\theta = 0$, $Q^{(\theta)} = Q$ has all its zeros on the real axis. By Proposition 24.4, all real zeros for each $Q^{(\theta)}$ are simple. If some $Q^{(\theta)}$ has non-real zeros and $\tilde{\theta} = \inf\{\theta \mid Q^{(\theta)}$ has a non-real zero$\}$, it is easy to see that two complex zeros of $Q^{(\tilde{\theta}+\epsilon_n)}$ must coalesce to a zero of $Q^{(\tilde{\theta})}$ which would be a double zero. Since this is impossible, we conclude that $Q^{(\theta=1)}$ has real simple zeros.

$Q^{(\theta=1)}(z)$ is some $q_{t_0}(z)$ for the original problem. The zeros of $q_t(z)$ are continuous in t in $\mathbb{R}$, all real zeros of q_t for t real are simple (by Proposition 24.4) and q_{t_0} has only real zeros. By repeating the above argument, each $q_t(z)$ has only real zeros. □

Proof of Theorem 24.3 By the discussion following the statement of the theorem, we can suppose the Loewner matrix, $L^{(e)}$, is strictly positive and $f_t = p_t/q_t$ has $\deg(q_t) = n$. By Proposition 24.5, q_t has only simple real zeros so by a partial fraction expansion

$$f_t(z) = b_t + \sum_{j=1}^{n} \frac{r_j(t)}{z - x_j(t)}$$

with b_t real. By (23.27), (24.7), and Proposition 24.4, $r_j(t) < 0$. It follows that $f_t(z)$ is a rational Herglotz function of degree n. □

Notes and Historical Remarks This chapter follows the discussion in Donoghue [81].

Chapter 25
Loewner's Theorem via Rational Interpolation: Loewner's Proof

Our main goal in this chapter is to prove the following:

Theorem 25.1 *Fix $n \geq 1$. Let $x_1, \dots, x_{2n+1}$ be $2n + 1$ distinct points in (a, b). Let f be a function on $(a, b) \subset \mathbb{R}$ for which every $m \times m$ extended Loewner matrix, $m \leq n$, $L^{(e)}(\lambda_1, \dots, \lambda_m; \mu_1, \dots, \mu_m; f)$, $\lambda_1 < \mu_1 < \cdots < \lambda_m < \mu_m$ among the x's has strictly positive determinant. Then there exists a unique rational function, h, of degree exactly n obeying*

$$h(x_j) = f(x_j), \qquad j = 1, \dots, 2n + 1 \tag{25.1}$$

Moreover, h is a Herglotz function with no poles in $[x_-, x_+]$ where $x_- = \min_j x_j$; $x_+ = \max_j x_j$.

Remark Recall (Theorem 5.26) that when $f \in \mathcal{M}_n(a, b)$, $\det(L^{(e)}) \geq 0$ and that by Lemma 5.17, any f is a limit of f_n with strictly positive $\det(L^{(e)})$.

Before turning to the proof of this result, we want to note that it immediately implies the hard part of Loewner's theorem:

Corollary 25.2 (Loewner's Proof of His Theorem) *Let $f : (a, b) \to \mathbb{R}$ lie in $\mathcal{M}_\infty(a, b)$. Then f is the restriction to (a, b) of a function analytic on $(\mathbb{C}\setminus\mathbb{R})\cup(a, b)$ which is Herglotz on $\mathbb{C}_+$.*

Proof As noted, there exist g's with $\det(L^{(e)}(\lambda_1, \dots, \lambda_m; \mu_1, \dots, \mu_m; g)) > 0$ for all λ, μ with $a < \lambda_1 < \mu_1 < \cdots < \lambda_m < \mu_m < b$ (see Lemma 5.17). Since $\det(L^{(e)}(\lambda_1, \dots, \lambda_m; \mu_1, \dots, \mu_m; \theta g + (1 - \theta)f))$ is a polynomial in θ which has non-negative values for $\theta \in [0, 1]$ and is non-vanishing at $\theta = 1$, $\det(L^{(e)}(\lambda_1, \dots, \lambda_m; \mu_1, \dots, \mu_m; \theta g + (1 - \theta)f)) > 0$ for all but finitely many θ's. Let S be a countable subset of (a, b). Then for all m and any λ, μ with $a < \lambda_1 < \mu_1 < \cdots < \lambda_m < \mu_m < b$ and with $\lambda_j, \mu_j \in S$, $\det(L^{(e)}(\lambda_1, \dots, \lambda_m; \mu_1, \dots, \mu_m; \theta g + (1 - \theta)f)) > 0$ for all but countably many θ, so we can find $\theta_m \downarrow 0$ so that all these determinants are positive.

B. Simon, *Loewner's Theorem on Monotone Matrix Functions*, Grundlehren der mathematischen Wissenschaften 354, https://doi.org/10.1007/978-3-030-22422-6_25

Let $\{y_j\}_{j=1}^{\infty}$ be this dense set, S, of distinct points in (a, b). Let $y_{n,+} = \max_{j \le n} y_j$, $y_{n,-} = \min_{j \le n} y_j$. Since the set is dense, $y_{n,+} \to b$ and $y_{n,-} \to a$ and they are monotone in n. Let $h_n(\theta_m, z)$ be the degree n rational Herglotz function with $h(y_j) = \theta g(y_j) + (1 - \theta_m) f(y_j)$; $j = 1, \dots, 2n + 1$ guaranteed by Theorem 25.1. Fix $N \ge 2$. For $n \ge N$, we have that $y_{n,-} < y_1, y_2 < y_{n,+}$. For $n \ge N$, h_n are Herglotz functions analytic on $(\mathbb{C} \setminus \mathbb{R}) \cup (y_{N,-}, y_{N,+})$, so by compactness and the fixed values at y_1, y_2, Theorem 3.9 implies that the h_n have limit points. Since any two limit points agree on a dense set in $(y_{N,-}, y_{N,+})$, all limit points agree, i.e., $h_n \to h_\infty$, a Herglotz function on $(\mathbb{C} \setminus \mathbb{R}) \cup (y_{N,-}, y_{N,+})$. Taking $N \to \infty$ and then $\theta_m \downarrow 0$ yields the desired result since f and h_∞ agree on a dense set and both are continuous. □

For general $f \in \mathcal{M}_n(a, b)$, we can find rational Herglotz approximations of degree at most n matching $2n - 1$, not $2n + 1$, points but with all poles outside (a, b).

Corollary 25.3 *Fix $n \ge 1$. Let f be a function on $(a, b) \subset \mathbb{R}$ which is monotone on $(n + 1) \times (n + 1)$ matrices (i.e., $f \in \mathcal{M}_{n+1}(a, b)$). Let $x_1, \dots, x_{2n+1}$ be $2n + 1$ distinct points in (a, b). Then there exists a rational Herglotz function, h, of degree at most $n + 1$ analytic on $(\mathbb{C} \setminus \mathbb{R}) \cup (a, b)$ obeying*

$$h(x_j) = f(x_j), \qquad j = 1, \dots, 2n + 1 \tag{25.2}$$

Proof Suppose first $\det(L^{(e)}(\lambda_1, \dots, \lambda_m; \mu_1, \dots, \mu_m; f)) > 0$ for any λ_j, μ_j among $\{x_j\}_{j=1}^{2n+1}$ or $\{a + 1/m, b - 1/m\}_{m=m_0}^{\infty}$ where $m_0 > 1/(b - a)$. For large m, consider the set with $2n + 3$ points obtained by adding $a + 1/m$ and $b - 1/m$ to $\{x_j\}_{j=1}^{2n+1}$. By Theorem 25.1, there exists a Herglotz function, h_m, obeying (25.1) at this $2n + 3$ point set with no poles in $(a + 1/m, b - 1/m)$. Using compactness and taking m to infinity, we get the required function (which is a rational function of degree at most $n + 1$).

The same $\theta g + (1 - \theta) f$ method used in the proof of Corollary 25.2 extends the result to general f. □

Remark The reader may wonder why we don't use the $\theta g + (1 - \theta) f$ trick to extend Theorem 25.1 to degenerate f's. But in taking limits, poles of h can approach x_- and/or x_+ causing those points to be exceptional, so $h(x_\pm)$ may not equal $f(x_\pm)$. One can only be sure of getting $2n - 1$ points this way consistent with Corollary 25.3.

The proof of Theorem 25.1 is a little involved. We'll first restrict ourselves to the non-degenerate case, where for each $m \le n$ and $\{\lambda_j, \mu_j\}_{j=1}^m$ with $a < \lambda_1 < \mu_1 < \lambda_2 < \cdots < \lambda_m < \mu_m < b$, the extended Loewner matrix $L^{(e)}(\lambda_1, \dots, \lambda_m; \mu_1, \dots, \mu_m)$ has a strictly positive determinant. Note in this non-degenerate case for every $2n$ point subset of $\{x_j\}_{j=1}^{2n+1}$, the interpolation problem is type 1, so by Theorem 23.10, there is a unique solution of (25.1) which is a rational function of degree n. We need to prove this function is a Herglotz function with no poles in $[x_-, x_+]$.

The steps in the proof are the following:

(1) Prove the result when $n = 1$. This will start an induction.
(2) Suppose $x_1 < x_2 < \cdots < x_{2n+1}$. Assuming the result for the $2n - 1$ point set with x_2 and x_{2n+1} dropped, we'll find a q_{t_1} denominator for the $2n$ problem at $[x_1, \dots, x_{2n}]$ with n simple zeros on $\mathbb{R}$ distinct from the x_j's and so that the Wronskian $W(y) < 0$ at each of these zeros.
(3) Using the existence of q_{t_1}, we'll prove that the Wronskian is negative on all of $\mathbb{R}$.
(4) We'll prove that the unique degree n solution of (25.1) is a Herglotz function with no poles in $[x_1, x_{2n+1}]$.

We begin with the case $n = 1$ with $x_1 < x_2 < x_3$ and $f(x_1) < f(x_2) < f(x_3)$. $L^{(e)}$ is a 1×1 matrix so positivity of the determinant is equivalent to $x_i < x_j \Rightarrow f(x_i) < f(x_j)$. Thus $f(x_1) < f(x_2) < f(x_3)$.

Proposition 25.4 *Let* $x_1 < x_2 < x_3$ *and* $f(x_1) < f(x_2) < f(x_3)$. *Then there is a degree* 1 *rational function, h, with* (25.1). *It is Herglotz with no poles in* $[x_1, x_3]$.

Proof As noted $f(x_i) \neq f(x_j)$ implies each $\{x_i, x_j\}$ two point interpolation is type 1, so by Theorem 23.10, there is a unique degree 1 rational function, $h = p/q$ obeying (25.1). Reality of h implies p and q are real on $\mathbb{R}$.

If $\deg(q) = 0$, h is linear, $h(z) = az + b$. Clearly, $a = \frac{f(x_2)-f(x_1)}{x_2-x_1} > 0$, so h is Herglotz.

Otherwise,

$$h(z) = b + \frac{r}{z - z_0} \tag{25.3}$$

with b, z_0 real and $z_0 \neq x_j$, $j = 1, 2, 3$ so we need to prove that $z_0 \notin [x_1, x_3]$ and $r < 0$. Suppose $z_0 \notin [x_1, x_2]$. Then h is monotone between x_1 and x_2 if and only if $r < 0$ so $f(x_2) > f(x_1) \Rightarrow r < 0$. If it were true that $z_0 \in [x_2, x_3]$, then $f(x_3) < b < f(x_2)$, so $f(x_3) > f(x_2) \Rightarrow z_0 \notin [x_2, x_3]$. A similar analysis works if we suppose a priori that $z_0 \notin [x_2, x_3]$. □

Remark It is illuminating to see what happens if $f(x_1) = f(x_2) < f(x_3)$ and we replace $f(x_2)$ by $f(x_2) + \epsilon$ and take $\epsilon \downarrow 0$. For example, suppose $f(0) = f(1) = 0$; $f(2) = 1$. The function, f_ϵ with ϵ added to $f(1)$ has the form $f_\epsilon(z) = b_\epsilon + \frac{c_\epsilon}{z - z_\epsilon}$. Explicit calculation shows $b_\epsilon \uparrow 0$, $c_\epsilon \uparrow 0$, and $z_\epsilon \downarrow 2$. The limiting function is the constant 0. $x_3 = 2$ is an exceptional point!

Now we turn to the second step. Given a $2n+1$ point set $x_1 < x_2 < \cdots < x_{2n+1}$, consider the $2n-1$ point set with x_2 and x_{2n+1} dropped. By an induction hypothesis, we can find a degree $n - 1$ rational Herglotz function, $h_0 = \sigma/\tau$ with no poles in $[x_1, x_{2n}]$ and so that (25.1) holds (with h_0 replacing h) for $j = 1, 3, 4, \dots, 2n$. Let $p(x) = (x - x_2)\sigma(x)$, $q(x) = (x - x_2)\tau(x)$. Then (p,q) is an exceptional solution for the linear problem with the $2n$ points $\{x_j\}_{j=1}^{2n}$. Thus q is a q_{t_0}. For now, assume $\deg(\tau) = n - 1$ (i.e., h_0 doesn't have a linear term in its Herglotz representation).

By the induction hypothesis:

$$\frac{p_{t_0}(z)}{q_{t_0}(z)} = \frac{\sigma(z)}{\tau(z)} = b + \sum_{j=1}^{n-1} \frac{r_j}{z - z_j} \tag{25.4}$$

where $r_j < 0$, $b \in \mathbb{R}$, and $z_j \in \mathbb{R} \setminus [x_1, x_{2n}]$. For t near t_0, the zeros of q_t will be near those of $q_{t_0}(z) = (z - x_2)\tau(z)$ and the residues at the poles other than the one near x_2 will be close to the r_j and so negative. So let $\{z_j(t)\}_{j=1}^{n-1}$ be the zeros near $\{z_j\}_{j=1}^{n-1}$ and let $z_0(t)$ be the zero near x_2. Write

$$f_t(z) = h_t(z) + \frac{r_0(t)}{z - z_0(t)}; \quad h_t(z) = b_t + \sum_{j=1}^{n-1} \frac{r_j(t)}{z - z_j(t)} \tag{25.5}$$

If $\deg(\tau) = n - 2$ (i.e., h_0 has linear growth at ∞), for t near t_0, $q_t(x)$ will have a zero, call it $z_{n-1}(t)$ near infinity in addition to the zeros near the $n - 2$ zeros of q_{t_0} and the one near x_2. We claim the residue at $z_{n-1}(t)$ is negative. For if the linear term is az with $a > 0$, then for t near t_0 this has the form

$$\frac{(a + C\epsilon)z}{1 + D\epsilon z} + \mathrm{O}(\epsilon^2) = \frac{\left(\frac{a}{D\epsilon} + \frac{C}{D}\right) z}{z - z_{n-1}(t)} + \mathrm{O}(\epsilon^2)$$

where $z_{n-1}(t) = -\frac{1}{D\epsilon} + \mathrm{O}(1)$ and $r_{n-1} = -\frac{a}{(D\epsilon)^2} < 0$. So the form of h is the same as the $\deg(\tau) = n - 1$ case.

Also, pick z_n different from $\{z_j(t)\}_{j=1}^{n-1}$ in $\mathbb{R} \setminus [x_1, x_{2n}]$ and define:

$$\tilde{h}_t(z) = h_t(z) - \frac{1}{z - z_n} \tag{25.6}$$

Proposition 25.5 *For t near but different from t_0:*

$$W(z_0(t)) < 0 \tag{25.7}$$

Remark On $\mathbb{R} \setminus [x_1, x_{2n}]$, we have that $\delta(z) = \prod_{j=1}^{2n}(z - x_j)$ is positive, so, by (23.27) and $r_j(t) < 0$, we also have the $W(z_j(t)) < 0$ for $j = 1, \ldots, n - 1$. Thus for t near but distinct from t_0, we have that q_t has all its zeros on $\mathbb{R}$, all are simple and at the zeros, W is negative.

Proof $\tilde{h}_t$ is a Herglotz function, so its Loewner determinant is positive. Thus by (23.47) and (23.49) and $r = -1$, we have that

$$0 < \det(L^{(e)}(\tilde{h}_t)) = -r\chi(z_n) = \frac{C\tau(z_n)^2}{\delta(z_n)} \tag{25.8}$$

Since $\delta(z_n) > 0$, we conclude that $C > 0$.

On the other hand, $L^{(e)}(h_t)$ is the extended Loewner matrix of our problem which is positive by the nondegeneracy hypothesis. Therefore by (23.47) and (23.49),

$$0 < -Cr(z_0(t))\tau(z_0)^2/\delta(z_0) \tag{25.9}$$

We conclude that $r(z_0(t))/\delta(z_0(t) < 0$. Since (23.27) says that $r(z_0(t))/\delta(z_0(t) = 1/W(z_0(t)$, we obtain (25.7). □

We now turn to step 3. Without loss, we can suppose that the q_t we showed had simple zeros at which W is negative is q_∞. We can also suppose that $\deg(q_\infty) = n$ since exactly one q_t has lower degree and we proved the result on zeros for a range of t's.

Proposition 25.6 *Consider a 2n point interpolation problem of degree n with real data. Suppose* $\deg(q_\infty) = n$ *and that* q_∞ *has n real simple zeros,* $z_1, \dots, z_n$ *with* $W(z_j) < 0$. *Then* $W(x) < 0$ *for all real x.*

Proof Find (p_0, q_0) another real-valued linearly independent solution so that they form a normalized pair. Let

$$T(z) = -q_0(z)/q_\infty(z) \tag{25.10}$$

Then T has poles at $\{z_j\}_{j=1}^n$ with residues $r_j = -q_0(z_j)/q'_\infty(z_j) = W(z_j)/q'_\infty(z_j)^2 < 0$ since $W(z_j) = q'_0(z_j)q_\infty(z_j) - q_0(z_j)q'_\infty(z_j) = -q_0(z_j)q'_\infty(z_j)$ because $q_\infty(z_j) = 0$. Thus

$$T(z) = b + \sum_{j=1}^{n} \frac{r_j}{z - z_j} \tag{25.11}$$

with b real and $r_j < 0$. It follows that T is a Herglotz function. In particular

$$T'(x) = -\sum_{j=1}^{n} \frac{r_j}{(x - z_j)^2} > 0 \tag{25.12}$$

for all $x \in \mathbb{R} \setminus \{z_j\}_{j=1}^n$

But

$$T'(x) = \frac{-q_\infty(x)q'_0(x) + q'_\infty(x)q_0(x)}{q_\infty(x)^2} = \frac{-W(x)}{q_\infty(x)^2} \tag{25.13}$$

so for $x \neq z_j$, $T'(x) > 0 \Rightarrow W(x) < 0$. By hypothesis, for all j, $W(z_j) < 0$. □

We turn now to step 4, completing the proof of Theorem 25.1:

Proposition 25.7 *Suppose Theorem 25.1 holds for non-degenerate $f \in \mathcal{M}_{n-1}(a,b)$ and sets with $2n - 1$ points. Then it holds for non-degenerate $f \in \mathcal{M}_n(a,b)$ and sets with $2n+1$ points.*

Proof Since we found one q_{t_1} with all real zeros and $W(x) < 0$, the same argument as in the proof of Proposition 24.5 shows that all q_t have only real zeros, each of which is a simple zero. Relabel the x's so that $x_1 < x_2 < \cdots < x_{2n+1}$.

Since the problem is non-degenerate, there is, by Theorem 23.10, a unique rational function of degree n, h, obeying (25.1).

By Theorem 23.10, h is an $f_{t_0}^{(0)}$ for an interpolation problem with points $[x_1, \dots, x_{2n}]$ and an $f_{t_1}^{(1)}$ for an interpolation problem with points $[x_2, \dots, x_{2n+1}]$. By the analysis in the last two propositions, both problems have Wronskians, $W_0(x) < 0$ and $W_1(x) < 0$ for all $x \in \mathbb{R}$. Moreover, since we are dealing with normalized δ's,

$$\frac{\delta_0(x)}{\delta_1(x)} = \frac{x - x_1}{x - x_{2n+1}} < 0 \text{ if } x \in [x_1, x_{2n+1}] \tag{25.14}$$

If there were a pole at $y \in [x_1, x_{2n+1}]$ its residue, r, will obey $r = \delta_0(y)/W_0(y) = \delta_1(y)/W_1(y)$. Since $W_j(y) < 0$ and (25.14) holds, one equation implies $r > 0$ and the other that $r < 0$. This contradiction implies there are no poles in $[x_1, x_{2n+1}]$.

In $\mathbb{R} \setminus [x_1, x_{2n+1}]$, $\delta_0(x) > 0$, $\delta_1(x) > 0$ so by (23.27), all poles have negative residues. Therefore, if $\deg(q_t) = n$, the ratio is a Herglotz function. If $\deg(q_t) = n - 1$, by slightly perturbing the α's we get a limit of problems with denominator of degree n and so see that even in this case, the function is Herglotz. □

Notes and Historical Remarks The discussion here has common features with the presentation of Donoghue [81] and follows the discussion in Loewner [209].

Chapter 26
The Moment Problem and the Bendat–Sherman Proof

The two proofs so far have involved approximation by rational functions, and that will also be true of the proofs in Chapters 31 and 32. The measures then arise as limits of the point measures associated to rational functions. The other proofs (see Chapters 27, 28, 29, 30, 33, and 34) involve constructing the measure in some more direct way (actually, our presentation of the Korányi proof in Chapter 27 is via rational functions but his original proof of which this is a variant does not use rational approximation). This chapter is our first proof that doesn't use finite approximation. Here we'll use the moment problem on a finite interval. In fact, there are proofs of the existence of solutions to the moment problem that go through finite approximation but most do not.

Suppose we have a representation of the form (1.13) for $f \in \mathcal{M}_\infty(-1, 1)$. Then f is C^∞ and

$$\frac{f^{(n+1)}(0)}{(n+1)!} = \int_{-1}^{1} \lambda^n \, d\nu(-\lambda) \tag{26.1}$$

since

$$\frac{x}{1+\lambda x} = \sum_{n=0}^{\infty} (-\lambda)^n x^{n+1} \tag{26.2}$$

and

$$f(x) = \sum_{n=0}^{\infty} \frac{f^{(n)}(0)}{n!} x^n \tag{26.3}$$

Of course, we do not have (1.13); the whole point is to prove that. But, while we do not know f is analytic (i.e., that (26.3) holds), we do know (see Theorem 5.2) that f is C^∞, so $f^{(n)}(0)$ makes sense. Thus, it is natural to try to first find μ where

B. Simon, *Loewner's Theorem on Monotone Matrix Functions*, Grundlehren der mathematischen Wissenschaften 354, https://doi.org/10.1007/978-3-030-22422-6_26

(26.1) holds and then try to prove the series converges and to the limit f. This two-step process, which we will do in opposite order, is exactly the strategy of Bendat–Sherman [22]. First, prove analyticity on $\mathbb{D}$, the unit disk, and then solve the moment problem.

The model of analyticity is the following result of Bernstein:

Theorem 26.1 (Bernstein's Theorem) *Let f be a C^∞ function on $(-1, 1)$ with*

$$f^{(k)}(x) \geq 0 \tag{26.4}$$

on all of $(-1, 1)$. Then f is the restriction to $(-1, 1)$ of a function analytic on the unit disk $\mathbb{D}$.

Proof Let $(T_{n,y}f)(x)$ be the Taylor approximation of order n about y,

$$(T_{n,y}f)(x) = \sum_{j=0}^{n} \frac{f^{(j)}(y)}{j!}\,(x-y)^j \tag{26.5}$$

Taylor's theorem with remainder plus the intermediate value theorem says

$$f(x) - (T_{n,y}f)(x) = \frac{f^{(n+1)}}{(n+1)!}\,(\xi_{n,x,y})x - y)^{n+1} \tag{26.6}$$

for some $\xi_{n,x,y} \in [y, x]$ (resp. $[x, y]$) if $x \geq y$ (resp. $x \leq y$).

In particular, by (26.4) taking $x > 0 = y$,

$$\sum_{j=1}^{n} \frac{f^{(j)}(0)}{j!}\,x^j \leq f(x) \tag{26.7}$$

Since the left side is monotone in n, we see for $0 < x < 1$ that

$$\sum_{j=0}^{\infty} \frac{f^{(j)}(0)}{j!}\,x^j < \infty \tag{26.8}$$

which guarantees that

$$\limsup_{n\to\infty} \left(\frac{f^{(n)}(0)}{n!}\right)^{1/n} \leq 1 \tag{26.9}$$

and that

$$g(z) = \sum_{n=0}^{\infty} \frac{f^{(n)}(0)}{n!}\,z^n \tag{26.10}$$

converges for $z \in \mathbb{D}$ and define an analytic function there.

The same argument, thinking of $f(x_0+x)$ defined on $(x_0-(1-|x_0|), x_0+(1-|x_0|))$ shows that

$$\limsup_{n\to\infty}\left(\frac{f^{(n)}(x_0)}{n!}\right)^{1/n} \le (1-|x_0|)^{-1} \tag{26.11}$$

If $|x| < \frac{1}{2}$, this shows

$$|x^n| \sup_{|w|<x}\left|\frac{f^{(n+1)}(w)}{(n+1)!}\right| \to 0 \tag{26.12}$$

as $n \to \infty$. Thus, (26.6) (with $y = 0$) shows that $f(x) = g(x)$ for $x \in (-\frac{1}{2}, \frac{1}{2})$; in particular, f is real analytic on $(-\frac{1}{2}, \frac{1}{2})$. By the same argument, f is real analytic on $(x_0 - \frac{1}{2}(1-|x_0|), x_0 + \frac{1}{2}(1-|x_0|))$, so f is real analytic on $(-1, 1)$, that is, analytic in a neighborhood of $(-1, 1)$. Since it agrees with g on $(-\frac{1}{2}, \frac{1}{2})$, it can be continued to $\mathbb{D}$. □

Remark The above argument used uniqueness of analytic continuity to piece together the analytic functions given by the Taylor series near x_0 in radius $\frac{1}{2}(1-|x_0|)$ to see that f must agree with g.

Returning to $\mathcal{M}_\infty(-1, 1)$, we do not have (26.4) for all k but only for k odd (by Proposition 5.20). We will control k even by using:

Lemma 26.2 *For any C^2 function f on an interval containing $[-\delta, \delta]$, we have*

$$|f'(0)| \le 2\delta^{-1} \sup_{|y|\le\delta}|f(y)| + \tfrac{1}{2}\delta \sup_{|y|\le\delta}|f''(y)| \tag{26.13}$$

Proof By Taylor's theorem with remainder,

$$\begin{aligned}|f(\delta) - f(0) - \delta f'(0)| &= \left|\int_0^\delta (\delta - y) f''(y)\,dy\right| \\ &\le \tfrac{1}{2}\delta^2 \sup_{|y|\le\delta}|f''(y)|\end{aligned}$$

which implies that

$$|f'(0)| \le \delta^{-1}[|f(\delta)| + |f(0)|] + \tfrac{1}{2}\delta^2 \sup_{|y|\le\delta}|f''(y)|$$

□

Theorem 26.3 (Bernstein–Boas Theorem) *Let f be a C^∞ function on $(-1, 1)$ so that (26.4) holds for k odd. Then f is the restriction to $(-1, 1)$ of a function analytic on the unit disk $\mathbb{D}$. In particular, for any $R > 1$,*

$$\left|\frac{f^{(n)}(0)}{n!}\right| \leq C_R R^n \tag{26.14}$$

Proof Equation (26.14) follows by a standard Cauchy estimate (writing Taylor coefficients as contour integrals over the circle of radius R^{-1}, so C_R can be taken to be $\sup_{|z|<R^{-1}}|f(z)|$).

Let $g(x) = f(x) - f(-x)$, so $g^{(k)}(x) \geq 0$ for $k = 1, 3, 5, \dots$ and $g^{(2k)}(0) = 0$. By (26.7), we have that

$$\sum_{j=0}^{k-1} \frac{g^{(2j+1)}(0)}{(2j+1)!} x^{2j+1} \leq g(x) \tag{26.15}$$

where we used $g^{(2k)}(0) = 0$. Since $f^{(2j+1)}(0) = \frac{1}{2} g^{(2j+1)}(0)$, we see that

$$\sum_{j=0}^{\infty} \frac{f^{(2j+1)}(0)}{(2j+1)!} x^{2j+1} < \infty \tag{26.16}$$

so

$$\limsup_{j\to\infty} \left(\frac{f^{(2j+1)}(0)}{(2j+1)!}\right)^{1/2j+1} \leq 1 \tag{26.17}$$

The same argument (to get a bound at x_0, we use $g(x) = f(x) - f(2x_0 - x)$ in $(x_0 - (1 - |x_0|), x_0 + (1 - |x_0|))$) shows that

$$\limsup_{j\to\infty} \left(\frac{f^{(2j+1)}(x_0)}{(2j+1)!}\right)^{1/2j+1} \leq (1 - |x_0|)^{-1} \tag{26.18}$$

If we apply (26.13) to $f^{(2j)} \equiv (f^{(2j-1)})'$, we first see that

$$\limsup_{j\to\infty} \left(\frac{f^{(2j)}(x_0)}{(2j+1)!}\right)^{1/2j} \leq (1 - |x_0| - \delta)^{-1} \tag{26.19}$$

for each $\delta < 1 - |x_0|$, and then taking δ to zero, that (26.11) holds.

The argument now follows exactly the one in the proof of Theorem 26.1. □

Corollary 26.4 *Let $f \in \mathcal{M}_\infty(-1, 1)$. Then f has an analytic continuation to $\mathbb{D}$, and if*

$$c_n \equiv \frac{f^{(n+1)}(0)}{n!} \tag{26.20}$$

then for any $R > 1$,

$$|c_n| \leq C_R R^n \tag{26.21}$$

Remark One nice feature of this proof is that it makes clear the leap from C^∞ to real analytic. The key is positivity of $f^{(2k-1)}(x)$.

We are now ready to look at the moment problem result relevant to this case.

Theorem 26.5 (Hausdorff Moment Problem) *Let* $\{c_n\}_{n=0}^{\infty}$ *be a sequence of real numbers. Then there exists a measure* $d\mu$ *on* $[-R, R]$ *with*

$$c_n = \int_{-R}^{R} x^n \, d\mu(x) \tag{26.22}$$

if and only if

(a) *Equation* (26.21) *holds for that fixed* R.
(b) *The* $n \times n$ *Hankel matrix*

$$H_{ij}^{(n)} = c_{i+j-2} \qquad 1 \leq i, j \leq n \tag{26.23}$$

is positive for each n.

Moreover, the measure $d\mu$ *in* (26.22) *is unique.*

Proof If (26.22) holds and

$$P(z) = \sum_{j=0}^{n} a_j x^j \tag{26.24}$$

then

$$0 \leq \langle P, P \rangle_{L^2(d\mu)} = \sum_{j,k=0}^{n} \bar{a}_j a_k c_{j+k} = \sum_{j,k=1}^{n+1} \bar{a}_{j-1} a_{k-1} H_{jk}^{(n+1)} \tag{26.25}$$

so $H^{(n)} \geq 0$. Moreover, (26.22) implies

$$|c_n| \leq R^n \mu([-R, R]) \tag{26.26}$$

so (26.21) holds. Thus, (a) and (b) are necessary conditions.

Moreover, if P has the form (26.24),

$$\int P(x)\,d\mu = \sum_{j=0}^{n} a_j c_j \tag{26.27}$$

so any solution of (26.22) determines $\int P(x)\,d\mu$, and so, by Weierstrass' theorem [325, Theorem 2.4.2], $\int f(x)\,d\mu$ for any continuous f, and so $d\mu$ is unique. Thus, we only need to show that (a) + (b) $\Rightarrow$ existence of $d\mu$ obeying (26.22).

For any polynomial P, define

$$\ell(P) = \sum_{j=0}^{n} a_j c_j \tag{26.28}$$

Suppose we show that (a) and (b) imply that

$$P(x) \geq 0 \text{ for } x \in [-R, R] \Rightarrow \ell(P) \geq 0 \tag{26.29}$$

If $\|\cdot\|_\infty$ is the L^∞ norm on $[-R, R]$, $\|P_\infty\| \pm P \geq 0$ on $[-R, R]$ for any real polynomial, so

$$|\ell(P)| \leq \|P_\infty\| c_0 \tag{26.30}$$

since $c_0 = \ell(1)$. By the Weierstrass approximation theorem, ℓ extends to a function, $\tilde{\ell}$, on $C([-R, R])$ and, by (26.29),

$$f(x) > 0 \text{ on } [-R, R] \Rightarrow \tilde{\ell}(f) \geq 0$$

and so, writing $f = \lim_{\varepsilon \downarrow 0} f + \varepsilon 1$,

$$f(x) \geq 0 \text{ on } [-R, R] \Rightarrow \tilde{\ell}(f) \geq 0 \tag{26.31}$$

It follows, by the Riesz–Markov theorem [325, Theorem 4.5.4], that

$$\tilde{\ell}(f) = \int f(x)\,d\mu(x) \tag{26.32}$$

so taking $f = x^n$, we obtain (26.22). Thus, we only need to prove (26.29).

By the lemma below, any $P(x) \geq 0$ on $[-R, R]$ can be written as a finite sum of four types of terms:

$$Q(x)^2, \quad (R - x)Q(x)^2, \quad (R + x)Q(x)^2, \quad (R^2 - x^2)Q(x)^2 \tag{26.33}$$

where Q is a real polynomial. By (26.25),

$$\ell(Q^2) \geq 0 \tag{26.34}$$

for any real Q. Thus, by the Schwarz inequality for the inner product $\langle P, Q \rangle_\ell = \ell(PQ)$ for real polynomials,

$$|\ell(PQ)| \leq \ell(P^2)^{1/2} \ell(Q^2)^{1/2} \tag{26.35}$$

It follows that for any Q,

$$\begin{aligned} |\ell(xQ^2)| &\leq \ell(x^2 Q^2)^{1/2} \ell(Q^2)^{1/2} \\ &\leq \ell(x^4 Q^2)^{1/4} \ell(Q^2)^{3/4} \\ &\leq \ell(x^{2^n} Q^2)^{1/2^n} \ell(Q^2)^{1-1/2^n} \end{aligned} \tag{26.36}$$

Note that if $Q^2 = \sum_{j=0}^n b_j x^j$, then

$$|\ell(x^{2^n} Q^2)| \leq \sum_{j=0}^n |b_j| \, |c_{2^n+j}| \leq d R^{2^n}$$

by (26.21). Here d is a constant depending on Q but not on n. Thus,

$$\limsup_{n \to \infty} |\ell(x^{2^n} Q^2)|^{1/2^n} \leq R \tag{26.37}$$

so (26.26) becomes

$$|\ell(xQ^2)| \leq R \ell(Q^2)$$

that is,

$$\ell((R \pm x) Q^2) \geq 0 \tag{26.38}$$

Similarly,

$$\begin{aligned} \ell(x^2 Q^2) &\leq \ell(x^{22^n} Q^2)^{1/2^n} \ell(Q^2)^{1-1/2^n} \\ &\leq R^2 \ell(Q^2) \end{aligned}$$

so

$$\ell((R^2 - x^2) Q^2) \geq 0 \tag{26.39}$$

Equations (26.34), (26.38), (26.39), and the lemma below imply (26.29). □

Lemma 26.6 *Any polynomial P with*

$$P(x) \geq 0 \, for \, x \in [-R, R] \tag{26.40}$$

is a finite sum of terms of the form (26.33).

Proof Equation (26.40) implies that P is real, so if z_0 is a zero in $\mathbb{C}\backslash\mathbb{R}$, so is $\bar{z}_0$ with the same multiplicity. Any zero in $[-R, R]$ has to have even multiplicity. Thus,

$$P(x) = C \prod_{z_j \in \mathbb{C}_+} |x - z_j|^2 \prod_{x_j \in [-R,R]} (x - x_j)^2 \prod_{y_j \in (-\infty,R)} (x - y_j) \prod_{y_j \in (R,\infty)} (w_j - x) \tag{26.41}$$

for suitable sets of zeros in the sets shown and $C > 0$. Since each $z_j = a_j + ib_j$ has

$$|x - z_j|^2 = (x - a_j)^2 + b_j^2 \tag{26.42}$$

P is a sum of terms of the form

$$Q^2 \prod_{y_j \in (-\infty,R)} (x - y_j) \prod_{w_j \in (R,\infty)} (w_j - x)$$

Write $w_j - x = (w_j - R) + (R - x)$ and $x - y_j = (-R - y_j) + (R + x)$ and expand so that P is a finite sum of terms of the form

$$Q^2(x + R)^n (R - x)^\ell$$

For n, ℓ both even, this is a Q^2, for both odd $Q^2(R^2 - x^2)$ and for the other cases $Q^2(x + R)$ or $Q^2(R - x)$. □

We are ready to prove the hard part of Loewner's theorem:

Theorem 26.7 *If $f \in \mathcal{M}_\infty(-1, 1)$, then there is a measure $d\nu$ on $[-1, 1]$ so that*

$$f(x) = f(0) + \int \frac{x}{1 + \lambda x} \, d\nu(x) \tag{26.43}$$

Proof Let c_n be given by (26.20) for $n \geq 0$. By Corollary 26.4, c_n obeys (26.21) for each $R > 1$. By (5.38) and Theorem 5.18, $H_{ij}^{(n)}$ given by (26.23) is positive. Thus, there exists a measure $d\mu_R$ supported on $[-R, R]$ so

$$c_n = \int_{-R}^{R} \lambda^n \, d\mu_R(\lambda) \tag{26.44}$$

By uniqueness, the $d\nu_R$'s all agree, and so is a measure $d\nu$ supported on $[-1, 1]$ with $(d\nu(x) = d\mu(-x))$

$$c_n = (-1)^n \int_{-1}^{1} \lambda^n \, d\nu(x) \tag{26.45}$$

Thus, the Taylor series for f sums for $|z| < 1$ to

$$\begin{aligned}\sum_{n=0}^{\infty} \frac{f^{(n)}(0)z^n}{n!} &= f(0) + \sum_{n=1}^{\infty} z(-z)^n \int_{-1}^{1} \lambda^n \, d\nu(x) \\ &= f(0) + \int_{-1}^{1} \frac{z}{1+\lambda z} \, d\nu(\lambda)\end{aligned}$$

where the sum and integral can be interchanged since $\sum_{n=0}^{\infty}(-\lambda z)^n$ sums to $(1 + z\lambda)^{-1}$ uniformly in $|z| \le 1 - \delta$, $|\lambda| \le 1$.

By Corollary 26.4, the Taylor series converges to $f(x)$. □

Notes and Historical Remarks The idea of proving Loewner's theorem by combining the Bernstein–Boas theorem and the moment problem is due to Bendat–Sherman [22]. There is one significant simplification in our approach here. They use the Hamburger moment problem, namely they exploit the theorem that the positivity of $H^{(n)}$ given by (26.23) (without (26.21)) implies the existence of a measure $d\mu$ on $\mathbb{R}$ so $\int |x|^n \, d\mu < \infty$ for all n and

$$c_n = \int x^n \, d\mu(x) \tag{26.46}$$

Then they use (26.21) for n even to conclude that $d\mu$ is supported on $[-R, R]$.

The proof of existence of solutions to the Hamburger moment problem is not extremely hard (see Akhiezer [5], Simon [322] or [325, Sections 4.17 and 5.6] and [329, Section 7.7]), but it is more difficult than Theorem 26.5. Because the interval is not a priori bounded, one cannot use the Weierstrass theorem and, indeed, uniqueness may fail if only condition (b) of Theorem 26.5 holds.

Bernstein's theorem (Theorem 26.1) is due to Bernstein [26, 27] and is usually stated for functions with $(-1)^n f^{(n)}(x) \ge 0$ ("completely monotone functions"). By $x \to -x$, the two results are equivalent. The extension to Theorem 26.3 using Lemma 26.2 is due to Boas [42], whose result is more general (allowing, e.g., cases where one assumes $f^{(3n)}(x) \ge 0$) and so his argument is more complicated. In particular, he cannot just use $f(x) - f(-x)$ as we do.

Donoghue [81] has an alternate approach that looks at $\mathcal{M}_\infty(-1, \infty)$ instead of $\mathcal{M}_\infty(-1, 1)$ that lets him avoid needing Boas' extension. He begins by supposing $f \le 0$ on $(-1, \infty)$, which can always be arranged by first restricting to $(-1 + \varepsilon, \infty)$ so f is bounded below, and then composing f with a suitable fractional linear transformation, T, that takes $\mathbb{C}_+ \to \mathbb{C}_+$ and (a, ∞) to $(-1, 0)$ ($T(z) = 1/(a-1-z)$ works).

Since $f' > 0$ and convex, if it were ever increasing, f', and so f, would go to $+\infty$. Thus, f' is decreasing, that is, $f'' \leq 0$. By iterating, $g = f'$ obeys $(-1)^n g^{(n)} \geq 0$ on $(-1, \infty)$, so by Bernstein's theorem (applied on each $(-1, K)$ with $K \to \infty$), g is the restriction of an analytic function on $\operatorname{Re} z > -1$. Then Donoghue invokes the moment problem to prove a representation of the form (26.43) to get a Herglotz extension to $\mathbb{C}_+ \cup \mathbb{C}_-$. Instead of this argument of Donoghue, one can appeal to Corollary 10.3 to see that Bernstein's theorem applies.

We use the Weierstrass theorem [325, Theorem 2.4.2] on density of the polynomials in the continuous functions, so we should note that our arguments in Proposition 5.4 using Bernstein polynomials provide a proof of this fact; see [325, Section 2.4].

The moment problem, that is, given c_n, determining when there is a measure so that (26.46) holds, was a central problem in analysis from 1895 to 1925. The key early paper was by Stieltjes [335], who solved the problem when μ is supported on $[0, \infty)$. The general problem is called the Hamburger moment problem after Hamburger [130]. Understanding the moment problem resulted in special cases of the Riesz–Markov theorem, the Hahn–Banach theorem, and the spectral theorem that motivated the general results. See Akhiezer [5] or Simon [322] for further historical references.

The moment problem on a bounded interval is named after Hausdorff [143]. His key realization in this case is that the positivity of $\det(H^{(n)})$ which are nonlinear in c_n can be replaced by linear conditions. For example, if the interval is $[0, 1]$, his conditions are that

$$\sum_{j=0}^{n} \binom{n}{j} (-1)^j c_{m+j} \geq 0 \tag{26.47}$$

for all $m, n = 0, 1, 2, \ldots$. If the c's are moments of a measure, $d\mu$, on $[\infty, 1]$, then

$$\text{LHS of (26.47)} = \int x^m (1-x)^n \, d\mu(x)$$

so (26.47) holds. If ℓ is defined by (26.28), we see that (26.47) implies $\ell(x^m(1 - x)^n) \geq 0$. Bernstein polynomials then imply $\ell(P) \geq 0$ for any positive polynomial P and show that (26.47) is sufficient for a measure, $d\mu$, to exist; see [325, Section 4.17].

Interestingly enough, we do not use the Hausdorff conditions but instead use the Hamburger conditions supplemented by (26.21).

Chapter 27
Hilbert Space Methods and the Korányi Proof

The moral of Chapter 17 and many other situations is that given a positive matrix, use it to form an inner product. In this chapter, following Korányi [186], we will do this with the Loewner matrix. By Theorem 20.1, we need only prove a representation of the form (20.2) for $f \in \widetilde{\mathcal{M}}_\infty(-1, 1)$, that is, we can suppose the Loewner matrix is strictly positive and f is C^∞ in a neighborhood of $[-1, 1]$. Fix $f \in \widetilde{\mathcal{M}}_\infty(-1-1)$ in the remainder of this chapter.

Given $-1 \leq x_1 < x_2 < \cdots < x_n \leq 1$, define an inner product on $\mathbb{C}^n$ by

$$\langle \alpha, \beta \rangle_L = \sum_{j=1}^{n} \bar{\alpha}_j \beta_k L_{jk}(x_1, \ldots, x_n; f) \tag{27.1}$$

We are going to begin by trying to construct a vector $\varphi \in \mathbb{C}^n$ and self-adjoint matrix A on $\mathbb{C}^n$ so that

$$\langle \varphi, (A - x_j)^{-1} \varphi \rangle_L = f(x_j) \tag{27.2}$$

and

$$\|(A - x_j)^{-1} \varphi\|_L \leq f'(x_j) \tag{27.3}$$

We will want

$$(A - x_j)^{-1} \varphi = \delta_j \tag{27.4}$$

the vector in $\mathbb{C}^n$ with coordinates δ_{jk}.

If (27.4) holds, then, by a simple calculation using (27.2) and (27.4)

B. Simon, *Loewner's Theorem on Monotone Matrix Functions*, Grundlehren der mathematischen Wissenschaften 354, https://doi.org/10.1007/978-3-030-22422-6_27

$$\langle \delta_j, A\delta_k \rangle_L = x_k \langle \delta_j, \delta_k \rangle_L + \langle \delta_j, \varphi \rangle_L$$

$$= \begin{cases} x_j f'(x_j) + f(x_j) & j = k \\ \frac{x_k f(x_j) - x_k f(x_k)}{x_j - x_k} & j \neq k \end{cases} \tag{27.5}$$

It will turn out that x_j might be an eigenvalue of A, but that is not a problem if φ is orthogonal to the eigenspace! Thus, we will only use (27.2)–(27.5) for motivation.

Proposition 27.1 *There exists a unique vector φ in $\mathbb{C}^n$ so that*

$$\langle \delta_j, \varphi \rangle_L = f(x_j) \tag{27.6}$$

and a unique self-adjoint operator A on $\mathbb{C}^n$ so that (27.5) *holds. They are related by*

$$(A - x_j)\delta_j = \varphi \tag{27.7}$$

Proof Since L is strictly positive and δ_j is a basis for $\mathbb{C}^n$, (27.6) defines a unique vector and (27.5) a real symmetric matrix. By running (27.5) backwards, we see

$$\langle \delta_j, (A - x_k)\delta_k \rangle_L = f(x_j) = \langle \delta_j, \varphi \rangle_L$$

so (27.7) holds. □

Theorem 27.2 *Let P_j be the projection onto* $\mathrm{Ran}(A - x_j)$. *Then $A - x_j$ is a bijection of* $\mathrm{Ran}\, P_j$ *to itself. We let $(A - x_j)^{-1} P_j$ be the inverse of this operator on* $\mathrm{Ran}\, P_j$ *and* 0 *on* $\mathrm{Ran}(1 - P_j)$. *Then*

$$(1 - P_j)\varphi = 0 \qquad P_j \delta_j = (A - x_j)^{-1} P_j \varphi \tag{27.8}$$

and

$$\|(A - x_j)^{-1} P_j \varphi\|_L^2 \leq \|\delta_j\|_L^2 \tag{27.9}$$

There exists a measure μ on $\mathbb{R}$ supported by at most n points so that $\mu(\{x_j\}) = 0$ for all j and

$$\text{(i)} \qquad f(x_j) = \int (x - x_j)^{-1} \, d\mu(x) \tag{27.10}$$

$$\text{(ii)} \qquad f'(x_j) \geq \int (x - x_j)^{-2} \, d\mu(x) \tag{27.11}$$

Proof For any self-adjoint matrix and real λ, $(A - \lambda)$ maps $\mathbb{C}^n = \ker(A - \lambda) \oplus \mathrm{Ran}(A - \lambda)$, each summand to itself, being zero on the first factor and a bijection on the second.

Since (27.7) holds, $\varphi \in \operatorname{Ran}(A - x_j)$, that is, $(1 - P_j)\varphi = 0$. Applying P_j to (27.7) and using the fact that $(A - x_j)$ is a bijection for $\operatorname{Ran}(A - x_j)$ to itself yields the second equation (27.8). Equation (27.9) is then just $\|P_j\delta\| \le \|\delta\|$, that is, the orthogonal projection is a contraction.

Let ψ_α be an orthonormal basis of eigenfunctions for A with $A\psi_\alpha = \lambda_\alpha \psi_\alpha$. Let $\tilde{P}_\alpha$ be the projection onto $\operatorname{Ran}(A - \lambda_\alpha)$. Let μ be the point measure supported on $\{\lambda_\alpha\}_{\alpha=1}^n$ (or a proper subset!) with

$$\mu(\{\lambda_\alpha\}) = |\langle \psi_\alpha, \varphi\rangle|^2 \tag{27.12}$$

If x_j is an eigenvalue, say λ_β, then $\tilde{P}_\beta = (1 - P_\beta)$, so (27.8) implies $\varphi \in \operatorname{Ran}(\tilde{P}_\beta)^\perp$, that is, $\langle \psi_j, \varphi\rangle = 0$, so $\mu(\{x_j\}) = 0$. Of course, if x_j is not an eigenvalue, then $\mu(\{x_j\}) = 0$.

By (27.6),

$$\begin{aligned} f(x_j) &= \langle \delta_j, \varphi\rangle \\ &= \langle P_j\delta_j, \varphi\rangle \qquad (27.13) \\ &= \langle \varphi, (A - x_j)^{-1} P_j \varphi\rangle \qquad (27.14) \end{aligned}$$

is given by (27.10). In this calculation, (27.13) follows from $(1 - P_j)\varphi = 0$ and (27.14) from (27.8).

We will start dropping the subscript L on $\langle \cdot, \cdot \rangle_L$ and $\| \cdot \|_L$. By (27.1), $\|\delta_j\|^2 = f'(x_j)$, so (27.11) is just (27.9). □

Now the argument is essentially a variant of the one in Chapter 20. We will pick $n = 2\,2^m + 1$, and x_j to be $j/2^m$ $(j = -2^m, -2^m + 1, \dots, 2^m)$ and μ_m to be the corresponding measure.

Proposition 27.3 *We have*

(i) $$\mu_m\left(\left[\frac{j}{2^m} - \frac{1}{2^{m+1}}, \frac{j}{2^m} + \frac{1}{2^{m+1}}\right]\right) \le 2^{-2m-2} f'\left(\frac{j}{2^m}\right) \tag{27.15}$$

(ii) $$\mu_m([-1, 1]) \le 2^{-m-1}(1 + 2^{-m-1})\|f'\|_\infty \tag{27.16}$$

Proof Clearly,

$$2^{2m+2}\mu_m\left(\left[\frac{j}{2^m} - \frac{1}{2^{m+1}}, \frac{j}{2^m} + \frac{1}{2^{m+1}}\right]\right) \le \int \frac{d\mu_m}{|x - j2^{-m}|^2}$$

so (27.7) follows from (27.11). Summing over j, (27.15) implies (27.16). □

Let $d\tilde{\mu}_m$ be $d\mu_m$ restricted to $\mathbb{R}\setminus[-1,1]$ and define

$$f_m(x) = \int \frac{d\tilde{\mu}_m(y)}{y - x} \tag{27.17}$$

Proposition 27.4 *We have that*

$$\text{(i)} \quad \left| f_m\left(\frac{j}{2^m}\right) - f\left(\frac{j}{2^m}\right)\right| \le c\, 2^{-(m+1)/2}, \quad \textit{for } j = -2^m, -2^m + 1, \ldots, 2^m \tag{27.18}$$

$$\text{(ii)} \quad f_m'(0) \le f'(0) \tag{27.19}$$

Proof

(i) Let $x_{j,m} = j/2^m$. Then

$$\begin{aligned} |f_m(x_{j,m}) - f(x_{j,m})| &\le \int_{[-1,1]} \frac{d\mu_m(y)}{|y - x_{j_m}|} \\ &\le \left(\int \frac{d\mu_m(y)}{|y - x_{j_m}|^2}\right)^{1/2} \mu_m([-1,1])^{1/2} \end{aligned} \tag{27.20}$$

by the Schwarz inequality. By (27.11), the first term in (27.11) is bounded by $\|f'\|_\infty^{1/2}$ and by (27.16), the second by $C2^{-(m+1)/2}$.

(ii)

$$f_m'(0) = \int \frac{d\tilde{\mu}_m(y)}{|y|^2} \le \int \frac{d\mu_m(y)}{|y|^2} \le f'(0) \tag{27.21}$$

by (27.11).

□

We can now prove Loewner's theorem (given Theorem 20.1).

Theorem 27.5 *Let* $f \in \widetilde{\mathfrak{M}}_\infty(-1,1)$. *Then there exists a measure* $d\nu$ *on* $[-1,1]$ *so that*

$$f(x) = f(0) + \int \frac{x}{1 + \lambda x}\, d\nu(\lambda) \tag{27.22}$$

Proof By Theorem 3.8,

$$f_m(x) = f_m(0) + \int \frac{x}{1 + \lambda x}\, d\nu_m(\lambda) \tag{27.23}$$

By (27.19),

$$\int d\nu_m(\lambda) \leq f'(0) \tag{27.24}$$

so the measures are uniformly bounded. By (27.18), $\lim_{m\to\infty} f_m(j/2^\ell)$ exists for each ℓ, so by Lemma 1.4 and (27.24), $d\nu_m$ has a weak limit $d\nu$ for which (27.22) holds for $x = j/2^\ell$. Since these x's are dense and both sides of (27.22) are continuous, (27.22) holds. □

Notes and Historical Remarks Korányi [186] provided a Hilbert space proof of Loewner's theorem closely related to our proof here. He does not use finite approximation but directly looks at all $x \in (-1, 1)$ and defines a real Hermitian operator on an infinite dimensional Hilbert space. This operator may be unbounded, so he needs to use von Neumann's theorem on self-adjoint extensions [329, Section 7.4] and the spectral theorem for unbounded operators—big guns we can avoid by finite approximation. He then shows that his spectral measure in the form (27.10) has no support on $(-1, 1)$ by a quadratic estimate. By doing things in the opposite order (eliminating $[-1, 1)$ before passing to the limit), we need some explicit estimates that make this proof look even more like the one in Chapter 20 than Korányi's original proof.

Chapter 28
The Krein–Milman Theorem and Hansen's Variant of the Hansen–Pedersen Proof

In this chapter, we will present a proof of Loewner's theorem due to Hansen–Pedersen [136] that relies on the Krein–Milman theorem; we follow a variant of Hansen [133]. Both variants rely on some a priori bounds on functions lying in some $\mathcal{M}_\infty(a, b)$ for suitable (a, b) to prove compactness of a suitably normalized subset, both use cleverly chosen mappings of the subset of $\mathcal{M}_\infty$ to itself to identify the extreme points and both exploit some relations between matrix monotone and matrix convex functions. The Hansen variant has $(a, b) = (0, \infty)$, while the original proof used $(a, b) = (-1, 1)$. Because the connection between $\mathcal{C}_\infty$ and $\mathcal{M}_\infty$ is direct for $(0, \infty)$, the use of those ideas is simpler in the latter variant and some other technicalities are also easier in this second proof which is why we follow it.

We begin by recalling the Krein–Milman theorem. While the correct framework is the theory of locally convex spaces, in our application, the topology will be a weak topology, so it suffices to consider such topological vector spaces, which we will do (see further discussion on this point in the Notes).

A *non-degenerate pairing* or *dual pair* is a pair of vector spaces, X and Y, over $\mathbb{C}$ and a bilinear map $x, y \mapsto \langle y, x\rangle$ of $X \times Y$ into $\mathbb{C}$ which is non-degenerate in that for any $x \in X\backslash\{0\}$, there is $y \in Y$ with $\langle y, x\rangle \neq 0$, and for any $y \in Y\backslash\{0\}$, there is $x \in X$ with $\langle y, x\rangle \neq 0$.

Thus, $x \mapsto \langle y, x\rangle = \ell_y(x)$ says each y defines a linear function on X and the latter half of the nondegeneracy theory says the map of Y into $(X^*)_{\text{alg}}$, the set of all linear functionals on X, is injective. So we can view Y as a set of linear functions on X which separates points.

The weak topology, $\sigma(X, Y)$, on X associated to the dual pair X, Y is the weakest topology in which each ℓ_y is continuous, that is, it is the topology generated by the open sets $\{x \mid |\langle y, x\rangle| \leq 1\}$ and their translates. A net x_α converges to x if and only if $\langle y, x_\alpha\rangle \to \langle y, x\rangle$ for each y.

A convex set, S, is a subset of a vector space, X, in which $x, y \in S$ and $\theta \in [0, 1]$ implies $\theta x + (1 - \theta)y \in S$. By induction, it follows that if

B. Simon, *Loewner's Theorem on Monotone Matrix Functions*, Grundlehren der mathematischen Wissenschaften 354, https://doi.org/10.1007/978-3-030-22422-6_28

$$\theta_1, \ldots, \theta_n \geq 0, \ \sum_{j=1}^{n} \theta_j = 1, \ x_1, \ldots, x_n \in S \Rightarrow \sum_{j=1}^{n} \theta_j x_j \in S \tag{28.1}$$

Sums like those in (28.1) are called *convex linear combinations* of the x's. Given $U \in S$, the set of all such sums with $x_j \in U$ is called the *convex hull* of U, and its closure the *closed convex hull.*

A point $x \in S$, a convex set, is called an *extreme point* of S if it is an interior point of no nontrivial line segment, that is, $x = \theta y + (1 - \theta)z$, $0 < \theta < 1$ with $y, z \in S$ implies $x = y = z$. This notion is nontrivial already in $\mathbb{R}^2$ (or $\mathbb{C}$). The open unit disk has no extreme points; the closed disk has its boundary as its extreme points; and a closed triangle has its vertices as extreme points. The closed unit ball in $L^1([0, 1], dx)$ is an example of a closed bounded convex subset in the infinite dimensional case with no extreme points. The Krein–Milman theorem says that compactness rather than merely closed and bounded implies lots of extreme points.

Theorem 28.1 (Krein–Milman Theorem) *Let X, Y be a dual pair and $S \subset X$ a convex subset which is compact in the $\sigma(X, Y)$ topology. Then any point in S is a limit of convex combinations of extreme points of S, that is, the closed convex hull of the extreme points is all of S.*

Remarks

1. In particular, S has extreme points. Indeed, once one knows that there are extreme points for any closed convex subset of X, the density result is not hard.
2. In the Notes, we will give a sketch of the proof.

Limits of point sums tend to be integrals, so this suggests that in nice cases, points in S are integrals of extreme points. One implementation of this idea is Choquet theory; we will instead focus on an easy version that works when the extreme points, $\mathcal{E}(S)$, are themselves closed.

We need to begin with the notion of integrating points in S. We will suppose that S is metrizable, so we can use Baire measure theory. S is metrizable if and only if there exists a countable family $y_n \in Y$ so that $x_\alpha \to x$ (*all* in S) if and only if $\langle y_n, x_\alpha \rangle \to \langle y_n, x \rangle$ for each n.

Proposition 28.2 (Existence of Barycenters) *Let X, Y be a dual pair. Let $S \subset X$ be a convex subset of X which is metrizable and compact in the $\sigma(X, Y)$ topology. Then for any probability measure, $\mu \in \mathcal{M}_{+,1}(S)$, there exists a point $B(\mu)$ in S (called the* barycenter *of μ) so that for all $y \in Y$,*

$$\langle y, B(\mu) \rangle = \int_S \langle y, x \rangle \, d\mu(x) \tag{28.2}$$

Remark It is easy to see that $\mu \mapsto B(\mu)$ is continuous if $\mathcal{M}$ is given the weak topology.

Proof Uniqueness of B is obvious because if there are two points z_1, z_2 so that for $j = 1, 2$,

$$\langle y, z_j \rangle = \int_S \langle y, z \rangle \, d\mu(x)$$

then $\langle y, z_1 - z_2 \rangle = 0$ for all y which, by nondegeneracy, implies $z_1 = z_2$.

For existence, we start by noting that if $\mu_x(f) = f(x)$, then measures of the form $\sum_{j=1}^n \theta_j \mu_{x_j}$, with $\sum_{j=1}^n \theta_j = 1, \theta_j \geq 0$, are dense in $\mathcal{M}_{+,1}(S)$ in the weak topology (i.e., $\sigma(\mathcal{M}(S), C(S))$ topology). To see this, use the Krein–Milman theorem! $\mathcal{M}_{+,1}$ is compact and the extreme points are $\{\mu_x\}_{x \in S}$.

For μ of this finite pure point form, define

$$B\left(\sum_{j=1}^n \theta_j \mu_{x_j}\right) = \sum_{j=1}^n \theta_j x_j \tag{28.3}$$

and note that (28.2) is obvious. If μ is any measure, by the last paragraph, there exists $\mu_n \to \mu$ weakly where $\mu_n = \sum_{j=1}^n \theta_j^{(n)} \mu_{x_j^{(n)}}$. By compactness, $B(\mu_n)$ given by (28.3) has a weak limit z. Then

$$\begin{aligned}
\langle y, z \rangle &= \lim_{n(k)} \langle y, B(\mu_{n(k)}) \rangle \\
&= \lim_{n(k)} \int_S \langle y, x \rangle \, d\mu_{n(k)} \\
&= \int_S \langle y, x \rangle \, d\mu
\end{aligned}$$

□

Under one additional condition, we can easily see that any point is an integral of extreme points:

Theorem 28.3 (Strong Krein–Milman Theorem) *Let X, Y be a dual pair. Let $S \subset X$ be a convex subset of X which is metrizable and compact in the $\sigma(X, Y)$ topology. Suppose the set of extreme points, $\mathcal{E}(S)$, is closed in S. Then for any $x \in S$, there is a probability measure on $\mathcal{E}$, $\mu \in \mathcal{M}_{+,1}(\mathcal{E}(S))$ so that $x = B(\mu)$.*

Proof Fix $x_0 \in S$. By the Krein–Milman theorem, there exist points $x_j^{(n)} \in \mathcal{E}(S)$ and reals $\{\theta_j^{(n)}\}_{1 \leq j \leq n < \infty}$ with $\theta_j^{(n)} \geq 0$ and $\sum_{j=1}^n \theta_j^{(n)} = 1$ so that

$$\sum_{j=1}^n \theta_j^{(n)} x_j^{(n)} \to x_0$$

in the $\sigma(X, Y)$ topology.

In $\mathcal{M}_{+,1}(\mathcal{E}(S))$, let

$$d\mu_n = \sum_{j=1}^{n} \theta_j^{(n)} d\mu_{x_j^{(n)}}$$

By compactness of $\mathcal{M}_{+,1}(\mathcal{E}(S))$, $d\mu_n$ has a weak limit μ, and clearly,

$$\begin{aligned}\int \langle y, x\rangle \, d\mu(x) &= \lim \sum_{j=1}^{n} \left\langle y, \sum \theta_j^{(n)} x_j^{(n)} \right\rangle \\ &= \langle y_0, x_0 \rangle\end{aligned}$$

□

The key to the Hansen variant proof is a pair of theorems about a particular subset of $\mathcal{M}_\infty(0,\infty)$. Recall that in Chapter 10 we defined

$$\mathcal{P} = \{f \in \mathcal{M}_\infty(0,\infty) \mid f(t) > 0 \text{ for all } t \in (0,\infty)\} \tag{28.4}$$

Here we focus on

$$\mathcal{P}_1 = \{f \in \mathcal{P} \mid f(1) = 1\} \tag{28.5}$$

The two key results are

Theorem 28.4 (Hansen–Pedersen [136], Hansen [133]) *Let $Y = \{\sum_{j=1}^{m} a_j \delta_{x_j} \mid x_j \in (0,\infty)\}$ where $\delta_{x_j}(f) = f(x_j)$. Then $(C(0,\infty), Y)$ is a dual pair and $\mathcal{P}_1$ is compact and convex in the $\sigma(C(0,\infty), Y)$ topology. Convergence in $\mathcal{P}_1$ is equivalent to convergence of $\{\delta_x\}_{x\in\mathbb{Q}\cap(0,\infty)}$ (so $\mathcal{P}_1$ is metrizable) and to convergence uniformly on each $[\varepsilon, \varepsilon^{-1}]$ for any $\varepsilon > 0$.*

Theorem 28.5 (Hansen–Pedersen [136], Hansen [133]) *For $\lambda \in [0,1]$, define*

$$\varphi_\lambda(t) = \frac{t}{\lambda + (1-\lambda)t} \tag{28.6}$$

Then each $\varphi_\lambda \in \mathcal{P}_1$, the set of (28.5), *and $\{\varphi_\lambda\}_{\lambda\in[0,1)}$ is precisely the set of extreme points of $\mathcal{P}_1$.*

Remark These two results, as stated, are from [133]. Earlier, [136] had analogous results but for a suitable subset of $\mathcal{M}_\infty(-1,1)$, not $\mathcal{M}_\infty(0,\infty)$. As usual, the hard part of Loewner's theorem for one set implies it for the other.

These two results and the strong Krein–Milman theorem imply the hard part of Loewner's theorem.

Theorem 28.6 *If $\mathcal{P}_1$ is given by (28.5), then for any $f \in \mathcal{P}_1$, there is a probability measure $d\nu$ on $[0, 1]$ so that*

$$f(t) = \int \frac{t}{\lambda + (1-\lambda)t} \, d\nu(\lambda) \tag{28.7}$$

Remarks

1. If f is nonconstant and in $\mathcal{M}_\infty(0, \infty)$, then for any $s > 0$

$$g_s(t) = \frac{f(t+s) - f(s)}{f(1+s) - f(s)} \tag{28.8}$$

lies in $\mathcal{P}_1$, so by (28.7), $f(t)$ is real analytic on (s, ∞) with an analytic continuation to $\mathbb{C}_+$ with $\operatorname{Im} f(z) > 0$ there, so, since s is arbitrary, we have the hard half of Loewner's theorem.
2. Equation (28.7) is, of course, (3.59).

Proof Assuming Theorems 28.4 and 28.5 $\{\varphi_\lambda\}_{\lambda \in [0,1]}$ is closed in the topology of pointwise convergence and the usual topology on $[0, 1]$ is the same as this topology of pointwise convergence. Thus, by Theorem 28.3, $f = B(\nu)$ for some probability measure ν on $[0, 1]$. Equation (28.7) is then just (28.2) for $y = \delta_t$. □

To prove Theorem 28.4, we need some preliminary inequalities.

Proposition 28.7 *Let $f \in \mathcal{P}_1$. Then*

(a) For all $s > t > 0$, we have that

$$0 < f(t) \le f(s) \le f(t) \left(\frac{s}{t}\right) \tag{28.9}$$

(b) For all $t > 0$, we have that

$$f(t) \le 1 + t \tag{28.10}$$

(c) For all $s > t > 0$, we have that

$$0 \le f(s) - f(t) \le (1 + t^{-1})(s - t) \tag{28.11}$$

Proof

(a) By Theorem 10.2, f is a scalar concave function, so for $s > t$, $f(s)$ lies below the straight line from $(0, \alpha)$ through $(t, f(t))$ where $\alpha = \lim_{u \downarrow 0} f(u)$, which exists since f is scalar monotone and bounded below. That is

$$s > t \Rightarrow f(s) \le \frac{f(t) - \alpha}{t}(s-t) + f(t)$$
$$\le \frac{f(t)}{t}(s-t) + f(t) = f(t)\left(\frac{s}{t}\right) \tag{28.12}$$

since $\alpha \ge 0$.

(b) Set $t = 1$ to see if $s \ge 1$, then $f(s) \le s$. If $0 \le t \le 1$, then by scalar monotonicity, $f(t) \le f(1) = 1$ so (28.10) holds since $\max(1, t) \le 1 + t$.

(c) By (28.9),

$$0 \le f(s) - f(t) \le f(t)\left[\frac{s}{t} - 1\right]$$
$$= \frac{f(t)}{t}(s-t) \le \frac{1+t}{t}(s-t) \tag{28.13}$$

which is (28.11).

□

Proof of Theorem 28.4 We need the Arzelà–Ascoli theorem [325, Theorem 2.3.14]. By (28.10)/(28.11), $\mathcal{P}_1$ is uniformly equicontinuous and uniformly bounded on each $[\varepsilon, \varepsilon^{-1}]$, so pointwise convergence is equivalent to uniform convergence on each such set and to pointwise convergence at $\mathbb{Q} \cap (0, \infty)$ (so the topology is metrizable) and $\mathcal{P}_1$ is compact by the above mentioned theorem. □

Next, we turn to proving that each of the $\{\varphi_\lambda\}_{\lambda \in [0,1]}$ are extreme points of $\mathcal{P}_1$. The key will be the extremal property, (a), of the following:

Proposition 28.8

(a) For any $f \in \mathcal{P}_1$, let $\lambda = f'(1)$. Then for all $t \in (0, \infty)$, we have that

$$f(t) \ge \varphi_\lambda(t) \tag{28.14}$$

(b) For any $t \in (0,1) \cup (1, \infty)$, we have that $\lambda \mapsto \varphi_\lambda(t)$ is strictly convex, i.e., if $\lambda_1 \ne \lambda_2$, $\theta \in (0,1)$, and $\lambda = \theta\lambda_1 + (1-\theta)\lambda_2$, then

$$\varphi_\lambda(t) < \theta\varphi_{\lambda_1}(t) + (1-\theta)\varphi_{\lambda_2}(t) \tag{28.15}$$

Proof

(a) We use the function, $f^\sharp(t) = t/f(t)$, of (10.20). By Theorem 10.11, $f \in \mathcal{P}_1 \Rightarrow f^\sharp \in \mathcal{P}_1$. By simple calculations, $f^\sharp(1) = 1$, $(f^\sharp)'(t) = f(t)^{-2}(f(t) - tf'(t))$ so $(f^\sharp)'(1) = 1 - \lambda$. Since $f^\sharp$ is scalar convex (by Theorem 10.2), we conclude that

$$f^\sharp(t) \le \lambda + (1-\lambda)t \tag{28.16}$$

which implies that $f(t) \ge t/[\lambda + (1-\lambda)t] = \varphi_\lambda(t)$.

(b) By a direct calculation

$$\frac{\partial^2}{\partial\lambda^2}\varphi_\lambda(t) = \frac{2t(1-t)^2}{(\lambda + (1-\lambda)t)^3} > 0 \tag{28.17}$$

if $t \neq 1$, which implies strict convexity.

□

Theorem 28.9 *Each φ_λ ($\lambda \in [0, 1]$) is an extreme point of $\mathcal{P}_1$.*

Proof Fix λ. Suppose that for some $\theta \in (0, 1)$, $f, g \in \mathcal{P}_1$, we have that

$$\varphi_\lambda(t) = \theta f(t) + (1-\theta)g(t) \tag{28.18}$$

Let

$$\lambda_1 = f'(1), \lambda_2 = g'(1) \Rightarrow \lambda = \theta\lambda_1 + (1-\theta)\lambda_2 \tag{28.19}$$

Let $t \neq 1$. Then

$$\begin{aligned}\varphi_\lambda(t) &= \theta f(t) + (1-\theta)g(t) \\ &\geq \theta\varphi_{\lambda_1}(t) + (1-\theta)\varphi_{\lambda_2}(t) \qquad (28.20)\\ &\geq \varphi_\lambda(t) \qquad (28.21)\end{aligned}$$

where (28.20) uses (28.14) and (28.21) uses (28.15). Clearly, we must have equality in (28.20) and (28.21). The latter implies that $\lambda_1 = \lambda_2$ (since, by (28.15), the inequality is strict if $\lambda_1 \neq \lambda_2$). The former implies that for all $t \neq 1$ (and so, by continuity, for all t), we have that $f(t) = \varphi_{\lambda_1}(t)$, $g(t) = \varphi_{\lambda_2}(t)$. Thus $f = g = \varphi_\lambda$ implying that φ_λ is an extreme point. □

What remains is to show that any extreme point is a φ_λ. We begin with

Proposition 28.10 *For and $f \in \mathcal{P}_1$, we have that*

$$0 \leq f'(1) \leq 1 \tag{28.22}$$

Moreover,

$$f'(1) = 0 \Rightarrow f \textit{ is constant;} (= \varphi_0); \quad f'(1) = 1 \Rightarrow f(t) = t (= \varphi_1)$$

In particular, the only extreme points with $f'(1) = 0, 1$ are φ_0, φ_1.

Proof By the calculation in the proof of Proposition 28.8 (a)

$$f'(1) + (f^\sharp)'(1) = 1 \tag{28.23}$$

so, since $f'(1) \geq 0$, we see (28.22) holds.

If f is nonconstant, $f'(1) > 0$ by Theorem 14.1 so $f'(1) = 0 \Rightarrow f$ is constant. Similarly, $f'(1) = 1 \Rightarrow (f^\sharp)(t) = 1 \Rightarrow f(t) = 1^\sharp(t) = t$.

This shows that φ_0, φ_1 are the only points with $f'(1) = 0, 1$, so the only possible extreme points. □

If $f \in \mathcal{P}_1, f \neq \varphi_0, \varphi_1$, we define two functions

$$Tf(t) = \frac{t}{f'(1)} \begin{cases} \dfrac{f(t)-1}{t-1}, & t \neq 1 \\ f'(1), & t = 1 \end{cases} \tag{28.24}$$

$$Sf(t) = \frac{1}{1-f'(1)} \begin{cases} \dfrac{f(t)-t}{1-t}, & t \neq 1 \\ 1-f'(1), & t = 1 \end{cases} \tag{28.25}$$

These definitions are made so that with $\theta = f'(1)$, we have that

$$\begin{aligned} \theta(Tf)(t) + (1-\theta)(Sf)(t) &= t\left(\frac{f(t)-1}{t-1}\right) + \left(\frac{f(t)-t}{1-t}\right) \\ &= f(t) \end{aligned} \tag{28.26}$$

Lemma 28.11 *If $f \in \mathcal{P}_1, f \neq \varphi_0, \varphi_1$, then $Tf, Sf \in \mathcal{P}_1$*

Proof Let

$$h(t) = \frac{1}{f'(1)} \frac{f(t)-1}{t-1}; \qquad g(t) = h(t^{-1}) \tag{28.27}$$

Since f is matrix concave, by Theorem 9.1, $-h$ is matrix monotone, so h is order reversing on positive matrices. Since $t \mapsto t^{-1}$ is also order reversing on strictly positive matrices, we see that $g \in \mathcal{P}_1$. By a direct computation

$$Tf(t) = th(t) = g^*(t) \tag{28.28}$$

is in $\mathcal{P}_1$ by Theorem 10.11.

By a direct computation $Sf = (Tf^*)^*$, so, by Theorem 10.11 again, $Sf \in \mathcal{P}_1$ also. □

Proof of Theorem 28.5 Let f be an extreme point of $\mathcal{P}_1$. If $f'(1) = 0, 1$, then $f = \varphi_0, \varphi_1$ as we've seen. If $0 < f'(1) < 1$, we have (28.26) with $\theta = f'(1)$. Since f is an extreme point and $\theta \in (0, 1)$, we must have that $f = Tf$. Solving for $f(t)$ yields $f(t) = t/(t + \lambda(t-1)) = \varphi_\lambda(t)$. □

There is another proof that each φ_λ is an extreme point due to Boutet de Monvel [45] that is illuminating and that proves more:

Theorem 28.12 *Each function* $\{\varphi_\lambda(x)\}_{0\le\lambda\le 1}$ *is an extreme point of*

$$\{f \in \mathcal{M}_n(0,\infty) \mid f(1) = 1\} = \mathcal{P}_1^{(n)} \tag{28.29}$$

for each $n \ge 2$.

Proof By a direct computation,

$$\frac{\varphi_\lambda(t) - \varphi_\lambda(s)}{t-s} = \frac{1}{(1+\lambda t)(1+\lambda s)} \tag{28.30}$$

so

$$L_2(t,s;\varphi_\lambda) = ((1+\lambda s)^{-1}(1+\lambda t)^{-1})\begin{pmatrix}(1+\lambda s)^{-1} \\ (1+\lambda t)^{-1}\end{pmatrix} \tag{28.31}$$

is rank one.

Suppose $\varphi_\lambda = \theta g + (1-\theta) f$ with $f, g \in \mathcal{P}_1^{(n)}$ and $0 < \theta < 1$. Since the only positive matrices less than a rank one matrix are multiples of that rank one matrix,

$$L_2(1,x;g) = cL_2(1,x;\varphi_\lambda) \tag{28.32}$$

But $L_2(1,x;h)_{11} = 1$ for all $h \in \mathcal{P}_1^{(n)}$, so $c = 1$ in (28.32) and thus $g'(x) = \varphi'_\lambda(x)$ looking at the 22 entries. Thus $g = \varphi_\lambda$ and so φ_λ is an extreme point. □

Corollary 28.13 $\mathcal{M}_\infty(0,\infty)$ *is a face of* $\mathcal{M}_n(0,\infty)$ *for each* n.

Proof This is immediate from the fact that any extreme point of $\mathcal{M}_\infty$ is an extreme point of $\mathcal{M}_n$. □

Open Question Is $\mathcal{M}_{n+1}(0,\infty)$ a face of $\mathcal{M}_n(0,\infty)$?

Notes and Historical Remarks The Krein–Milman theorem (Theorem 28.1) is due to Krein–Milman [192]. The now standard proof that we sketch below is due to Kelley [180]. The strong Krein–Milman theorem (Theorem 28.3) is due to Choquet [60] and is a precursor of his more general Choquet theory also discussed below. It is a remarkable fact that most of the integral representation theorems of classical analysis can be viewed as versions of the strong Krein–Milman theorem, including Bernstein's theorem on completely monotone functions on $[0,\infty)$, Bochner's theorem, the Herglotz representation theorem (see Armitage [18] and Holland [164]), and the Levy–Khintchine formula. Details for many of these can be found in Choquet's three-volume series [61] and Simon [324].

The Kelley proof goes as follows (see also [325, Section 5.11] : A *face* of a compact convex set, S, is a nonempty closed convex subset, F, so that $x, y \in S$ and

$\theta x + (1-\theta)y \in F$ for $0 < \theta < 1$ implies $x, y \in F$. An extreme point is precisely a one-point face. Any chain of decreasing faces $\{F_\alpha\}_{\alpha \in I}$, $F_\alpha \subset F_\beta$, if $\alpha > \beta$, has a lower bound, namely $\bigcap_\alpha F_\alpha$ which is nonempty by compactness. Thus, by Zorn's lemma, minimal faces exist.

If F is a minimal face and F has more than one point, say $x_0 \neq x_1$, find $\ell \in Y$ so $\ell(x_0) \neq \ell(x_1)$. Let $F_1 = \{x \in F \mid \ell(x) = \sup_{y \in F} \ell(y)\}$. Then F_1 is a face. It is nonempty by compactness and $F_1 \subsetneq F$ since not both of x_0, x_1 lie in F. This contradicts minimality, so minimal faces are single points, hence extreme points. Thus, $\mathcal{E}(S) \neq \emptyset$.

Let $\tilde{S}$ be the closed convex hull of $\mathcal{E}(S)$. If $\tilde{S} \neq S$, pick $x_0 \in S \backslash \tilde{S}$, By separating hyperplane theorems [325, Section 5.10] since $\tilde{S}$ is convex, there is $\ell \in Y$, so $\ell(x_0) > \sup_{y \in \tilde{S}} \ell(y)$. $F = \{x \in S \mid \ell(x) = \sup_{y \in S} \ell(y)\}$ is a face disjoint from $\tilde{S}$. But $\mathcal{E}(F) \neq \emptyset$, so let x_1 be an extreme point of F. By a simple argument, it is an extreme point of S but, by construction, it is not in $\tilde{S}$! This contradiction proves the theorem.

There are various ways to see that there is no loss in looking only at the weak topology. For a general compact convex subset, S, of locally convex vector spaces, X, S is automatically compact in the $\sigma(X, X^*)$ topology. Thus, the map of S with the original topology to S with the $\sigma(X, X^*)$ topology is a continuous bijection between compact sets, hence a homeomorphism. That is, we have shown that on any compact convex subset of a locally convex space, the restriction of the regular topology is equal to the restriction of the weak topology.

In fact, if X is separable, it is enough to have the result for subsets of ℓ^2 in the weak topology; see Phelps [272], who has remarked that any separable convex subset of a locally convex space is affinely homeomorphic to a compact convex subset of ℓ^2.

The strong Krein–Milman theorem has a slight generalization with the same proof: In any compact convex set S, for any $x \in S$, there is a measure on $\overline{\mathcal{E}(S)}$, the closure of the extreme points, whose barycenter is μ. It is actually this result that Hansen–Pedersen use (see below). There are interesting examples where $\overline{\mathcal{E}(S)}$ is all of S, so this theorem may say nothing. For example, when one has an indeterminate moment problem, the extreme points are dense; see [322]. Also, there is a class of complexes called Poulsen simplices discussed in Poulsen [275] and Lindenstrauss, Olsen, and Sternfeld [205], where the extreme points are dense.

If $\mathcal{E}(S)$ is not closed but is still a Baire subset of S, one can ask if, given $x \in S$, there is a measure μ on S with $B(\mu) = x$ and $\mu(S\backslash\{0\}) = 0$. The answer is yes when S is metrizable, in which case, $\mathcal{E}(S)$ is always a Baire set (indeed, a Baire G_δ). This is the content of Choquet's theorem. Its extensions, uniqueness issues, etc., are the subject of Choquet theory and discussed in Choquet's three books [61].

A compact, convex set where the Choquet representing measure is unique is called a *simplex*. If also $\mathcal{E}(S)$ is closed, it is called a *Brauer simplex*. Our set, S, is a Brauer simplex.

The proof in this chapter is mainly that in the Hansen's [133]. The strategy follows a much earlier paper of Hansen–Pedersen [136] in that one finds extreme points by constructing maps $f \mapsto Tf, Sf$ so that f is a convex combination of the two which implies that if f is an extreme point, one must have $f = Tf$. In [136], they deal with a subset of $\mathcal{M}_\infty(-1, 1)$, not $\mathcal{M}_\infty(0, \infty)$ and instead of building Tf out of $f \mapsto f^*$, they use the maps $f \mapsto x^{-1}(x \pm 1)f$ which they show map $\mathcal{M}_\infty(-1, 1)$ to itself. The technicalities in [133] are simpler, in part, because of the direct relation of $\mathcal{C}_\infty(0, \infty)$ and $-\mathcal{M}_\infty(0, \infty)$. Bhatia[33] and Hiai–Petz[157] have textbook expositions of the original Hansen–Pedersen proof.

The one major new element here is our direct proof that each φ_λ is an extreme point (Theorem 28.12). Hansen–Pedersen and Hansen prove this indirectly as follows: He proves (and we follow his proof) that $\{\varphi_\lambda\}$ are the only possible extreme points. By using the variant of the strong Krein–Milman theorem above, one still gets existence of integral representations. Uniqueness of representation (i.e., Theorem 1.5) implies that φ_λ is an extreme point.

The argument that the φ_λ are extreme in $\mathcal{P}_1$ because the Loewner matrix is rank one is due to Boutet de Monvel [45]. That this implies they are a face in $\mathcal{M}_n$ is an immediate consequence, although $\mathcal{M}_n$ is not discussed in his note. It follows from (28.30) that all the Loewner matrices for φ_λ (i.e., $L_k(x_1, \ldots, x_k; \varphi_t)$) are rank one.

There is another illuminating way of seeing that the φ_λ's are all extreme points, knowing that they are the only possible extreme points. First, one notes that φ_0, φ_1 are extreme points.

For the other φ_λ's, one proceeds as follows: Consider the cone generated by $\mathcal{P}_1$, that is, $\mathcal{P}$. Let Q be a conformal map of taking $(0, \infty)$ monotonically to itself. Then

$$\mathcal{Q}(f) = f \circ Q$$

maps $\mathcal{P}$ invertibly to itself, so it takes extreme rays to extreme rays. A simple calculation shows that $\mathcal{Q}(\varphi_\lambda)$ is a multiple of some other φ_μ so φ_λ is an extreme point of $\mathcal{P}_1$ if and only if φ_μ is extreme. The orbit of any λ_0 under the set of such maps is all of $(0, 1)$, so either all φ_λ are extreme points of $\mathcal{P}_1$ or only φ_0, φ_1 are. But if they are the only extreme points, then $\mathcal{P}_1$ is one-dimensional, which it is not!

The Hansen–Pedersen proof has the usual drawback of Krein–Milman-type proofs: Miracles are unexplained! The magical element of the hard part of Loewner's theorem is that matrix monotonicity implies analyticity. In this proof, the reason that happens is that each extreme point is analytic (with some kind of uniformity), but there is no explanation why this is so. In fact, the proof only requires f to be C^1, so there is not even an explanation why monotone matrix functions are C^∞.

Chapter 29
Positive Functions and Sparr's Proof

The key to many proofs of Loewner's theorem is finding a measure. Indeed, to prove (1.13), we want a measure $d\nu$ with

$$f(x) - f(0) = \int_{-1}^{1} \frac{x}{1 + \lambda x}\, d\nu(\lambda) \tag{29.1}$$

and a measure is basically a positive functional on $C(X)$. The guts of proving such a ν exists is clearly proving for any $x_j \in [-1, 1]$ and $a_j \in \mathbb{R}$, $j = 1, \dots, k$, we have

$$\sum_{j=1}^{k} a_j \frac{x_j}{1 + \lambda x_j} \geq 0 \text{ for } \lambda \in [-1, 1] \Rightarrow \sum_{j=1}^{k} a_j (f(x) - f(0)) \geq 0 \tag{29.2}$$

This is essentially the strategy of a proof of Sparr [332], except that he works with $\mathcal{M}_\infty(0, \infty)$ and with slightly less general a's than $a \in \mathbb{R}^k$. Indeed, to get rid of the pesky $f(0)$ term, it is clearly convenient to assume $\sum_{j=1}^{k} a_j = 0$.

Let $x \in (0, \infty)$ and $\lambda \in [-\infty, 0]$ ($-\infty$ allowed) and define

$$\varphi_x(\lambda) = \frac{1 + x\lambda}{\lambda - x} \tag{29.3}$$

a reasonable function, given Theorem 3.4. For $\lambda = -\infty$, we interpret

$$\varphi_x(-\infty) \equiv x \tag{29.4}$$

By a Sparr function, we mean a continuous function g on $[-\infty, 0]$ of the form

$$g = \sum_{j=1}^{2n} a_j \varphi_{x_j} \tag{29.5}$$

B. Simon, *Loewner's Theorem on Monotone Matrix Functions*, Grundlehren der mathematischen Wissenschaften 354, https://doi.org/10.1007/978-3-030-22422-6_29

for some distinct $x_1, \dots, x_{2n}$ in $(0, \infty)$ and $(a_1, \dots, a_{2n}) \in \mathbb{R}^{2n}$ with

$$\sum_{j=1}^{2n} a_j = 0 \tag{29.6}$$

We call n the order of f. Given (29.2) and (3.33), the key to Sparr's proof is

Theorem 29.1 (Sparr's Lemma) *If $f \in \mathcal{M}_{n+1}(0, \infty)$ and g is a Sparr function of order n given by (29.5) with $g(\lambda) \geq 0$ for $\lambda \in [-\infty, 0]$, then*

$$\sum_{j=1}^{2n} a_j f(x_j) \geq 0 \tag{29.7}$$

We will first show that this implies Loewner's theorem and then turn to the proof of Sparr's lemma.

Proposition 29.2

(i) *For any $x_1, \dots, x_{2n}$ distinct, $\{\varphi_{x_j}\}_{j=1}^{2n}$ are linearly independent in $C([-\infty, 1])$.*
(ii) *If $x < y$, then $\varphi_y - \varphi_x$ is strictly positive on $C([-\infty, 1])$.*
(iii) *The Sparr functions are dense in $C([-\infty, 1])$.*
(iv) *The strictly positive Sparr functions are dense in the non-negative functions in $C([-\infty, 1])$*

Proof

(i) Suppose that

$$\sum_{j=1}^{2n} a_j \varphi_{x_j}(\lambda) = 0$$

for all $\lambda \in [-\infty, 0]$. Then, since the function on the left is analytic on $\mathbb{C} \setminus \{x_j\}_{j=1}^{2n}$, the function is zero. But the residue of the pole at $\lambda = x_j$ is $a_j(1 + x_j^2)$, so we conclude that each $a_j = 0$.
(ii) By a direct calculation,

$$\varphi_y(\lambda) - \varphi_x(\lambda) = \frac{(y - x)(1 + \lambda^2)}{(\lambda - y)(\lambda - x)} \tag{29.8}$$

This function is $y - x$ at $-\infty$ and non-vanishing on $(-\infty, 0]$. By compactness of $[-\infty, 0]$ and continuity, the minimum is strictly positive.
(iii) Let $\mathcal{F}$ be the set of Sparr functions. By (29.8),

$$\frac{\varphi_y - \varphi_{x_0}}{y - x_0} \rightarrow \frac{(1 + \lambda^2)}{(\lambda - x_0)^2} \tag{29.9}$$

uniformly on $[-\infty, 0]$ as $y \downarrow x_0$. Since the left side of (29.9) are Sparr functions, the right side lies in $\overline{\mathcal{F}}$. Taking successive derivatives (checking convergence uniformly of the differences), we see

$$\eta_n(\lambda) = \frac{(1+\lambda^2)}{(\lambda-1)^{2+n}} \tag{29.10}$$

lies in $\overline{\mathcal{F}}$ for each $n = 0, 1, 2, \ldots$.

Let $M : [-\infty, 0] \to [0, 1]$ by

$$M(\lambda) = y = \frac{\lambda}{\lambda - 1} \tag{29.11}$$

whose inverse is

$$M^{-1}(y) = \frac{y}{y-1} \tag{29.12}$$

and $\lambda - 1 = (y-1)^{-1}$. Moreover,

$$1 + \lambda^2 = \frac{y^2 + (y-1)^2}{(y-1)^2}$$

Thus,

$$(\eta_n \circ M^{-1})(y) = [y^2 + (1-y)^2](y-1)^n \equiv Q_n(y)$$

Since $[y^2 + (1-y)^2]$ is strictly positive, the Weierstrass theorem [325, Theorem 2.4.2] implies the linear combinations of the Q_n are dense in $\mathbb{C}([0, 1])$ (to approximate h by linear combinations of the Q_n's, approximate $h/[y^2 + (1-y)^2]$ by polynomials). Since $G \mapsto G \circ M^{-1}$ is an isometry of $\mathbb{C}([-\infty, 0])$ on $\mathbb{C}([0, 1])$, totality of the Q_n's implies totality of the η_n's, that is, $\overline{\mathcal{F}} = \mathbb{C}([-\infty, 0])$.

(iv) If $f \geq 0$ on $[-\infty, 0]$, given n, by (iii), find a Sparr function g_n with $\|g_n - (f + \frac{1}{n})\| \leq \frac{1}{n}$. Since $f \geq 0$, we have that g_n is strictly positive and $\|g_n - f\| \leq \frac{2}{n}$. □

Assuming Theorem 29.1, we have a proof of the hard part of Loewner's theorem given Theorem 2.4:

Theorem 29.3 *Let $f \in \mathcal{M}_\infty(0, \infty)$. Then there is $C \in \mathbb{R}$ and a measure $d\nu$ on $[-\infty, 0]$ so that*

$$f(x) = C + \int \varphi_x(\lambda)\, d\nu(\lambda) \tag{29.13}$$

Proof Define for g of the form (29.5),

$$\ell(g) = \sum_{j=1}^{2n} a_j f(x_j) \tag{29.14}$$

which is well-defined by (i) of Proposition 29.2. By Sparr's lemma,

$$g \geq 0 \Rightarrow \ell(g) \geq 0 \tag{29.15}$$

Let

$$\beta_0 = \min_{\lambda \in [-\infty, 0]} [\varphi_2(\lambda) - \varphi_1(\lambda)] > 0 \tag{29.16}$$

by Proposition 29.2(ii). For any Sparr function g,

$$\|g\|(\varphi_2 - \varphi_1) \pm \beta_0 g \geq 0 \tag{29.17}$$

so

$$|\ell(g)| \leq \beta_0^{-1} \ell(\varphi_2 - \varphi_1)\|g\|$$

Thus, by Proposition 29.2(iii), ℓ extends to $\overline{\mathfrak{F}} = \mathbb{C}([-\infty, 0])$ and is positive by (29.15). Therefore, by Proposition 29.2(iv) the Riesz–Markov theorem [325, Section 4.5], there is a positive measure $d\nu$ so that $\ell(g) = \int g(\lambda)\, d\nu(\lambda)$ for any Sparr function g. Since $\varphi_x - \varphi_1 \in \mathfrak{F}$, we see that

$$f(x) - f(1) = \int (\varphi_x(\lambda) - \varphi_1(\lambda))\, d\nu(\lambda) \tag{29.18}$$

which means (29.13) holds with

$$C = f(1) - \int \varphi_1(\lambda)\, d\nu(\lambda)$$

□

That leaves the proof of Sparr's lemma. We should begin by explaining in general terms how matrix monotonicity will be brought in. Let B be the diagonal matrix with entries $x_1, x_2, \ldots, x_n$, and A a real symmetric matrix with eigenvalues $y_1, \ldots, y_n$. So if $\tilde{A}$ is the diagonal matrix with entries $y_1, \ldots, y_n$, then

$$A = U^* \tilde{A} U$$

for some orthogonal U. Thus,

$$A \leq B \Leftrightarrow \forall \zeta \in \mathbb{C}^n \quad \sum_{j=1}^{n} (U\zeta)_j^2 y_j \leq \sum_{j=1}^{n} \zeta_j^2 x_j \tag{29.19}$$

$$f(A) \leq f(B) \Leftrightarrow \forall \zeta \in \mathbb{C}^n \quad \sum_{j=1}^{n} (U\zeta_j)^2 f(y_j) \leq \sum_{j=1}^{n} \zeta_j^2 f(x_j) \tag{29.20}$$

The goal will be to associate matrices A and B obeying (29.19) to a large enough class of positive Sparr functions, g, and deduce $\ell(g) \geq 0$ (ℓ given by (29.14)) from (29.20).

As a useful preliminary, we note that if $\sum_{j=1}^{2n} a_j = 0$, then

$$\begin{aligned}
\sum_{j=1}^{n} a_j \varphi_{x_j}(\lambda) &= \sum_{j=1}^{n} a_j \left(\frac{1 + x_j \lambda}{\lambda - x_j} \right) \\
&= \sum_{j=1}^{2n} a_j \left[\left(\lambda + \frac{1}{\lambda} \right) \frac{x_j}{\lambda - x_j} + \frac{1}{\lambda} \right] \\
&= \left(\lambda + \frac{1}{\lambda} \right) \sum_{j=1}^{2n} \frac{a_j x_j}{\lambda - x_j}
\end{aligned} \tag{29.21}$$

Since $\lambda + \frac{1}{\lambda} < 0$ on $(-\infty, 0)$ and a Sparr function is positive on $[0, \infty]$ if and only if it is positive on $(0, \infty)$, we see that Sparr's lemma is equivalent to a result about negative functions for $\lambda \in (-\infty, 0)$. It is notationally easier to speak about positive polynomials on $(0, \infty)$ rather than $(-\infty, 0)$, so we replace λ by $-\lambda$ and use $(-\lambda - x_j)^{-1} = -(\lambda + x_j)$ to change negative functions back to positive ones. Thus, Sparr's lemma is equivalent to

Theorem 29.1′ If $f \in \mathcal{M}_{n+1}(0, \infty)$ and $a_j \in \mathbb{R}$, $j = 1, \ldots, 2n$ with $\sum_{j=1}^{2n} a_j = 0$, and if for some $0 < x_1 < \cdots < x_{2n}$

$$\sum_{j=1}^{2n} \frac{a_j x_j}{\lambda + x_j} \geq 0 \text{ for } \lambda \in (0, \infty) \tag{29.22}$$

then

$$\sum_{j=1}^{2n} a_j f(x_j) \geq 0 \tag{29.23}$$

Henceforth, we fix $0 < x_1 < \cdots < x_{2n}$ and let

$$W(\lambda) = \prod_{j=1}^{2n} (\lambda + x_j) \tag{29.24}$$

and write

$$\sum_{j=1}^{2n} \frac{a_j x_j}{\lambda + x_j} = \frac{P(\lambda; a_j, x_j)}{W(\lambda)} \tag{29.25}$$

where

$$P(\lambda; a_j, x_j) = \sum_{j=1}^{2n} a_j x_j \prod_{k \neq j} (\lambda + x_k) \tag{29.26}$$

is a polynomial of degree $2n - 1$. Given any such P, we can recover the a_j's by

$$a_j = x_j^{-1} P(-x_j) \prod_{k \neq j} (x_k - x_j)^{-1} = x_j^{-1} P(-x_j) W'(-x_j)^{-1} \tag{29.27}$$

From this point of view, (29.22) is equivalent to

$$P(\lambda) \geq 0 \quad \text{for } \lambda \in (0, \infty) \tag{29.28}$$

and $\sum_{j=1}^{2n} a_j = 0$ is equivalent to

$$P(0) = 0 \tag{29.29}$$

We thus see that Theorem 29.1$'$ is equivalent to

Theorem 29.1$''$ If $f \in \mathcal{M}_{n+1}(0, \infty)$, $0 < x_1 < \cdots < x_{2n}$, then for any polynomial P obeying (29.28) and (29.29), we have that (29.23) holds where a_j is given by (29.27).

The following result on polynomials on $(0, \infty)$ is an analog of Lemma 26.6 proven in precisely the same way:

Lemma 29.4 *Any polynomial of degree at most* $2n - 1$ *obeying* (29.28) *and* (29.29) *is a finite sum of the form* $\lambda Q_1(\lambda)^2$ *and* $Q_2(\lambda)^2$ *where*

$$\deg Q_1 \le n-1 \qquad \deg Q_2 \le n-1 \qquad Q_2(0)=0 \tag{29.30}$$

The map from P to (29.23) is linear, so it suffices to prove Theorem 29.1″ for the special cases $P = \lambda Q_1^2$ and $P = Q_2^2$ where (29.30) holds. We will need the following well-known piece of algebra:

Proposition 29.5 *Let $z_1, \ldots, z_n$ be distinct points in $\mathbb{C}$. Let $\mathcal{P}_{n-1}$ be the space of polynomials of degree $n-1$. The map $\mathcal{E}_{z_1,\ldots,z_n} : P \mapsto (P(z_1), \ldots, P(z_n))$ of $\mathcal{P}_{n-1}$ to $\mathbb{C}^n$ is a bijection.*

Proof Let $P(z) = \sum_{j=0}^{n-1} c_j z^j$, then $\mathcal{E}$ is a linear map whose determinant is

$$\begin{vmatrix} 1 & 1 & \ldots & 1 \\ z_1 & z_2 & \ldots & z_n \\ \vdots & \vdots & & \vdots \\ z_1^{n-1} & z_2^{n-1} & \ldots & z_n^{n-1} \end{vmatrix} \tag{29.31}$$

which is well-known (Vandermonde determinant) to be $\prod_{i<j}(z_i - z_j) \neq 0$. □

Remark If $x_1, \ldots, x_n$ are real, $\mathcal{E}$ maps real polynomials bijectively to $\mathbb{R}^n$.

Here is Theorem 29.1″, for the case $\lambda Q_1(\lambda)^2$:

Theorem 29.6 *Fix $0 < x_1 < \cdots < x_{2n}$. Let $f \in \mathcal{M}_n(0,\infty)$. For any polynomial Q_1 of degree $n-1$, let $P(\lambda) = \lambda Q_1(\lambda)^2$ and let $a_1, \ldots, a_{2n}$ be given by (29.27). Then (29.23) holds.*

Remark Note that here we only need $\mathcal{M}_n$, not $\mathcal{M}_{n+1}$.

Proof Rename the x's to be $0 < y_1 < x_1 < y_2 < \cdots$. This choice is natural since $-W'(-y_j) < 0$, $-W'(-x_j) > 0$, which means $\lambda Q^2/W$ has negative residues at $\lambda = -y_j$ and positive at $-x_j$. Indeed, we can write

$$\frac{\lambda Q^2(\lambda)}{W} = -\sum_1^n \eta_j^2 \frac{y_j}{\lambda + y_j} + \sum_1^n \zeta_j^2 \frac{x_j}{\lambda + x_j} \tag{29.32}$$

where

$$\zeta_j = \frac{Q(-x_j)}{(-W'(-x_j))^{1/2}} \qquad \eta_j = \frac{Q(-y_j)}{(W_j'(-y_j))^{1/2}} \tag{29.33}$$

Define $U : \mathbb{R}^n \to \mathbb{R}^n$ by

$$U = M_1 \mathcal{E}_{-y_1,-y_2,\ldots,-y_n} \mathcal{E}_{-x_1,\ldots,-x_n}^{-1} M_2^{-1} \tag{29.34}$$

where M_2 is the diagonal matrix with diagonal elements $((-W'(-x_1))^{-1/2}, (-W'(-x_2))^{-1/2}, \dots)$, and M_1 the diagonal matrix elements $((W'(-y_1))^{-1/2}, (W'(-y_2))^{-1/2}, \dots)$. U encodes (29.33).

We make three claims:

(A) As ζ_j runs through all $\mathbb{R}^n$,

$$-\sum_{j=1}^{n} (U\zeta)_j^2 \frac{y_j}{\lambda + y_j} + \sum_{j=1}^{n} \zeta_j^2 \frac{x_j}{\lambda + x_j} \tag{29.35}$$

runs through all choices of $\lambda Q^2(\lambda)/W(\lambda)$.

(B) U is orthogonal for all choices of $0 < y_1 < x_1 < y_2 < x_2 < \dots < x_{2n}$.

(C) If

$$A = \begin{pmatrix} x_1 & & 0 \\ & \ddots & \\ 0 & & x_n \end{pmatrix} \qquad B = U^{-1} \begin{pmatrix} y_1 & & \\ & \ddots & \\ & & y_n \end{pmatrix} U \tag{29.36}$$

then $B \leq A$.

Given these claims, we can complete the proof since $B \leq A$ implies $f(B) \leq f(A)$ and thus $\langle \zeta, f(A)\zeta \rangle - \langle \zeta, f(B)\zeta \rangle \geq 0$, that is,

$$-\sum_{j=1}^{n} (U\zeta_j)^2 f(y_j) + \sum_{j=1}^{n} \zeta_j^2 f(x_j) \geq 0 \tag{29.37}$$

which is (29.23). Thus, we only need to prove the claims.

To prove (A), we need only note that $\mathcal{E}_{x_1,\dots,x_n}^{-1} M_2^{-1}$ maps $\mathbb{R}^n$ bijectively to $\mathcal{P}_{n-1}$, that is, all Q_1's.

To prove (B), note that $\lambda Q^2(\lambda)/W\big|_{\lambda=0} = 0$, so $\sum_{j=1}^n \zeta_j^2 = \sum_{j=1}^n \eta_j^2$, that is, U is orthogonal!

To prove (C), note that since $Q^2 \geq 0$, $\lim_{\lambda\to\infty} \lambda^2 Q^2(\lambda) \geq 0$, which implies $\sum \zeta_j^2 x_j \geq \sum \eta_j^2 y_j$, that is, $U^{-1}BU \leq A$. □

For the remaining case, we need the following:

Lemma 29.7 *If $f \in \mathcal{M}_2(0,\infty)$, then*

$$\lim_{x\to 0} x^2 f(x) = 0 \tag{29.38}$$

$$\lim_{x\to\infty} x^{-2} f(x) = 0 \tag{29.39}$$

Proof These are trivial if f is constant. If not, $(f')^{-1/2}$ is positive and concave (see Theorem 14.1), so

$$(f')^{-1/2}(x) \geq (f'(1))^{-1/2} x \qquad x \text{ in } (0,1) \tag{29.40}$$

$$(f')^{-1/2}(x) \geq (f'(1))^{-1/2} \qquad x \text{ in } (1,\infty) \tag{29.41}$$

(for if it is ever smaller, concavity implies it goes negative).

Equation (29.40) says

$$f(1) - \frac{f'(1)}{x} \leq f(x) \leq f(1) \qquad x \text{ in } (0,1)$$

implying (29.38). Equation (29.41) says that

$$f(1) \leq f(x) \leq f(1) + xf'(1) \qquad x \text{ in } (1,\infty)$$

which implies (29.39). □

The following completes the proof of Sparr's lemma, and so his proof of Loewner's theorem:

Theorem 29.8 *Fix* $0 < x_1 < \cdots < x_{2n}$. *Let* $f \in \mathcal{M}_{n+1}(0,\infty)$. *For any polynomial* Q_2 *of degree* $n-1$, *let* $P(\lambda) = Q_2(\lambda)^2$ *and let* $a_1, \ldots, a_{2n}$ *be given by* (29.27). *Suppose* $Q_2(0) = 0$. *Then* (29.23) *holds.*

Proof Pick δ, R with $\delta < x_1$ and $x_{2n} < R$. Let $\widetilde{Q}_1 = \sqrt{R}\, Q_2$ and $\widetilde{W}$ given by

$$\widetilde{W}(\lambda) = (\lambda+\delta)W(\lambda)(\lambda+R) \tag{29.42}$$

so

$$\begin{aligned}\frac{\lambda \widetilde{Q}_1(\lambda)^2}{\widetilde{W}(\lambda)} &= \left(\frac{\lambda}{\lambda+\delta}\right)\frac{Q_2(\lambda)^2}{W(\lambda)}\,\frac{R}{\lambda+R} \\ &= \sum_{j=1}^{2n} a_j(\delta,R)\,\frac{x_j}{\lambda+x_j} + a_0(\delta,R)\frac{\delta}{\lambda+\delta} + a_{2n+1}(\delta,R)\,\frac{R}{\lambda+R}\end{aligned} \tag{29.43}$$

Since $f \in \mathcal{M}_{n+1}$ and $\deg \widetilde{Q}_1 \leq n$, we can use Theorem 29.6 to conclude that

$$a_0(\delta,R)f(\delta) + a_{2n+1}(\delta,R)f(R) + \sum_{j=1}^{2n} a_j(\delta,R)f(x_j) \geq 0 \tag{29.44}$$

Since $\lambda\widetilde{Q}_1^2/\widetilde{W} \to Q_2^2/W$ as $\delta \downarrow 0$, $R \to \infty$, and $a_j(\delta,R)$ are residues of poles, we see

$$\lim_{\substack{\delta\downarrow 0\\ R\to\infty}} a_j(\delta,R) = a_j \qquad (1 \leq j \leq 2n) \tag{29.45}$$

Now,

$$a_0(\delta, R) = -\delta^{-1}\delta \frac{Q_1(-\delta)^2}{W(-\delta)} \frac{R}{R-\delta}$$

which, since $Q_2(0) = 0$, has

$$|a_0(\delta, R)| \le C\delta^2 \tag{29.46}$$

uniformly in δ small and R large.

Similarly,

$$\begin{aligned} a_{2n+1}(\delta, R) &= \frac{-R}{\delta - R} \frac{Q_2(R)^2}{W(R)} \\ &= O(R^{-2}) \end{aligned} \tag{29.47}$$

uniformly in δ small and R large since $\deg Q_2^2 \le 2n-2 \le 2n = \deg W(R)$.

By Lemma 29.7 and (29.46)/(29.47), the contributions of a_0 and a_{2n+1} in (29.44) go to zero as $\delta \downarrow 0$, $R \to \infty$. Thus, (29.44) and (29.45) imply (29.23). □

Notes and Historical Remarks The proof in this chapter follows Sparr [332] fairly closely. The one major difference is that he uses a refined Hahn–Banach theorem (to deal with positive extensions) to extend the functional ℓ to $\mathbb{C}([-\infty, 0])$. We instead prove density so that the big gun of the Hahn–Banach theorem (including infinite induction) is unnecessary and shown to be somewhat inelegant!

Schilling, Song, and Vondraček [305] have a presentation of Sparr's proof with some simplifications. Their book is on *Bernstein functions*, that is, functions f : $(0, \infty) \mapsto \mathbb{R}$ which are C^∞ with $(-1)^n f^{(n)}(x) \ge 0$ for all $n = 1, 2, \dots$. The big Bernstein theorem (the little theorem says that such functions are the restriction of functions analytic on the right half plane and that follows from Theorem 26.1) says that such functions have a representation of the form:

$$f(x) = a + bx + \int_{(0,\infty)} (1 - e^{-xs})\, d\mu(s) \tag{29.48}$$

where μ is a positive measure with $\int \min(1, s)\, d\mu(s) < \infty$. A *complete Bernstein function* is one for which $d\mu(s) = q(s)ds$ where q is the Laplace transform of a positive measure. It is known that such functions are exactly the restriction to $(0, \infty)$ of functions analytic on $\mathbb{C} \setminus (-\infty, 0]$ which are Herglotz, so for Schilling et al., Loewner's theorem says that $\mathcal{M}_\infty(0, \infty)$ is precisely the complete Bernstein functions.

Sparr was motivated by the connection of Loewner's theorem to interpolation theory (see Chapter 15).

In a note added in proof, Sparr remarks that it can be shown that his map U of (29.34) is the same as the map constructed implicitly by Loewner [209] in his proof of Theorem 5.26. Here is what Sparr means.

What U determines are the matrix elements between the eigenvectors $\psi^{(k)}$ of B and $\eta^{(j)}$ of A. By (29.36) and $\eta_j = \delta_j$ the vector with 1 in row j and 0 elsewhere, $U\psi_k = \delta_k$, that is,

$$\begin{aligned} U_{kj} &= \langle \delta_k, U\delta_j \rangle \\ &= \langle U^{-1}\delta_k, \delta_j \rangle \\ &= \langle \psi_k, \delta_j \rangle \end{aligned} \tag{29.49}$$

On the other hand, by (29.34), we can compute $U\delta_j$ as follows:

$$M_2^{-1}\delta_j = \left[-\prod_{\ell \neq j} (x_\ell - x_j) \prod_p (y_p - x_j) \right]^{1/2} \delta_j \equiv c\delta_j \tag{29.50}$$

Now $\mathcal{E}^{-1}_{-x_1,\dots,x_n} c\delta_j$ is a polynomial that is 0 at $-x_\ell$ for $\ell \neq j$, and c at $-x_j$, that is,

$$Q(\lambda) = \frac{c \prod_{\ell \neq j} (\lambda + x_\ell)}{\prod_{\ell \neq j} (x_\ell - x_j)} \tag{29.51}$$

Taking the inner product with δ_j we see that we need to evaluate $Q(\lambda)$ at $\lambda = -y_k$ and multiply by $W'(-y_k)^{-1/2}$, that is,

$$U_{kj} = \prod_\ell (x_\ell - y_k)^{-1/2} \prod_{j \neq k} (y_p - y_k)^{-1/2} c \prod_{\ell \neq j} (x_\ell - y_k) \prod_{\ell \neq j} (x_\ell - x_j)^{-1} \tag{29.52}$$

Plugging the value of c in this formula shows that

$$U_{kj} = \left[-\prod_{\ell \neq j} (x_\ell - x_j) \prod_{p \neq k} (y_p - y_k) \right]^{-1/2} \left[-\prod_p (y_p - x_j) \prod_\ell (x_\ell - y_k) \right]^{1/2} (y_k - x_j)^{-1} \tag{29.53}$$

where $(y_k - x_j)^{-1}$ is there because (29.53) has a product over all p and all ℓ, while c has $\prod_p (y_p - x_j)^{-1/2}$, and the first factor in (29.52) has $\prod_\ell (x_\ell - y_k)^{-1/2}$ but the second factor has $\prod_{\ell \neq j} (x_\ell - y_k)$.

In comparison, using the notation from this chapter for the formulae in Proposition 5.25 (so $\lambda_j^{\text{Sec. 5}} = y_j$, $\mu_j^{\text{Sec. 5}} = x_j$, and the roles of (A, η) and (B, φ) are reversed), we find

$$\langle \varphi, \psi_k^{(n)} \rangle = \left[\frac{\prod_\ell (x_\ell - y_k)}{\prod_{p \neq k} (y_p - y_j)} \right]^{1/2} \tag{29.54}$$

and, by symmetry,

$$\langle \varphi, \eta_j^{(n)} \rangle = \left[\frac{-\prod_p (y_p - x_j)}{\prod_{\ell \neq j} (x_\ell - x_j)} \right]^{1/2} \tag{29.55}$$

and, by (5.103),

$$\langle \psi_k, \eta_j \rangle = \frac{\langle \psi_k, \varphi \rangle \langle \varphi, \eta_j \rangle}{(x_j - y_k)} \tag{29.56}$$

which, by (29.49), is consistent with (29.53).

What we have shown is that the pair (A, B) Sparr constructs, where A has eigenvalues y_j and B has eigenvalues x_j, has $B - A$ rank one! This pair is the same as what Loewner used in proving Theorem 5.27.

However, the uses made by Loewner and Sparr seem to be rather different.

Chapter 30
Ameur's Proof Using Quadratic Interpolation

This chapter relies on the theory of quadratic interpolation as presented in Chapter 15. In particular, we will use K_2 defined by (15.11) and (15.19), that is,

$$K_2(t\varphi) = \left\langle \varphi, \frac{tA}{1+tA} \varphi \right\rangle^{1/2} \tag{30.1}$$

Our goal will be to prove Ameur's form of Donoghue's Exact Interpolation Theorem (Theorem 15.12):

Theorem 30.1 (Ameur's Form of Donoghue's Theorem) *$h \in \mathcal{I}_\infty$ if and only if there is a (positive) measure, $d\rho_h$, with*

$$\langle \varphi, h(A)\varphi \rangle = \int \left(1 + \frac{1}{t}\right) K_2(t, \varphi)\, d\rho_h(t) \tag{30.2}$$

for all finite matrices, $A \geq 0$.

Remarks

1. If $A\varphi = \lambda\varphi$, (30.2) implies that

$$h(\lambda) = \int (1+t) \left(\frac{\lambda}{1+\lambda t}\right) d\rho_h(t) \tag{30.3}$$

   showing that any $h \in \mathcal{I}_\infty$ is a Herglotz function.
2. The proof will not make use of any ideas involving matrix monotone functions. That said, since we know that $\mathcal{M}_\infty(0, \infty) \subset \mathcal{I}_\infty$ (by Theorem 15.4 or better, by Proposition 15.5), we see that any $h \in \mathcal{M}_\infty(0, \infty)$ is Herglotz. Thus, this provides another proof of the hard part of Loewner's theorem.

B. Simon, *Loewner's Theorem on Monotone Matrix Functions*, Grundlehren der mathematischen Wissenschaften 354, https://doi.org/10.1007/978-3-030-22422-6_30

Since each $K_2(t, \cdot)$ is an exact interpolation norm (Theorem 15.10), if h has a representation of the form (30.2), then $T^*T \le \mathbf{1}$ & $T^*AT \le A \Rightarrow T^*h(A)T \le h(A)$, i.e., $h \in \mathfrak{I}_\infty$. Thus we need only focus on the other direction (which is the half that implies the hard part of Loewner's theorem). The key to proving this direction is:

Theorem 30.2 (Ameur's Lemma) *Let $A \ge 0$ on $\mathbb{C}^n$ and $\varphi, \psi \in \mathbb{C}^n$ so that for some $a < 1$, we have that*

$$K_2(t, \psi) \le a K_2(t, \varphi) \tag{30.4}$$

for all $t \in [0, \infty]$. Then there exists $T \in \mathcal{L}(\mathbb{C}^n)$ with $T^*T \le \mathbf{1}$ & $T^*AT \le A$ so that

$$\psi = T\varphi \tag{30.5}$$

Remark Conversely, if (30.5) holds for T obeying $T^*T \le \mathbf{1}$ & $T^*AT \le A$, since $K_2(t, \cdot)$ is an exact interpolation norm, (30.4) holds with $a = 1$.

The proof of Theorem 30.1 from Theorem 30.2 relies on the functions

$$e_\lambda(t) = \frac{(1+t)\lambda}{1+t\lambda} \tag{30.6}$$

defined for $t \in [0, \infty]$ and $\lambda \in (0, \infty)$.

Lemma 30.3 *Finite linear combinations of $\{e_\lambda\}_{\lambda \in (0,\infty)}$ are dense in $C([0, \infty])$ in $\|\cdot\|_\infty$. The set is finitely independent (i.e., for $\{\lambda_j\}_{j=1}^n$ distinct, $\sum_{j=1}^n a_j e_{\lambda_j}(t) = 0$ for all $t \in [0, \infty] \Rightarrow a_1 = a_2 = \cdots = a_n = 0$).*

Proof Let $\mathfrak{F}$ be the set of finite linear combinations of $\{e_\lambda\}_{\lambda \in (0,\infty)}$. Note that

$$e_{\lambda=1}(t) \equiv 1 \tag{30.7}$$

Thus

$$e_\lambda(t) - e_1(t) = \frac{\lambda - 1}{1 + t\lambda} \tag{30.8}$$

so

$$f_\lambda(t) = (1 + t\lambda)^{-1} \tag{30.9}$$

lies in $\mathfrak{F}$ for $\lambda \ne 1$ and $f_{\lambda=1}$ lies in $\overline{\mathfrak{F}}$. By an easy calculation,

$$\frac{d^k}{d\lambda^k} f_\lambda(t) = (-1)^k \frac{k!t^k}{(1+t\lambda)^{k+1}} \tag{30.10}$$

Since the limits defining the derivative converge uniformly in t, $q_k(t) \equiv t^k/(1+t)^{k+1}$ lies in $\overline{\mathcal{F}}$.

$s(t) \equiv t/(1+t)$ maps $[0,\infty]$ bijectively to $[0,1]$. It follows from the Weierstrass theorem [325, Section 2.4] that any continuous function, $\ell(t)$, on $[0,\infty]$ is a uniform limit of polynomials in s. If $p \in C([0,\infty])$ has support in $(0,\infty)$, $\ell(t) = (1+t)p(t)$ is a uniform limit of polynomials in s, so since $(1+t)^{-1}$ is bounded, p is a uniform limit of linear combinations of $(1+t)^{-1}s(t)^k = q_k(t)$, i.e., $p \in \overline{\mathcal{F}}$. It follows that if $p \in C([0,\infty])$ with $p(0) = p(\infty) = 0$, then $p \in \overline{\mathcal{F}}$. Since $f_1(0) \neq 0 = f_1(\infty)$ and $e_1(0) = e_1(\infty) = 1$, any $f \in C([0,\infty])$ is a limit of linear combinations of p with $p(0) = p(\infty) = 0$, e_1 and f_1 so $\overline{\mathcal{F}} = C([0,\infty])$.

Suppose that

$$w(t) = \sum_{j=1}^{n} a_j e_{\lambda_j}(t) \tag{30.11}$$

is identically 0 for $t \in [0,\infty]$. $w(t)$ is a rational function for $t \in \mathbb{C} \setminus \{-\lambda_j^{-1}\}_{j=1}^{n}$ so this rational function is identically zero. Since

$$(\lambda_j - 1)a_j = \lim_{t \to -\lambda_j^{-1}} (1 + t\lambda_j) w(t)$$

we see that for all $\lambda_j \neq 1$, we have that $a_j = 0$. If some $\lambda_{j_0} = 1$, we have that $w(t) = a_{j_0}$ so $a_{j_0} = 0$ also. □

Proof of Theorem 30.1 Given Theorem 30.2 We already noted that any h of the form (30.2) lies in $\mathcal{I}_\infty$, so we need only prove the converse. Let $h \in \mathcal{I}_\infty$. Let $\mathcal{F}$ be the finite linear combinations of $\{e_\lambda\}_{\lambda \in (0,\infty)}$. Define $\phi : \mathcal{F} \to \mathbb{C}$ by

$$\phi\Big(\sum_{j=1}^{n} a_j e_{\lambda_j}\Big) = \sum_{j=1}^{n} a_j h(\lambda_j) \tag{30.12}$$

Since $\{e_\lambda\}_{\lambda \in (0,\infty)}$ are finitely independent, ϕ is well-defined.

Given $\{\lambda_j\}_{j=1}^{n}$ and $\{a_j\}_{j=1}^{n}$, with $w(t) \equiv \sum_{j=1}^{n} a_j e_{\lambda_j}(t) \geq 0$ for all $t \in [0,\infty]$, define, for $j = 1, \ldots, n$

$$\varphi_j = \sqrt{(a_j)_+} \qquad \psi_j = \sqrt{-(a_j)_-} \tag{30.13}$$

where $x_+ = \max(x, 0)$, $x_- = -\max(-x, 0)$. Thus

$$a_j = |\varphi_j|^2 - |\psi_j|^2 \tag{30.14}$$

so $w(t) \geq 0$ implies that, when A is the diagonal matrix with eigenvalue, λ_j

$$\begin{aligned}\left(1+\frac{1}{t}\right) K_2(t,\psi) &= \sum_{j=1}^n |\psi_j|^2 e_{\lambda_j}(t) \\ &\leq \sum_{j=1}^n |\varphi_j|^2 e_{\lambda_j}(t) \\ &= \left(1+\frac{1}{t}\right) K_2(t,\varphi)\end{aligned}$$

Since $\inf_{t\in[0,\infty]} \left(1+\frac{1}{t}\right) K_2(t,\varphi) > 0$ and we just saw that (30.4) holds for $a = 1$, by Theorem 30.2, for every $\epsilon \in (0,1)$, there is an operator T_ϵ, with $T_\epsilon^* T_\epsilon \leq \mathbf{1}$, $T_\epsilon^* A T_\epsilon \leq A$, and $T_\epsilon \varphi = (1-\epsilon)\psi$.

Since h is an interpolation function, we conclude that $(1-\epsilon)^2 \langle \psi, h(A)\psi\rangle \leq \langle \varphi, h(A)\varphi\rangle$ so

$$\sum_{\{j \mid a_j > 0\}} a_j h(\lambda_j) + (1-\epsilon)^2 \sum_{\{j \mid a_j < 0\}} a_j h(\lambda_j) \geq 0$$

Taking $\epsilon \downarrow 0$, we conclude that

$$w \in \mathcal{F},\ w \geq 0 \Rightarrow \phi(w) \geq 0 \tag{30.15}$$

where the middle inequality is just $w(t) \geq 0$. Since $\mathbf{1} \in \mathcal{F}$ (with $\phi(\mathbf{1}) = h(1)$) and $\|w\|_\infty \pm w \geq 0$ for w real, we conclude that $|\phi(w)| \leq h(1)\|w\|_\infty$ so ϕ extends to a positive functional on $\overline{\mathcal{F}} = C([0,\infty])$. It follows from the Riesz–Markov theorem (see [325, Section 4.5]) that there is a measure, $d\rho_h$ with

$$h(\lambda) = \int e_\lambda(t) d\rho_h(t) \tag{30.16}$$

which is (30.3). □

The remainder of this chapter will be devoted to Ameur's proof of Theorem 30.2 which has several ad hoc constructions. Given $t > 0$, $\{\lambda_j\}_{j=1}^n$ in $\mathbb{R}_+^n$, and $\varphi \in \mathbb{C}^n$, define

$$K_{2,\lambda}(t,\varphi) = \left(\sum_{j=1}^n |\varphi_j|^2 \frac{t\lambda_j}{1+t\lambda_j}\right)^{1/2} \tag{30.17}$$

We first note that we can suppose that A is diagonal. Moreover, we can suppose that A has only simple eigenvalues, for if A does not, it is a limit of A's that are, so it is easy to go from that special case using the fact that $\{T \in \mathcal{L}(\mathbb{C}^n) \mid T^*T \leq 1\}$ is compact and $T^*AT \leq A$ is continuous in A. So we suppose that A has eigenvalues $\lambda_1 < \lambda_2 < \cdots < \lambda_n$.

Given $a < 1$ in (30.4), pick $\rho > 1$ so that $a\rho < 1$ and $\rho\lambda_j < \lambda_{j+1}$ for $j = 1, \ldots, n$. Thus for $t \in (0, \infty)$

$$K_{2,\boldsymbol{\lambda}}(t^{-1}, \psi) \leq a K_{2,\boldsymbol{\lambda}}(t^{-1}, \varphi) < \rho^{-1} K_{2,\boldsymbol{\lambda}}(t^{-1}, \varphi) \tag{30.18}$$

We've shifted from t to t^{-1} to be able to deal later with polynomials in t rather than t^{-1}.

Note next that if $\varphi, \psi \in \mathbb{C}^n$ and $T \in \mathcal{L}(\mathbb{C}^n)$ and if $\varphi = |\varphi_j| \exp^{i\kappa_j}$, $\psi_j = |\psi_j| \exp^{i\eta_j}$ and if $T^{(0)}_{jk} = \exp^{-i\eta_j} T_{jk} \exp^{i\kappa_j}$, then $T\varphi = \psi \Leftrightarrow T^{(0)}|\varphi| = |\psi|$. Thus, we can suppose $\varphi \geq 0$, $\psi \geq 0$ and get the general case from the special case. Since, by decreasing ρ slightly, we can perturb, φ, ψ so that

$$\varphi_j > 0, \quad \psi_j > 0 \tag{30.19}$$

for all j and take limits. So we henceforth suppose that (30.19) holds.

Define $\beta_j = \lambda_j$, $\alpha_j = \rho\beta_j$ so

$$0 < \beta_1 < \alpha_1 < \beta_2 < \cdots < \beta_n < \alpha_n \tag{30.20}$$

Since $\frac{t\beta_j}{1+t\beta_j} \leq \frac{t\alpha_j}{1+t\alpha_j} \leq \frac{\rho t\beta_j}{1+t\beta_j}$, we have that

$$K_{2,\boldsymbol{\beta}}(t, \eta)^2 \leq K_{2,\boldsymbol{\alpha}}(t, \eta)^2 \leq \rho K_{2,\boldsymbol{\beta}}(t, \eta)^2 \tag{30.21}$$

for any $\eta \in \mathbb{C}^n$. In particular, by (30.18) and (30.21),

$$K_{2,\boldsymbol{\alpha}}(t^{-1}, \psi)^2 \leq K_{2,\boldsymbol{\beta}}(t^{-1}, \varphi)^2 \tag{30.22}$$

We define polynomials of degree n and $2n$ by

$$L_{\boldsymbol{\beta}} = \prod_{j=1}^{n} (t + \beta_j), \quad L_{\boldsymbol{\alpha}} = \prod_{j=1}^{n} (t + \alpha_j), \quad L(t) = L_{\boldsymbol{\alpha}}(t) L_{\boldsymbol{\beta}}(t) \tag{30.23}$$

Of course,

$$t > 0 \Rightarrow L_{\boldsymbol{\beta}}(t) > 0, \; L_{\boldsymbol{\beta}}(t) > 0, \; L(t) > 0 \tag{30.24}$$

Note that by (30.20), we have that

$$L'(-\beta_j) > 0, \qquad L'(-\alpha_j) < 0 \tag{30.25}$$

Define the polynomial P of degree at most $2n - 1$ by

$$\frac{P(t)}{L(t)} = K_{2,\beta}(t^{-1}, \phi)^2 - K_{2,\alpha}(t^{-1}, \psi)^2$$
$$= \sum_{j=1}^{n} \varphi_j^2 \frac{\beta_j}{t + \beta_j} - \sum_{j=1}^{n} \psi_j^2 \frac{\alpha_j}{t + \alpha_j} \tag{30.26}$$

By (30.22) and (30.24), $P(t) > 0$ if $t \geq 0$. By (30.26)

$$P(-\beta_j) = \varphi_j^2 \beta_j L'(-\beta_j), \qquad P(-\alpha_j) = -\psi_j^2 \alpha_j L'(-\alpha_j) \tag{30.27}$$

so by (30.25), we have that

$$P(-\beta_j) > 0, \quad P(-\alpha_j) > 0, \quad L'(-\beta_j)P(-\beta_j) > 0, \quad L'(-\alpha_j)P(-\alpha_j) < 0 \tag{30.28}$$

By slightly wiggling the values of α and β (and then taking limits), we can suppose that P has exact degree $2n - 1$ and only simple zeros.

Lemma 30.4 *The* $2n - 1$ *zeros of* P *consist of* $n - m$ *zeros in* $\mathbb{C}_+$, $\{-c_j\}_{j=1}^{n-m}$ ($m \in \{0, 1, \ldots, n\}$), *their complex conjugates and* $2m - 1$ *real negative zeros* $\{-\delta_j\}_{j=1}^{m}$ *and* $\{-\gamma_j\}_{j=1}^{m-1}$ *where the leftmost zero is* $-\delta_1$ *with* $\delta_1 > \alpha_n$ *and*

$$L(-\delta_j)P'(-\delta_j) > 0, \qquad L(-\gamma_j)P'(-\gamma_j) < 0 \tag{30.29}$$

Proof Since P is real, its complex roots come in conjugate pairs. Since $P(x) > 0$ for $x \geq 0$ and $\deg(P)$ is odd. $P(x) \to -\infty$ as $x \to -\infty$. Since $P(-\alpha_n) > 0$ (by (30.28)), the leftmost zero, $-\delta_1$ has $-\delta_1 < -\alpha_n$. Since $L(t) > 0$ for $t < -\alpha_n$ and $P'(-\delta_1) > 0$, (30.29) holds for $-\delta_1$. Let $H(x) = L(x)P(x)$. It has simple zeros, so that $H'(x)$ alternates signs at zeros. Since we have just seen that $H'(x) > 0$ at the leftmost zero, there are $n + m$ zeros where $H'(x) > 0$ and $n + m - 1$ where $H'(x) < 0$. By (30.25) and (30.28), $H'(-\alpha_j) < 0$, $H'(-\beta_j) > 0$, $j = 1, \ldots, n$ so among the zeros of P, we have that $H'(x) > 0$ for m points $\{-\delta_j\}_{j=1}^{m}$ and $H'(x) < 0$ for $m-1$ points $\{-\gamma_j\}_{j=1}^{m}$. Since $P(x_0) = 0 \Rightarrow H'(x_0) = L(x_0)P'(x_0)$, we obtain (30.29). □

Define

$$L_\delta = \prod_{j=1}^{n}(t + \delta_j), \qquad L_c = \prod_{j=1}^{n}(t + c_j) \tag{30.30}$$

Recall (30.28) that $L'(-\beta_j)P(-\beta_j) > 0$. Given $\eta \in \mathbb{C}^n$, there is by Theorem 22.2 a unique polynomial, $Q(z)$ of degree $2n - 1$ so that

$$\eta_j = \frac{Q(-\beta_j)}{[\beta_j L'(-\beta_j)P(-\beta_j)]^{1/2}}, \quad Q(z) = 0 \,\text{for}\, z \in \{-\delta_j\}_{j=1}^{M} \cup \{-c_j\}_{j=1}^{n-m} \tag{30.31}$$

(alternatively, $Q(z) = L_{\boldsymbol{\delta}} L_{\boldsymbol{c}} q(z)$ with $\deg(q) = n - 1$). By (22.9), Q is linear in η so $\eta \mapsto Q$ defines a map T_1 from $\mathbb{C}^n$ to the polynomials of degree $2n - 1$.

Recall that $L'(-\alpha_j)P(-\alpha_j) < 0$, $L(-\gamma_j)P'(-\gamma_j) < 0$. Given Q, a polynomial of degree $2n - 1$, define

$$\kappa_j = \frac{Q(-\alpha_j)}{[-\alpha_j L'(-\alpha_j)P(-\alpha_j)]^{1/2}}, \quad \chi_j = \frac{Q(-\gamma_j)}{[-\gamma L(-\gamma_j)P'(-\gamma_j)]^{1/2}} \tag{30.32}$$

Define $\overline{Q}(z) = \overline{Q(\bar{z})}$. We can consider a partial fraction expansion of $\frac{\overline{Q}(z)Q(z)}{L(z)P(z)}$. Since $2\deg(Q) \leq 4n-2 = \deg(LP)-1$, it is a sum of pole terms. Since $\overline{Q}Q(t) = 0$ for $t \in \{-c_j, -\overline{c_j}\}_{j=1}^{n-m} \cup \{-\delta_j\}_{j=1}^{m}$, there are pole terms only at $\{\alpha_j, \beta_j\}_{j=1}^{n} \cup \{\gamma_j\}_{j=1}^{m-1}$. By (30.31)/(30.32), we have for $t \geq 0$ that

$$\frac{|Q(t)|^2}{L(t)P(t)} = \sum_{j=1}^{n} |\eta_j|^2 \frac{\beta_j}{t+\beta_j} - \sum_{j=1}^{n} |\kappa_j|^2 \frac{\alpha_j}{t+\alpha_j} - \sum_{j=1}^{m-1} |\chi_j|^2 \frac{\gamma_j}{t+\gamma_j} \tag{30.33}$$

Proof of Theorem 30.2 The map $T_2 : Q \mapsto \kappa = \{\kappa_j\}_{j=1}^{n}$ given by (30.32) is a linear map of all polynomials of degree $2n-1$ to $\mathbb{C}^n$. We claim that $T \equiv T_2 T_1 : \mathbb{C}^n \to \mathbb{C}^n$ has the required properties for the theorem. T_1 is defined just below (30.31) and T_2 at the start of this proof.

By (30.27) and $P(-\beta_j) > 0$, we have that

$$\frac{P(-\beta_j)}{[\beta_j L'(-\beta_j)P(-\beta_j)]^{1/2}} = \sqrt{\frac{P(-\beta_j)}{\beta_j L'(-\beta_j)}} = \varphi_j \tag{30.34}$$

Also $P(-\delta_j) = P(-c_j) = 0$. By the uniqueness of interpolating polynomials, $T_1\varphi = P$. By (30.27) again, $T_2 P = \psi$. Thus $T\varphi = T_2(T_2\varphi) = \psi$.

Let $\eta \in \mathbb{C}^n$ be arbitrary and $\kappa = T\eta$. Then, by (30.33), for $t \in (0, \infty)$

$$\begin{aligned} K_{2,\boldsymbol{\beta}}(t^{-1}, \eta)^2 - K_{2,\boldsymbol{\alpha}}(t^{-1}, T\eta)^2 &= \sum_{j=1}^{n} |\eta_j|^2 \frac{\beta_j}{t+\beta_j} - \sum_{j=1}^{n} |\kappa_j|^2 \frac{\alpha_j}{t+\alpha_j} \\ &= \frac{|Q(t)|^2}{L(t)P(t)} + \sum_{j=1}^{m-1} |\chi_j|^2 \frac{\gamma_j}{t+\gamma_j} > 0 \end{aligned}$$

By (30.21)

$$K_{2,\boldsymbol{\beta}}(t^{-1}, T\eta)^2 \leq K_{2,\boldsymbol{\alpha}}(t^{-1}, T\eta)^2 \leq K_{2,\boldsymbol{\beta}}(t^{-1}, \eta)^2 \tag{30.35}$$

Taking $t \downarrow 0$, we get $T^*T \le 1$ and multiplying by t taking $t \to \infty$, we get that $T^*AT \le A$. □

Notes and Historical Remarks This chapter follows Ameur's proof in his paper [7] with some details from his thesis [6] available online. His calculations clearly owe something to Sparr [332]. Like Sparr, he doesn't prove that the span of the trial functions is dense but appeals to the Hahn–Banach theorem. As we did with Sparr's proof, we instead prove a density result.

Ameur's paper [8] deals with the arguments needed to prove Donoghue's result in the infinite dimensional case. From our point of view, rather than prove Theorem 30.1 directly in the infinite dimensional case, one notes that if h is an interpolation function in infinite dimensions, by picking A finite rank, and applying the finite dimensional result, one concludes that h is Herglotz. Conversely, as we've seen, Herglotz functions are infinite dimensional interpolation functions since each $K_2(t, \cdot)$ is such a function.

There is a sense in which Ameur's papers while they do have an elegant proof of Theorem 30.1 do not have a new proof of the hard part of Loewner's theorem. For his proof that $\mathcal{M}_\infty(0,\infty) \subset \mathcal{I}_\infty$ uses a result (Theorem 25.1) that immediately implies Loewner's theorem. However, given Boutet's direct proof that $\mathcal{M}_\infty(0,\infty) \subset \mathcal{I}_\infty$ (see Proposition 15.5), we can regard this chapter as providing another proof of the hard part of Loewner's theorem.

Ameur has a recent review[9] and extension of his work on interpolation and Loewner's theorem.

Chapter 31
One-Point Continued Fractions: The Wigner–von Neumann Proof

In this chapter and the next, we provide proofs of the hard part of Loewner's theorem that rely on continued fractions and a key lemma on the Loewner matrix under stripping off the front of the fraction. Of course, continued fractions are a kind of rational approximation so these proofs are not unconnected to that in Chapter 25, but its reliance on the key lemma makes the details different. Here is the key lemma:

Lemma 31.1 (Wigner–von Neumann Lemma) *Let f be a C^3 function on a bounded interval $I = (a, b) \subset \mathbb{R}$ with $f'(x) > 0$ on I. For $x_0 \in (a, b)$, define*

$$g_{x_0}(x) = -\frac{1}{f(x) - f(x_0)} + \frac{1}{f'(x_0)(x - x_0)} \tag{31.1}$$

If $f \in \mathcal{M}_{n+1}(a, b)$, then $g_{x_0} \in \mathcal{M}_n(a, b)$

Remarks

1. Since f is strictly monotone, $g_{x_0}(x)$ is well-defined for $x \neq x_0$. Since $f(x) - f(x_0) = f'(x_0)(x - x_0) + \frac{1}{2} f''(x_0)(x - x_0)^2 + O((x - x_0)^3)$, a simple calculation shows that

$$g_{x_0}(x) = \frac{f''(x_0)}{2f'(x_0)^2} + O((x - x_0)) \tag{31.2}$$

   so g_{x_0} has a continuous extension to all of I including x_0.
2. As we'll explain in the Notes, this lemma has a kind of converse.
3. If f is a scalar monotone function, $-(f(x) - f(x_0))^{-1}$ is monotone on (a, x_0) and on (x_0, b) but has a pole at x_0. The second term cancels the pole so there is hope of restoring global monotonicity.

Proof As easy argument shows that, as extended by setting $g_{x_0}(x_0) = \frac{f''(x_0)}{2f'(x_0)^2}$, $g_{x_0} \in C^1$ given that $f \in C^3$. Let $x_0, x_1, \ldots, x_n$ be distinct points in (a, b) and let

B. Simon, *Loewner's Theorem on Monotone Matrix Functions*, Grundlehren der mathematischen Wissenschaften 354, https://doi.org/10.1007/978-3-030-22422-6_31

$$L_{ij} = [x_i, x_j; f] \quad i, j = 0, \dots, n \tag{31.3}$$

A straightforward calculation shows that

$$[x_i, x_j; g_{x_0}] = \frac{L_{00}L_{ij} - L_{i0}L_{0j}}{f'(x_0)(f(x_i) - f(x_0))(f(x_j) - f(x_0))} \tag{31.4}$$

Since $f'(x_0) > 0$ and $\{M_{ij}\}_{1 \le i,j \le n} > 0 \Rightarrow \{\zeta_i M_{ij} \zeta_j\}_{1 \le i,j \le n} \ge 0$ for any real, non-zero $\{\zeta_i\}_{i=1}^n$, it suffices to prove that $\{M_{ij}\}_{i,j=1,\dots,n} \equiv \{L_{00}L_{ij} - L_{i0}L_{0j}\}_{i,j=1,\dots,n}$ is a positive matrix if $\{L_{ij}\}_{i,j=0,\dots,n}$ is a positive matrix.

If L is strictly positive, $L_{00} > 0$. Given $(\zeta_1, \dots, \zeta_n) \in \mathbb{C}^n$, set $\zeta_0 = -(\sum_{j=1}^n L_{j0}\zeta_j)/L_{00}$ and compute that

$$L_{00} \sum_{i,j=0,\dots,n} \zeta_i L_{ij} \zeta_j = \sum_{i,j=1,\dots,n} \zeta_i M_{ij} \zeta_j$$

(since the $(ij) = (00)$ term gives $\sum_{i,j=1,\dots,n} \zeta_i L_{i0} L_{0j} \zeta_j$ and the $0i$ and $j0$ terms each give minus this sum). □

To exploit this, we first suppose without real loss that $f(0) = 0$ and write (31.1) for $x_0 = 0$ as

$$f(x) = \frac{1}{q_1(x) - h_1(x)} \tag{31.5}$$

where

$$q_1(x) = \frac{1}{f'(0)x} - \frac{f''(0)}{2f'(0)^2} \qquad h_1(x) = g_0(x) - \frac{f''(0)}{2f'(0)^2}$$

Suppose now that $f \in \mathcal{M}_\infty(-1, 1)$. By the Wigner–von Neumann Lemma, $h_1 \in \mathcal{M}_\infty(-1, 1)$ also and we can iterate multiple times. That is, we define inductively:

$$q_{n+1}(x) = \frac{1}{h_n'(0)x} - h_n(0) \tag{31.6}$$

$$h_{n+1}(x) = q_{n+1}(x) - \frac{1}{h_n(x)} \tag{31.7}$$

so that

$$h_n(x) = \frac{1}{q_{n+1}(x) - h_{n+1}(x)} \tag{31.8}$$

Thus, by the Wigner–von Neumann Lemma

$$h_{n+1} \in \mathcal{M}_\infty(-1,1) \qquad h_{n+1}(0) = 0 \tag{31.9}$$

This implies an exact finite continued fraction representation

$$f(x) = \cfrac{1}{q_1(x) - \cfrac{1}{q_2(x) - \cfrac{1}{q_3(x) - \dots - \cfrac{1}{q_n(x) - h_n(x)}}}} \tag{31.10}$$

$f_n(x)$, for $x \in (-1,1)$, is defined to be the function given by taking the continued fraction in (31.10) with h_n replaced by 0. In this definition, it might happen some unraveling of the fraction has a zero value at some x. We continue interpreting $1/0 = \infty$ and $1/\infty = 0$ which corresponds to viewing continued fractions as composition of fractional linear transformation and view such transformation as maps of $\widehat{\mathbb{C}} \to \widehat{\mathbb{C}}$.

Put more formally, we define, for $x \in (-1,1)$

$$\tau_{j,x}(w) = \frac{1}{q_j(x) - w} \tag{31.11}$$

which we note is a sensible definition for all $x \in \mathbb{C}$ since $q_j(x)$ is a rational function of x whose only pole is at $x = 0$ (where we interpret $\tau_{j,x=0}(w) = 0$ for all $w \in \mathbb{C}$). Equation (31.10) and the definition of f_n can then be written as

$$f(x) = \tau_{1,x}(\tau_{2,x}(\dots \tau_{n,x}(h_n(x))\dots) \qquad f_n(x) = \tau_{1,x}(\tau_{2,x}(\dots \tau_{n,x}(0)\dots) \tag{31.12}$$

Lemma 31.2

(a) *$f_n(x)$ is the ratio of a polynomial in $1/x$ of degree $n-1$ divided by one of degree n.*
(b) $\tau_{j,x}(w) = h'_{j-1}(0)x + O_w(x^2)$ *where the error is uniformly* $\mathrm{O}(\mathrm{x}^2)$ *for w in compact subsets of $\mathbb{C}$.*
(c) $\mathrm{Im}(w) \geq 0$, $\mathrm{Im}(z) > 0 \Rightarrow \mathrm{Im}(\tau_{j,z}(w)) > 0$.
(d) *$f_n(z)$ is a rational function of z with* $\mathrm{Im}(f_n(z)) > 0$ *on* $\mathbb{C}_+$.

Remark (a) says that $f_n(x) - f(x) = \mathrm{O}(x^2)$ and it is not hard to show that $f_n(x) - f(x) = \mathrm{O}(x^{2n+1})$ so f_n is the $[n,n]$-Padé approximation. This shows that this proof is not unrelated to the Loewner and Bendat–Sherman proofs.

Proof

(a) $q_j(x)$ is 1 divided by a linear function of $1/x$, so this follows by induction from (31.11) and (31.12).
(b) This is immediate from (31.8) and (31.11).
(c) Since $h'_{j-1}(0) > 0$, we have that $\mathrm{Im}(w) \geq 0, \mathrm{Im}\, z > 0 \Rightarrow \mathrm{Im}(q_j(z)) < 0 \Rightarrow \mathrm{Im}(q_j(z) - w) < 0 \Rightarrow \mathrm{Im}(\tau_{j,z}(w)) > 0$.
(d) Follows from (c) and (31.12).

□

Proposition 31.3 *Let $f \in \mathcal{M}_\infty(-1, 1)$. Let $f(0) = 0$ and set $h_0 \equiv f$. Then for $x \in (0, 1)$ and $j = 1, 2, \dots$*

(a) $0 < h_j(x) < q_j(x)$
(b) $0 < w < w' < q_j(x) \Rightarrow 0 < \tau_{j,x}(0) < \tau_{j,x}(w) < \tau_{j,x}(w')$
(c) *If $0 \leq w \leq h_j(x)$, then no denominators vanish in forming $\tau_{1,x}(\tau_{2,x}(\dots \tau_{j,x}(w)\dots)$*
(d) $0 \leq f_j(x) \leq f_{j+1}(x) \leq f(x)$

For $x \in (-1, 0)$, reverse the signs of all inequalities.

Proof

(a) By hypothesis, $h_0(0) = 0$ and by definition $h_j(0) = 0$. Thus, for $x \in (0, 1)$, by monotonicity, $0 < h_j(x)$ and $0 < (h_{j-1}(x))^{-1} = q_j(x) - h_j(x)$.
(b) $\tau_{j,x}(0) = 1/q_j(x) > 0$ and $y \mapsto 1/(q_j(x) - y)$ is monotone increasing in y for $y \in [0, q_j(x))$.
(c) If $w = h_j(x)$, the successive results of applying $\tau_{m,x}$ give us $h_{j-\ell}(x), \ell = 1, \dots$ so by (a), the denominators don't vanish. By repeated use of (b), the intermediate values are in $[0, h_{j-\ell}(x)]$ if $w < h_j(x)$.
(d) By (b), $0 < \tau_{j+1,x}(0) < h_j(x) = \tau_{j+1,x}(h_{j+1}(x))$. Applying $\tau_{1,x}(\tau_{2,x}(\dots \tau_{j,x}(\cdot)\dots)$ and using the monotonicity of (b) yields $0 \leq f_j(x) \leq f_{j+1}(x) \leq f(x)$.

To get the results for $(-1, 0)$, we apply the above results to $-f(-x)$. □

Theorem 31.4 *Let $f \in \mathcal{M}_\infty(-1, 1)$. Then $f_n(z)$ is a rational Herglotz function all of whose poles lie in $\mathbb{R} \setminus (-1, 1)$. As $n \to \infty$, $f_n(z)$ converges to a limiting function, $f_\infty(z)$, uniformly for z in compact subsets of $\mathbb{C} \setminus (-\infty, -1] \cup [1, \infty)$. This function is a Herglotz function analytic on $(-1, 1)$.*

Proof By Lemma 31.2 (d), each f_n is a rational Herglotz function and by Proposition 31.3 (c), there are no poles in $(-1, 1)$. By the monotonicity and boundedness in Proposition 31.3 (d), f_n has a limit on $(-1, 1)$. Note that $f_n(0) = f(0)$, $f'_n(0) = f'(0)$. By Theorem 3.9, and the fact that, by monotonicity, all possible limit points agree on $(-1, 1)$, we get the claimed convergence. □

Clearly, to complete the proof of the hard part of Loewner's theorem, we need to know that $f_\infty(x) = f(x)$ for all $x \in (-1, 1)$. By a trick, we'll see that we only need this for $|x| < \delta$ for some $\delta > 0$. To prove that, we'll exploit the Euler–Wallis formulae (discussed in Chapter 21) to which we now turn. These take the form (since for $j \geq 1$, we have that $B_j = -1,\ A_j = -q_j(w)$):

$$Q_n(x) = -Q_{n-2}(x) - q_n(x)Q_{n-1}(x) \tag{31.13}$$

$$P_n(x) = -P_{n-2}(x) - q_n(x)P_{n-1}(x) \tag{31.14}$$

$$Q_{-1} = P_0 = 1 \qquad\qquad Q_0 = P_{-1} = 0$$

Theorem 31.5 *We have that*

(a) $(-x)^n P_n(x)$ *and* $(-x)^{n-1} Q_n(x)$ *are non-negative on* $(-1, 0) \cup (0, 1)$ *for* $n = 1, 2, \ldots$

(b)

$$\tau_{1,x}(\tau_{2,x}(\ldots \tau_{n,x}(w) \ldots) = \frac{wQ_{n-1}(x) + Q_n(x)}{wP_{n-1}(x) + P_n(x)} \tag{31.15}$$

(c)

$$f_n(x) = \frac{Q_n(x)}{P_n(x)} \qquad f(z) = \frac{h_n(x)Q_{n-1}(x) + Q_n(x)}{h_n(x)P_{n-1}(x) + P_n(x)} \tag{31.16}$$

(d)

$$P_n(x)Q_{n-1}(x) - Q_n(x)P_{n-1}(x) = 1 \tag{31.17}$$

(e)

$$f_n(x) - f_{n-1}(x) = -(P_n(x)P_{n-1}(x))^{-1} \tag{31.18}$$

(f)

$$\begin{aligned} f(x) - f_n(x) &= \frac{h_n(x)}{(P_n(x))(h_n(x)P_{n-1}(x) + P_n(x))} \\ &= \frac{1}{(P_n(x))(h_n(x)^{-1}P_n(x) + P_{n-1}(x))} \end{aligned} \tag{31.19}$$

Remark For $x \in (0, 1)$, the signs of P_n alternate, so (31.18) implies again the monotonicity of f_n.

Proof

(a) is immediate from (31.14) and the sign of $q_n(x)$.

(b) Define $W_{j,x} = \begin{pmatrix} 0 & -1 \\ 1 & -q_j(x) \end{pmatrix}$ so

$$\begin{pmatrix} \tau_{j,x}(w) \\ 1 \end{pmatrix} = \frac{1}{-q_j(x) + w} W_{j,x} \begin{pmatrix} w \\ 1 \end{pmatrix}$$

Thus, with $\left(\begin{smallmatrix} a \\ b \end{smallmatrix}\right) \doteq \left(\begin{smallmatrix} c \\ d \end{smallmatrix}\right) \Leftrightarrow \exists\lambda$ with $c = \lambda a,\ d = \lambda b$, and $T_n = W_1 \dots W_n$, we have that

$$T_n \begin{pmatrix} w \\ 1 \end{pmatrix} \doteq \begin{pmatrix} \tau_{n,x}(w) \\ 1 \end{pmatrix} \tag{31.20}$$

Thus $T_n = T_{n-1} \left(\begin{smallmatrix} 0 & -1 \\ 1 & -q_n(x) \end{smallmatrix}\right) \Rightarrow T_n = \left(\begin{smallmatrix} Q_{n-1} & Q_n \\ P_{n-1} & P_n \end{smallmatrix}\right)$, where P, Q obey (31.14). Equation (31.15) is (31.20).

(c) is just $w = 0$ in (31.15).

(d) $\det(W_j) = \det \left(\begin{smallmatrix} 0 & -1 \\ 1 & -q_j(x) \end{smallmatrix}\right) = 1$ so $\det T_n = 1$ which implies (31.17).

(e) $f_n - f_{n-1} = \frac{P_{n-1}Q_n - P_n Q_{n-1}}{P_n P_{n-1}} = \frac{-1}{P_n P_{n-1}}$

(f) is immediate from (31.16) and the same calculation as in (e).

□

The basic idea behind the convergence proof is to note that $q_n(x)$ is very large for small x, so q_n dominates h_n for such x and thus we expect $P_n \sim q_1 \dots q_n$. Thus, we expect P_n to be large and the convergence from (31.18) to be (better than) geometric. To implement this, we need to control the product $h_1'(0)h_2'(0)\dots h_n'(0)$. If this were very large, it could cancel the small x effect—i.e., make how small x needs to be shrunk as $n \to \infty$. The key will be an upper bound $h_j'(0)h_{j-1}'(0) < C$ for some universal constant.

Besides this bound, we need to make explicit the ideas that for small x, $q_j(x) - h_j(x) \sim (h_{j-1}'(0)x)^{-1}$ and that $h_j(x) \sim h_j'(0)x$. Since the precise values of the constants aren't important, we'll often use specific values while making no attempt to optimize them.

Proposition 31.6 *Write*

$$q_j(x) = \frac{1}{h_{j-1}'(0)x} - c_j \tag{31.21}$$

Then

(a)

$$|c_j| \leq \frac{3}{2h_{j-1}'(0)} \tag{31.22}$$

(b)

$$0 < x \leq \frac{1}{3} \Rightarrow \frac{1}{2h_{j-1}'(0)x} < q_j(x) < \frac{3}{2h_{j-1}'(0)x} \tag{31.23}$$

(c) *We have that for all* $j = 1, 2, \ldots$

$$h'_j(0)h'_{j-1}(0) \le \tfrac{27}{2} \tag{31.24}$$

(d) *For all* $j = 1, 2, \ldots$ *and* $0 < x < \frac{1}{3}$, *we have that*

$$q_{j+1}(x)q_j(x) \ge \frac{1}{54x^2} \tag{31.25}$$

(e)

$$0 < x < \frac{1}{18} \Rightarrow q_j(x) - h_j(x) \ge \frac{1}{4h'_{j-1}(0)x} \tag{31.26}$$

(f)

$$0 < x < \frac{1}{18} \Rightarrow h_j(x) \le 4h'_j(0)x \tag{31.27}$$

Proof

(a) $\pm q_j(\pm\frac{2}{3}) \ge 0 \Rightarrow \pm\left[\frac{\pm 3}{2h'_{j-1}(0)} - c_j\right] \ge 0 \Rightarrow$ (31.22).

(b) $0 < x \le \frac{1}{3} \Rightarrow |c_j| \le \frac{1}{2h'_{j-1}(0)x} \Rightarrow$ (31.23).

(c) By $\pm(q_j(x) - h_j(x)) \ge 0$ for $\pm x \ge 0$, Proposition 14.10 and $h_j \in \mathcal{M}_\infty(-1, 1)$

$$\begin{aligned} h'_j(0) &\le \tfrac{3}{2}[h_j(\tfrac{1}{3}) - h_j(-\tfrac{1}{3})] \le \tfrac{3}{2}[q_j(\tfrac{1}{3}) - q_j(-\tfrac{1}{3})] \\ &\le \frac{3}{2}2\left[\frac{9}{2h'_{j-1}(0)}\right] \end{aligned}$$

by (31.23). This proves (31.24).

(d) Equation (31.25) is immediate from (31.23) and (31.24).

(e) Since h_j is monotone, we have that $h_j(x) \le h_j(6x) \le q_j(6x)$ if $x \in (0, \frac{1}{6})$. If $0 < 6x < \frac{1}{3}$, then by (31.23),

$$\begin{aligned} q_j(x) - h_j(x) &\ge \frac{1}{2h'_j(0)x} - \frac{3}{2h'_j(0)(6x)} \\ &= \frac{1}{4h'_j(0)x} \end{aligned}$$

(f) Immediate from (31.26) and (31.8) (for $n = j-1$ and then replacing $j-1$ by j).

□

Proposition 31.7

(a) *There exist* $r_1 > 0$ *and* C, *so that*

$$|x| \leq r_1 \Rightarrow \frac{P_{n+1}(x)}{P_{n-1}(x)} \geq \frac{1}{Cx^2} \tag{31.28}$$

(b) *There exist* $r_2 > 0$ *and* N *so that for all* $x \in (0, r_2)$ *and* $n > N$, *either*

$$\left| P_{n-1} + \frac{P_n}{h_n} \right| > 1 \text{ or } \left| P_n + \frac{P_{n+1}}{h_{n+1}} \right| > 1 \tag{31.29}$$

(c) *For some* $r_3 > 0$, $f_\infty(x) = f(x)$ *for all* $x \in (0, r_3)$.

Remarks

1. The r_j are not independent of f, but depend on small n data but the various constants can be picked universally.
2. While the sign of P_n is n dependent, the ratio in (31.28) is not.

Proof

(a) Multiplying $P_{n+1} + P_{n-1} = -q_{n+1}P_n$ by $P_n + P_{n-2} = -q_n P_{n-1}$ yields

$$P_{n+1} + P_{n-1} = \frac{q_n q_{n+1} P_n P_{n-1}}{P_n + P_{n-2}}$$

or

$$\frac{P_{n+1}}{P_{n-1}} = \frac{q_n q_{n+1}}{\left(1 + \frac{P_{n-2}}{P_n}\right)} - 1 \tag{31.30}$$

Pick r_{01} so that $|P_2(x)P_0(x)^{-1}| > 1$ for $0 < x < r_{01}$, possible since $P_0 \equiv 1$ and $P_2(x) \to \infty$ as $x \downarrow 0$. Using (31.23) and (31.24), if also $0 < x < \frac{1}{3}$, then, so long as $|P_{n-2}(x)P_n(x)^{-1}| < 1$,

$$\frac{q_n q_{n+1}}{\left(1 + \frac{P_{n-2}}{P_n}\right)} - 1 \geq \frac{1}{108x^2} - 1 \geq \frac{1}{109x^2} \tag{31.31}$$

if we also pick $|x| < r_{02}$ for small r_{02}. Pick $r_{03} = \min(\frac{1}{3}, r_{01}, r_{02})$ and $C \geq 109$ so that (31.28) holds for $n = 1$, if $|x| < r_{03}$. Finally, pick $r_1 \leq r_{03}$ so that also $Cr_1^{-2} < 1$. Then an easy induction using (31.31) proves that (31.28) holds.

(b) Suppose that for some n and x that (31.29) fails. Then

$$\left| \frac{1}{h_n(x)} \right| \leq \frac{1}{|P_n(x)|} + \frac{|P_{n-1}(x)|}{|P_n(x)|}$$

$$\left|\frac{1}{h_{n+1}(x)}\right| \le \frac{1}{|P_{n+1}(x)|} + \frac{|P_n(x)|}{|P_{n+1}(x)|} = \left(1 + \frac{1}{|P_n(x)|}\right)\frac{|P_n(x)|}{|P_{n+1}(x)|}$$

so

$$\frac{1}{|h_n(x)||h_{n+1}(x)|} \le \left(\frac{1}{|P_{n+1}(x)|} + \frac{|P_{n-1}(x)|}{|P_{n+1}(x)|}\right)\left(1 + \frac{1}{|P_n(x)|}\right) \tag{31.32}$$

By induction and (31.28), $|P_n(x)|^{-1} \to 0$ as $n \to \infty$ uniformly on $[0, r_{20}]$ for r_{20} so that $Cr_{20}^2 \le \frac{1}{2}$. By further shrinking and using (31.28) again, we can find r_{21} and N_0 so that $0 < x < r_{21}$ and $n \ge N_0$ imply that RHS of (31.32) is less than $1/2$. On the other hand, by (31.26) and (31.27), we can be sure that LHS of (31.32) is bigger than 1 if $|x| < r_{22}$. Taking $r_2 = \min(r_{21}, r_{22})$ and $N = N_0$ yields a contradiction.

(c) By (31.19), (b) says that for each $x \in (0, r_2)$ and $n \ge N$, either $|f(x) - f_n(x)| \le |P_n(x)|^{-1}$ or $|f(x) - f_{n+1}(x)| \le |P_{n+1}(x)|^{-1}$. Since $f_n \le f_{n+1} \le f$, we see that in either event

$$|f(x) - f_{n+1}(x)| \le |P_n(x)|^{-1} + |P_{n+1}(x)|^{-1} \tag{31.33}$$

As noted, $|P_n(x)|^{-1} \to 0$ uniformly in small x, proving (c).

□

With one more trick, we can complete this proof of the hard part of Loewner's theorem:

Theorem 31.8 *Any* $f \in \mathcal{M}_\infty(-1, 1)$ *is the restriction to* $(-1, 1)$ *of a function analytic in* $\mathbb{C} \setminus (-\infty, -1] \cup [1, \infty)$, *real on* $(-1, 1)$ *and with* $\operatorname{Im} f(z)$ *if* $z \in \mathbb{C}_+$.

Proof The function f_∞ of Theorem 31.4 has the requisite properties so we need only show that $f_\infty(x) = f(x)$ for all $x \in (-1, 1)$. Proposition 31.7 says that this is true for $|x| < r_3$. In particular, any $f \in \mathcal{M}_\infty(-1, 1)$ is real analytic near $x = 0$. But then, by scaling, if $f \in \mathcal{M}_\infty(a, b)$, then f is real analytic near $\frac{1}{2}(a+b)$. Since $x_0 \in (-1, 1)$ and $f \in \mathcal{M}_\infty(-1, 1)$ imply that $f \in \mathcal{M}_\infty(x_0 - (1 - |x_0|), x_0 + (1 - |x_0|))$, any such f is real analytic on all $(-1, 1)$. Since it agrees with f_∞ near 0, it agrees on all of $(-1, 1)$. □

Notes and Historical Remarks This proof appeared in a paper of Wigner and von Neumann [363] with the title *Significance of Loewner's theorem in the quantum theory of collisions* but mathematicians reading the paper didn't realize that there was a proof because of the title and the fact the first few pages were talk that might make a quick reader think that it really didn't have any serious discussion of theorems. This paper was written at a period when von Neumann was spending most of time in Washington working as a member of the Atomic Energy Commission. He was spending weekends in Princeton and presumably discussed the paper with Wigner. In particular, I'm guessing that he put Wigner in touch with H.S. Wall (an expert on continued fractions, author of [359]). It is likely von Neumann knew of

Loewner's work although in an earlier paper, Wigner thanks Sherman for telling him of that work. It is clear from the style that Wigner wrote the paper and he put his name first even though alphabetically, W comes after both N and v!

If one reads the paper, one finds the proof we give here including the rather involved convergence argument—we have fleshed out some of the details. Wigner quotes the classic book of Perron [267] on continued fractions saying that the convergence arguments are based on Jensen's proof [172] of van Vleck's theorem [356]. (I'd guess, using other continued fraction techniques, one could simplify their argument.)

What we call the Wigner–von Neumann lemma, their paper calls the Lemma of Schiffer and Bargmann, although it appears that Schiffer and Bargmann had little to do with it! Rather, in an earlier paper [361], Wigner noted a result that a meromorphic Herglotz function remains such a function if a pole term is removed. He attributes that result to Schiffer and Bargmann (although it is an immediate consequence of the Herglotz representation).

Nayak [233] proved a converse to the Wigner–von Neumann lemma: If g_{x_0} is given by (31.1) and $g_{x_0} \in \mathcal{M}_n(a, b)$ for *all* $x_0 \in (a, b)$, then $f \in \mathcal{M}_{n+1}(a, b)$.

Wigner had no interest in matrix monotone functions. Instead, in the early 1950s he got very interested in meromorphic Herglotz functions which he viewed as a model of quantum collisions. He wrote three mathematical papers on the subject besides the one with von Neumann [360–362]. It seems clear that he was unaware of the Herglotz representation theorem which was not so well-known then. Many of the results in these three papers are easy consequences of that result. The positivity of the Loewner matrix arose from some physical axioms and what really interested him is that assuming the function was defined on all of $\mathbb{R}$ except for some possible poles, these axioms implied the meromorphic Herglotz functions he was so fond of. This all fit in well with work on dispersion relations which was significant in the elementary particle physics of the era.

In the early years of the current century, I gave in many places a colloquium talk called "The Lost Proof of Loewner's Theorem". In 1964, Loewner [211] gave a talk before an AMS national meeting which appeared in the Bulletin. He focused on his two great pieces of work, neither well appreciated at the time, mainly on Loewner evolution. But he had a section on the monotone matrix theorem where he quoted his proof and mentioned those of Korányi and of Bendat–Sherman. The Wigner–von Neumann paper appears in the bibliography but is not quoted in the paper at all! In his 1974 book, Donoghue [81] says his main goal is to expose the three existing proofs of Loewner's theorem (namely Loewner, Korányi, and Bendat–Sherman). In the Notes, he says: "An important study of Loewner's theorem from the viewpoint of the physicist can be found in" and he quotes Wigner and von Neumann. Bhatia [33] exposes the Hansen–Pedersen proof and mentions several others. About Wigner–von Neumann, he writes: "Among several applications of these ideas, there are two we should mention here . . . They also occur in problems related to elementary particles, see, for example,. . . ."

When I became interested in Loewner's theorem about 2000, I found these references and was then shocked to discover a full fledged and apparently lost proof of the theorem in [363]. In my talk, I gave these quotes but as referring to a "mystery paper." The punch line of part of my talk was then: "You might think the mystery paper appeared in some obscure journal or worse, a physics journal by obscure authors." I then pause for dramatic effect and said "The journal was the Annals of Mathematics. The authors were Wigner and von Neumann!" I still find it remarkable that despite lots of people working on this theorem, this proof was not known to many mathematicians working in this area!

The complicated part of this proof is to show that f_∞, the limit of continued fractions, agrees with f. However, if one is willing to borrow the first part of the Bendat–Sherman proof (see Chapter 26), one can avoid this argument. It can be shown (see below) that for all $k = 0, 1, \dots$

$$f_\infty^{(k)}(0) = f^{(k)}(0) \tag{31.34}$$

The first half of the Bendat–Sherman proof (i.e., the Bernstein–Boas theorem, Theorem 26.3) shows that f is real analytic on $(-1, 1)$. Since f_∞ is also real analytic, (31.34) immediately shows that $f = f_\infty$.

To prove (31.34), we begin by noting that is it easy to see that

$$f_n^{(k)}(0) = f^{(k)}(0); \qquad k = 1, \dots, 2n \tag{31.35}$$

(it even holds for $k = 2n+1$). Since f_n is a meromorphic Herglotz function analytic on $(-1, 1)$, we have for $x \in (-1, 1)$ that

$$f_n(x) = \sum_{j=1}^{n} \lambda_j^{(n)} (y_j^{(n)} - x)^{-1} \tag{31.36}$$

where $\lambda_j^{(n)} > 0$ and $y_j^{(n)}$ is real with $|y_j^{(n)}| > 1$, so for such x

$$\frac{f_n^{(k)}(x)}{k!} = \sum_{j=1}^{n} \lambda_j^{(n)} (y_j^{(n)} - x)^{-k-1} \tag{31.37}$$

Since $|y_j^{(n)}| > 1$, we have for $n \geq 1$, $k \geq 1$, and $|x| < 1/2$ that

$$\frac{|f_n^{(k)}(x)|}{k!} \leq 2^{k+1} \sum_{j=1}^{n} \lambda_j^{(n)} (y_j^{(n)})^{-2} = 2^{k+1} f'(0) \tag{31.38}$$

since $f_n'(0) = f'(0)$ for $n \geq 1$. It follows from Taylor's theorem with remainder that for K fixed, $|x| < 1/2$ and n with $K < 2n$, we have that

$$\left| f_n(x) - \sum_{j=1}^{n} \frac{f^{(j)}(0)}{k!} x^j \right| \leq \frac{|2x|^{K+1}}{(K+1)!} f'(0) \tag{31.39}$$

which proves (31.34) by taking $n \to \infty$ and then $x \to 0$.

Chapter 32
Multipoint Continued Fractions: A New Proof

This chapter has a variant of the proof of the last one that we regard as a new proof of the hard part of Loewner's theorem. The proof of the last chapter was fairly straightforward but the details of the proof of convergence to the right limit were involved—one might even say tedious. The variant here is a simple change in the strategy but one where the convergence will be effortless—no estimates will be needed. The Wigner–von Neumann lemma has a base point about which the continued fraction is expanded. In their proof, to get the approximation at step n, we repeated the process n times at 0. In our new proof, at step n, we'll only make the expansion once at each point but we'll do it at each of the dyadic rationals $j/2^n$, $j \in \mathbb{Z}$ which also lie in $(-1, 1)$. The construction will be such that $f_n(j/2^n) = f(j/2^n)$ so convergence at the dense set of dyadic rationals is immediate and that is enough to get weak convergence of the measures in a Herglotz representation.

It is likely that as in the Wigner–von Neumann analysis, the Herglotz representations, f_n, obtained by truncating the continued fractions have no poles in $(-1, 1)$ but we'll settle for proving that there are no poles in $(-1 + 2^{-n}, 1 - 2^{-n})$ which suffices for our needs. Here is what we'll prove:

Theorem 32.1 *Let $f \in \mathcal{M}_\infty(-1, 1)$ be a C^1 function whose Loewner matrices are all strictly positive. Let (the funny labeling is intended because of the inductive construction we use later)*

$$-1 < x_m < x_1 < \cdots < x_{m-1} < 1 \tag{32.1}$$

Let $\varphi(x)$ be the function obtained by truncating the Wigner–von Neumann continued fraction obtained by successively expansion at $x_1, \ldots, x_m$. Then

(1) *φ is a rational function of degree m*
(2) *We have that*

B. Simon, *Loewner's Theorem on Monotone Matrix Functions*, Grundlehren der mathematischen Wissenschaften 354, https://doi.org/10.1007/978-3-030-22422-6_32

$$\varphi(x_j) = f(x_j), \qquad \varphi'(x_j) = f'(x_j); \qquad j = 1, \dots, m \tag{32.2}$$

(3) *φ is a Herglotz function with exactly m poles each with strictly negative residues.*
(4) *The poles are all outside $[x_m, x_{m-1}]$.*

Proof of Loewner's Theorem Given Theorem 32.1 Without loss of generality, we can suppose that f has strictly positive Loewner matrices since we can add a small multiple of log. For each n=1,2,..., let $m = 2^{n+1} - 1$ and let $x_1, \dots, x_m$ be a reordering of $\{j/2^n\}_{j=-2^n+1}^{2^n-1}$ obeying (32.1) and let f_n be the φ associated to this set of points. Then, in the notation of Theorem 3.9, the f_n are Herglotz functions in $\mathcal{H}(-1 + 1/2^n, 1 - 1/2^n) \equiv \mathcal{H}_n$. By that theorem, for each ℓ, $\{f_n\}_{n \geq \ell}$ lie in a compact subset of $\mathcal{H}_\ell$ since $f_n(0) = f(0), \quad f_n'(0) = f'(0)$.

It follows that there are limit points in $\mathcal{H}(-1, 1)$. All limit points, f_∞, clearly have $f_\infty(j/2^n) = f(j/2^n)$ for all dyadic rationals in $(-1, 1)$. Thus, by continuity of f and density of the dyadic rationals, the limit exists and equals f, so $f \in \mathcal{H}(-1, 1)$. □

Remark As advertised, we only need to prove properties of the f_n; the convergence result is effortless and needs no estimates.

We begin the proof of Theorem 32.1 with some notation. Let $k_0(x) = f(x)$ and inductively define for $j = 0, \dots, m-1$

$$p_{j+1}(x) = \frac{1}{k_j'(x_{j+1})(x - x_{j+1})} - \frac{k_j''(x_{j+1})}{2k_j'(x_{j+1})^2} \tag{32.3}$$

$$k_{j+1} = -\frac{1}{k_j(x) - k_j(x_{j+1})} + p_{j+1}(x) \tag{32.4}$$

so

$$k_j(x) = k_j(x_{j+1}) + \frac{1}{p_{j+1}(x) - k_{j+1}(x)} \tag{32.5}$$

By the Wigner–von Neumann Lemma (Lemma 31.1), $f \in \mathcal{M}_\infty(-1, 1) \Rightarrow k_j \in \mathcal{M}_\infty(-1, 1)$ for $j = 1, 2, \dots, m$.

Define fractional linear transformations $w \mapsto \sigma_{j,x}(w)$ for $j = 1, \dots, n; x \in (-1, 1)$ by

$$\sigma_{j,x}(w) = k_{j-1}(x_j) + \frac{1}{p_j(x) - w} \tag{32.6}$$

so (32.5) becomes

$$\sigma_{j,x}(k_j(x)) = k_{j-1}(x) \tag{32.7}$$

Moreover, since $p_j(x)$ is real for $x \in \mathbb{R} \setminus \{x_{j+1}\}$, we have that

$$\operatorname{Im} w > 0 \Rightarrow \operatorname{Im} \sigma_{j,x}(w) > 0 \tag{32.8}$$

Thus, for $j = 1, 2, \dots, n$

$$f(x) = \sigma_{1,x}(\sigma_{2,x}(\dots \sigma_{j,x}(k_j(x)))) \tag{32.9}$$

We will use the fact that for any $w \in \mathbb{C}$ (but not necessarily $w = \infty$) and any C^1 function, g, with $g(x_j) = w$, we have that

$$\sigma_{j,x_j}(w) = k_{j-1}(w); \qquad \frac{d}{dx}\sigma_{j,x}(g(x)\Big|_{x=x_j} = k'_{j-1}(x) \tag{32.10}$$

This implies that for any such g

$$h(x) = \sigma_{1,x}(\sigma_{2,x}(\dots \sigma_{j,x}(g(x))) \Rightarrow h(x_j) = f(x_j),\ h'(x_j) = f'(x_j) \tag{32.11}$$

The truncated continued fractions are

$$f^{(j)}(x) = \sigma_{1,x}(\sigma_{2,x}(\dots \sigma_{j,x}(0))) \tag{32.12}$$

We begin with

Proposition 32.2 *Let $f \in \mathcal{M}_\infty(-1, 1)$, $-1 < x_2 < x_1 < 1$ with f not a rational function of degree 2. Then $f^{(2)}(x)$ is a rational function of degree two whose two poles are in $\mathbb{R} \setminus [x_2, x_1]$ and with*

$$f^{(2)}(x_j) = f(x_j),\ [f^{(2)}]'(x_j) = f'(x_j); \quad j = 1, 2 \tag{32.13}$$

Remark The key is that (32.13) holds which requires that the poles of $f^{(1)}$ not be at the x_j. Once one has that one can use the same continuity argument we'll use to prove Theorem 32.1 to show that the poles lie in $\mathbb{R} \setminus [x_2, x_1]$. However, since a clean, crisp calculation cleanses the soul and can be illuminating, we give explicit formulae.

Proof Without loss, we can suppose that $f(x_1) = 0$. We set

$$f(x_2) = \gamma < 0, \quad f'(x_1) = \alpha > 0, \quad f'(x_2) = \beta > 0 \tag{32.14}$$

If $\tilde{k}_1$ is the k_1 resulting by taking $k_2 \equiv 0$ in (32.5), then

$$\tilde{k}_1(x) = \zeta_1 + \frac{1}{\frac{1}{\beta_1(x-x_2)}} = \zeta_1 + \beta_1(x - x_2) \tag{32.15}$$

where since $k_1 \in \mathcal{M}_\infty(-1, 1)$ (by the Wigner–von Neumann lemma) and k_1 is not constant (if it were $f^{(2)}$ would be rational of degree 2)

$$\beta_1 = k_1'(x_2) > 0, \qquad \zeta_1 = k_1(x_2) \tag{32.16}$$

Thus, for suitable γ_1

$$f^{(2)}(x) = \frac{\alpha(x - x_1)}{P_2(x)}; \qquad P_2(x) = 1+\gamma_1(x-x_1)-\beta_2(x-x_1)(x-x_2) \tag{32.17}$$

where

$$\beta_2 = \alpha\beta_1 > 0 \tag{32.18}$$

Since $\tilde{k}_1(x_1) \neq \infty$ we can use (32.11) to see that

$$\gamma = f^{(2)} = \frac{\alpha(x_2 - x_1)}{P_2(x_2)} \Rightarrow P_2(x_2) = \frac{\alpha(x_2 - x_1)}{\gamma} > 0 \tag{32.19}$$

since $\alpha > 0$, $\gamma < 0$, $x_2 - x_1 < 0$. Obviously $P_2(x_1) = 1$ so since $\beta_2 > 0$, we have that

$$P_2(x_1) > 0, \quad P_2(x_2) > 0, \quad P_2(\pm\infty) = -\infty \tag{32.20}$$

so P_2 has its two zeros, one each in $(-\infty, x_2)$ and (x_1, ∞) so the poles of $f^{(2)}$ lie outside $[x_1, x_2]$ as claimed. □

Proof of Theorem 32.1 Fix f. Let $\varphi(x; x_1, \dots, x_n)$ defined for x_j's obeying (32.1) be the function described in the theorem. By construction, φ is a ratio of polynomials of degree at most n. The denominator is seen to be $\prod_{j=1}^n (x - x_j)$ plus a lower order polynomial so of degree exactly n. It follows that φ is bounded at infinity. By iterating (32.8), we see that φ is Herglotz.

Suppose that φ obeys (32.2) so, in particular, the Loewner matrices $L_n(x_1, x_2, \dots, x_n; f)$ and $L_n(x_1, x_2, \dots, x_n; \varphi)$ agree, so, in particular the one for φ is strictly positive. Then φ has a representation as a sum of at most n simple poles on the real axis with strictly negative residues plus its value at infinity. If there were fewer than n poles, the Loewner matrix wouldn't be strictly positive so we conclude there are exactly n poles, each with strictly negative residue. We have thus proven (1) and, once we have (2), we have also proven (3).

By Proposition 32.2, (32.9) and (32.10), we have that (32.2) holds for $j = m - 1, m$ no matter what the value of the other x_j's. In the limit as all the x_j's go to zero, the function φ goes to the function constructed in the last chapter. We actually know that this limit has no poles in $(-1, 1)$ but we don't need to use that fact whose proof is not hard but does need some work. Rather, all we need is that this limit φ is finite at 0 so it has no poles in a small neighborhood of 0. It is not hard to see that as a map of the Riemann sphere to itself, $\varphi(\cdot; x_1, \dots, x_n)$ as a composition of fractional linear transformations is continuous in the x_j's. It follows by continuity that for distinct x_j's, there are no poles in $[x_n, x_{n-1}]$ for φ nor of the iterating continued fractions when all the $|x_j|$ are small. In particular, (2) holds for such small $|x_j|$.

As we move the x_j's, the poles move continuously, although a priori, it might happen a residue vanishes at some values of the parameters. We know that (32.2) holds as x_{m-1}, x_m for all values of the parameters so there can't be a pole with a non-zero residue at those points. If for some value of the x_j's, all the poles of φ and the continued fraction that build it up lie outside $[x_m, x_{m-1}]$, then as noted above, (2) holds for that φ. If a pole tried to move from outside $[x_m, x_{m-1}]$ to inside, it would have to move through an endpoint. That can only happen if the residue vanished. But by continuity, (2) would also hold at the critical point and, as we've seen, that would imply that the residues are all non-zero. This implies the poles can't move inside and proves (4) in all cases which implies (2). □

Notes and Historical Remarks The proof in this chapter is new. It is an obvious way to try and improve the Wigner–von Neumann approach given work on multipoint Padé approximants. Perhaps if their proof hadn't been "lost," this proof might have been found earlier. We remark these are not quite multipoint Padé approximants because a ratio of two degree n polynomials needs $2n + 1$ parameters and (32.2) is only $2n$ conditions.

Chapter 33
Hardy Spaces and the Rosenblum–Rovnyak Proof

In this chapter, we present a proof of the hard part of Loewner's theorem due to Rosenblum and Rovnyak based on their general theorem that appeared as Theorem 19.8. Since H^2 and H^∞ are central to their machine, I've emphasized the role of Hardy spaces in titling the chapter.

In thinking about applying Theorem 19.8, there are two potential puzzles that actually solve each other. First Loewner's theorem asserts the existence of a map from $\mathbb{C}_+$ to $\mathbb{C}_+$, while Theorem 19.8 yields maps from $\mathbb{D}$ to $\mathbb{D}$. Secondly, the hypothesis of the hard part of Loewner's theorem (at least after one proves Theorem 5.9) involves a single function, f, with a linear inequality on f but the Rosenblum–Rovnyak proof needs two functions, β and γ, and a quadratic inequality between them. In dealing with the change of domain, we used conformal maps of $\mathbb{C}_+$ to $\mathbb{D}$ already in Chapters 16 and 19. We can do something similar with ranges, i.e., if f maps $\mathbb{C}_+$ to $\mathbb{C}_+$, then $\omega = \frac{f-i}{f+i}$ maps $\mathbb{C}_+$ to $\mathbb{D}$ and one can go in the other direction from information on $\mathrm{Ran}(\omega)$ to $\mathrm{Ran}(f)$. Thus, we'll take $\beta = f - i$ and $\gamma = f + i$ and note that if f is real, then

$$\overline{\gamma(x)}\gamma(y) - \overline{\beta(x)}\beta(y) = 4i[f(x) - f(y)]$$

so a quadratic inequality on β and γ will be equivalent to a linear inequality on f!

Because of the application we have in mind, we'll make the simplifying assumption that the initial functions are continuous on some open set, $\Delta \subset \mathbb{R}$, although Rosenblum and Rovnyak only require measurable functions on Borel sets, Δ. Also, while Theorem 19.8 has if and only if arguments, we'll only go in the direction needed for the hard part of Loewner's theorem.

We start with a warmup: interpolation from an open subset of $\partial\mathbb{D}$. We'll let $d\sigma(u) = \frac{d\theta}{2\pi}$ when $u = e^{i\theta}$.

B. Simon, *Loewner's Theorem on Monotone Matrix Functions*, Grundlehren der mathematischen Wissenschaften 354, https://doi.org/10.1007/978-3-030-22422-6_33

Theorem 33.1 *Let $\Xi \subset \partial\mathbb{D}$ be an open subset and let b,c be complex-valued bounded continuous functions on Ξ with c non-vanishing. Suppose that for all bounded continuous functions, h, on Ξ, we have that*

$$\lim_{r\uparrow 1} \int_{u,v\in\Xi} \frac{\overline{c(u)}c(v) - \overline{b(u)}b(v)}{1 - r^2\bar{u}v} \overline{h(u)}h(v)\, d\sigma(u)\, d\sigma(v) \geq 0 \tag{33.1}$$

Then there is a function, $w \in H^\infty(\mathbb{D})$ with $\|w\|_\infty \leq 1$ so that for every $u_0 \in \Xi$, $\lim_{r\uparrow 1} w(ru_0) \equiv w(u_0)$ exists. Moreover, w so defined is continuous on $\mathbb{D} \cup \Xi$ and so that for all $u_0 \in \Xi$

$$w(u_0)c(u_0) = b(u_0) \tag{33.2}$$

Remarks

1. Our proof will show that the limit in (33.1) exists for all bounded continuous b, c, h on Ξ.
2. As we'll note in our proof, the space K of the proof of Theorem 19.8 is all of $H^2(\mathbb{D})$, so that one doesn't need commutant lifting at all!

Proof We use the framework of Theorem 19.8. We let V be the set of bounded continuous functions on Ξ, so, by hypothesis, $b, c \in V$. We let V act on itself as linear functionals via

$$\ell_h(g) = \int_\Xi h(u)g(u)\, d\sigma(u) \tag{33.3}$$

(no complex conjugate since Theorem 19.8 is stated in terms of real, not complex conjugate, duality). We let A be the operator on V of multiplication by u, the variable on $\Xi \subset \mathcal{D}$. $\mathcal{D}$ will be $\{\ell_h \mid h \in V\}$.

Thus

$$\ell_h(A^\ell c) = \langle \overline{u^\ell}, hc\rangle \tag{33.4}$$

are the negative Fourier coefficients of hc viewed as a function on $\partial\mathbb{D}$ that vanishes on $\partial\mathbb{D} \setminus \Xi$. Since $hc \in L^2(\partial\mathbb{D}, d\sigma)$, we have (19.51), i.e., $\sum_{j=0}^\infty |\ell_h(A^j c)|^2 < \infty$ for all $h \in \mathcal{D}$. The function f_ℓ of (19.55) is

$$\begin{aligned} f_\ell(z) &= \sum_{j=0}^\infty z^j \int_\Xi u^j h(u)c(u)\, d\sigma(u) \\ &= \int_\Xi \frac{h(u)c(u)}{1 - uz}\, d\sigma(u) \end{aligned} \tag{33.5}$$

where the sum and integral can be interchanged by any easy estimate since $|z| < 1$. (We note parenthetically that $\{f_\ell\}$ is dense in $H^2(\mathbb{D})$, for changing u to $\bar{u}$ in (33.5) shows that f_ℓ is $P_+(g_\ell)$ where $g_\ell(u) = h(\bar{u})c(\bar{u})$. It follows that any $q \in H^2$ orthogonal to all f_ℓ obeys $0 = \langle q, P_+ g_\ell\rangle = \langle P_+ q, g_\ell\rangle = \langle q, g_\ell\rangle$ so q vanishes on $\overline{\Xi}$ and so is zero.)

To check (19.52), we compute

$$\begin{aligned}\sum_{j=0}^{\infty} |\ell_h(A^j c)|^2 &= \lim_{r\uparrow 1} \sum_{j=0}^{\infty} |\ell_h(r^j A^j c)|^2\\ &= \lim_{r\uparrow 1} \sum_{j=0}^{\infty} \left[\int_\Xi \overline{(ru)}^j \overline{h(u)c(u)}\, d\sigma(u)\right]\left[\int_\Xi (rv)^j h(v)c(v)\, d\sigma(v)\right]\\ &= \lim_{r\uparrow 1} \int_{u,v\in\Xi} \frac{\overline{c(u)}c(v)}{1-r^2\bar{u}v}\overline{h(u)}h(v)\, d\sigma(u)d\sigma(v)\end{aligned}$$

Thus (19.52) is equivalent to (33.1) and Theorem 19.8 is applicable.

It follows there is a function, $w \in H^\infty$ with $\|w\|_\infty \le 1$ so that if $w(z) = \sum_{j=0}^\infty w_j z^j$, then for all bounded continuous h on Ξ,

$$\begin{aligned}\int_\Xi h(u)b(u)\, d\sigma(v) &= \sum_{j=0}^{\infty} w_j \lim_{r\uparrow 1} \int_\Xi (ru)^j h(u)c(u)\, d\sigma(u)\\ &= \lim_{r\uparrow 1} \int_\Xi w(ru)h(u)c(u)\, d\sigma(u)\\ &= \int_\Xi w(u)h(u)c(u)\, d\sigma(u)\end{aligned}$$

in terms of the boundary values of w. Since $\{h\}$ is dense in $L^2(\Xi, d\sigma)$, we conclude that for a.e. $u \in \Xi$, we have that $b(u) = w(u)c(u)$.

Thus, since c is non-vanishing, the boundary values of w are a.e. equal to the continuous function bc^{-1} on Ξ. Since Ξ is open, a simple argument using the Poisson kernel shows that w extends to a continuous function on $\mathbb{D} \cup \Xi$ which is equal to bc^{-1} everywhere on Ξ. □

We transfer this from $\Delta \subset \partial\mathbb{D}$ to $\Delta \subset \mathbb{R}$ using the maps (19.16).

Theorem 33.2 *Let $\Delta \subset \mathbb{R}$ be a bounded, open subset and let β,γ be complex-valued bounded continuous functions on Δ with γ non-vanishing. Suppose that for all bounded continuous functions, φ, on Δ, we have that*

$$\lim_{r\uparrow 1} \int_{x,y\in\Delta} \frac{(-i)\left[\overline{\gamma(x)}\gamma(y) - \overline{\beta(x)}\beta(y)\right]}{(1+r^2)(x-y) - i(1-r^2)(1+xy)}\overline{\varphi(x)}\varphi(y)\, dx\, dy \ge 0 \qquad (33.6)$$

Then there is a function, $\omega \in H^\infty(\mathbb{C}_+)$ with $\|\omega\|_\infty \leq 1$ so that ω has a continuous extension to $\mathbb{C}_+ \cup \Delta$ and so that for all $x_0 \in \Delta$

$$\omega(x_0)\gamma(x_0) = \beta(x_0) \tag{33.7}$$

Remarks

1. One might prefer to have $\lim_{\epsilon \downarrow 0}$ with $(1 - r^2)(1 + xy)$ replaced by ϵ which is the same as approximating $\mathbb{R}$ with lines parallel it rather than the images of constant radius circles under the conformal map of $\mathbb{D}$ to $\mathbb{C}_+$. This is the result in the book of Rosenblum–Rovnyak [289]—they prove it by mimicking the proof of Theorem 33.1 rather than using the result itself with conformal mapping as we do below.
2. If β and γ are complex conjugates on Δ, then $|\omega(x)| = 1$ on Δ which, by a reflection principle, implies that ω has a meromorphic continuation to the lower half plane.

Proof Use the restriction of the map (19.16) to $\mathbb{R}$, i.e.,

$$u(x) = \frac{i - x}{i + x} \qquad v(y) = \frac{i - y}{i + y} \tag{33.8}$$

Given γ, β, φ, define c, b, h on $\boldsymbol{\zeta}[\Delta]$ by ($\boldsymbol{\zeta}$ given by (19.16))

$$c(u(x)) = \gamma(x)(i + x)^{-1} \quad b(u(x)) = \beta(x)(i + x)^{-1}$$

$$h(u(x)) = \varphi(x) \tag{33.9}$$

Taking into account that $d\sigma(u(x)) = \pi^{-1}(1 + x^2)^{-1}$ (see (19.18)), the integral in (33.6) becomes

$$\pi^2 \int_{u,v \in \Xi} \frac{\left[\overline{c(u)}c(v) - \overline{b(u)}b(v)\right] \overline{h(u)}h(v)\, d\sigma(u)\, d\sigma(v)}{i(-i + x)^{-1}(i + y)^{-1}\left[(1 + r^2)(x - y) + i(1 - r^2)(1 + xy)\right]} \tag{33.10}$$

Next compute

$$\begin{aligned}(1 - r^2\overline{u(x)}v(y))(-i + x)(i + y) &= (-i + x)(i + y) - r^2(-i - x)(i - y) \\ &= i\left[(1 + r^2)(x - y) - i(1 - r^2)(1 + xy)\right]\end{aligned} \tag{33.11}$$

Thus the integral in (33.10) is π^2 times that in (33.1) so the positivity in (33.6) implies (33.1). Equation (33.2) becomes (33.7) if we define $\omega(z) = w(\boldsymbol{\zeta}(z))$. $\square$

This allows the following proof of the hard half of one version of Loewner's theorem for general bounded open sets (Theorem 35.2). $\mathcal{M}_\infty^{(r)}(\Delta)$ is defined at the start of Chapter 35.

Theorem 33.3 *Let Δ be an open, bounded subset of $\mathbb{R}$. Let $f \in \mathcal{M}_\infty^{(r)}(\Delta)$. Then f is the restriction to Δ of an analytic function on $(\mathbb{C}\backslash\mathbb{R})\cup\Delta$ which has* $\mathrm{Im}(f(z)) > 0$ *if* $\mathrm{Im}(z) > 0$.

Proof Let $\Delta_\epsilon = \{x \in \mathbb{R} \mid \mathrm{dist}(x, \mathbb{R}\setminus\Delta) > \epsilon\}$. f is bounded and continuous on Δ_ϵ so if we prove the theorem for $f \restriction \Delta_\epsilon$, we get it general. That is, without loss we can suppose that f is bounded and C^1 on Δ with bounded derivative.

Let φ be a bounded continuous function on Δ. We claim that

$$\int_{x,y\in\Delta} \frac{f(x)-f(y)}{x-y}\overline{\varphi(x)}\varphi(y)\,dx\,dy \geq 0 \tag{33.12}$$

Proposition 35.7 says that the Riemann sums approximating this are non-negative.

Let $\gamma(x) = f(x)+i$, $\beta(x) = f(x)-i$. Then, since f is real on Δ, we have that

$$\overline{\gamma(x)}\gamma(y) - \overline{\beta(x)}\beta(y) = 2i(f(x)-f(y)) \tag{33.13}$$

Since f is C^1 with bounded derivatives, we can take the limit (33.6) inside the integral which is two times (33.12) so (33.12) implies (33.6). By Theorem 33.2, there is ω continuous from $\mathbb{C}_+\cup\Delta$ to $\overline{\mathbb{D}}$ and analytic on $\mathbb{C}_+$, so that for all $x \in \Delta$

$$f(x) - i = \omega(x)(f(x)+i) \tag{33.14}$$

Solving for f, we see that

$$f(x) = i\frac{1+\omega(x)}{1-\omega(x)} \tag{33.15}$$

Since $\zeta \to i\frac{1+\zeta}{1-\zeta}$ maps $\mathbb{D}$ to $\mathbb{C}_+$ and $\|\omega\|_\infty \leq 1$, f obeys the required condition on $\mathbb{C}_+$. Since f is real on Δ, we can analytically extend f to $\mathbb{C}_-$ by $f(\bar{z}) = \overline{f(z)}$. $\square$

Notes and Historical Remarks Between 1974 and 1980, Rosenblum and Rovnyak wrote several works [286–289] on interpolation and on analytic continuation from boundary data. Each had a proof of a theorem going from (33.12) to the same conclusion as Theorem 33.3. The proofs were variants of each other and we give the version from [289]. They emphasized that they could deal with arbitrary Borel sets $\Delta \subset \mathbb{R}$.

Some of these works state Loewner's theorem in terms of the restricted Hilbert transform. The *Hilbert transform* can be defined for any $g \in L^p(\mathbb{R})$, $1 \leq p < \infty$ by

$$Hg(x) = \lim_{\epsilon \downarrow 0} \frac{1}{\pi} \int_{|x-t|>\epsilon} g(t) \frac{dt}{t-x} \tag{33.16}$$

where the limit exists for a.e. x. Moreover, for a.e. x and such g, one has that

$$Hg(x) = \lim_{\epsilon \downarrow 0} \frac{1}{\pi} \int_{-\infty}^{\infty} \operatorname{Re}\left[\frac{1}{t-x-i\epsilon}\right] g(t)dt \tag{33.17}$$

For $g \in L^2(\mathbb{R})$, we have that for a.e. k

$$\widehat{Hf}(x) = \begin{cases} -i\hat{f}(k), & k > 0 \\ i\hat{f}(k), & k < 0 \end{cases} \tag{33.18}$$

so, in particular, H is a unitary map. One also has that for any $g, h \in L^2(\mathbb{R})$

$$\lim_{\epsilon \downarrow 0} \frac{1}{\pi} \int \frac{\overline{h(x)}g(y)}{x-y+i\epsilon} \, dx \, dy = \langle h, Hg - ig \rangle \tag{33.19}$$

(This is an L^2 version of the formula that weakly as distributions, $(x+i\epsilon)^{-1} \to \mathcal{P}(\frac{1}{x}) - i\pi\delta(x)$ as $\epsilon \downarrow 0$.) Proofs of these facts can be found in [185] or [328, Section 5.9].

If P_Δ is the orthogonal projection onto $L^2(\Delta, dx) \subset L^2(\mathbb{R}, dx)$, one defines $H_\Delta = P_\Delta H P_\Delta$, the *restricted Hilbert transform.* This is not unitary but is a contraction and is skew-adjoint. From (33.19), we see that for a real-valued, C^1 function, f, one has that

$$\frac{1}{\pi} \int_{x,y\in\Delta} \frac{f(x)-f(y)}{x-y} \overline{\varphi(x)}\varphi(y) \, dx \, dy = \langle \varphi, (fH_\Delta - H_\Delta f)\varphi \rangle \tag{33.20}$$

(33.12) can be interpreted in terms of positivity of $[f, H_\Delta]$ and that is how Rosenblum and Rovnyak stated Loewner's theorem in their first paper on the subject [286].

Chapter 34
Mellin Transforms: Boutet de Monvel's Proof

In this chapter, we'll see that Mellin transforms are a natural approach to proving the hard part of Loewner's theorem. We've already seen several reformulations of the integral representation for matrix monotone functions on $(0, \infty)$. Here is one more:

Theorem 34.1 *Let $f \in \mathcal{M}_\infty(0, \infty)$ have $f(0) = 0$, $f'(0) < \infty$, and $\lim_{x\to\infty} f(x) < \infty$. Then for a measure, ν on $(0, \infty)$ with $\int d\nu(\lambda) < \infty$ and $\int \lambda^{-1} d\nu(\lambda) < \infty$, we have that*

$$f(x) = \int \frac{x}{x+\lambda} d\nu(\lambda) \tag{34.1}$$

Remarks

1. By $f(0)$ and $f'(0)$, we mean $\lim_{t\downarrow 0} f(t)$ and $\lim_{t\downarrow 0} f'(t)$, respectively, since f is only defined on $(0, \infty)$. These limits exist (and also, $\lim_{x\to\infty} f(t)$) since f is monotone increasing and f' monotone decreasing (since $f''(x) \leq 0$; see Corollary 14.5).
2. In the proof in this chapter, we'll not use directly that f is matrix monotone, but rather, in the language of Chapter 15, that f is an interpolation function in the infinitesimal form of Theorem 15.9. Since any function on $(0, \infty)$ with a Herglotz representation is an interpolation function in non-infinitesimal form (see the proof of Theorem 15.12), this theorem provides a converse to Theorem 15.9.

The proof of this theorem is the main result of this chapter. We first show that the conditions that f is bounded and $f'(0)$ is finite are unimportant in that, given this result, one can get that any $f \in \mathcal{M}_\infty(0, \infty)$ is Herglotz.

Proof That Theorem 34.1 Implies the Hard Part of Loewner's Theorem Note that

B. Simon, *Loewner's Theorem on Monotone Matrix Functions*, Grundlehren der mathematischen Wissenschaften 354, https://doi.org/10.1007/978-3-030-22422-6_34

$$\operatorname{Im}\left(\frac{z}{z+\lambda}\right) = \frac{\operatorname{Im}(z(\bar{z}+\lambda))}{|z+\lambda|^2} = \frac{\lambda \operatorname{Im} z}{|z+\lambda|^2}$$

so the theorem implies that any $f \in \mathcal{M}_\infty(0,\infty)$ with $f'(0) < \infty$ and f bounded is the restriction of a Herglotz function to $(0,\infty)$. Thus, it suffices to prove that any $f \in \mathcal{M}_\infty(0,\infty)$ with $f(0) = 0$ is a pointwise limit of such functions.

Since $A \mapsto -A^{-1}$ is matrix monotone on positive operators (and also on negative operators), if f is strictly positive and matrix monotone, so is $-f(x)^{-1}$. Thus if f, g are positive and matrix monotone functions on $(0,\infty)$, so is $(f^{-1}+g^{-1})^{-1} = fg/(f+g)$. Let $g(x) = x(x+1)^{-1}$ and $f_n(x) = nf(x)g(x)/(f(x)+ng(x))$. Then $f_n(x) \le f(x)$ and converges pointwise to $f(x)$ as $n \to \infty$. Moreover, $f_n(x) \le ng(x) \le n$ so f_n is bounded, vanishes at 0 and $\lim_{x\downarrow 0} x^{-1} f_n(x) \le ng'(0) = n$ so $f_n(0) < \infty$. This provides the required approximations. □

The key to Boutet de Monvel's proof of Theorem 34.1 is to observe that (34.1) is a convolution. It doesn't look like one because we don't mean additive convolution but multiplicative convolution. That is, if f, g are functions on, and ν a measure on $(0,\infty)$, we define (assuming the integrals converge!):

$$f \wedge g(x) = \int_0^\infty f(xy^{-1}) g(y) \frac{dy}{y} \tag{34.2}$$

$$g \wedge \nu(x) = \int g(xy^{-1}) d\nu(y) \tag{34.3}$$

In (34.2), dy/y occurs as Haar measure (see [325, Section 4.19]) on $(0,\infty)$ under the group operation $(\lambda,\alpha) \mapsto \lambda\alpha$. If we associate $f \in L^1(\mathbb{R}, dx/x)$ with the measure $f(x)dx/x$, (34.2) and (34.3) are consistent. Because dy/y is invariant not only under multiplication but also under $y \mapsto y^{-1}$, it is easy to see that

$$f \wedge g(x) = \int_0^\infty f(y) g(xy^{-1}) \frac{dy}{y} = g \wedge f(x)$$

Note that (34.1) can be rewritten

$$f(x) = \int \frac{x\lambda^{-1}}{1+x\lambda^{-1}} d\nu(\lambda) = g \wedge \nu(x) \tag{34.4}$$

where

$$g(y) = \frac{y}{1+y} \tag{34.5}$$

Just as additive convolution is turned into multiplication by Fourier transform on the additive group of reals, $g \wedge \nu$ is turned to multiplication by Fourier analysis on $\mathbb{R}_+$ as a multiplicative group (see [329, Section 6.9] for Fourier analysis on general locally compact abelian groups—of course the group under discussion here is topologically isomorphic to the additive group of reals, so, as we'll see, one can reduce everything to conventional Fourier transform). This transform, known as *Mellin transform*, is formally defined by

$$\tilde{f}(s) = \int_0^\infty x^{s-1} f(x)dx \tag{34.6}$$

$$\tilde{\nu}(s) = \int x^s d\nu(x) \tag{34.7}$$

(consistent with associating f with the measure $f(x)dx/x$). So our plan is to first present a lightning course on Mellin transforms and show for a suitable choice of bounded operators, A and B, on $L^2(\mathbb{R}, dx/x)$, that (15.8) implies that (with g given by (34.5)) $\tilde{f}(s)/\tilde{g}(s)$ is a function of positive type in Bochner's sense so Bochner's theorem implies (34.1).

Given a positive measure, ν on $(0, \infty)$, we define the *interval of convergence*, I_ν, to be the set of those $c \in \mathbb{R}$ with $\int x^c d\nu(x) < \infty$. For a function, f, I_f is those $c \in \mathbb{R}$ so that $\int x^{c-1}|f(x)|dx < \infty$. If f or ν has compact support in $(0, \infty)$, then $I_f = \mathbb{R}$. Since $a < c < b \Rightarrow x^c \leq x^a + x^b$ (by considering $x \leq 1$ and $x \geq 1$) I_f and I_ν are intervals which can be open, closed, or half open on either side.

The *fundamental strip*, S_f (resp. S_ν), is $\{z \in \mathbb{C} \mid \operatorname{Re} z \in I_f \text{ (resp. } I_\nu)\}$. It follows from the dominated convergence theorem [325, Theorem 4.6.3] that $\tilde{f}(z)$ and $\tilde{\nu}(z)$ are defined and continuous on S_f or S_ν and (by [326, Theorem 3.1.6]) that they are analytic in the interiors of the fundamental strip.

For the inverse, we note the formal formula

$$f(x) = \frac{1}{2\pi i} \int_{c-i\infty}^{c+i\infty} \tilde{f}(s)x^{-s} ds \tag{34.8}$$

for c in the interior of I_f. If the Mellin transform obeys $|\tilde{f}(s)| \leq C(1 + |s|)^{-1-\epsilon}$ for some $\epsilon > 0$ (which is true if f is C^2 with suitable integral conditions on the derivative), then the integral in (34.8) is independent of c in the strip (by the Cauchy integral formula) and (34.8) recovers f.

There is a distributional approach to Mellin transform. Given real a, b with $a < b$, let $\mathcal{T}(a, b)$ be the set of C^∞ functions, φ on $(0, \infty)$ with

$$\int_0^1 x^{k-a}|\varphi^{(k)}(x)|dx < \infty \qquad \int_1^\infty x^{k-b}|\varphi^{(k)}(x)|dx < \infty \tag{34.9}$$

for $k = 0, 1, \ldots$ with the obvious seminorms. Notice that for $a < \operatorname{Re}(s) < b$, $x^{s-1} \in \mathcal{T}(a, b)$. $\mathcal{T}'(a, b)$ is the set of continuous linear functionals on $\mathcal{T}(a, b)$. If $T \in \mathcal{T}'(a, b)$, one defines

$$\tilde{T}(s) = T(t^{s-1}) \tag{34.10}$$

for $a < \text{Re}(s) < b$.

The main feature of Mellin transform for us is that it turns convolution into multiplication.

Theorem 34.2 *If $S_g \cap S_\nu$ has a nonempty interior, then $S_{g\wedge\nu} \supset S_g \cap S_\nu$ and for $s \in S_g \cap S_\nu$*

$$\widetilde{g \wedge \nu}(s) = \tilde{g}(s)\tilde{\nu}(s) \tag{34.11}$$

Remark If $d\nu(x) = f(x)dx/x$, we see that $\widetilde{f \wedge g}(s) = \tilde{f}(s)\tilde{g}(s)$.

Proof We have that

$$\begin{aligned}\widetilde{g \wedge \nu}(s) &= \int \left[\int g(y^{-1}x)d\nu(y)\right] x^{s-1}dx \\ &= \int\int g(w)w^{s-1}y^s d\nu(y)dw \\ &= \tilde{g}(s)\tilde{\nu}(s)\end{aligned}$$

because $dx = ydw$ if $x = wy$. □

An alternate approach to inverse Mellin transforms comes by noting the connection to Fourier transforms on $\mathbb{R}$ (discussed in detail in [325, Chapter 6]) which we define by

$$\hat{h}(k) = (2\pi)^{-1/2} \int e^{-iky}h(y)dy \tag{34.12}$$

Given f on $(0, \infty)$, map $(-\infty, \infty)$ to $(0, \infty)$ by $y \mapsto e^y = x$, so $x^{-1}dx = e^{-y}(e^y dy) = dy$ and if

$$h(y) = f(e^y) \tag{34.13}$$

then

$$\tilde{f}(-ik) = (2\pi)^{1/2}\hat{h}(k) \tag{34.14}$$

This allows the definition of Mellin transform on $L^2((0, \infty), dx/x)$ functions.

Given this and (see [325, equation (6.2.45)])

$$\widehat{fg} = (2\pi)^{-1/2}\hat{f} * \hat{g} \tag{34.15}$$

and (34.14), one obtains for functions f, g on $(0, \infty)$

$$\widetilde{fg} = (2\pi i)^{-1}\tilde{f} * \tilde{g} \tag{34.16}$$

where, for functions on $i\mathbb{R}$

$$(f * g)(s) = \int_{-i\infty}^{i\infty} f(s-t)g(t)dt \tag{34.17}$$

((34.16) has an i^{-1} because in (34.17), we have that $dt = id(\text{Im}(t))$). Originally, (34.16) holds for $f, g \in C_0^\infty(0,\infty)$, but then for $f, g \in L^2((0,\infty), dx/x)$ where $\tilde{f}, \tilde{g}$ are L^2 Mellin transforms and $\widetilde{fg}$ is an L^1 Mellin transform.

Equation (34.14) also implies that if $x^c f(x) \in L^1((0,\infty), dx/x)$ and $\tilde{f}(c + iy) = 0$ for all $y \in \mathbb{R}$, then f is a.e. 0. For if $g(x) = x^c f(x)$, then $\tilde{f}(c+iy) = \tilde{g}(iy)$ so it suffices to prove the result when $c = 0$. By (34.14), $\hat{h}(k) = 0$ for all real k. Since $\hat{}$ is injective on distributions (by the Fourier inversion formula; see [325, Theorem 6.2.13]), $h(y) = 0$ for a.e. y so $f(x) = 0$ for a.e. x. We have therefore proven

Theorem 34.3 *If $\int_0^\infty x^{c-1}|f(x)|dx < \infty$ for some $c \in \mathbb{R}$ and $\tilde{f}(c+iy) = 0$ for all $y \in \mathbb{R}$, then $f(x) = 0$ for a.e. x.*

Equation (34.14) also implies (by the Fourier inversion theorem [325, Theorem 6.2.12]) that if $f \in C_0^\infty(0,\infty)$, so that $\tilde{f}$ is an entire function, then (34.8) holds for any $c \in \mathbb{R}$.

This also implies a Plancherel's theorem [325, Theorem 6.2.14]

$$\frac{1}{2\pi i} \int_{-i\infty}^{i\infty} |\tilde{f}(s)|^2 \, ds = \int |f(x)|^2 \, \frac{dx}{x} \tag{34.18}$$

Initially, one proves this for $f \in C_0^\infty((0,\infty))$ and then uses a density argument to extend $\tilde{}$ to a unitary map of $L^2((0,\infty), dx/x)$ to $L^2(i\mathbb{R}, (2\pi i)^{-1} ds)$. Note that if $s = iy$, then $ds = idy$ so $i^{-1}ds$ is a positive measure. We will need a somewhat more general version of the last formula:

Theorem 34.4

(a) If $\int_0^\infty x^{-2c-1}|f(x)|^2 dx < \infty$ and $\int_0^\infty x^{2c-1}|g(x)|^2 dx < \infty$, then

$$\int_0^\infty f(x)g(x)\frac{dx}{x} = \frac{1}{2\pi i} \int_{c-i\infty}^{c+i\infty} \tilde{f}(-s)\tilde{g}(s)ds \tag{34.19}$$

(b) Suppose that

$$|f(x)| \le \frac{Cx}{1+x} \tag{34.20}$$

for some constant C and that for all $c \in (0, 1)$, $x^c g \in L^1(\mathbb{R}_+, \frac{dx}{x}) \cap L^\infty(\mathbb{R}_+, \frac{dx}{x})$ *and that* $|\tilde{g}(s)| \le C(1 + |s|)^{-2}$, $0 < \operatorname{Re} s < 1$. *Then* (34.19) *holds for all* $c \in (0, 1)$.

Remark The hypothesis on g in (b) is stronger than necessary but suffices for our needs.

Proof

(a) By replacing $f(x)$ by $x^{-c} f(x)$ and $g(x)$ by $x^c g(x)$, it suffices to prove (34.19) for $c = 0$. Let $h(x) = \overline{f(x)}$. Then

$$\overline{\tilde{h}(s)} = \tilde{\bar{h}}(\bar{s}) = \tilde{f}(-s) \tag{34.21}$$

if $s = iy$, $y \in \mathbb{R}$. Thus, (34.19) is the inner product version of (34.18) which follows by polarization (see [325, equation (3.1.8)]).

(b) Define

$$f_n(x) = \begin{cases} f(x), & \text{if } \frac{1}{n} \le |x| < n \\ 0, & \text{otherwise} \end{cases} \tag{34.22}$$

Since $x^{-c} f_n \in L^2$ for $0 < c < 1$, by (a), (34.19) holds if f is replaced by f_n and $c \in (0, 1)$.

As $n \to \infty$, we have that $f_n \to f$ pointwise, so since $\int_0^\infty \frac{x}{1+x} |g(x)| \frac{dx}{x} < \infty$ (because $\frac{x^{1/2}}{1+x} \frac{dx}{x}$ is in $L^1(dx)$ and $x^{1/2}|g(x)| \in L^\infty$), the integral on the left of (34.19) for f_n converges as $n \to \infty$ to the integral for f.

By (34.20), for $0 < \operatorname{Re}(s) < 1$,

$$|\tilde{f}_n(-s)| \le \int \frac{C|x^{-s}|}{1+x} dx < \infty$$

so since $\tilde{g} \in L^1$ and $\tilde{f}_n \to \tilde{f}$ pointwise (by dominated convergence and (34.20)), we get convergence of the right sides of (34.19). □

The final general result we'll need about Mellin transforms is Bochner's theorem (see [325, Theorem 6.6.6]) which for Fourier transforms on $\mathbb{R}$ says there is a finite (positive) measure μ on $\mathbb{R}$ with $\hat{\mu} = h$ if and only if h is continuous and for all n and any $\{k_j\}_{j=1}^n$ in $\mathbb{R}$ and all $\boldsymbol{\zeta} \in \mathbb{C}^n$

$$\sum_{i,j=1}^n \overline{\zeta_i} \zeta_j h(k_i - k_j) \ge 0 \tag{34.23}$$

The Mellin transform version of this is

Theorem 34.5 *Let $c \in \mathbb{R}$. Let f be a continuous function on $\{z \mid \mathrm{Re}(z) = c\}$. Then there is a measure, ν on $(0, \infty)$ with*

$$\int x^c d\nu(x) < \infty \quad \text{and} \quad \tilde{\nu}(s) = f(s) \tag{34.24}$$

if and only if for all n, any $\{z_j\}_{j=1}^n$ with $\mathrm{Re}(z_j) = c$ and all $\boldsymbol{\zeta} \in \mathbb{C}^n$

$$\sum_{i,j=1}^{n} \overline{\zeta_i}\zeta_j f\left(\frac{1}{2}(\bar{z_i} + z_j)\right) \geq 0 \tag{34.25}$$

Proof By setting $d\nu^\sharp(x) = x^c d\nu(x)$ and replacing z_j by $z_j - c$, it is easy to see we can suppose that $c = 0$. In that case, we can set up a bijection between finite measures, μ on $\mathbb{R}$ and ν on $(0, \infty)$ by

$$\int g(y) d\mu(y) = \int g(\log x) d\nu(x) \tag{34.26}$$

(equivalently $\int f(e^y) d\mu(y) = \int f(x) d\nu(x)$). Then

$$\hat{\mu}(k) = (2\pi)^{-1/2} \tilde{\nu}(-ik) \tag{34.27}$$

Given f on $i\mathbb{R}$, set $h(k) = f(-ik)$. If $\frac{1}{2}z_j = -ik_j$, (34.25) is the same as (34.23) and the result follows from Bochner's theorem. □

Example 34.6 Let $f(x) = x/(1 + x)$, so $I_f = (-1, 0)$. We claim that for $s \in S_f$, we have that

$$\tilde{f}(s) = -\frac{\pi}{\sin(\pi s)} \tag{34.28}$$

equivalently, if $g(x) = (1 + x)^{-1}$ so $\tilde{g}(s) = \tilde{f}(s - 1)$, then

$$\tilde{g}(s) = \frac{\pi}{\sin(\pi s)} \tag{34.29}$$

for $0 < \mathrm{Re}(s) < 1$. This is equivalent to

$$\int_0^\infty \frac{x^{a-1}}{1 + x} dx = \frac{\pi}{\sin(\pi a)} \tag{34.30}$$

For $0 < a < 1$, this is a standard textbook contour integral [326, Example 5.7.6, Example 5.7.8, or Problem 10 of Section 5.7].

Alternatively, the change of variables $y = x/(1+x)$ (so $x = y/(1 - y)$, $1+x = 1/(1 - y)$, $dx = (1 - y)^{-2} dy$) shows that

$$\text{LHS of (34.30)} = \int_0^1 y^{a-1}(1-y)^{-a}dy$$
$$= \mathrm{B}(a, 1-a) \tag{34.31}$$
$$= \frac{\Gamma(a)\Gamma(1-a)}{\Gamma(1)} \tag{34.32}$$
$$= \frac{\pi}{\sin(\pi a)} \tag{34.33}$$

where (34.31) is the beta function (see [326, Section 9.6]), (34.32) is Euler's formula for the beta integral [326, Equation (9.6.44)], and (34.33) is the Euler reflection formula [326, Equation (9.6.25)]. □

Now define an operator, B, on $L^2(\mathbb{R}, dx/x)$ by

$$\widetilde{Bu}(s) = \tan(\pi s)\tilde{u}(s) \tag{34.34}$$

We'll prove, for suitable A, that $\mathrm{Re}(B) \geq 0$, $\mathrm{Re}(AB) \geq 0$ and use Theorem 15.9. At first sight, one thinks that $\tan(\pi s)$ is real and unbounded so that B is a self-adjoint, unbounded operator. But remember that s is in $i\mathbb{R}$, not in $\mathbb{R}$ and that

$$\tan(i\pi y) = i \tanh(\pi y) \tag{34.35}$$

Since $|\tanh(y)| \leq 1$, B is a bounded skew-adjoint operator (so $T = e^{\beta B}$ is unitary for $\beta > 0$); thus, $\mathrm{Re}(B) = \frac{1}{2}(B + B^*) = 0$. For our calculations, we'll need the space, $\mathbb{V}$, of finite linear combinations of functions $v \in L^2(i\mathbb{R}, d(\mathrm{Im}\, s))$ of the form $v(s) = e^{\alpha(s-s_0)^2}$ for some $\alpha > 0$ and $s_0 \in \mathbb{C}$. Here is the key calculation:

Proposition 34.7 *Let $v \in \mathbb{V}$. Let $u \in L^2(\mathbb{R}, dx/x)$ be given by*

$$\tilde{u}(s) = \cos(\pi s)v(s) \tag{34.36}$$

Then, if

$$g(x) = \overline{u(x)}(Bu)(x) + \overline{Bu(x)}u(x) \tag{34.37}$$

we have that

$$\tilde{g}(s) = (2\pi i)^{-1}\sin(\pi s)(v^* * v)(s) \tag{34.38}$$

where convolution is given by (34.17) and where for any function, F, on the strip $a < \mathrm{Re}\, s < b$

$$F^*(s) = \overline{F(\bar{s})} \tag{34.39}$$

Remark For $s = iy$, $\cos(\pi s)$ grows as $e^{|s|}$ but since $|v(s)| \leq Ce^{-\alpha|s|^2}$, $\tilde{u}$, $\widetilde{Bu} \in \mathcal{S}(\mathbb{R})$ so u decays rapidly at 0 and ∞ so that $\tilde{g}$ is an entire function. We'll prove (34.38) for $s \in i\mathbb{R}$, but, by analyticity, it holds for any complex s.

Proof By (34.34) and (34.36)

$$\widetilde{Bu}(s) = \sin(\pi s)v(s) \tag{34.40}$$

Moreover, for any $h \in L^1((0, \infty), dx/x)$ and $s \in i\mathbb{R}$

$$\tilde{\bar{h}}(s) = \int x^{s-1}\bar{h}(x)dx = \overline{\int x^{\bar{s}-1}h(x)dx} = \tilde{h}^*(s) \tag{34.41}$$

Therefore, since $F^*(s) = F(s)$ for $F(s) = \cos(\pi s)$ or $\sin(\pi s)$, by (34.16), with g given by (34.37),

$$\begin{aligned}\tilde{g}(s) &= (2\pi i)^{-1}\left\{[\cos(\pi s)v^*] * [\sin(\pi s)v] + [\sin(\pi s)v^*] * [\cos(\pi s)v]\right\} \\ &= [4\pi(i)^2]^{-1}[e^{i\pi s}v^* * e^{i\pi s}v - e^{-i\pi s}v^* * e^{-i\pi s}v]\end{aligned} \tag{34.42}$$

because the cross terms like $e^{i\pi s}v^* * e^{-i\pi s}v$ cancel between the two terms.

For any p, q with $e^{\pi|s|}p, e^{\pi|s|}q \in L^1(i\mathbb{R}, d\,\mathrm{Im}\, s)$, (all $+$ or all $-$)

$$\begin{aligned}(e^{\pm i\pi s}p * e^{\pm i\pi s}q)(s) &= (2\pi i)^{-1}\int e^{\pm i\pi(s-t)}e^{\pm i\pi t}p(s-t)q(t)dt \\ &= e^{\pm i\pi s}(p * q)(s)\end{aligned} \tag{34.43}$$

so

$$\begin{aligned}\tilde{g}(s) &= [4\pi(i)^2]^{-1}[e^{i\pi s} - e^{-i\pi s}](v^* * v)(s) \\ &= (2\pi i)^{-1}\sin(\pi s)(v^* * v)(s)\end{aligned}$$

□

Proof of Theorem 34.1 First, let $h_\epsilon(x) = x(1 + \epsilon x)^{-1}$, let $h \equiv h_{\epsilon=1}$, and let A_ϵ be the multiplication operator on $L^2((0, \infty), dx/x)$ by h_ϵ. By scaling

$$\begin{aligned}\tilde{h}_\epsilon(s) &= \int \frac{x^{1+s}}{1+\epsilon x}\frac{dx}{x} = \epsilon^{-1-s}\tilde{h}(s) \\ &= -\epsilon^{-1-s}\frac{\pi}{\sin(\pi s)}\end{aligned} \tag{34.44}$$

Thus, by (34.19), with u given by (34.36) for $v \in \mathbb{V}$ and g given by (34.37)

$$
\begin{aligned}
\langle u, (A_\epsilon B + B^* A_\epsilon)u\rangle &= \int h_\epsilon(x)g(x)\,\frac{dx}{x} \\
&= \frac{1}{2\pi i}\int_{c-i\infty}^{c+i\infty} \tilde{h}_\epsilon(-s)\tilde{g}(s)\,ds \text{ for } 0 < c < 1 \\
&= \frac{\pi}{(2\pi i)^2}\int \epsilon^{-1+s}(v^* * v)(s)\,ds \text{ (by (34.38))} \\
&= \frac{\pi\epsilon^{-1}}{(2\pi i)^2}\int (\epsilon^s v^* * \epsilon^s v)(s)ds \qquad (34.45) \\
&= \frac{\pi\epsilon^{-1}}{(2\pi)^2}\int \epsilon^{iy}\bar{v}(-iy)\epsilon^{c+i(x-y)}v(ix-iy+c)\,dx\,dy \\
&= \frac{\pi\epsilon^{-1}}{(2\pi)^2}\int \epsilon^{-iu}\bar{v}(iu)\epsilon^{c+iw}v(c+iw)\,du\,dw \\
&= \frac{\pi\epsilon^{-1}}{(2\pi)^2}\left|\int \epsilon^{iu}v(u)\,du\right|^2 \geq 0 \qquad (34.46)
\end{aligned}
$$

where we make the change of variables $s = ix + c; t = iy$ in the line after (34.45) and $w = x - y; u = -y$ in the line after that. In the above, we used that $\int_{c-i\infty}^{c+i\infty} \epsilon^z v(z)\,dz$ is c independent by the decay and analyticity of v and the Cauchy integral formula. Thus $\mathrm{Re}(A_\epsilon B) \geq 0,\ \mathrm{Re}(B) = 0$.

Since f is matrix monotone, it is an interpolation function (by Theorem 15.4) so Theorem 15.9 implies that (with g given by (34.37))

$$\int f\left(\frac{x}{1+\epsilon x}\right) g(x)\frac{dx}{x} \geq 0 \qquad (34.47)$$

Since $|f(x)| \leq Cx(1+x)^{-1}$ and g is bounded and decays rapidly at 0 and ∞, $fg \in L^1((0,\infty), dx/x)$. Since f is monotone and positive $0 \leq f(x/1+\epsilon x) \leq f(x)$, the monotone convergence theorem implies that

$$\int f(x)g(x)\frac{dx}{x} \geq 0 \qquad (34.48)$$

By (34.19) and (34.38)

$$(2\pi i)^{-2}\int_{c-i\infty}^{c+i\infty} \tilde{f}(-s)\sin(\pi s)(v^* * v)(s)\,ds \geq 0 \qquad (34.49)$$

for $0 < c < 1$. For $\beta > 0$ and $s_0 \in \mathbb{C}$, let

$$v_{\beta,s_0}(s) = (4\pi\beta)^{-1/2}\exp\left(\frac{(s-s_0)^2}{\beta}\right) \qquad (34.50)$$

If $\operatorname{Re} s_0 = d$, $v_{\beta,s_0}(s) \to \delta(s - s_0)$ as $\beta \downarrow 0$ in the sense that for any continuous function $q(s)$ on $\{s \mid \operatorname{Re} s = d\}$ with $|q(s)| \le C_1 e^{C_2|s|}$, one has that $i^{-1} \int_{d-i\infty}^{d+i\infty} v_{\beta,s_0}(s) q(s)\, ds \to q(s_0)$.

If z_i, z_j have $\operatorname{Re} z_i = \operatorname{Re} z_j = c$, then as $\beta \downarrow 0$,

$$i^{-2} \int_{c-i\infty}^{c+i\infty} q(s)(v^*_{\beta,\frac{1}{2}z_i} * v_{\beta,\frac{1}{2}z_j})(s)\, ds \to q(\tfrac{1}{2}(\bar{z}_i + z_j)) \tag{34.51}$$

To see this, one uses the decay of the v_{β,s_0} and the Cauchy integral formula to rewrite

$$(v^*_{\beta,s_0} * v_{\beta,s_1})(s) = \int_{\frac{1}{2}c-i\infty}^{\frac{1}{2}c+i\infty} v^*_{\beta,s_0}(t) v_{\beta,s_1}(s-t)\, dt \tag{34.52}$$

Given $\{z_j\}_{j=1}^n$ with $\operatorname{Re} z_j = c \in (0,1)$ and

$$q(s) \equiv \tilde{f}(-s) \sin(\pi s) \tag{34.53}$$

(34.49) with $v(s) = \sum_{j=1}^n \zeta_j v_{\beta,z_j}(s)$ and (34.51) imply that

$$(2\pi)^{-2} \sum_{i,j=1}^n q(\tfrac{1}{2}(\bar{z}_i + z_j)) \bar{\zeta}_i \zeta_j \ge 0 \tag{34.54}$$

Applying Bochner's theorem in the form Theorem 34.5, there is a measure μ on $(0,\infty)$ with

$$\int x^s\, d\mu(x) < \infty, \quad 0 < s < 1 \tag{34.55}$$

so that $\tilde{\mu}(s) = \tilde{f}(-s)\sin(\pi s)/\pi$. If $d\nu(x) = d\mu(x^{-1})$ (i.e., for all $\psi \in C_0^\infty(0,\infty)$, $\int \psi(x) d\nu(x) = \int \psi(x^{-1}) d\mu(x)$), then (since $\tilde{\nu}(s) = \tilde{\mu}(-s)$)

$$\int x^s\, d\nu(s) < \infty,\ -1 < s < 0 \text{ and } \tilde{\nu}(s) = -\tilde{f}(s)\sin(\pi s)/\pi \tag{34.56}$$

Let $t \in (0,1)$. Since $|f(x)| \le Cx(1+x)^{-1}$ by assumption, we have that

$$|\tilde{f}(-t)| \le C \int x^{-t}(1+x)^{-1}\, dx = C \frac{\pi}{\sin(\pi t)} \tag{34.57}$$

By (34.56),

$$\overline{\lim_{t \downarrow 0}}\, \nu(-t) \le C < \infty \tag{34.58}$$

Thus, by the monotone convergence theorem

$$\nu((1,\infty)) = \lim_{t\downarrow 0} \int_1^\infty x^{-t} d\nu(t) \leq \overline{\lim_{t\downarrow 0}}\, \tilde{\nu}(t) < \infty$$

Since $\int x^{-1/2} d\nu(x) < \infty$, we have that $\nu((0,1)) < \infty$, so $\nu((0,\infty)) < \infty$.

Therefore, we can define a continuous function, h,

$$h(x) = \int \frac{x}{x+\lambda}\, d\nu(\lambda) \tag{34.59}$$

by a convergent integral (since $|x/(x+\lambda)| \leq 1$). By Theorem 34.2 and (34.28), $I_h \supset (-1,0)$ and for $-1 < \operatorname{Re} s < 0$,

$$\tilde{h}(s) = \tilde{\nu}(s)\left[\frac{-\pi}{\sin(\pi s)}\right] = \tilde{f}(s) \tag{34.60}$$

on account of (34.56). By Theorem 34.3, $f = h$ proving (34.1).

By the monotone convergence theorem

$$\lim_{x\downarrow 0} x^{-1} f(x) = \int \lambda^{-1}\, d\nu(\lambda) \tag{34.61}$$

so, by hypothesis, $\int \lambda^{-1}\, d\nu(\lambda) < \infty$ as well as $\int d\nu(\lambda) < \infty$. □

Notes and Historical Remarks We have emphasized the Mellin transform as the novel aspect of the proof of this chapter but one could argue that a central part of most proofs is the construction of a measure and from that point of view, the novel element here is the use of Bochner's theorem. It is a standard aspect of Bochner's theorem (see [325, Proposition 6.6.8 and Problem 3 of Section 6.6]) that the pointwise sum positivity is equivalent to an integral condition.

Mellin transforms are named after two papers of Mellin [219, 220]. Before Mellin, they were used implicitly in number theory where they remain an important tool because of the connection to Dirichlet series. Two clear references are Bertrand et al. [29] which is a chapter of a book on various transforms and Zagier [369] which is the appendix of a set of lectures on number theory. [246] is a table of Mellin transforms. Such tables were once important tools in some areas of applied mathematics and theoretical science but with the rise of computer symbolic mathematics programs, they have lost their significance. Perry [268] applies Mellin transforms to quantum scattering theory and includes a nice discussion of the tool. The naturalness of Mellin transforms in scattering theory is connected with the fact that the operator $A = \frac{1}{2}(x \cdot p + p \cdot x)$ is positive when states are going outwards and negative when they are going inwards. The quantum version of A where "p" means multiplication in Fourier space (i.e., $-i\nabla$) is the generator of dilations whose eigenfunctions are powers of x. Thus Mellin transform is an eigenfunction expansion for A.

The discussion here follows an unpublished note of Louis Boutet de Monvel [45] which appears as an appendix to this book. We have fleshed out some of the details. As I mentioned in the Notes to Chapter 31, around 2005, I gave a talk at many math departments entitled "The Lost Proof of Loewner's Theorem". One place I gave it was at the branch of the University of Paris known as "Paris 6" or "Jussieu" (more formally the *Université Pierre et Marie Curie*). The talk was at the invitation of Anne Berthier–Boutet de Monvel, a mathematical physicist who I've known for many years. Afterwards, I was approached by her husband, Louis, who I didn't know personally although I knew his reputation for brilliance and a reluctance to publish.

He told me that he was particularly interested in my talk because he'd long known what I call the hard part of Loewner's theorem that he had his own proof and had assumed the result must be known but he hadn't heard of Loewner's work. On June 6, 2005, he sent an email which read:

Dear Barry,

Here is the note about matrix-monotone functions I mentioned last Thursday. Initially I wrote it for Bourbaki (who never read it), towards 1989. I have corrected some misprints and incorrections, but the note was never meant to be published so there may remain some; the initial note also contained 1.5 *pages of remarks about interpolation which are not really relevant). I was quite excited by your colloquium talk because I did not know the Loewner reference, and although I knew that the interpolation result must have been well known I did not know where it is accurately proved.*

Best regards,
Louis

Included was a three page note in French. There was an exchange of brief emails including a remark from me that I didn't think Krein–Milman was needed which resulted in a brief two page note from him in English.

I found these notes truly remarkable. They were filled with observations about the problem—a few were known in 2005 although some of those were not in 1989. For example, his direct proof that $\mathcal{M}_{2n}(0,\infty) \subset \mathcal{I}_n$ (Proposition 15.5) was new; indeed, so far as I know all published proofs that $\mathcal{M}_\infty(0,\infty) \subset \mathcal{I}_\infty$ depend on Loewner's theorem or something equivalent to it. He had the form of the extreme points independently of Hansen–Pedersen (although several years later). And, notably, his Mellin transform approach was new. This proof is one of the few that doesn't rely on the positivity of the Loewner matrix. Anne tells me it was originally intended to be part of a Bourbaki book (Louis was a member for a period) but the project fell through and he put the note away unpublished.

Unfortunately, Louis passed away on Dec. 25, 2014 at the age of 73. With the concurrence of his widow, it was decided to publish his notes as an appendix to this book.

Chapter 35
Loewner's Theorem for General Open Sets

In this chapter, we discuss versions of Loewner's theorem that replace (a, b) by a more general open subset of $\mathbb{R}$. Let $\Omega \subset \mathbb{R}$ be an open subset of $\mathbb{R}$. We define $\mathcal{M}_n(\Omega)$ in the expected way, namely for $f : \Omega \to \mathbb{R}$, we say that $f \in \mathcal{M}_n(\Omega)$ if and only if for all pairs of self-adjoint $n \times n$ matrices A and B with $\sigma(A) \subset \Omega$, $\sigma(B) \subset \Omega$, and $A \leq B$, we have that $f(A) \leq f(B)$.

There is actually a second very natural class of functions. Ω as an open set is uniquely a countable union of disjoint open intervals, $\{I_j\}_{j=1}^N$ (where $N \in \{1, 2, \dots\} \cup \{\infty\}$), each of which is a connected component of Ω. We say that a pair of $n \times n$ matrices, A, B is a *restricted pair* for Ω if and only if we have that $\sigma(A) \subset \Omega$, $\sigma(B) \subset \Omega$, $A \leq B$ and for each j, A and B have the same number of eigenvalues in I_j. We say that $f \in \mathcal{M}_n^{(r)}(\Omega)$ if for all restricted pairs (A, B), we have that $f(A) \leq f(B)$.

As usual, we define

$$\mathcal{M}_\infty(\Omega) = \cap_{n=1}^\infty \mathcal{M}_n(\Omega), \qquad \mathcal{M}_\infty^{(r)}(\Omega) = \cap_{n=1}^\infty \mathcal{M}_n^{(r)}(\Omega)$$

Here are the main results on these classes which generalize Theorems 1.7 and 1.8.

Theorem 35.1 *Let $\Omega \subset \mathbb{R}$ be an open set and let Ω^{cvx} be its convex hull. Let $f \in \mathcal{M}_\infty(\Omega)$. Then f is the restriction to Ω of a function $\tilde{f} \in \mathcal{M}_\infty(\Omega^{cvx})$.*

Remarks

1. In particular, $\tilde{f}$ is real analytic on Ω^{cvx} and $\tilde{f}$ is uniquely determined as the analytic continuation of f to Ω^{cvx}.
2. Obviously, any $\tilde{f} \in \mathcal{M}_\infty(\Omega^{cvx})$ has $f = \tilde{f} \restriction \Omega \in \mathcal{M}_\infty(\Omega)$ so this has an if and only if form.
3. Suppose we know this theorem when $\Omega = (a, b) \cup (c, d)$ with $a < b < c < d$. Then given any open Ω, the special case implies we can analytically continue between any two I_j and so to all of Ω^{cvx}. And by Loewner's theorem for (a, b),

B. Simon, *Loewner's Theorem on Monotone Matrix Functions*, Grundlehren der mathematischen Wissenschaften 354, https://doi.org/10.1007/978-3-030-22422-6_35

this continuation has a Herglotz continuation to the upper half plane. Thus, it suffices to prove the result in the two interval case, which is Theorem 1.7.

Theorem 35.2 *Let $\Omega \subset \mathbb{R}$ be an open set and let $f \in \mathcal{M}_\infty^{(r)}(\Omega)$. Then there is a function, g, analytic on $\mathbb{C}_+ \cup \mathbb{C}_- \cup \Omega$ which is Herglotz and which has $g \restriction \Omega = f$.*

Remarks

1. If $(a, b) \cup (c, d) \subset \Omega$ and $a < b < e < c < d$ with $e \notin \Omega$, then, by Corollary 35.6 below, $f(x) = (e - x)^{-1}$ lies in $\mathcal{M}_\infty^{(r)}(\Omega)$ but not in $\mathcal{M}_\infty(\Omega)$ by these two theorems, so $\mathcal{M}_\infty(\Omega)$ is strictly smaller than $\mathcal{M}_\infty^{(r)}(\Omega)$.
2. Corollary 35.6 below says that for any real $e \notin \Omega$, $f(x) = (e - x)^{-1}$ lies in $\mathcal{M}_\infty^{(r)}(\Omega)$, so by the Herglotz representation, any g analytic on $\mathbb{C}_+ \cup \mathbb{C}_- \cup \Omega$ which is Herglotz has $g \restriction \Omega = f \in \mathcal{M}_\infty^{(r)}(\Omega)$. Thus, there is a converse to this theorem.
3. For each component, I_j of Ω, a pair of $n \times n$ matrices with spectrum in I_j and $A \leq B$ is a restricted pair. Thus, if $f \in \mathcal{M}_\infty^{(r)}$, Loewner's theorem implies that there is a Herglotz function, f_j whose restriction to I_j agrees with f there. The point of this theorem is that the f_j's are equal to each other. Thus, it suffices to prove this result in the case there Ω is the union of two disjoint open intervals. By translation and scaling (and shrinking one component) we can suppose that

$$\Omega = (-1, -b) \cup (b, 1) \tag{35.1}$$

for some $b \in (0, 1)$. Moreover, by slightly shrinking the intervals, we can suppose that f and all its derivatives are bounded on Ω and, as usual, we can add $\epsilon \log((3 - x)/(2 - x))$ and take ϵ to 0, so that once we show Loewner matrices are positive, we can assume they are strictly positive.

One can ask whether $f \in \mathcal{M}_\infty(\Omega)$ or $f \in \mathcal{M}_\infty^{(r)}(\Omega)$ says something about pairs of self-adjoint operators with $\sigma(A) \subset \Omega$, $\sigma(B) \subset \Omega$, and $A \leq B$. The $\mathcal{M}_\infty(\Omega)$ extension is trivial. Either by approximation or by using Theorem 35.1, we immediately see

Theorem 35.3 *Let $f \in \mathcal{M}_\infty(\Omega)$ and let A, B be self-adjoint operators with $\sigma(A) \subset \Omega$, $\sigma(B) \subset \Omega$, and $A \leq B$. Then $f(A) \leq f(B)$.*

At first sight, it seems hopeless to define a restricted pair when (A, B) operate on an infinite dimensional space since typically the spectrum in each I_j is not a finite set of points. However, we have the following:

Proposition 35.4 *Let Ω be an open subset of $\mathbb{R}$. Let A, B be a pair of self-adjoint $n \times n$ matrices with $\sigma(A) \subset \Omega$, $\sigma(B) \subset \Omega$, and $A \leq B$. Then (A, B) is a restricted pair for Ω if and only if for all $\theta \in [0, 1]$, $A(\theta) = (1-\theta)A + \theta B$ has $\sigma(A(\theta)) \subset \Omega$.*

Proof If $A(\theta)$ has $\sigma(A(\theta)) \subset \Omega$ for all θ, all the eigenvalues which move monotonically and continuously must stay in Ω so the number of eigenvalues in any I_j is the same for A and B.

Conversely, if (A, B) is a restricted pair, any C with $A \leq C \leq B$ has its eigenvalues between those of A and B and so all in Ω. Since $A \leq A(\theta) \leq B$, we see that $\sigma(A(\theta)) \subset \Omega$. □

This suggests the right infinite dimensional version of Theorem 35.2:

Theorem 35.5 *Let Ω be an open subset of $\mathbb{R}$. Let A, B be a pair of self-adjoint operators on a separable Hilbert space, $\mathcal{H}$. Suppose that for all $\theta \in [0, 1]$, $A(\theta) = (1 - \theta)A + \theta B$ has $\sigma(A(\theta)) \subset \Omega$ and that $A \leq B$. Let $f \in \mathcal{M}_\infty^{(r)}(\Omega)$. Then $f(A) \leq f(B)$.*

Proof (Given Theorem 35.2) We can't mimic the proof of Theorem 2.7 because $\sigma(PAP \restriction \text{Ran } P)$ may not lie in Ω (but only in Ω^{cvx}). Instead, by Theorem 35.2 and the Herglotz representation theorem, it suffices to prove that

$$(\lambda - A)^{-1} \leq (\lambda - B)^{-1} \tag{35.2}$$

for all $\lambda \in \mathbb{R} \setminus \Omega$. So fix such a λ. Since $\sigma(A(\theta)) \subset \Omega$, $g(\theta) \equiv (\lambda - A(\theta))^{-1}$ is a real analytic function of θ. Moreover, by a simple calculation,

$$g'(\theta) = (\lambda - A(\theta))^{-1}(B - A)(\lambda - A(\theta))^{-1} \tag{35.3}$$

Since $(\lambda - A(\theta))^{-1}$ is self-adjoint and $B - A \geq 0$, we see that $g'(\theta) \geq 0$ proving (35.2). □

The above proof immediately implies:

Corollary 35.6 *For any open $\Omega \subset \mathbb{R}$ and all $e \in \mathbb{R} \setminus \Omega$, $f(x) = (e - x)^{-1} \in \mathcal{M}_\infty^{(r)}(\Omega)$.*

We'll first prove Theorem 35.2 whose proof is close to the spirit of many proofs of Loewner's theorem since one can relate it to the positivity of Loewner matrices.

Proposition 35.7 (See also Theorem 5.27) *Let Ω be an open subset of $\mathbb{R}$ and f a C^1 function on Ω. Then $f \in \mathcal{M}_n^{(r)}(\Omega)$ if and only if for any $x_1 < \cdots < x_n$ with all $x_j \in \Omega$, the Loewner matrix, $L_n(x_1, \ldots, x_n; f) \geq 0$.*

Proof The proof is identical to that of Theorem 5.1. The limitation to restricted pairs comes from the fact that we need $\sigma(A(\theta)) \subset \Omega$, for the Loewner matrix that enters has $x_1, \ldots, x_n$ as eigenvalues of $A(\theta)$. □

Proof of Theorem 35.2 As noted earlier, it suffices to consider $\Omega = (-1, -b) \cup (b, 1)$ for $0 < b < 1$ and matrices where f and its derivatives are bounded and, given Proposition 35.7, where the Loewner matrices are strictly positive.

We can now use the method of Chapter 20. For each n, let $\{z_\ell\}_{\ell=1}^{m(n,b)}$ be the points $\frac{j}{2^n} + i\frac{\epsilon_n}{2^n}$ where $\frac{j}{2^n} \in \Omega$, $0 \leq \epsilon_n < 1$ and ϵ_n so small that the matrix M of (20.6) is positive. The rest of the proof is unchanged from the arguments in Chapter 20. The only difference is that we learn that $(\lambda - z)^{-1}$ terms are absent in the limit only for $\lambda \in \Omega$ (rather than $\lambda \in [-1, 1]$). □

We will give two proofs of Theorem 35.1—first, the original proof of Šmul'jan and Chandler and then a new proof of Heinävaara[149]. As a preliminary to the first proof of Theorem 35.1, we need

Theorem 35.8 (Šmul'jan's Theorem) *Let f be a function defined on $\Omega = \{0\} \cup (1, \beta)$ for some $\beta > 1$, so that for any n, if A, B are any $n \times n$ self-adjoint matrices with $\sigma(A) \subset \Omega$, $\sigma(B) \subset \Omega$, and $A \leq B$, then $f(A) \leq f(B)$. Then f has an analytic continuation to* $(0, \beta)$.

Remarks

1. By Loewner's theorem, it then follows that $f \in \mathcal{M}_\infty(0, \beta)$.
2. Analyticity is not claimed for f at $x = 0$ and may not hold. It can be shown though that $f(0) \leq \lim_{x \downarrow 0} f(x)$ and that if $f \in \mathcal{M}_\infty(0, \beta)$ with this condition on $f(0)$, then f obeys the hypothesis of the theorem.

Proof By replacing $f(x)$ by $f(x)+\alpha+\gamma x$ for suitable α and $\gamma \geq 0$, we can arrange first that $f(0) = 0$ and then (since taking $A = 0$ and $B = x\mathbf{1}$ with $x \in (1, \beta)$ implies $f(x) \geq f(0)$) taking $\gamma > 0$ if need be that

$$f(0) = 0, \qquad f(x) > 0 \text{ for } x \in (1, \beta) \tag{35.4}$$

Define the function $g : (\beta^{-1}, 1) \to \mathbb{R}$ by

$$g(x) = f(x^{-1})^{-1} \tag{35.5}$$

The idea will be to show that $g \in \mathcal{M}_\infty(\beta^{-1}, 1) \cap [-\mathcal{C}_\infty(\beta^{-1}, 1)]$ and then apply Theorem 10.9 to see that g has an analytic continuation to (β^{-1}, ∞) which will imply that f has an analytic continuation to $(0, \beta)$.

Since $x \mapsto x^{-1}$ is antimonotone for $x \in (0, \infty)$ (i.e., $0 < A \leq B \leq \mathbf{1} \Rightarrow \mathbf{1} \leq B^{-1} \leq A^{-1}$), we see that g is monotone (i.e., $\beta^{-1} \leq A \leq B \leq \mathbf{1} \Rightarrow \mathbf{1} \leq B^{-1} \leq A^{-1} \leq \beta \Rightarrow 0 < f(B^{-1}) \leq f(A^{-1}) \Rightarrow g(A) \leq g(B)$).

Fix C self-adjoint with

$$\sigma(C) \subset (\beta^{-1}, 1) \tag{35.6}$$

and P a self-adjoint projection. By the Davis condition [(1) $\iff$ (4) in Theorem 11.1], to prove that $-g \in \mathcal{C}_\infty(\beta^{-1}, 1)$, it suffices to prove that for all such C and P, we have that

$$Pg(PCP)P \geq Pg(C)P \tag{35.7}$$

We'll use Theorem 13.6 and the map $N \mapsto N^\sharp$ of (13.13). Let $B = C^{-1}$ so $\sigma(B) \subset (1, \beta)$. By (13.14), we have that $B^\sharp \oplus 0 \leq B$. Since $\sigma(B^{-1}) \subset (\beta^{-1}, 1)$, we have that $\sigma((B^{-1})_P) \subset (\beta^{-1}, 1)$ so $\sigma(B^\sharp) \subset (1, \beta)$. Therefore $\sigma(B^\sharp \oplus 0) \subset \Omega$. By the hypothesis on f and $B^\sharp \oplus 0 \leq B$, we see that $f(B^\sharp) \oplus 0 \leq f(B)$, so by

Theorem 13.6 again

$$f(B^\sharp) \leq f(B)^\sharp \tag{35.8}$$

Since $B = C^{-1}$, we have that $f(B) = g(C)^{-1}$. By the definition of $N^\sharp$, we have that $(g(C)^{-1})^\sharp = [g(C)_P]^{-1}$, so

$$f(B)^\sharp = [g(C)_P]^{-1} \tag{35.9}$$

On the other hand, $(B^\sharp)^{-1} = (B^{-1})_P = C_P$, so $f(B^\sharp) = g((B^\sharp)^{-1})^{-1} = g(C_P)^{-1}$. Therefore, (35.8) says that

$$g(C_P)^{-1} \leq [g(C)_P]^{-1} \Rightarrow g(C)_P \leq g(C_P)$$

which is (35.7). Thus we have shown that $g \in \mathcal{M}_\infty(\beta^{-1}, 1) \cap [-\mathcal{C}_\infty(\beta^{-1}, 1)]$ so by Theorem 10.9 we see that g has an analytic continuation to (β^{-1}, ∞). Since g is scalar monotone, it is everywhere strictly positive so f has an analytic continuation to $(0, \beta)$. □

First Proof of Theorem 35.1 As noted in the remarks after the theorem, it suffices to consider the case $\Omega = (a, b) \cup (c, d)$ with $a < b < c < d$. By translation and scaling, we can suppose that $a < 0 < b$ and that $c = 1$. Then any $f \in \mathcal{M}_\infty(\Omega)$ obeys the hypotheses of Šmul'jan's theorem and therefore f has an analytic continuation to $(0, d)$ and so to (a, d), since by the single interval Loewner's theorem, we know that f is real analytic on Ω. By Loewner's theorem for (a, b), we also know that f has a Herglotz continuation to the upper half plane. Thus, by the easy half of Loewner's theorem, $f \in \mathcal{M}_\infty(a, d)$. □

We begin our second proof of Theorem 35.1 with

Theorem 35.9 *Let $\Omega \subset \mathbb{R}$ be open and $\lambda_1 < \mu_1 < \cdots < \lambda_n < \mu_n$ all in Ω. If $f \in \mathcal{M}_n(\Omega)$, then for all $q \in \mathcal{P}_{n-1}$, the real polynomials of degree at most $n-1$, we have that*

$$[\lambda_1, \mu_1, \ldots, \lambda_n, \mu_n; q^2 f] \geq 0 \tag{35.10}$$

Remark Given the relationship of positivity of Loewner and multipoint Loewner matrices and positivity of divided differences of $q^2 f$ proven in Proposition 8.3 and (8.14) and given Theorem 5.26, this theorem is an analog of Theorem 5.18 for extended Loewner matrices instead of Loewner matrices. Put differently, (35.10) is a kind of positivity for a multipoint extended Loewner matrix.

To prove this theorem, we will extend the machinery used to prove Theorem 5.26, explicitly Proposition 5.25. Recall that there, for any $\lambda_1 < \mu_1 < \lambda_2 \cdots < \lambda_n < \mu_n$, we constructed real symmetric $n \times n$ matrices, A and B and $\varphi \in \mathbb{R}^n$ so that

$$B - A = \langle \varphi, \cdot \rangle \varphi \tag{35.11}$$

and orthonormal basis for $\mathbb{R}^n$, $\{\eta^{(j)}\}_{j=1}^n$, $\{\psi^{(j)}\}_{j=1}^n$ so that

$$A\eta^{(j)} = \lambda_j \eta^{(j)}; \qquad B\psi^{(j)} = \mu_j \psi^{(j)} \tag{35.12}$$

Moreover, for all j

$$\langle \varphi, \eta^{(j)} \rangle > 0; \qquad \langle \varphi, \psi^{(j)} \rangle > 0 \tag{35.13}$$

By the second resolvent formula (see (5.86)) with $P_\varphi = \langle \varphi, \cdot \rangle \varphi$ (we use P even though it is only a projection if φ is a unit vector), we have that

$$(A - z)^{-1} = (B - z)^{-1} + (B - z)^{-1} P_\varphi (A - z)^{-1} \tag{35.14}$$

$$= (B - z)^{-1} \left(\mathbf{1} + P_\varphi (A - z)^{-1} \right) \tag{35.15}$$

If C is rank 1, $\det(\mathbf{1} + C) = 1 + \operatorname{Tr}(C)$, so (35.15) implies that

$$\det(B - z) = \det(A - z) \left(1 + \langle \varphi, (A - z)^{-1} \varphi \rangle \right) \tag{35.16}$$

(35.14) also implies that

$$(A - z)^{-1} \varphi = \left(1 + \langle \varphi, (A - z)^{-1} \varphi \rangle \right) (B - z)^{-1} \varphi$$

so, by (35.16)

$$\det(z - B)(z - B)^{-1} \varphi = \det(z - A)(z - A)^{-1} \varphi \tag{35.17}$$

As a final preliminary, we note that by an eigenvector expansion, for any $\xi \in \mathbb{R}^n$

$$\langle \varphi, (A - z)^{-1} \xi \rangle = \sum_{j=1}^n \frac{\langle \varphi, \eta^{(j)} \rangle \langle \eta^{(j)}, \xi \rangle}{\lambda_j - z} \tag{35.18}$$

so if we define

$$q_j(z) = \prod_{\ell \neq j} (\lambda_j - z) \tag{35.19}$$

then

$$\det(A - z) \langle \varphi, (A - z)^{-1} \xi \rangle = \sum_{j=1}^n \langle \varphi, \eta^{(j)} \rangle \langle \eta^{(j)}, \xi \rangle q_j(z) \equiv q_\xi(z) \tag{35.20}$$

Proof of Theorem 35.9 For each $q \in \mathcal{P}_{n-1}$, we'll find $\xi_q \in \mathbb{R}^n$ so that for any continuous function f on Ω

$$\langle \xi_q, [f(B) - f(A)]\xi_q \rangle = [\lambda_1, \mu_1, \dots, \lambda_n, \mu_n; q^2 f] \tag{35.21}$$

If f is monotone, we get (35.10) since $B \geq A$.

As usual, we can suppose that f is entire analytic by extending f to $\mathbb{R}$, convoluting with Gaussians and taking limits. Thus we can write $f(C)$ as a contour integral of $(z - C)^{-1}$ as in (5.55). We compute for $\xi \in \mathbb{R}^n$

$$\langle \xi, [(z - B)^{-1} - (z - A)^{-1}]\xi \rangle = \langle \xi, (z - B)^{-1}\varphi \rangle \langle \varphi, (z - A)^{-1}\xi \rangle \tag{35.22}$$

$$= \frac{\det(z - A)}{\det(z - B)} \langle \xi, (z - A)^{-1}\varphi \rangle \langle \varphi, (z - A)^{-1}\xi \rangle \tag{35.23}$$

$$= \frac{q_\xi(z)^2}{\det(z - A)\det(z - B)} \tag{35.24}$$

where $q_\xi(z) \in \mathcal{P}_{n-1}$. Equation (35.22) is the second resolvent formula, (35.23) uses (35.17), and we get (35.24) by using (35.20). q_ξ is real because all $\langle \varphi, \eta^{(j)} \rangle$ and $\langle \xi, \eta^{(j)} \rangle$ are real. Thus

$$\langle \xi, [f(B) - f(A)]\xi \rangle = \frac{1}{2\pi i} \oint \frac{f(z) q_\xi(z)^2 \, dz}{\prod_{j=1}^n (z - \lambda_j)(z - \mu_j)})$$

$$= [\lambda_1, \mu_1, \dots, \mu_n; q_\xi^2 f]$$

In the proof of Proposition 8.3 we showed that $\{q_j(z)\}_{j=1}^n$ is a basis for $\mathcal{P}_{n-1}$. Since all $\langle \varphi, \eta^{(j)} \rangle > 0$, the map $\xi \mapsto q_\xi$ is onto $\mathcal{P}_{n-1}$. Thus, given q, we can find ξ so that $q_\xi = q$. This proves the assertion in the first sentence of the proof. □

To complete the second proof of Theorem 35.1, we need three lemmas.

Lemma 35.10 *Fix $a < b$ in $\mathbb{R}$ and $\epsilon, \delta > 0$. Then there exists n, $q \in \mathcal{P}_{n-1}$ and distinct $x_1, \dots, x_{2n} \in (a - \epsilon, a) \cup (b, b + \epsilon)$ so that the function $r(x) = q(x)^2 / \prod_{j=1}^{2n} (x - x_j)$ obeys*

$$r(x) \leq -1 \text{ on } (a, b) \tag{35.25}$$

$$0 \leq r(x) \leq \frac{\delta}{1 + x^2} \text{ on } (-\infty, a - \epsilon) \cup (b + \epsilon, \infty) \tag{35.26}$$

Proof Start with

$$r_0(x) = \frac{1}{(x - a)(b - x)} \tag{35.27}$$

and note that $r_0(x) < 0$ on $\mathbb{R} \setminus [a, b]$ and

$$r_0(x) \geq \frac{4}{(b-a)^2} \text{ if } x \in [a, b] \tag{35.28}$$

For any $M > 0$, $r_M(x) \equiv M + r(x)$ has two zeros, $\alpha_M \in (-\infty, a)$ and $\beta_M \in (b, \infty)$ with $\alpha_M \to a$, $\beta_M \to b$ as $M \to \infty$. Thus for suitable M_0, we have that for all $M > M_0$, $\alpha_M \in (a - \epsilon, a)$, $\beta_M \in (b, b + \epsilon)$.

Since r_M changes sign at α_M, a, b, β_M and goes to M at ∞, we have that

$$r_M \geq M + \frac{4}{(b-a)^2} \text{ if } x \in [a, b] \tag{35.29}$$

$$0 \leq r_M(x) \leq M \text{ if } x \in \mathbb{R} \setminus (a - \epsilon, b + \epsilon) \tag{35.30}$$

Moreover,

$$r_M(x) = \frac{[\sqrt{M}(x - \alpha_M)(x - \beta_M)]^2}{(x - \alpha_M)(x - \beta_M)(x - a)(x - b)} \tag{35.31}$$

since this has the right zeros and poles and asymptotics at ∞.

For any N, define

$$r_{M,N} = -r_0(x) \frac{(b-a)^2}{2} \left(\frac{r_M(x)}{M + \frac{4}{(b-a)^2}} \right)^N \tag{35.32}$$

By (35.28) and (35.29), we see that

$$r_{M,N}(x) \leq -2 \text{ if } x \in [a, b] \tag{35.33}$$

By (35.27) and (35.31), we have that, if $x \in \mathbb{R} \setminus (a - \epsilon, b + \epsilon)$

$$0 \leq r_{M,N}(x) \leq \frac{D(\epsilon)}{(1 + x^2)} \left(\frac{M}{M + \frac{4}{(b-a)^2}} \right)^N \tag{35.34}$$

where

$$D(\epsilon) \equiv \sup_{x \in \mathbb{R} \setminus (a - \epsilon, b + \epsilon)} \frac{(b-a)^2}{2} [-(1 + x^2) r_0(x)] < \infty$$

is finite since $-(1 - x^2) r(x) \to 1$ as $x \to \pm\infty$. And, by (35.31), for a constant $D_{N,M}$

$$r_{M,N}(x) = D_{N,M} \frac{(x-\alpha_M)^{2N}(x-\beta_M)^{2N}}{(x-\alpha_M)^N(x-\beta_M)^N(x-a)^{N+1}(x-b)^{N+1}} \tag{35.35}$$

Fix $M > M_0$. By picking N large enough, we can be sure that the quantity on the right of (35.34) is less than $\frac{1}{2}\delta(1+x^2)^{-1}$. Then we can move the $4N+2$ zeros in the denominator of (35.35) into distinct points in $(a-\epsilon, a) \cup (b, b+\epsilon)$ so that the new function has -1 on the right side of (35.33) and $\delta(1+x)^{-1}$ in (35.34). If we then take $n = 2N+1$ and $q(x) = (x-\alpha_M)^N(x-\beta_M)^N$ and the x_j as above, we get the required function. □

Lemma 35.11 *Fix $\lambda \in \mathbb{C}$. Let P be a polynomial of degree at most $m-1$ and $x_1, \dots, x_m$, m distinct points. Let $h(x) = P(x)(\lambda - x)^{-1}$. Then*

$$[x_1, \dots, x_m; h] = P(\lambda) \prod_{j=1}^{m} (\lambda - x_j)^{-1} \tag{35.36}$$

Proof Let $g = P(\lambda)(\lambda - x)^{-1}$. By (5.54), which this lemma generalizes, (35.36) holds if h is replaced by g, so we need to show that

$$[x_1, \dots, x_m; h - g] = 0 \tag{35.37}$$

$h - g$ is a polynomial of degree at most $m - 2$ so (35.37) follows from Lemma 22.3(d). □

Lemma 35.12 *Let $f(x)$ be a function on an open set $\Omega \subset \mathbb{R}$ of the form*

$$f(x) = \alpha + \beta x + \int \left(\frac{1}{\lambda - x} - \frac{1}{1+\lambda^2} \right) d\nu(\lambda) \tag{35.38}$$

where ν is a measure supported on $\mathbb{R} \setminus \Omega$ and obeying (3.42). Fix $m \geq 2$ and P a polynomial of degree at most $m-2$. Then for any distinct points $x_1, \dots, x_m$ in Ω, we have that

$$[x_1, \dots, x_m; fP] = \beta A + \int \frac{P(\lambda)}{(\lambda - x_1)\dots(\lambda - x_m)} d\nu(\lambda) \tag{35.39}$$

where $P(x) = AX^{m-2} +$ plus lower order; equivalently

$$A = \lim_{x \to \infty} \frac{(1+x^2)P(x)}{\prod_{j=1}^{m}(x - x_j)} \tag{35.40}$$

Proof If $d\nu$ is a finite measure, the result is immediate from Corollary 22.4 and (35.36).

For a general f, let ν_n be the measure obtained by restricting ν to $[-n, n]$ and f_n the function given by (35.38) with $d\nu$ replaced by $d\nu_n$. Then $f_n \to f$ pointwise so $[x_1, \dots, x_m; f_n P] \to [x_1 \dots, x_m; fP]$. Since (3.42) holds and

$|P(\lambda)/(\lambda - x_1)\dots(\lambda - x_m)| \leq C(1+\lambda^2)^{-1}$ for $\lambda \in \mathbb{R} \setminus \Omega$, we can take the limit in the integral. □

Second Proof of Theorem 35.1 By Theorem 35.2, there exists $\alpha, \beta \geq 0$ and a measure obeying (3.42) and $\nu(\Omega) = 0$ so that f is given by (35.38). We need to show that for every $a, b \in \Omega$ we have that

$$\nu((a,b)) = 0 \tag{35.41}$$

By Theorem 35.9 and (35.39), we have that for any $q \in \mathcal{P}_{n-1}$ and $x_1, \dots, x_{2n} \in \Omega$

$$\beta A + \int \frac{q(\lambda)^2}{(\lambda - x_1)\dots(\lambda - x_{2n})}\, d\nu(\lambda) \geq 0$$

Now pick $\epsilon > 0$ so that $\Xi \equiv [a-\epsilon, a] \cup [b, b+\epsilon] \subset \Omega$ and then n, $q \in \mathcal{P}_{n-1}$ and $x_1, \dots, x_{2n} \in \Xi$ so that $r(x) = q(x)^2/\prod_{j=1}^{2n}(x - x_j)$ obeys (35.25) and (35.26). By (35.26), we have that A given by (35.40) has $0 \leq A \leq \delta$.

It follows that

$$\nu((a,b)) \leq \delta \left[\beta + \int \frac{d\nu(x)}{1+x^2}\right] \tag{35.42}$$

Since δ is arbitrary, (35.41) holds. □

Remark The proof of (35.10) only needs monotonicity when $B - A$ is rank one. It is easy to see that if $\Omega = (a, b)$, then monotonicity under rank 1 positive perturbations implies it for all positive perturbations since the intermediate operators have eigenvalues in Ω but this is not obvious when Ω has holes. But taking $\mu_j \downarrow \lambda_j$ and using the analog of Proposition 8.3 shows that (35.10) implies that $f \in \mathcal{M}_\infty^{(r)}(\Omega)$ and then the above argument shows that $f \in \mathcal{M}_\infty$, so monotonicity under rank one perturbations does imply monotonicity under general positive perturbations after all.

Notes and Historical Remarks Theorem 35.2 was found by Rosenblum–Rovnyak [287, 288] whose proof appears as Theorem 33.2. From their point of view, $\mathcal{M}^{(r)}$ was natural since they were studying positive Loewner matrices rather than matrix monotonicity. The proof we give here is new. It would be an interesting exercise to figure out which of our eleven proofs of Loewner's theorem extend to get this generalization.

Theorem 35.8 and the subtle, beautiful proof we give are due to J. L. Šmul'jan [331]. Theorem 35.1 appeared first in Chandler [58]. Chandler was a student of Rosenblum and this result was from his thesis. There are two puzzles here given that, as we've seen, it is only a few lines to go from Theorem 35.8 to Theorem 35.1. First it is surprising that Theorem 35.1 isn't in Šmul'jan [331] and that it took 11 years for it to be found. Secondly, while Chandler's paper is only 4 pages long, it is unnecessarily complicated in that it relies on first using Theorem 35.2 which is not needed.

Donoghue was an expert on applying the Loewner rational approximation machine and he found proofs using this machine of both Theorem 35.1 [82] and Theorem 35.2 [83].

The second proof we give of Theorem 35.1 is due to Heinävaara[149].

Theorem 35.5 seems to be new but is an elementary observation.

Part III
Applications and Extensions

This brief final part discusses a selection of applications of Loewner's theorem. Most applications involve the use of monotonicity of specific functions like $x \mapsto x^\alpha;\ 0 < \alpha \leq 1$. These involve all areas of analysis and don't shed particular light on Loewner's theorem. Instead, here we focus on results connected to either the hard part of Loewner's theorem or the equivalence. We study three topics:

(1) *Operator Means* One issue is how to generalize the geometric mean $a, b \mapsto \sqrt{ab}$ when a and b are replaced by finite matrices. Chapter 37 has an axiomatic approach to operator means and proves there is a one–one correspondence between such means and positive matrix monotone functions on $[0, \infty)$ with $f(1) = 1$.
(2) *Quantum Strong Subadditivity* We discuss several proofs of a basic inequality of Lieb and Ruskai that is used in quantum statistical mechanics and quantum information theory. These proofs rely on operator means and/or matrix concave functions.
(3) *Unitarily Invariant Norm Inequalities* After Chapter 42 which presents the theory of unitarily invariant norms, Chapter 43 proves results that relate $|||f(A) - f(B)|||$ and $|||f(|A - B|)|||$ for unitarily invariant norms $|||\cdot|||$.

Chapter 36
Operator Means, I: Basics and Examples

If $a, b > 0$, there are various "means" that have turned out to be useful, in particular, the *arithmetic mean*, $\frac{1}{2}(a + b)$, the *harmonic mean*, $2(a^{-1} + b^{-1})^{-1}$, and the *geometric mean*, $\sqrt{ab}$. In this chapter, we'll begin by considering analogs when the positive reals a, b are replaced by positive matrices A, B. We'll then present a set of axioms that captures what is an operator mean. In the next chapter, we'll prove a remarkable result that sets a $1-1$ correspondence between such operator means and matrix monotone functions on $(0, \infty)$. This is why this subject is a part of this book.

We define the *arithmetic mean* of two operators in the obvious way as $\frac{1}{2}(A+B)$. As a preliminary to our discussion of the harmonic mean, we recall that using the notion of Schur complement, we were able to prove a criteria for a block matrix to be positive. Theorem 13.3(a) can be rewritten

Theorem 36.1 *Let $A, B, X \in \mathcal{L}(\mathcal{H})$ with $A \geq 0$, $B \geq 0$ with B invertible. Then*

$$\begin{pmatrix} A & X \\ X^* & B \end{pmatrix} \geq 0 \Leftrightarrow XB^{-1}X^* \leq A \tag{36.1}$$

Corollary 36.2 *The function $F(X, B) \equiv (X, B) \mapsto XB^{-1}X^*$ is jointly convex from pairs of arbitrary matrices, X and positive invertible matrices, B. The function $(A, B, X) \mapsto A - XB^{-1}X^*$ is jointly concave.*

Proof The second statement follows from the first. Given (X_1, B_1) and (X_2, B_2), by the Theorem, we have that $\begin{pmatrix} X_jB_j^{-1}X_j^* & X_j \\ X_j^* & B_j \end{pmatrix} \geq 0$, so for any $t \in (0, 1)$, we have that

$$\begin{pmatrix} tX_1B_1^{-1}X_1^* + (1-t)X_2B_2^{-1}X_2^* & tX_1 + (1-t)X_2 \\ tX_1^* + (1-t)X_2* & tB_1 + (1-t)B_2 \end{pmatrix} \geq 0$$

B. Simon, *Loewner's Theorem on Monotone Matrix Functions*, Grundlehren der mathematischen Wissenschaften 354, https://doi.org/10.1007/978-3-030-22422-6_36

By the Theorem again, $F(tX_1+(1-t)X_2, tB_1+(1-t)B_2) \leq tF(X_1, B_1)+(1-t)F(X_2, B_2)$, proving convexity. □

Given two strictly positive operators, A and B, we define their *parallel sum* by

$$A : B \equiv (A^{-1} + B^{-1})^{-1} \tag{36.2}$$

The *harmonic mean*, $A!B \equiv 2A : B$ is twice this but the parallel sum is used more often in the literature. We compute:

$$\begin{aligned}(A^{-1} + B^{-1})^{-1} &= (A^{-1}(A + B)B^{-1})^{-1} \\ &= B(A + B)^{-1}A \\ &= B(A + B)^{-1}[(A + B) - B] \\ &= B - B(A + B)^{-1}B\end{aligned}$$

We thus have the first statement of

Theorem 36.3 *The parallel sum $A : B$ obeys*

(a)

$$A : B = B - B(A + B)^{-1}B \tag{36.3}$$

$$= A - A(A + B)^{-1}A \tag{36.4}$$

(b) $A, B \mapsto A : B$ *is jointly monotone increasing and jointly concave*
(c) $A : B \leq A, \quad A : B \leq B$
(d) *A:B is the maximal positive operator Y so that*

$$\begin{pmatrix} A & A \\ A & A + B \end{pmatrix} \geq \begin{pmatrix} Y & 0 \\ 0 & 0 \end{pmatrix} \tag{36.5}$$

(e)

$$\text{If } A \leq B, \text{ then } A \leq A!B \leq B \tag{36.6}$$

Remarks

1. From (36.2), $A : B$ is clearly in symmetric in A and B but this is obscured in (36.3)/(36.4) alone
2. Thus $A : B$ is the Schur complement $C\backslash(A + B)$ for $C = \begin{pmatrix} A & A \\ A & A+B \end{pmatrix}$ or $C = \begin{pmatrix} B & B \\ B & A+B \end{pmatrix}$
3. (e) is something one expects of any mean
4. Concavity implies that

$$\tfrac{1}{2}\,[A : B + C : D] \leq \left(\tfrac{1}{2}(A + C) : \tfrac{1}{2}(B + D)\right)$$

Multiplying by 2, we obtain that for any positive operators:

$$A : B + C : D \leq (A + C) : (B + D) \tag{36.7}$$

It is not hard to see that this equality implies (and so is equivalent) to concavity.

Proof

(a) We proved (36.3) and (36.4) then follows by symmetry.
(b) As we've seen (Theorem 4.1), $A \mapsto -A^{-1}$ is monotone so $A, B \mapsto -(-A^{-1} - B^{-1})^{-1}$ is also. Concavity follows from (36.4) and the concavity result in Corollary 36.2.
(c) is immediate from (36.3)/(36.4).
d) By Theorem 36.1, (36.5) holds if and only if $A - Y \geq A(A + B)^{-1}A$ if and only if $Y \leq A : B$ which is what we claim.
(e) $A \leq B \Rightarrow 2A \leq A + B \leq 2B \Rightarrow \frac{1}{2}B^{-1} \leq (A + B)^{-1} \leq \frac{1}{2}A^{-1} \Rightarrow A(A + B)^{-1}A \leq \frac{1}{2}A$ and $\frac{1}{2}B \leq B(A + B)^{-1}B \Rightarrow$ (36.6).

□

Next, we want to remark that if $A, B \geq 0$, we can define $A : B$ even if A and/or B is not invertible because $\lim_{\epsilon \downarrow 0}(A + \epsilon\mathbf{1}) : (B + \epsilon\mathbf{1})$ exists (in the strong operator topology) by a simple argument—for example, $A(A + B + 2\epsilon\mathbf{1})^{-1}A$ is increasing and bounded in norm by $\|A\|$. The same argument (monotonicity follows from monotonicity of $A, B \mapsto A : B$) proves that (even in the infinite dimensional case):

Theorem 36.4 *If $A_n \downarrow A \geq 0$ and $B_n \downarrow B \geq 0$ (convergence in the strong operator topology), then $A_n : B_n \downarrow A : B$.*

Lastly, we want to prove some useful inequalities that depend on

Lemma 36.5 *For every $\varphi \in \mathcal{H}$, we have that*

$$\langle \varphi, A : B\varphi \rangle = \inf\{\langle \psi, A\psi \rangle + \langle \eta, B\eta \rangle \mid \varphi = \psi + \eta\} \tag{36.8}$$

Proof By taking limits, we can suppose that A and B are invertible. In that case, by (36.3):

$$\begin{aligned}
&\langle \psi, A\psi \rangle + \langle (\varphi - \psi), B(\varphi - \psi) \rangle - \langle \varphi, (A : B)\varphi \rangle \\
&= \langle \varphi, B\varphi \rangle + \langle \psi, (A + B)\psi \rangle - 2\,\mathrm{Re}\langle \psi, B\varphi \rangle - \langle \varphi, (A : B)\varphi \rangle \\
&= \langle \varphi, B(A + B)^{-1}B\varphi \rangle + \langle \psi, (A + B)\psi \rangle - 2\,\mathrm{Re}\langle \psi, B\varphi \rangle \\
&= \|(A + B)^{-1/2}B\varphi\|^2 + \|(A + B)^{1/2}\psi\|^2 - \\
&\qquad 2\,\mathrm{Re}\langle (A + B)^{1/2}\psi, (A + B)^{-1/2}B\varphi \rangle
\end{aligned}$$

which is always positive by the Schwarz inequality with equality if $\psi = (A+B)^{-1}B\varphi$. □

Proposition 36.6 *We have the following for all positive A and B:*

(a) *For every* $S \in \mathcal{L}(\mathcal{H})$, $S^*(A:B)S \le (S^*AS):(S^*BS)$
(b) $A:B+C:D \le (A+B):(C+D)$

Remark Our proof of (b) below is an alternate proof of (36.7) and so of concavity.

Proof

(a) When $\varphi = \psi + \eta$, by the lemma:

$$\begin{aligned}\langle \varphi, S^*(A:B)S\varphi\rangle &= \langle S\varphi, (A:B)S\varphi\rangle \\ &\le \langle S\psi, AS\psi\rangle + \langle S\eta, BS\eta\rangle \\ &= \langle \psi, S^*AS\psi\rangle + \langle \eta, S^*BS\eta\rangle\end{aligned}$$

so by the lemma again, we get (a).
(b) When $\varphi = \psi + \eta$, by the lemma:

$$\begin{aligned}\langle \varphi, [A:B+C:D]\varphi\rangle &\le \langle \psi, A\psi\rangle + \langle \eta, B\eta\rangle + \langle \psi, C\psi\rangle + \langle \eta, D\eta\rangle \\ &= \langle \psi, (A+C)\psi\rangle + \langle \eta, (B+D)\eta\rangle\end{aligned}$$

so by the lemma again, we get (b).

□

This completes most of what we want to say about parallel sums and harmonic means. We turn next to geometric means. It is interesting to try various simple guesses like $\frac{1}{2}\left[(AB)^{1/2} + (BA)^{1/2}\right]$ and see that they lack positivity or monotonicity. We'll take a definition out of the hat and show it has all the properties one might like. Given A and B positive with A invertible, we define their *geometric mean* to be

$$A\#B = A^{1/2}\left[A^{-1/2}BA^{-1/2}\right]^{1/2}A^{1/2} \tag{36.9}$$

This at least has the necessary property that $(\alpha A)\#B = A\#(\alpha B) = \alpha^{1/2}A\#B$ for positive α but it doesn't seem to be symmetric in A and B. One way of seeing that it is to note that:

Proposition 36.7 $X = A\#B$ *solves*

$$XA^{-1}X = B \tag{36.10}$$

Moreover, if $Y \ge 0$ *obeys* $YA^{-1}Y \le B$, *then*

$$Y \le A\#B \tag{36.11}$$

In particular, $A\#B$ is the unique positive solution of (36.10).

Proof We note first that

$$(A\#B)A^{-1}(A\#B) = A^{1/2}\left[\sqrt{A^{-1/2}BA^{-1/2}}\right]^2 A^{1/2} = B$$

On the other hand, if $YA^{-1}Y \leq B$, then $(A^{-1/2}YA^{-1/2})^2 \leq A^{-1/2}BA^{-1/2}$, so, since the square root is matrix monotone (see Chapter 4), $A^{-1/2}YA^{-1/2} \leq \sqrt{A^{-1/2}BA^{-1/2}}$, so (36.10) holds. □

Corollary 36.8 $A\#B = B\#A$

Proof $XB^{-1}X = A \Leftrightarrow B^{-1} = X^{-1}AX^{-1} \Leftrightarrow B = XA^{-1}X$. By the Proposition, $B\#A \leq A\#B$ and $A\#B \leq B\#A$, so we have equality. □

Theorem 36.9

(a) $X = A\#B$ *is the unique positive solution of* (36.10).
(b) *If* $X \geq 0$ *and obeys* $\left(\begin{smallmatrix} A & X \\ X & B \end{smallmatrix}\right) \geq 0$, *then* $X \leq A\#B$.

Proof

(a) is a restatement of the Proposition
(b) follows from Theorem 36.1 and the Proposition.

□

We want to conclude our discussion of the geometric mean with another proof of $A\#B = B\#A$ that is suggestive of the developments of the next chapter. We start with the Herglotz representation for $\sqrt{C}$ given in (4.5).

$$\begin{aligned}\sqrt{C} &= \frac{1}{\pi}\int_0^\infty w^{-1/2}C(C+w)^{-1}dw \\ &= \frac{1}{\pi}\int_0^\infty w^{1/2}(1+wC^{-1})^{-1}\frac{dw}{w}\end{aligned}$$

Thus, using the parallel sum,

$$A\#B = \frac{1}{\pi}\int_0^\infty w^{1/2}(A : w^{-1}B)\frac{dw}{w} \tag{36.12}$$

since $A^{1/2}(1+w(A^{1/2}B^{-1}A^{1/2}))^{-1}A^{1/2} = (A^{-1}+wB^{-1})^{-1} = (A : w^{-1}B)$.

Changing variables, $u = w^{-1}$, $\frac{dw}{w} = -\frac{du}{u}$, we see that:

$$\begin{aligned}A\#B &= \frac{1}{\pi}\int_0^\infty u^{1/2}u^{-1}(A : uB)\frac{du}{u} \\ &= \frac{1}{\pi}\int_0^\infty u^{1/2}(u^{-1}A : B)\frac{du}{u}\end{aligned} \tag{36.13}$$

Comparing, (36.12) and (36.13) and using the symmetry $A : B = B : A$ proves the claimed symmetry. We note that (36.12) and the properties of harmonic mean imply that $(A, B) \mapsto A\#B$ is jointly monotone and jointly concave.

Before leaving the subject of these three classical operator means, we note that in the next chapter, we'll prove that

$$A!B \leq A\#B \leq A\nabla B \equiv \tfrac{1}{2}(A + B) \tag{36.14}$$

$A\nabla B$ is, of course, the *arithmetic mean*.

As a final subject of this chapter, we'll state a set of reasonable axioms for general operator means, which in the next chapter, we'll show are natural one-one correspondence with a set of normalized matrix monotone functions. An *operator mean* is a map $(A, B) \mapsto A\sigma B$ from pairs of non-negative operators to non-negative operators which has the following four properties (A, B, C, D are general non-negative bounded operators):

(1) (Joint Monotonicity) $A \leq C$ and $B \leq D \Rightarrow A\sigma B \leq C\sigma D$
(2) (Transformer Inequality) $C(A\sigma B)C \leq (CAC)\sigma(CBC)$
(3) (Upper Semicontinuity) $A_n \downarrow A$ and $B_n \downarrow B \Rightarrow A_n\sigma B_n \downarrow A\sigma B$
(4) (Normalization) $\mathbf{1}\sigma\mathbf{1} = \mathbf{1}$

We end with several remarks. We'll normally suppose in our discussion that A and B are invertible. Letting $A_n = A + n^{-1}\mathbf{1}$ and $B_n = B + n^{-1}\mathbf{1}$ and using upper semicontinuity allows the general case. Second, we note that if C is invertible, property (2) implies that $C^{-1}[(CAC)\sigma(CBC)]C^{-1} \leq A\sigma B$ which means that

$$C \text{ invertible } \Rightarrow C(A\sigma B)C = (CAC)\sigma(CBC) \tag{36.15}$$

Third, this is a very broad definition: $A\sigma B \equiv A$ is an operator mean! Fourth, one of the results in the next chapter will be that every operator mean is jointly concave. Fifth, it is not hard to show that $A \leq B \Rightarrow A \leq A\sigma B \leq B$. Finally, we have proven above that the harmonic mean is an operator mean in this sense. In particular, parallel sum obeys axioms (1)–(3).

Notes and Historical Remarks Our discussion follows in part the lecture notes of Hiai [155] and the book of Bhatia [34]. Parallel sums go back to a paper of Anderson–Duffin [11] (see also [12]) who first proved monotonicity and joint concavity of parallel sums. The name comes from circuit theory and Kirchoff's law that if two circuits with resistance r_1 and r_2 are placed in parallel, the resulting resistance, r is given by $r^{-1} = r_1^{-1} + r_2^{-1}$.

Corollary 36.2 goes back at least to Kiefer [182] in 1959.

Geometric means for operators were first defined by Pusz–Woronowicz [278] and were developed especially by Ando [14, 15]. The axiom scheme we discuss is due to Kubo–Ando [193]. That monotonicity implies concavity can be thought of as a two variable version of Theorem 10.2.

Chapter 37
Operator Means, II: Kubo–Ando Theorem

In this chapter, we'll see a natural 1–1 correspondence between operators means $A, B \mapsto A\sigma B$ and non-negative matrix monotone functions, f on $(0, \infty)$, normalized by $f(1) = 1$. We begin by recalling the Herglotz representation for positive functions on $(0, \infty)$ in the form (3.65)

$$f(z) = a + bz + \int_0^\infty \frac{z(1+\lambda)}{z+\lambda} d\rho_f(\lambda) \tag{37.1}$$

$$f(1) = a + b + \int_0^\infty d\rho_f(\lambda) \tag{37.2}$$

where $a, b \geq 0$

The main theorem of this chapter is

Theorem 37.1 (Kubo–Ando Theorem) *Let $A, B \mapsto A\sigma B$ be an operator mean. Then $\mathbf{1}\sigma(t\mathbf{1})$ is a multiple of $\mathbf{1}$, i.e. for some function, $f : [0, \infty) \to [0, \infty)$*

$$\mathbf{1}\sigma(t\mathbf{1}) = f(t)\mathbf{1} \tag{37.3}$$

f is a matrix monotone function with $f(1) = 1$ and σ is given in terms of the representation (37.1) *as*

$$A\sigma B = aA + bB + \int_0^\infty \frac{1+\lambda}{\lambda} ((\lambda A) : B)\, d\rho_f(\lambda) \tag{37.4}$$

where $(\lambda A) : B$ is the parallel sum. Conversely, if f is a non-negative matrix monotone function with $f(1) = 1$ and is given by (37.1)*, then the right side of* (37.4) *defines an operator mean which obeys* (37.3)*. Moreover*

B. Simon, *Loewner's Theorem on Monotone Matrix Functions*, Grundlehren der mathematischen Wissenschaften 354, https://doi.org/10.1007/978-3-030-22422-6_37

(a) *If A is invertible, then*

$$A\sigma B = A^{1/2} f(A^{-1/2} B A^{-1/2}) A^{1/2} \tag{37.5}$$

(b) *For any A*

$$A\sigma A = A \tag{37.6}$$

We know that $(A + B) : (C + D) \geq (A : C) + (B : D)$ (see (36.7)) and that $A, B \mapsto (A : B)$ is jointly concave (Theorem 36.3 and Proposition 36.6). Since any σ is a positive superposition of $(\lambda A : B)$, we conclude from the Kubo–Ando Theorem:

Corollary 37.2 *For any operator mean, σ, we have that*

$$(A + B)\sigma(C + D) \geq (A\sigma C) + (B\sigma D) \tag{37.7}$$

and $(A, B) \mapsto A\sigma B$ is jointly concave.

Remark This is especially interesting as we'll see for the case of geometric mean. Joint concavity in that case also follows from Theorem 36.9 (b) by the same argument used to get joint convexity of $(X, B) \mapsto XB^{-1}X^*$ from Theorem 36.1 (see the proof of Corollary 36.2.)

We'll prove Theorem 37.1 in a number of steps:

Lemma 37.3 *Let σ be an operator mean and A, B be positive operators. For any orthogonal projection, P, commuting with A and B, we have that P commutes with $A \sigma B$ and*

$$[(AP)\sigma(BP)]P = (A\sigma B)P \tag{37.8}$$

Proof We begin by noting that

$$P(A\sigma B)P \leq (PAP)\sigma(PBP) \leq A\sigma B \tag{37.9}$$

The first inequality is the transformer inequality (axiom (2)) and the second follows from monotonicity (axiom (1)) since $\langle\varphi, PAP\varphi\rangle = \|PA^{1/2}\varphi\|^2 \leq \|A^{1/2}\varphi\|^2 \leq \langle\varphi, A\varphi\rangle$. Thus $C \equiv A\sigma B - P(A\sigma B)P \geq 0$. We compute

$$|C^{1/2}P|^2 = PCP = 0$$

so $CP = C^{1/2}(C^{1/2}P) = 0$, i.e. $(A\sigma B)P = P(A\sigma B)P$ and thus P commutes with $(A\sigma B)$. Since P also commutes with AP and BP, we see that P commutes with $(AP)\sigma(BP)$. Since $C \leq D$ and $PC = CP$, $PD = DP \Rightarrow PC \leq PD$, (37.9) implies (37.8). □

Proposition 37.4 *Let σ be an operator mean. Then $\mathbf{1}\sigma(t\mathbf{1})$ is a multiple, $f(t)\mathbf{1}$, of $\mathbf{1}$. For any A,*

$$\mathbf{1}\sigma A = f(A) \tag{37.10}$$

In particular, f is a matrix monotone function and $f(1) = 1$.

Proof $C = \alpha\mathbf{1}$ if and only if C commutes with every projection (since if P is the rank one projection onto multiples of φ, $(\mathbf{1}-P)CP = 0 \Rightarrow$ every φ is an eigenvector $\Rightarrow$ C is a multiple of $\mathbf{1}$) so, by the Lemma, (37.3) holds for a non-negative function f with $f(1) = 1$ (since $\mathbf{1}\sigma\mathbf{1} = \mathbf{1}$).

Now suppose that $A = \sum_{j=1}^{m} \alpha_j P_j$ where $\alpha_j \geq 0$ and $\{P_j\}_{j=1}^{m}$ are mutually orthogonal projections with $\sum_{j=1}^{m} P_j = \mathbf{1}$. Then

$$\mathbf{1}\sigma A = \sum_{j=1}^{m} (\mathbf{1}\sigma A) P_j \tag{37.11}$$

$$= \sum_{j=1}^{m} (P_j \sigma (A P_j)) P_j \tag{37.12}$$

$$= \sum_{j=1}^{m} (P_j \sigma (\alpha_j P_j)) P_j \tag{37.13}$$

$$= \sum_{j=1}^{m} (\mathbf{1}\sigma (\alpha_j \mathbf{1})) P_j \tag{37.14}$$

$$= \sum_{j=1}^{m} f(\alpha_j) P_j = f(A) \tag{37.15}$$

In the above, (37.11), (37.13), and (37.15) use $A = \sum_{j=1}^{m} \alpha_j P_j$ and $\mathbf{1} = \sum_{j=1}^{m} P_j$ while (37.12) and (37.14) use (37.8).

By the spectral theorem, for any $A \geq 0$, there exist A_n, each a finite sum of the form $A_n = \sum_{j=1}^{m_n} \alpha_j^{(m)} P_j^{(m)}$, so that $A_n \downarrow A$. Thus by axiom (3), (37.10) for the special A's implies it for all $A \geq 0$.

Monotonicity of $A \mapsto \mathbf{1}\sigma A$ in A (axiom (1)) implies matrix monotonicity of f.

□

Proposition 37.5 *For any operator mean, σ, define f by* (37.3). *Then for A invertible*

$$A\sigma B = A^{1/2} f(A^{-1/2} B A^{-1/2}) A^{1/2} \tag{37.16}$$

Moreover, (37.4) *holds.*

Proof With $C = A^{1/2}$, which is invertible if A is, we have, by (36.15), that $C(\mathbf{1}\sigma(A^{-1/2}BA^{-1/2}))C = A\sigma B$ which is (37.16).

Notice next that if B is invertible

$$f(B) = a + bB + \int_0^\infty \frac{B}{B + \lambda\mathbf{1}}\,(1+\lambda)\,d\rho_f(\lambda) \tag{37.17}$$

and that $(\lambda\mathbf{1} : B) = (B^{-1}+\lambda^{-1}\mathbf{1})^{-1} = \left[(B+\lambda\mathbf{1})B^{-1}\lambda^{-1}\right]^{-1} = \lambda\frac{B}{B+\lambda\mathbf{1}}$, so (37.17) says that

$$f(B) = a + bB + \int_0^\infty \frac{1+\lambda}{\lambda}\,[(\lambda\mathbf{1}) : B]\,d\rho_f(\lambda)$$

proving (37.4) when $A = \mathbf{1}$ and B is invertible.

If A is also invertible, using $A^{1/2}\left[\lambda\mathbf{1} : (A^{-1/2}BA^{-1/2})\right]A^{1/2} = (\lambda A : B)$, we get (37.4) from (37.16) for A, B invertible. Using $A\sigma B = \lim_{n\to\infty}(A + n^{-1}\mathbf{1})\sigma(B + n^{-1}\mathbf{1})$, we get the result for general A, B. □

The following completes the proof of Theorem 37.1.

Proposition 37.6 *Let f be a non-negative function on $(0,\infty)$ with $f(1) = 1$. Suppose that f has the form* (37.1) *and that $A\sigma B$ is defined by* (37.4)*. Then σ is an operator mean obeying* (37.3)

Proof Since $A, B \mapsto (\lambda A) : B$ obeys axioms (1)–(3) for operator mean and it is easy to prove uniform bounds on the integrals, $A\sigma B$ defined by (37.4) also obeys axioms (1)–(3). Moreover, since $(\lambda\mathbf{1} : t\mathbf{1}) = \frac{\lambda t}{\lambda+t}$, we have that

$$\mathbf{1}\sigma(t\mathbf{1}) = a\mathbf{1} + at\mathbf{1} + \int_0^\infty \frac{(1+\lambda)t}{\lambda+t}\,\mathbf{1}\,d\rho_f(\lambda) = f(t)\mathbf{1}$$

Since $f(1) = 1$, σ obeys axiom (4). □

We turn next to the question of when $A\sigma B = B\sigma A$. Define the *transpose*, σ', of σ by $A\sigma' B = B\sigma A$. If $\sigma' = \sigma$ we say that σ is *symmetric*. Let $\tilde{f}$ be the function associated to σ'. Since $\alpha(A\sigma B) = (\alpha A)\sigma(\alpha B)$ for $\alpha > 0$, we see that

$$\begin{aligned}\tilde{f}(t)\mathbf{1} &= \mathbf{1}\sigma'(t\mathbf{1}) = (t\mathbf{1})\sigma\mathbf{1}\\ &= t\{t^{-1}[(t\mathbf{1})\sigma\mathbf{1}]\} = t(\mathbf{1}\sigma(t^{-1}\mathbf{1}))\\ &= tf(t^{-1})\mathbf{1}\end{aligned}$$

We have thus proven

Theorem 37.7 *The function, $\tilde{f}$, associated to σ', the transpose of σ is related to the function, f, corresponding to σ by*

$$\tilde{f}(t) = tf(t^{-1}) \tag{37.18}$$

In particular, σ is symmetric if and only if

$$tf(t^{-1}) = f(t) \tag{37.19}$$

Example 37.8

(a) (Arithmetic Mean) $A\nabla B \equiv \frac{1}{2}(A + B)$. Thus $f(t)\mathbf{1} = \mathbf{1}\nabla(t\mathbf{1}) = \frac{1}{2}(1 + t)\mathbf{1}$, so $f(t) = \frac{1}{2}(1 + t)$. This obeys (37.19) as it must since the mean is clearly symmetric.

(b) (Harmonic Mean) $A!B = 2(A^{-1} + B^{-1})^{-1}$. Thus $f(t)\mathbf{1} = 2(t^{-1} + 1)^{-1}\mathbf{1} = \frac{2t}{1+t}\mathbf{1}$, so

$$f(t) = \frac{2t}{1+t}$$

which also obeys (37.19) as it must.

(c) (Power Means including Geometric Mean) $x \mapsto x^\alpha$ is matrix monotone for $0 < \alpha < 1$ (see Theorem 4.1) so $A\#_\alpha B = A^{1/2}(A^{-1/2}BA^{-1/2})^\alpha A^{1/2}$ is an operator mean. By an easy calculation $f(t) = t^\alpha$ and, thus, $\tilde{f}(t) = t^{1-\alpha}$. $\#_\alpha$ is symmetric if and only if $\alpha = 1/2$. This is, of course, the geometric mean.

□

Our final result on this subject is a universal inequality:

Theorem 37.9 *Let σ be a symmetric mean. Then*

$$A!B \leq A\sigma B \leq A\nabla B \tag{37.20}$$

In particular, (36.14) *holds.*

Proof Taking the derivative of (37.19) at $t = 1$ yields $f(1) - f'(1) = f'(1) \Rightarrow f'(1) = \frac{1}{2}f(1) = \frac{1}{2}$. Thus the tangent to f at $t = 1$ is $\frac{1}{2}(1+t)$. Since f is concave, we conclude that

$$f(t) \leq \tfrac{1}{2}(1 + t) \tag{37.21}$$

Given a matrix monotone function, f,

$$g(t) = f(t^{-1})^{-1} \tag{37.22}$$

is also matrix monotone since $t \mapsto t^{-1}$ is operator order reversing. (Note that $A\sigma_g B = (A^{-1}\sigma_f B^{-1})^{-1}$.) Moreover (37.19) for $f \Rightarrow tg(t)^{-1} = g(t^{-1})^{-1} \Rightarrow tg(t^{-1}) = g(t)$ which is (37.19) for g. Therefore, if f is symmetric, (37.21) holds for g, i.e. $g(t) \leq \frac{1}{2}(1 + t) \Rightarrow f(t) \geq 2(1 + t^{-1})^{-1} = 2t/(1 + t)$.

Therefore, we have proven that f symmetric implies that

$$\frac{2t}{1+t} \leq f(t) \leq \tfrac{1}{2}(1+t) \tag{37.23}$$

Plugging into the representation (37.4) yields (37.20). □

Notes and Historical Remarks The results in this chapter are from the seminal paper of Kubo–Ando [193]. Our presentation follows in part the lecture notes of Hiai [155].

We've seen that if f defines an operator mean, so does $t \mapsto tf(t^{-1})$ and this defines an involution between operator means. Hiai et al. [156] note that $t \mapsto 1/f(t^{-1})$ defines another involution between operator means.

Chapter 38
Lieb Concavity and Lieb–Ruskai Strong Subadditivity Theorems, I: Basics

The next four chapters concern some important matrix inequalities originally of significance in quantum statistical mechanics and, more recently, in quantum information theory. In this chapter, we'll state the two basic results and show that the second implies the first, while in the subsequent chapters we'll provide three different proofs of the second. This material is here because the theory of matrix convex and of matrix monotone functions will be central: one proof depends on the Hansen–Jensen–Pederson theorem and the other two on joint concavity of harmonic and of geometric means. We'll state and prove the results for finite matrices although they easily extend to the infinite dimensional case and even to general von Neumann algebras with traces (see the Notes).

We begin with a discussion of entropy. In classical mechanics, in a system with a countable set of configurations, a statistical mechanical state is a set of probabilities $\{p_j\}_{j=1}^{\infty}$ where p_j is the probability of configuration j. Thus

$$p_j \geq 0; \qquad \sum_{j=1}^{\infty} p_j = 1 \tag{38.1}$$

The *classical entropy* of the state is given by

$$S(p) = -\sum_{j=1}^{\infty} p_j \log p_j \equiv \sum_{j=1}^{\infty} h(p_j) \tag{38.2}$$

where $h(x) = -x \log x$ is positive (for $x \in [0, 1]$) and concave. Here we interpret $H(0) = \lim_{t \downarrow 0} h(t) = 0$.

The *quantum entropy* is defined on a quantum density matrix, i.e., a trace class operator, ρ (see [329, Chapter 3] for a discussion of trace class), obeying

$$\rho \geq 0; \qquad \mathrm{Tr}(\rho) = 1 \tag{38.3}$$

B. Simon, *Loewner's Theorem on Monotone Matrix Functions*, Grundlehren der mathematischen Wissenschaften 354, https://doi.org/10.1007/978-3-030-22422-6_38

clearly an analog of (38.1). Given a Hilbert space, $\mathcal{H}$, and a ρ obeying (38.3), one defines a linear function on $\mathcal{L}(\mathcal{H})$ by $\ell_\rho(B) = \mathrm{Tr}(\rho B)$. It obeys

$$\ell_\rho(\mathbf{1}) = 1, \qquad B \geq 0 \Rightarrow \ell_\rho(B) \geq 0 \tag{38.4}$$

In the finite dimensional case, every ℓ obeying (38.4) is an ℓ_ρ. In the infinite dimensional case, we either require $B \mapsto \ell(B)$ be continuous in the weak topology defined by the $\{\ell_\rho\}$ or only look at ℓ on the compact operators and demand $\|\cdot\|$-continuity. This is the analog of the Riesz–Markov theorem in classical measure theory [325, Section 4.5].

One defines the entropy of ρ (aka *von Neumann entropy* or *quantum entropy*) by

$$S(\rho) = -\mathrm{Tr}(\rho \log \rho) \tag{38.5}$$

Thus ρ has eigenvalues $\{p_j\}_{j=1}^\infty$, $S(\rho)$ is given by (38.2).

Classically, one often considers $X_1 \times X_2$, i.e., configurations which are ordered pairs of two simpler configurations (e.g., two particles). A state of $X_1 \times X_2$ is then given by $\{p_{ij}\}_{1 \leq i,j < \infty}$. One defines induced probabilities on the subsystems by

$$p_i^{(1)} = \sum_j p_{ij}, \qquad p_j^{(2)} = \sum_i p_{ij} \tag{38.6}$$

Especially if we are dealing with $X_1 \times \cdots \times X_k$, it is useful to write $p^{(12\ldots k)}$ for the original probability. *Subadditivity* of the entropy says that

$$S(p^{(12)}) \leq S(p^{(1)}) + S(p^{(2)}) \tag{38.7}$$

(rather than prove this now, we'll see it follows from the Lieb–Ruskai SSA theorem below). *Strong subadditivity* of the classical entropy says that on $X_1 \times X_2 \times X_3$, one has that

$$S(p^{(123)}) + S(p^{(2)}) \leq S(p^{(12)}) + S(p^{(23)}) \tag{38.8}$$

For the quantum case, we need to define the analog of $X_1 \times X_2$ and of the induced probabilities of (38.6). X_j is now a Hilbert space, $\mathcal{H}_j$ and $X_1 \times \cdots \times X_k$ is replaced by the tensor product Hilbert space $\mathcal{H}_1 \otimes \cdots \otimes \mathcal{H}_k$ (see [325, Section 3.8] or [329, Section 1.3] for more on tensor products).

Let $\mathcal{S}_{+,1}(\mathcal{H})$ be the set of trace class ρ on $\mathcal{L}(\mathcal{H})$ obeying (38.3) (aka quantum states). If $J = \{1, \ldots, k\}$ and $K \subset J$ has $K = \{i_1, \ldots, i_\ell\}$, $i_1 < i_2 < \cdots < i_\ell$, we set $\mathcal{H}^K = \mathcal{H}_{i_1} \otimes \cdots \otimes \mathcal{H}_{i_\ell}$. So in natural way,

$$\mathcal{H}^J = \mathcal{H}^K \otimes \mathcal{H}^{J \setminus K} \tag{38.9}$$

There is a natural map, $\mathrm{Tr}^{J\setminus K} : \mathcal{S}_{+,1}(\mathcal{H}^J) \mapsto \mathcal{S}_{+,1}(\mathcal{H}^K)$ (here, $J \setminus K$ is set difference, i.e., the complement of K in J) defined by the requirements that for any $\rho \in \mathcal{S}_{+,1}(\mathcal{H}^J)$ and any $A \in \mathcal{L}(\mathcal{H}^K)$:

$$\mathrm{Tr}_{\mathcal{H}^K}(\mathrm{Tr}^{J\setminus K}(\rho)A) = \mathrm{Tr}_{\mathcal{H}^J}(\rho(A \otimes \mathbf{1})) \tag{38.10}$$

It is called the *partial trace*. Equivalently if $\{e_j^{J\setminus K}\}$ is a basis for $\mathcal{H}^{J\setminus K}$ and $\varphi, \psi \in \mathcal{H}^K$, then

$$\langle \varphi, \mathrm{Tr}^{J\setminus K}(\rho)\psi\rangle = \sum_j \langle \varphi \otimes e_j^{J\setminus K}, \rho\left(\psi \otimes e_j^{J\setminus K}\right)\rangle \tag{38.11}$$

Given $\rho^{(123)} \in \mathcal{S}_{+,1}(\mathcal{H}_1 \otimes \mathcal{H}_2 \otimes \mathcal{H}_3)$, we define

$$\rho^{(12)} = \mathrm{Tr}^3(\rho^{(123)}), \quad \rho^{(23)} = \mathrm{Tr}^1(\rho^{(123)}), \quad \rho^{(2)} = \mathrm{Tr}^{13}(\rho^{(123)}) \tag{38.12}$$

Then the first major result of this chapter is

Theorem 38.1 (Lieb–Ruskai SSA of Quantum Entropy) *For any finite dimensional Hilbert spaces, $\mathcal{H}_1, \mathcal{H}_2, \mathcal{H}_3$ and $\rho^{(123)} \in \mathcal{S}_{+,1}(\mathcal{H}_1 \otimes \mathcal{H}_2 \otimes \mathcal{H}_3)$, we have that*

$$S(\rho^{(123)}) + S(\rho^{(2)}) \le S(\rho^{(12)}) + S(\rho^{(23)}) \tag{38.13}$$

Remarks

1. This is also true in infinite dimensions and even in some von Neumann algebra contexts; see the Notes.
2. Taking $\mathcal{H}_2 = \mathbb{C}$, so $\rho^{(2)} = \mathbf{1} \in \mathbb{C}$, (38.13) yields $S(\rho^{(13)}) \le S(\rho^{(1)}) + S(\rho^{(3)})$, i.e., $SSA \Rightarrow SA$. If we demand the eigenvectors of $\rho^{(123)}$ be tensor products of basis vectors, (38.13) reduces to classical SSA so this theorem implies classical SSA and classical and quantum SA.

The other major result of this chapter is

Theorem 38.2 (Lieb Concavity, First Form) *Let K be a Hilbert–Schmidt operator on a Hilbert space, $\mathcal{H}$. Then, for all $p \in [0, 1]$*

$$(A, B) \mapsto Tr(A^p K^* B^{1-p} K) \tag{38.14}$$

is jointly concave in A and B on positive A, B in $\mathcal{L}(\mathcal{H})$.

Remarks

1. For discussion of Hilbert–Schmidt class and of trace in infinite dimensions, see [329, Chapter 3].

2. This remains true if K maps $\mathcal{H}_1$ to $\mathcal{H}_2$, $B \in \mathcal{L}(\mathcal{H}_2)$, $A \in \mathcal{L}(\mathcal{H}_1)$, Tr is $\mathrm{Tr}_{\mathcal{H}_1}$, and K^*K is trace class as a map of $\mathcal{H}_1$ to itself. Indeed, if $\mathcal{H} = \mathcal{H}_1 \oplus \mathcal{H}_2$, $\tilde{A} = A \oplus 0$, $\tilde{B} = 0 \oplus B$, $\tilde{K} = \begin{pmatrix} 0 & 0 \\ K & 0 \end{pmatrix}$, then $\mathrm{Tr}_{\mathcal{H}_1}(A^p K^* B^{1-p} K) = \mathrm{Tr}_{\mathcal{H}}(\tilde{A}^p \tilde{K}^* \tilde{B}^{1-p} \tilde{K})$ so the one Hilbert space case implies the two Hilbert space result.
3. If $\widetilde{\mathcal{H}} = \mathcal{H} \oplus \mathcal{H}$, $\tilde{A} = A \oplus B$, and $\tilde{K} = \begin{pmatrix} 0 & 0 \\ K & 0 \end{pmatrix}$, then $\mathrm{Tr}_{\mathcal{H}}(A^p K^* B^{1-p} K) = \mathrm{Tr}_{\widetilde{\mathcal{H}}}(\tilde{A}^p \tilde{K}^* \tilde{A}^{1-p} \tilde{K})$, so concavity when $A = B$ implies joint concavity when B is not necessarily equal to A.
4. The right side of (38.14) is positive since $\mathrm{Tr}(A^p K^* B^{1-p} K) = \mathrm{Tr}(C^* C)$ where $C = B^{(1-p)/2} K A^{p/2}$.

A major point of the rest of the current chapter is to show that Lieb Concavity implies Ruskai–Lieb SSA. In the next three chapters, we'll provide three different proofs of Lieb Concavity.

We also want a rephrasing of Lieb Concavity in terms of tensor products. Let $K : \mathbb{C}^n \to \mathbb{C}^m$ be given by a matrix, $\{K_{\alpha j}\}_{j=1}^{n}{}_{\alpha=1}^{m}$ so that

$$(K\varphi)_\alpha = \sum_{j=1}^{n} K_{\alpha j} \varphi_j \tag{38.15}$$

We can think of K also as an element of $\mathbb{C}^{nm} = \mathbb{C}^n \otimes \mathbb{C}^m$ by $K = \sum K_{\alpha j}(\delta_\alpha \otimes \delta_j)$ where δ_j is the vector with components $(\delta_j)_i = \delta_{ij}$.

If $A \in \mathcal{L}(\mathbb{C}^m)$ has matrix $\{A_{\alpha\beta}\}_{1 \le \alpha, \beta \le m}$ and $B \in \mathcal{L}(\mathbb{C}^n)$ has matrix $\{B_{ij}\}_{1 \le i, j \le n}$, then the operator $A \otimes B$ applied to K thought of as a vector in $\mathbb{C}^n \otimes \mathbb{C}^m$ obeys

$$[(A \otimes B)K]_{\alpha j} = \sum_{\beta k} A_{\alpha\beta} B_{jk} K_{\beta k} \tag{38.16}$$

It follows that

$$\begin{aligned} \langle K, (A \otimes B)K \rangle &= \sum_{\beta k} A_{\alpha\beta} K_{\beta k} (B^t)_{kj} \overline{K}_{\alpha j} \\ &= \mathrm{Tr}(A K B^t K^*) \end{aligned} \tag{38.17}$$

Thus Theorem 38.2 is equivalent to

Theorem 38.3 (Lieb Concavity, Second Form) *The map*

$$(A, B) \mapsto A^p \otimes B^{1-p} \tag{38.18}$$

from $\mathcal{L}_+(\mathbb{C}^m) \otimes \mathcal{L}_+(\mathbb{C}^n) \to \mathcal{L}(\mathbb{C}^m \otimes \mathbb{C}^n)$ (where $\mathcal{L}_+$ is positive matrices) is operator jointly concave.

The relation of Lieb Concavity to SSA goes through a function called the *relative entropy* defined on $\mathcal{L}_{++}(\mathbb{C}^n) \times \mathcal{L}_{++}(\mathbb{C}^n)$ (where $\mathcal{L}_{++}$ is the strictly positive matrices) by

$$S(A|B) = \text{Tr}(A \log A - A \log B) \tag{38.19}$$

Corollary 38.4 (Joint Convexity of the Relative Entropy) *Taking* $K = \mathbf{1}$ *in Lieb–Concavity, we see for* $p < 1$,

$$(A, B) \mapsto \frac{Tr(A^p B^{1-p}) - Tr(A)}{p - 1} \tag{38.20}$$

is jointly convex (since $p - 1 < 0$*). As* $p \uparrow 1$*, the ratio converges to* $S(A|B)$ *which we conclude is also jointly convex.*

Notice next that if ρ^{JK} is a density matrix on $\mathcal{H}^J \otimes \mathcal{H}^K$ and $\rho^J = \text{Tr}^K \rho^{JK}$, then, by the definition of partial trace

$$\text{Tr}(\rho^{JK} \log(\mathbf{1}_{\mathcal{H}_K} \otimes \rho^J)) = \text{Tr}(\rho^J \log(\rho^J)) \tag{38.21}$$

so

$$S(\rho^{JK} | \tfrac{1}{n_K} \mathbf{1}_{\mathcal{H}_K} \otimes \rho^J) = -S(\rho^{JK}) + S(\rho^J) - \log n_K \tag{38.22}$$

The key to proving SSA is that partial trace is a convex average:

Proposition 38.5

(a) Let $\mathcal{U}(n)$ *be the* $n \times n$ *unitary matrices and let* dU *be normalized Haar measure on* $\mathcal{U}(n)$*. Then for any* $A \in \mathcal{L}(\mathbb{C}^n)$,

$$\int UAU^{-1} dU = \frac{1}{n} Tr(A) \mathbf{1} \tag{38.23}$$

(b) Let $Tr^{(1)}$ *be the partial trace of* $\mathcal{L}(\mathbb{C}^m \otimes \mathbb{C}^n)$ *to* $\mathcal{L}(\mathbb{C}^m)$*. Then for any* $A \in \mathcal{L}(\mathbb{C}^m \otimes \mathbb{C}^n)$

$$\int (\mathbf{1} \otimes U) A (\mathbf{1} \otimes U)^{-1} dU = \frac{1}{n} \mathbf{1}_n \otimes Tr^{(1)}(A) \tag{38.24}$$

where U *runs through* $\mathcal{U}(m)$.

Remark See [325, Section 4.19] for a discussion of Haar measure.

Proof

(a) By the invariance of Haar measure, if B is the left side of (38.23), then $VBV^{-1} = B$ for any unitary, V. If φ is an eigenvector of B with eigenvalue

β, so is $V\varphi$ so $B = \beta\mathbf{1}$. Both sides of (38.23) have the same trace proving the equality.

(b) By the same argument, for any unitary V on $\mathbb{C}^m$, we have that $(\mathbf{1} \otimes V)B(\mathbf{1} \otimes V)^{-1} = B$ so $B = C \otimes \mathbf{1}$. Multiplying by $D \otimes \mathbf{1}$ and taking traces yields the formula.

□

Proof that Corollary 38.4 Implies Theorem 38.1 (And So That Theorem 38.2 ⇒ Theorem 38.1) If U_3 is a unitary on $\mathcal{H}_3 = \mathbb{C}^{n_3}$, then for density matrices on $\mathcal{H}_1 \otimes \mathcal{H}_2 \otimes \mathcal{H}_3$, one has that

$$\frac{1}{n_3}\rho^{(12)} \otimes \mathbf{1}_{\mathcal{H}_3} = \int U_3 \rho^{(123)} U_3^{-1} dU_3 \tag{38.25}$$

and, of course similarly as operators on $\mathcal{H}_2 \otimes \mathcal{H}_3$,

$$\frac{1}{n_3}\rho^{(2)} \otimes \mathbf{1}_{\mathcal{H}_3} = \int U_3 \rho^{(23)} U_3^{-1} dU_3 \tag{38.26}$$

Therefore by Corollary 38.4, with $\tilde{\rho} = n_1^{-1}\mathbf{1}_{\mathcal{H}_1} \otimes \rho$

$$\begin{aligned} S(\rho^{(12)}|\widetilde{\rho^{(2)}}) &\le \int S(U_3\rho^{(123)}U_3^{-1}|U_3\widetilde{\rho^{(23)}}U_3^{-1})dU \\ &= S(\rho^{(123)}|\widetilde{\rho^{(23)}}) \end{aligned} \tag{38.27}$$

by the unitary invariance of $S(A|B)$.

By (38.22)

$$S(\rho^{(2)}) - \log(n_1) - S(\rho^{(12)}) \le S(\rho^{(23)}) - \log(n_1) - S(\rho^{(123)})$$

which is (38.13). □

Notes and Historical Remarks Classical entropy has its roots in the thermodynamics of Clausius, Carnot, and Thompson (aka Lord Kelvin). The formula (38.2) in the form $S(p) = -k\sum_{j=1}^{n} p_j \log p_j$ is called the Boltzmann or Boltzmann–Gibbs–Shannon formula. Here k is Boltzmann's constant and the formula arose in foundational work on the statistical mechanical foundation of thermodynamics [43, 125]. Shannon [313] called $-\sum p_j \log_2 p_j$ the information or entropy content of a distribution. If there are 2^n states with each $p_j = 2^{-n}$, it requires n bits to describe the distribution and, of course, $n = -\sum p_j \log_2 p_j$ in this case.

von Neumann [238] wrote down the definition (38.5) of quantum entropy. In the 1960s, Lanford, Robinson, and Ruelle reexamined the mathematical foundations of statistical mechanics. In this context, the usefulness of classical SSA, which was used earlier, was emphasized by Robinson–Ruelle [285] in their study of the thermodynamic limit of the mean entropy. Lanford–Robinson [196] conjectured the quantum inequality in 1967 which was proven in 1973 by Lieb–Ruskai [202, 203].

Wigner–Yanese [364, 365] had defined an alternate entropy $S_p(\rho, K) = \frac{1}{2}\text{Tr}\left([\rho^p, K][\rho^{1-p}, K]\right) = \text{Tr}(\rho^p K \rho^{1-p} K) - \text{Tr}(\rho K^2)$. Concavity in ρ is equivalent to Lieb Concavity. Wigner–Yanese proved concavity in case $p = 1/2$. Because of unpublished remarks of Dyson, Lieb Concavity is sometimes called the Wigner–Yanese–Dyson conjecture.

The idea of getting joint convexity of the relative entropy by differentiating (38.20) as we do in Corollary 38.4 is due to Lindblad [204].

In his paper on Lieb Concavity [200], Lieb proved several inequalities equivalent to the main results including concavity of

$$C \mapsto \text{Tr}(\exp(K + \log C)) \tag{38.28}$$

for $C \geq 0$ and K fixed and self-adjoint. The original Lieb–Ruskai proof of SSA relied on (38.28). Our proof that Lieb Concavity implies SSA is due to Uhlmann [354] using his earlier idea of writing induced states on subsystems in terms of averaging unitaries over Haar measure [353].

We will present three proofs of SSA (via Lieb Concavity) in the next three chapters but there are many other proofs since the result has captured the imaginations of mathematicians, statistical physicists, and experts in quantum information theory. In particular, we mention Epstein's 1973 proof of Lieb Concavity [90] and directly of the concavity (38.28). He gets this by establishing a Herglotz representation and using the fact that, as we saw in Chapter 10, Herglotz functions which are analytic across $(0, \infty)$ are concave. So his proof has a connection to the ideas central to much of this book. An exposition of Epstein's proof can be found in the appendix and erratum of Ruskai [291]. We also note the recent work of Sutter et. al [336] who estimate the error in SSA.

Three lovely reviews of SSA including equivalences among various assertions are Lieb [201], Ruskai [291], and Vershynina et. al [357].

The original Lieb–Ruskai SSA paper [203] has an appendix by Simon showing how to get quantum entropy on Hilbert space as limits their finite dimensional approximate entropies and they use this to show how to go from SSA on finite matrices to the infinite dimensional case. Araki [17], Narnhofer–Thirring [232], and Petz [248, 270, 271] have extensions to the context of states on suitable von Neumann algebras.

Chapter 39
Lieb Concavity and Lieb–Ruskai Strong Subadditivity Theorems, II: Effros' Proof

In this chapter, we'll prove Lieb Concavity as a special case of the following theorem (we explain the name in the Notes):

Theorem 39.1 (EENG Theorem) *Let f be a real-valued function on $(0, \infty)$. For two strictly positive self-adjoint matrices, C and D define*

$$g(C, D) = D^{1/2} f(D^{-1/2} C D^{-1/2}) D^{1/2} \tag{39.1}$$

Then f is matrix convex if and only if $C, D \mapsto g(C, D)$ is jointly convex on the $n \times n$ matrices for each n.

Proof Taking $D = \mathbf{1}$, it is immediate that convexity of g implies the convexity for f.

For the converse, suppose that f is matrix convex. Given C_0, C_1, D_0, D_1 strictly positive matrices and $\theta \in [0, 1]$, let

$$C = (1-\theta)C_0 + \theta C_1 \qquad D = (1-\theta)D_0 + \theta D_1$$

$$T_0 = ((1-\theta)D_0)^{1/2} D^{-1/2} \qquad T_1 = (\theta D_1)^{1/2} D^{-1/2}$$

Then

$$T_0^* T_0 + T_1^* T_1 = \mathbf{1} \quad D^{1/2} T_j^* = \beta_j^{1/2} D_j^{1/2} \quad D^{-1/2} \beta_j^{1/2} = T_j^* D_j^{-1/2} \tag{39.2}$$

where $\beta_0 = 1 - \theta$ and $\beta_1 = \theta$. Thus

$$\begin{aligned} g(C, D) &= D^{1/2} f(D^{-1/2} C D^{-1/2}) D^{1/2} && (39.3) \\ &= D^{1/2} f(T_0^* D_0^{-1/2} C_0 D_0^{-1/2} T_0 + T_1^* D_1^{-1/2} C_1 D_1^{-1/2} T_1) D^{1/2} && (39.4) \\ &\le D^{1/2} T_0^* f(D_0^{-1/2} C_0 D_0^{-1/2}) T_0 D^{1/2} + \end{aligned}$$

B. Simon, *Loewner's Theorem on Monotone Matrix Functions*, Grundlehren der mathematischen Wissenschaften 354, https://doi.org/10.1007/978-3-030-22422-6_39

$$D^{1/2}T_1^* f(D_1^{-1/2}C_1D_1^{-1/2})T_1D^{1/2} \tag{39.5}$$

$$= (1-\theta)D_0^{1/2} f(D_0^{-1/2}C_0D_0^{-1/2})D_0^{1/2} + \theta D_1^{1/2} f(D_1^{-1/2}C_1D_1^{-1/2})D_1^{1/2} \tag{39.6}$$

$$= (1-\theta)g(C_0, D_0) + \theta g(C_1, D_1) \tag{39.7}$$

proving that g is jointly convex. Equation (39.3) and (39.7) are just the definition of g, (39.1). Equation (39.4) uses the relation among C, C_0, and C_1 and the third formula in (39.2). Equation (39.6) uses the second relation in (39.2). Finally, the critical inequality (39.5), given the first equation in (39.2) is just the Hansen–Jensen–Pedersen theorem, Theorem 11.1. □

To emphasize there is a concave function version, we state the following corollary in that setting (replacing f by $-f$, the proof is immediate from Theorem 39.1):

Corollary 39.2 (Effros' Theorem) *Let $\mathcal{A}_1$ and $\mathcal{A}_2$ be two convex sets of $n \times n$ matrices so that $C \in \mathcal{A}_1$, $D \in \mathcal{A}_2 \Rightarrow CD = DC$. Let f be a matrix concave function on strictly positive matrices. Define $g(C, D) = Df(CD^{-1})$ for $C \in \mathcal{A}_1$, $D \in \mathcal{A}_2$. Then g is jointly concave.*

First Proof of Theorem 38.2 As noted in Example 9.13, $A \mapsto A^\alpha$ is matrix concave if $0 \le \alpha \le 1$. Letting $\mathcal{A}_1$ (resp. $\mathcal{A}_2$) be the set of $A \otimes \mathbf{1}$ (resp. $\mathbf{1} \otimes B$), the corollary implies that $A, B \mapsto A^\alpha \otimes B^{1-\alpha}$ is jointly concave. □

Remarks

1. Similarly, since $A \mapsto A\log(A)$ is matrix convex, $A, B \mapsto A(\log(A) - \log(B))$ is matrix convex, which as we saw in Chapter 38 is just the derivative of Lieb Concavity.
2. Since $A \mapsto A^{1+\beta}$ is matrix convex if $0 \le \beta \le 1$, we also see that $A, B \mapsto A^{1+\beta} \otimes B^{-\beta}$ is jointly matrix convex. We discuss this fact, known as Ando's convexity theorem in the next chapter.

Notes and Historical Remarks In the theory of scalar convex functions, the notion and name "perspective" appeared first in [162]. If f is convex on a convex subset of $\mathbb{R}^n$, they defined the perspective of f on a convex subset of $\mathbb{R}^{n+1}$ by $g(x, t) = tf(x/t)$. Maréchal [214] developed these ideas and studied the generalization $x, t \mapsto h(t)f(x/h(t))$. In a seminal paper, Effros [88] realized that, for commuting families of operators, there was an operator analog which implied Lieb Concavity—in particular, he proved Corollary 39.2. Ebadian, Nikoufar, and Eshaghi–Gordji [87] then realized that using the definition (39.1) (used earlier, as we've seen, by Ando), Effros' proof worked in a non-commutative setting. We emphasize that the applications in this chapter only require Effros' commutative version. Effros–Hansen [89] found a set of axioms on $A, B \mapsto g(A, B)$ that are equivalent to (39.1) for a convex function f. For a different approach to many of the additional inequalities related to Theorem 39.1, see Carlen–Lieb [56] .

Chapter 40
Lieb Concavity and Lieb–Ruskai Strong Subadditivity Theorems, III: Ando's Proof

In Chapter 37, we showed that if f is a matrix monotone function on $(0, \infty)$, then $A, B \mapsto A^{1/2} f(A^{-1/2} B A^{-1/2}) A^{1/2}$ is jointly concave on pairs of positive operators (see Corollary 37.2). The discussion there supposes that f is non-negative but that is only to get that this function of two variables is an operator mean. The proof of the matrix convexity uses the joint concavity of harmonic mean and the positivity of the measure ρ_f in (37.4) which holds for any matrix monotone f. We therefore have:

Second Proof of Theorem 38.2 Since the map $C \mapsto C^\alpha$ is matrix monotone for $\alpha \in [0, 1]$ and $C^{1/2}(C^{-1/2} D C^{-1/2})^\alpha C^{1/2} = C^{1-\alpha} D^\alpha$ if C and D commute, we immediately get Lieb Concavity since $A \otimes \mathbf{1}$ and $\mathbf{1} \otimes B$ commute. □

Remark What is critical is the integral representation for f which for the mean $\#_\alpha$ takes the form

$$A \#_\alpha B = \pi^{-1} \sin(\alpha\pi) \int_0^\infty \lambda^{-1}((\lambda A) : B) \lambda^{\alpha-1} d\lambda \tag{40.1}$$

and the joint concavity of $C, D \mapsto C : D$. This integral converges because $\lambda^{-1}((\lambda A) : B) = A(\lambda A + B)^{-1} B$ is bounded for all λ and decays as $1/\lambda$ at infinity. This is the same integral representation that we used in (4.5).

We saw that $A \mapsto -A^{-\gamma}$ is matrix monotone for $\gamma \in [0, 1]$ (see the Remarks after Theorem 4.1) and when A and B commute, we have that

$$A^{1/2}(A^{-1/2} B A^{-1/2})^{-\gamma} A^{1/2} = A^{1+\gamma} B^{-\gamma}$$

we see that setting $\alpha = 1 + \gamma$:

Theorem 40.1 (Ando's Convexity Theorem) *Let $\alpha \in [1, 2]$. Then $A, B \mapsto A^\alpha \otimes B^{1-\alpha}$ is jointly convex.*

B. Simon, *Loewner's Theorem on Monotone Matrix Functions*, Grundlehren der mathematischen Wissenschaften 354, https://doi.org/10.1007/978-3-030-22422-6_40

Notes and Historical Remarks That the integral representation (40.1) plus joint concavity of the parallel sum (a result of Anderson–Duffin [11]) imply Lieb Concavity is a result of Ando [15]. The simplicity of the argument is somewhat hidden by the fact that it appears in a 38 page paper (in part because of other applications) and that the proof uses discussion from several different places in the paper. This paper also had his result that appears as Theorem 40.1. We note that this paper predated the general paper on operator means by Kubo–Ando [193] and, indeed, seems to have motivated it.

The same proof that shows Theorem 40.1 proves joint convexity when $\alpha \in [-1, 0]$ also. It was noted by Hasegawa [142] (see also Jencova–Ruskai [170]) that one can compactly express Ando's result, Lieb's result, and the just mentioned result by saying that

$$(A, B) \mapsto \frac{\mathrm{Tr}(A^{\alpha} B^{1-\alpha}) - \mathrm{Tr}(A)}{\alpha(\alpha - 1)} \tag{40.2}$$

is jointly matrix convex for all $\alpha \in [-1, 2]$.

Chapter 41
Lieb Concavity and Lieb–Ruskai Strong Subadditivity Theorems, IV: Aujla–Hansen–Uhlmann Proof

In this chapter we'll prove a generalization of Lieb Concavity (which was already in his original paper):

Theorem 41.1 *Let $p_1, \dots, p_k$ be real numbers obeying:*

$$p_j \geq 0 \qquad p_1 + \dots + p_k \leq 1 \tag{41.1}$$

Then $(A_1, \dots, A_k) \mapsto A_1^{p_1} \otimes A_2^{p_2} \otimes \dots \otimes A_k^{p_k}$ is jointly concave on the positive matrices.

Given convex sets, $\mathcal{A}_1, \dots, \mathcal{A}_k$ of positive operators so that $C_j \in \mathcal{A}_j$ are mutually commuting and $F : [0, \infty)^k \to \mathbb{R}$, we can define $F(C_1, \dots, C_k)$ by the spectral theorem (or alternately since we are focusing on finite matrices, by using the fact that there is an orthonormal basis of common eigenvector so that $F(C_1, \dots, C_k)$ has the same eigenvectors with $F(C_1, \dots, C_k)\psi = F(\lambda_1, \dots, \lambda_k)\psi$ when $C_j\psi = \lambda_j\psi$).

Theorem 41.2 *Let F and G be two functions from $[0, \infty)^k$ to $\mathbb{R}$. Define*

$$H(x_1, \dots, x_k) = \sqrt{F(x_1, \dots, x_k)G(x_1, \dots, x_k)} \tag{41.2}$$

Let $\mathcal{A}_1, \dots, \mathcal{A}_k$ be convex sets of mutually commuting operators. Suppose $F(C_1, \dots, C_k)$ and $G(C_1, \dots, C_k)$ are jointly concave functions when $C_j \in \mathcal{A}_j$. Then so is $H(C_1, \dots, C_k)$.

Proof The proof will depend on the fact that when X and Y commute, their geometric mean, $X\#Y = (XY)^{1/2}$ and in particular

$$H(C_1, \dots, C_k) = F(C_1, \dots, C_k)\#G(C_1, \dots, C_k) \tag{41.3}$$

B. Simon, *Loewner's Theorem on Monotone Matrix Functions*, Grundlehren der mathematischen Wissenschaften 354, https://doi.org/10.1007/978-3-030-22422-6_41

and on the fact that $A, B \mapsto A\#B$ is jointly concave which follows either from the integral representation in terms of parallel sums or from the maximum principle in Part (b) of Theorem 36.9.

Given $C_j, D_j \in \mathcal{A}_j$ and $\theta \in [0, 1]$, let $E_j = (1-\theta)C_j + \theta D_j$. Then:

$$H(E_1, \dots, E_k) = F(E_1, \dots, E_k)\#G(E_1, \dots, E_k) \tag{41.4}$$

$$\geq [(1-\theta)F(C_1, \dots, C_k) + \theta F(D_1, \dots, D_k)]\#$$
$$[(1-\theta)G(C_1, \dots, C_k) + \theta G(D_1, \dots, D_k)] \tag{41.5}$$

$$\geq (1-\theta)F(C_1, \dots, C_k)\#G(C_1, \dots, C_k)+$$
$$\theta F(D_1, \dots, D_k)\#G(D_1, \dots, D_k) \tag{41.6}$$

$$= (1-\theta)H(C_1, \dots, C_k) + \theta H(D_1, \dots, D_k) \tag{41.7}$$

proving joint matrix concavity of H. In the above, (41.4) and (41.7) depend on (41.3), and (41.5) follows from the assumed concavity of F and G and the monotonicity of $C, D \mapsto C\#D$ which for commuting C and D follows from the monotonicity of the square root. Finally, (41.6) follows from the joint concavity of $X, Y \mapsto X\#Y$. □

Given $p_1, \dots, p_k$, define $P_{p_1,\dots,p_k}(x_1, \dots, x_k) \equiv x_1^{p_1}x_2^{p_2}\dots x_k^{p_k}$. If $F = P_{p_1,\dots,p_k}$ and $G = P_{q_1,\dots,q_k}$, then $H = P_{\frac{1}{2}p_1+\frac{1}{2}q_1,\dots}$. Given that P_p is continuous in p, we conclude with $\mathcal{A}_1, \dots \mathcal{A}_k$ fixed that the set, S, of p for which the map is concave when the arguments lie in $\mathcal{A}_j$, is a convex set.

Third Proof of Theorem 38.2 Let $k = 2$, $\mathcal{A}_1 = \{A \otimes \mathbf{1}\}$, and $\mathcal{A}_2 = \{\mathbf{1} \otimes B\}$. The above set S contains $(1, 0)$ and $(0, 1)$ since the maps are then linear. Thus by convexity, it contains all $(1-\theta, \theta)$. □

Proof of Theorem 41.1 We take $\mathcal{A}_j$ to be the tensor products with an arbitrary positive X in slot j and all the other slots equal to $\mathbf{1}$. The set of p obeying (41.1) is a convex simplex so it suffices to show its $k+1$ vertices have concavity. k of these have on $p_j = 1$ and the other p's equal 0 and the last one has all $p_j = 0$. For each of these vertices, F_p is linear and so concave. □

Notes and Historical Remarks The idea of proving Lieb Concavity by showing the set, S, of p with F_p concave is a convex set appeared first in a 1977 paper of Uhlmann [354]. Instead of a general result about concavity of $C, D \mapsto C\#D$, Uhlmann uses a Schwarz inequality argument not unrelated to the Proof of Lemma 36.5; in the presentation of Simon [320, Section 8.4], his proof is half a page. Ando [14] found a similar argument. Aujla [20] also exploited the idea for $k = 2$ of using concavity of the geometric mean to get convexity of the set of p for which Lieb concavity held. He didn't quote the literature on geometric mean but instead found the standard proof of it which we presented in Chapter 36. Apparently unaware of Uhlmann or Aujla (but aware of Ando), Hansen [132] found the general Theorem 41.2. We follow his presentation here.

Chapter 42
Unitarily Invariant Norms and Rearrangement

In the next chapter, we will discuss some operator norm inequalities for matrix monotone functions and also some functions which are functional inverses of matrix monotone functions. The operator norms we'll give the proofs for are a large class of norms called unitarily invariant so the current chapter is a preliminary on the subject of such norms. For most of this chapter, we'll consider the norms on finite rank operators although, as we discuss in the Notes, there are also results for the natural completions to spaces of compact operators.

A *unitarily invariant norm* is a norm, $|||\cdot|||$, on finite rank operators, A, so that for all such A and all unitaries U and V, we have that

$$|||UAV||| = |||A||| \tag{42.1}$$

A norm is called a *symmetric operator norm* (or symmetric norm), if for all finite rank A and bounded operators T and S, we have that

$$|||TAV||| \leq \|T\| |||A||| \|V\| \tag{42.2}$$

We emphasize that on the right side of (42.2), the norms on T and V are operator norms, not $|||\cdot|||$. Writing $B = UAV$ and noting that $A = U^*BV^*$, we see that any symmetric norm is unitarily invariant. One result in this chapter will be the converse.

Any unitarily invariant norm, $|||\cdot|||$ induces a norm $\mathfrak{s}_{|||\cdot|||}$, on sequences of finite support by letting $D(a_1, \dots, a_n, 0, \dots)$ be the diagonal matrix with $D\delta_j = a_j\delta_j$ and setting

$$\mathfrak{s}_{|||\cdot|||}(a_1, \dots, a_n, 0, \dots) = |||D(a_1, \dots, a_n, 0, \dots)||| \tag{42.3}$$

Such norms on sequences are invariant under sign flips of the $a's$ (since if $b_j = \sigma_j a_j$ with each σ_j equal ± 1, then $D(b_1, \dots, b_n, 1, \dots) = D(\sigma_1, \dots, \sigma_n, 1, \dots)D(a_1, \dots, a_n, 1, \dots)$ and $D(\sigma_1, \dots, \sigma_n, 1, \dots)$ is unitary),

B. Simon, *Loewner's Theorem on Monotone Matrix Functions*, Grundlehren der mathematischen Wissenschaften 354, https://doi.org/10.1007/978-3-030-22422-6_42

and invariant under permutations (i.e., $\mathfrak{s}_{|||\cdot|||}(a_{\pi(j)}) = \mathfrak{s}_{|||\cdot|||}(a_j))$ since there is a unitary permutation matrix, U_π with $U_\pi D(a_j)U_\pi^{-1} = D(a_{\pi(j)})$. Norms on sequences with these two properties are called *symmetric sequence norms*.

If $\{\mu_j(A)\}_{j=1}^N$ are the singular values of A (defined at the end of Chapter 2), there are unitary operators U and V so that $A = UD(\mu_j(A))V$ (by diagonalizing $|A|$ and using the polar decomposition) and thus

$$|||A||| = \mathfrak{s}_{|||\cdot|||}(\mu_j(A)) \tag{42.4}$$

Conversely, given any symmetric sequence norm, $\mathfrak{s}$, we can define $|||\cdot|||$ by (42.4) although it is not obviously a norm. At the end of this chapter, we'll prove it is a norm so that (42.4) sets up a one–one correspondence between unitarily invariant norms and symmetric sequence norms.

Example 42.1 ($\mathcal{I}_p$ Norms) Define

$$\|A\|_p = \left(\sum_{j=1}^{N} |\mu_j(A)|^p\right)^{1/p} \tag{42.5}$$

A priori, this is not obviously a norm but it has the form (42.4) so the general theorem says it is a norm. One can also prove this fact by showing that $A \mapsto \mathrm{Tr}(|A|^p)$ is convex and using the argument in [325, Section 5.3] or by proving that $\|A\|_p = \sup\{|\mathrm{Tr}(BA)| \mid \|B\|_q \le 1\}$ or (essentially our proof of the general theorem) that $\|A\|_p = \sup\{(\sum |\langle\varphi_j, A\psi_j\rangle|^p)^{1/p}\}$ where the sup is over all pairs of orthonormal sequences or vectors. Equation (42.5) is called the $\mathcal{I}_p$ norm and the completion of the finite rank operators in this norm is realizable as a space of compact operators called the trace ideal, $\mathcal{I}_p$. For $p = 2$ this is the Hilbert–Schmidt operators and for $p = 1$, the trace class. Reference [329, Chapter 3] is on these ideals as well as substantial parts of the books of Gohberg–Krein [127] and Simon [320]. □

Example 42.2 (Ky Fan Norms) For each $k = 1, 2, \ldots,$ define

$$|||A|||^{(k)} = \sum_{j=1}^{k} \mu_j(A) \tag{42.6}$$

called the *Ky Fan norms*. That they are norms depends on the proposition below. The major result in this chapter, relevant in the next chapter is that $|||A||| \ge |||B|||$ for all unitarily invariant norms $|||\cdot|||$ if and only if $|||A|||^{(k)} \ge |||B|||^{(k)}$ for each of the Ky Fan norms. □

Lemma 42.3 *Let $\mu_1 \ge \mu_2 \ge \cdots \ge \mu_N \ge 0$ and $k \le N$. Suppose that $0 \le b_j \le 1$, $j = 1, \ldots, N$ with $\sum_{j=1}^N b_j \le k$. Then*

$$\sum_{j=1}^{N} b_j \mu_j \leq \sum_{j=1}^{k} \mu_j \tag{42.7}$$

Proof If $N = k$, the result is obvious from $0 \leq b_j \leq 1$. If $N > k$, take b_N and split in pieces as follows. Let $c_1 = \min(b_N, 1 - b_1)$. If $c_1 = b_N$, stop and set $c_2 = c_3 = \cdots = c_k = 0$. Otherwise, set $c_2 = \min(b_N - c_1, 1 - b_2)$. If $c_1 + c_2 = b_N$, stop and set $c_3 = \cdots = c_k = 0$. Otherwise let $c_3 = b_N - c_1 - c_2$. If we have not stopped at c_{k-1} or earlier, then $c_j = 1 - b_j$ for $j = 1, \ldots, k-1$ so $b_N - \sum_{j=1}^{k-1} c_j = b_N - (k-1) + \sum_{j=1}^{k-1} b_j \leq k - b_k - (k-1) \leq 1 - b_k$, so we can take $c_k = b_N - \sum_{k=1}^{k-1} c_j$ and have $c_k + b_k \leq 1$. Thus, we can define $\tilde{b}_j = b_j + c_j$, $j = 1, \ldots, k$, $\tilde{b}_N = 0$ and $\tilde{b}_j = b_j$ for $j = k+1, \ldots, N-1$ and get that

$$\sum_{j=1}^{N-1} \tilde{b}_j \mu_j = \sum_{j=1}^{N} b_j \mu_j + \sum_{j=1}^{k} c_j (\mu_j - \mu_N) \geq \sum_{j=1}^{N} b_j \mu_j$$

so we've reduced the situation to the case where $b_N = 0$. Iterating yields the result. □

Proposition 42.4 *We have that*

$$|||A|||^{(k)} = \sup\{|Tr(BA)| \mid \|B\| \leq 1, \ rank(B) = k\}$$

Proof Let $A = \sum_{j=1}^{N} \mu_j(A) \langle \varphi_j, \cdot \rangle \psi_j$ be the canonical decomposition of A (see Chapter 2). Let $B_0 = \sum_{j=1}^{k} \langle \psi_j, \cdot \rangle \varphi_j$. Then $\|B_0\| = 1$, $\text{rank}(B_0) = k$, and $\text{Tr}(B_0 A) = \sum_{j=1}^{k} \mu_j(A)$.

On the other hand if B is arbitrary with $\|B\| \leq 1$ and $\text{rank}(B) = k$, we have that $B = \sum_{\ell=1}^{k} \mu_\ell(B) \langle \eta_\ell, \cdot \rangle \kappa_\ell$ and

$$\begin{aligned} |\text{Tr}(BA)| &\leq \sum_{j=1}^{N} \sum_{\ell=1}^{k} \mu_j(A) \mu_\ell(B) |\langle \eta_\ell, \varphi_j \rangle| \, |\langle \psi_j, \kappa_\ell \rangle| \\ &\equiv \sum_{j=1}^{N} \mu_j(A) \beta_j \end{aligned} \tag{42.8}$$

where (since $0 \leq \mu_\ell(B) \leq \|B\| \leq 1$)

$$\beta_j \leq \sum_{\ell=1}^{k} |\langle \eta_\ell, \varphi_j \rangle| \, |\langle \psi_j, \kappa_\ell \rangle|$$

Therefore, by the Schwarz and Bessel inequalities

$$\begin{aligned} 0 \leq \beta_j &\leq \left(\sum_{\ell=1}^{k} |\langle \eta_\ell, \varphi_j \rangle|^2 \right)^{1/2} \left(\sum_{\ell=1}^{k} |\langle \psi_j, \kappa_\ell \rangle|^2 \right)^{1/2} \\ &\leq \|\varphi_j\| \|\psi_j\| \leq 1 \end{aligned}$$

while by the same inequalities

$$\begin{aligned} \sum_{j=1}^{N} \beta_j &\leq \sum_{\ell=1}^{k} \left(\sum_{j=1}^{N} |\langle \eta_\ell, \varphi_j \rangle|^2 \right)^{1/2} \left(\sum_{j=1}^{N} |\langle \psi_j, \kappa_\ell \rangle|^2 \right)^{1/2} \\ &\leq \sum_{\ell=1}^{k} \|\eta_\ell\| \|\kappa_\ell\| = k \end{aligned}$$

By the lemma and (42.8), $|\mathrm{Tr}(BA)| \leq \sum_{j=1}^{k} \mu_j(A)$. □

The rest of this chapter discusses and applies the theory of majorization and rearrangement inequalities. Given $(a_1, \ldots, a_n) \in \mathbb{R}^n$, we define $a^* \in \mathbb{R}^n_+ = \{x \mid x_j \geq 0\}$ by requiring that $a_1^* \geq a_2^* \geq \cdots \geq a_n^*$ and that $(a_1^*, \ldots, a_n^*)$ is a permutation of $(|a_1|, \ldots, |a_n|)$. Given $a, b \in \mathbb{R}^n$, we say that b *majorizes* a, written $a \prec b$, if and only if, for $k = 1, \ldots, n$,

$$\sum_{j=1}^{k} a_j^* \leq \sum_{j=1}^{k} b_j^* \tag{42.9}$$

Notice that $c_j^* = b_j^*$ if and only if for some permutation, π of $\{1, \ldots, n\}$ and $\sigma_j = \pm 1$ for each j, we have that $c_j = \sigma_j b_{\pi(j)}$.

Lemma 42.5 *Let $a, b \in \mathbb{R}^n$. Then*

$$\sum_{j=1}^{n} |a_j b_j| \leq \sum_{j=1}^{n} a_j^* b_j^* \tag{42.10}$$

Proof By renumbering, we can suppose that $|a_1| \geq |a_2| \geq \cdots \geq |a_n|$ so that $a_j^* = |a_j|$. Then

$$\sum_{j=1}^{n} |a_j b_j| = |a_n| \sum_{j=1}^{n} |b_j| + (|a_{n-1}| - |a_n|) \sum_{j=1}^{n-1} |b_j| + \cdots + (|a_2| - |a_1|)|b_1| \tag{42.11}$$

Clearly $\sum_{j=1}^{\ell} |b_j| \leq \sum_{j=1}^{\ell} b_j^*$, so using $|a_{\ell-1}| - |a_\ell| \geq 0$,

$$\text{RHS of (42.11)} \leq a_n^* \sum_{j=1}^{n} b_j^* + \cdots + (a_2^* - a_1^*) b_1^*$$

$$= \sum_{j=1}^{n} a_j^* b_j^*$$

□

Let $\mathcal{G}_n$ be the group of $2^n n!$ elements generated by permutations of the coordinates on an n-vector and arbitrary sign flips of the n coordinates. Then $c = \pi(a)$ for some $\pi \in \mathcal{G}_n \iff c^* = a^*$ (generically this is for a unique π—indeed the π is unique if and only if $a_1^* > a_2^* > \cdots > a_n^* > 0$ (strict inequality)). The key rearrangement theorem is

Theorem 42.6 (Markus–Mitjagin Theorem) *Given $a \in \mathbb{R}^n$, let $H(a)$ be the convex hull of $\{\pi(a)\}_{\pi \in \mathcal{G}}$, i.e., $b \in H(a) \iff \exists(\theta_1, \ldots, \theta_{m \equiv 2^n n!})$ so that $\sum_{j=1}^{m} \theta_j = 1$, $\theta_j \geq 0$ and $b = \sum_{\pi \in \mathcal{G}_n} \theta_\pi \pi(a)$. Then $b \prec a \iff b \in H(a)$.*

Proof Let $\ell_k(a) = a_1 + \cdots + a_k$. Then

$$L_k(a) \equiv a_1^* + \cdots + a_k^* = \sup_{\pi \in \mathcal{G}_n} \ell_k(\pi(a)) \tag{42.12}$$

is a sup of linear functions, so convex. Since $L_k(\pi(a)) = L_k(a)$, we conclude that $L_k(b) \leq L_k(a)$ for all $b \in H(a)$ proving that $b \in H(a) \Rightarrow b \prec a$.

If $b \notin H(a)$, by separating hyperplane theorems (see [325, Theorem 5.10.6]), there is a linear functional, ℓ, on $\mathbb{R}^n$ with $\ell(b) > sup_{\pi \in \mathcal{G}_n} \ell(\pi(a)) \geq 0$ by the $\pm$ symmetry. If $\ell(x) = \sum_{j=1}^{n} \ell_j x_j$ and $\ell^*(x) \equiv \sum_{j=1}^{n} \ell_j^* x_j$, then, by the lemma, $\ell^*(b^*) \geq |\ell(b)| = \ell(b)$. Moreover, there is $\pi \in \mathcal{G}_n$ so that $\ell^*(a^*) = \ell(\pi(a))$. It follows that

$$\ell^*(b^*) > \ell^*(a^*) \tag{42.13}$$

If $b \prec a$, we'd have that

$$\begin{aligned}
\ell^*(b^*) &= \ell_n^* \sum_{j=1}^{n} b_j^* + (\ell_{n-1}^* - \ell_n^*) \sum_{j=1}^{n-1} b_j^* + \cdots + (\ell_1^* - \ell_2^*) b_1^* \\
&\leq \ell_n^* \sum_{j=1}^{n} a_j^* + (\ell_{n-1}^* - \ell_n^*) \sum_{j=1}^{n-1} a_j^* + \cdots + (\ell_1^* - \ell_2^*) a_1^* \\
&= \ell^*(a^*)
\end{aligned}$$

Thus $[b \prec a \Rightarrow \text{ not } (b \notin H(a))] \Rightarrow b \in H(a)$. □

Corollary 42.7 *Let $g : \mathbb{R} \to \mathbb{R}$ be convex with $g(x) = g(-x)$. Let $a, b \in \mathbb{R}^n$. If $b \prec a$, and $g(c_1, \dots, c_n) \equiv (g(c_1), \dots, g(c_n))$, then $g(b) \prec g(a)$.*

Remark To apply this, it is useful to note that if h is continuous from $[0, \infty)$ to $\mathbb{R}$ and $g(x) = h(|x|)$ for $x \in \mathbb{R}$, then g is convex if and only if h is convex and monotone increasing.

Proof For each subset $S \subset \{1, \dots, n\}$, define $G_S : \mathbb{R}^n \to \mathbb{R}$ by

$$G_S(a) = \sum_{j \in S} g(a_j) \tag{42.14}$$

Then it is easy to see that G_S is convex, so Theorem 42.6 implies that if S has k elements, then

$$G_S(b) \le \max_{\pi \in \mathcal{G}_n} G_S(\pi(a)) = \sum_{j=1}^{k} (g(a))_j^* \tag{42.15}$$

Maximizing over all S with k elements shows that $\sum_{j=1}^{k} (g(b))_j^* \le \sum_{j=1}^{k} (g(a))_j^*$, so $g(b) \prec g(a)$. $\square$

The following corollaries to Theorem 42.6 are so important, we'll call them theorems:

Theorem 42.8 $|||A||| \ge |||B|||$ *for every unitarily invariant norm (and, in particular, for every symmetric norm) if and only if* $|||A|||^{(k)} \ge |||B|||^{(k)}$ *for each of the Ky Fan norms.*

Remark We'll prove this for A, B finite rank. However, since $(\mu_1, \dots, \mu_n, 0) = \frac{1}{2}[(\mu_1, \dots, \mu_n, \mu_{n+1}) + (\mu_1, \dots, \mu_n, -\mu_{n+1})]$, we see that $\mathfrak{s}_{|||\cdot|||}(\mu_1, \dots, \mu_n)$ is monotone increasing in n, so one can define $|||A||| \equiv \lim_{n\to\infty} \mathfrak{s}_{|||\cdot|||}(\mu_1, \dots, \mu_n)$ for any compact A although the limit may be infinite. The finite rank result then implies the result for arbitrary compact A and B.

Proof Clearly, the "only if" is trivial since $|||\cdot|||^{(k)}$ is a symmetric norm. For the other direction, note that $\mathfrak{s}_{|||\cdot|||}$ is a convex function with $\mathfrak{s}_{|||\cdot|||}(\pi a) = \mathfrak{s}_{|||\cdot|||}(a)$ for all $\pi \in \mathcal{G}_n$. It follows from Theorem 42.6 that $\mathfrak{s}_{|||\cdot|||}(b) \le \mathfrak{s}_{|||\cdot|||}(a)$ if $b \prec a$. Since $|||A|||^{(k)} \ge |||B|||^{(k)}$ for $k = 1, \dots, n$ is equivalent to $\mu_j(B) \prec \mu_j(A)$, and $|||A||| = \mathfrak{s}_{|||\cdot|||}(\mu_j(A))$, we have proven that $|||A|||^{(k)} \ge |||B|||^{(k)} \Rightarrow |||B||| \le |||A|||$. $\square$

Theorem 42.9 *Any unitarily invariant norm is a symmetric norm.*

Proof By replacing T by $T/\|T\|$ and S by $S/\|S\|$, it suffices to prove (42.2) the result when $\|T\| = \|S\| = 1$. By (2.14) and (2.15), $\mu_j(TAS) \le \mu_j(A)$ for such T and S. It follows that $|||TAS|||^{(k)} \le |||A|||^{(k)}$ for all k so Theorem 42.8 implies (42.2). $\square$

Next, we want to consider matrix maps on symmetrically normed spaces. We consider $\mathbb{C}^n$ with such a norm. An $n \times n$ complex matrix, $B = \{b_{ij}\}_{1 \le i,j \le n}$ is called *doubly substochastic* if and only if

$$\sum_{j=1}^{n} |b_{ij}| \le 1, \quad \sum_{j=1}^{n} |b_{ji}| \le 1 \text{ for } i = 1, \dots, n \tag{42.16}$$

Theorem 42.10 *Let $\mathfrak{s}$ be a symmetric norm on $\mathbb{R}^n$. For $v \in \mathbb{C}^n$, define $|v| \in \mathbb{R}^n$ by $|v|_j = |v_j|$. Then if B is doubly substochastic, we have that*

$$\mathfrak{s}(|Bv|) \le \mathfrak{s}(|v|) \tag{42.17}$$

Remarks

1. One enlightening way to prove this theorem is to show that the extreme points in the set of real doubly substochastic matrices are precisely matrices that permute and sign flips coordinates, that is, the matrices associated to the natural representation of $\mathcal{G}_n$ (so matrices with exactly one non-zero element in each row and each column and each matrix element 0, 1, or -1). With this fact, the theorem follows from the concavity of s.
2. Since $\mathfrak{s}(|w|) = \mathfrak{s}(w)$, (42.17) is equivalent to $\mathfrak{s}(Bv) \le \mathfrak{s}(v)$.

Proof By the argument in the proof of Theorem 42.8, it suffices to prove (42.17) for $\mathfrak{s} = \mathfrak{s}_k$ given by $\mathfrak{s}_k(a) = a_1^* + \cdots + a_k^*$. Clearly,

$$\mathfrak{s}_k(|Bv|) \le \sum_{j=1}^{n} \beta_j |v_j| \tag{42.18}$$

where

$$\beta_j = \sum_{i=1}^{k} |b_{ij}| \tag{42.19}$$

Since $\sum_{i=1}^{n} |b_{ij}| \le 1$, we have that $0 \le \beta_j \le 1$. Moreover,

$$\sum_{j}^{n} \beta_j = \sum_{i=1}^{k} \sum_{j=1}^{n} |b_{ij}| \le k \tag{42.20}$$

by $\sum_{j=1}^{n} |b_{ij}| \le 1$. By Lemma 42.3, and (42.18), we have that (42.17) holds. □

Given a Hilbert space, let $\mathcal{B}$ be the set of all orthonormal sets $\{\varphi_n\}_{n=1}^{\infty}$ (not necessarily a basis).

Lemma 42.11 *Let $\varphi, \psi, \eta, \kappa \in \mathcal{B}$. Then*

$$b_{ij} = \langle \eta_i, \psi_j \rangle \langle \varphi_j, \kappa_i \rangle \tag{42.21}$$

is a doubly substochastic matrix.

Proof By the Schwarz and Bessel inequalities

$$\sum_{i=1}^{\infty} |b_{ij}| \le \left(\sum_{i=1}^{\infty} |\langle \eta_i, \psi_j \rangle|^2 \right)^{1/2} \left(\sum_{i=1}^{\infty} |\langle \varphi_j, \kappa_i \rangle|^2 \right)^{1/2}$$
$$\le \|\psi_j\| \|\varphi_j\| = 1$$

Similarly, $\sum_{j=1}^{\infty} |b_{ij}| \le \|\eta_i\| \|\kappa_i\| = 1$. □

Proposition 42.12 *Let $\mathfrak{s}$ be a symmetric sequence norm. Then for any finite rank operator, A,*

$$\mathfrak{s}(\mu_j(A)) = \sup_{\eta, \kappa \in \mathcal{B}} \mathfrak{s}(\langle \eta_i, A\kappa_i \rangle) \tag{42.22}$$

Proof Let

$$A = \sum_{j=1}^{N} \mu_j(A) \langle \varphi_j, \cdot \rangle \psi_j \tag{42.23}$$

be the canonical expansion of A. Then

$$\langle \psi_i, A\varphi_i \rangle = \mu_i(A) \tag{42.24}$$

so LHS of (42.22) $\le$ RHS of (42.22).

On the other hand,

$$\langle \eta_i, A\kappa_i \rangle = \sum_{j=1}^{N} b_{ij} \mu_j(A) \tag{42.25}$$

where $b_{ij} = \langle \eta_i, \psi_j \rangle \langle \varphi_j, \kappa_i \rangle$. By Lemma 42.11 and Theorem 42.10

$$\mathfrak{s}(\langle \eta_i, A\kappa_i \rangle) \le \mathfrak{s}(\mu_j(A)) \tag{42.26}$$

proving RHS of (42.22) $\le$ LHS of (42.22). □

Theorem 42.13 *Let $\mathfrak{s}$ be a symmetric sequence norm. For A finite rank, let*

$$|||A|||_{\mathfrak{s}} = \mathfrak{s}(\mu_j(A)) \tag{42.27}$$

Then $|||\cdot|||_{\mathfrak{s}}$ is a unitarily invariant norm.

Remark Clearly $\mathfrak{s}_{|||\cdot|||_{\mathfrak{s}}} = \mathfrak{s}$, so this shows that there is a one–one correspondence between symmetric sequence norms and unitarily invariant norms.

Proof Since $\mu_j(UAV) = \mu_j(A)$, $|||\cdot|||_{\mathfrak{s}}$ is unitarily invariant. That (42.5) defines a norm (i.e., that the triangle inequality holds) follows from (42.22). □

Notes and Historical Remarks The Ky Fan norms are named after his work on the subject [92–94]. The granddaddy of $\mathcal{I}_p$ spaces is the Hilbert–Schmidt space when $p = 2$. It appeared not in terms of trace but as integral operators on L^2 spaces with square integrable integral kernels (which is known to be equivalent to finite $\mathcal{I}_2$ norm [329, Theorem 3.8.5]). It is named after the work of Hilbert [160] and Schmidt [306] (Schmidt was Hilbert's student). For many years there were theorems on operators which were products of two Hilbert Schmidt until Schatten–von Neumann [303] introduced the trace class using the trace that von Neumann introduced in his work on the foundations of quantum mechanics [238]. The theory was developed extensively by Schatten [300–304] (the second and third with von Neumann). An earlier related work is Calkin [52] (the subject of [320, Chapter 2]). The standard monographs on the theory of trace ideals are Schatten [302], Gohberg–Krein [127], and Simon [320]. In particular, these books discuss the infinite dimensional case; essentially one takes the completion of the finite rank operators although for some norms, one should also consider the space of compact operators for which $\sup_k\{\mathfrak{s}_{|||\cdot|||}(\mu_1(A), \dots, \mu_k(A), 0, 0, \dots)\} < \infty$ which may be strictly bigger than the completion. There is a rich duality theory for these norms as described in those books.

Consideration of unitarily invariant norms on finite matrices goes back to von Neumann [239] and was turned into an art in the above quoted works on trace ideals. Proposition 42.12 and the uses we make of it goes back to Simon [317].

There is a huge literature on majorization. Theorem 42.6 is due to Markus [215] with the now standard proof due to Mitjagin [225] although it should be emphasized that the connection of majorization and convex hulls goes back to earlier work by Hardy–Littlewood–Polya [140, 141] (with even earlier work described in detail in the book of Marshall–Olkin [216], a love poem to majorization). Another monograph treatment of majorization is Simon [324, Chapter 15].

Chapter 43
Unitarily Invariant Norm Inequalities

In this chapter, we'll prove two beautiful results of Ando [16]:

Theorem 43.1 *Let $f \in \mathcal{M}_\infty(0, \infty)$ with $f \geq 0$. Then for all self-adjoint operators, $A, B \geq 0$, and any unitarily invariant norm, $|||\cdot|||$, one has that*

$$|||f(A) - f(B)||| \leq |||f(|A - B|)||| \tag{43.1}$$

Theorem 43.2 *Let g be a continuous, monotone, increasing bijection of $[0, \infty)$ to itself so that it compositional inverse, $f \in \mathcal{M}_\infty(0, \infty)$. Then for all self-adjoint operators, $A, B \geq 0$, and any unitarily invariant norm, $|||\cdot|||$, one has that*

$$|||g(|A - B|)||| \leq |||g(A) - g(B)||| \tag{43.2}$$

Remarks

1. We intend this and will prove it for the finite dimensional case but it can be shown easily to extend to various infinite dimensional cases. For example, if A, B are finite rank on a Hilbert space, or, for Theorem 43.1, for any f with $f'(0) < \infty$ and both $|||A|||$ and $|||B|||$ finite. And Theorem 43.2 for any g with $g'(0) > 0$ if A and B are trace class.
2. By Theorem 42.8, it suffices to prove this when $|||\cdot|||$ is a Ky Fan norm, $|||\cdot|||^{(k)}$, which is all we'll consider below.

Since $x \mapsto x^p$ lies in $\mathcal{M}_\infty(0, \infty)$ for $0 < p \leq 1$ as does $x \mapsto \log(x + 1)$ (see Theorem 4.1), we have that

Corollary 43.3 *For any $A, B \geq 0$ and any unitarily invariant norm $|||\cdot|||$, we have that*

$$\text{(a)} \quad \left|\left|\left|A^p - B^p\right|\right|\right| \leq \left|\left|\left||A - B|^p\right|\right|\right| \; \textit{for}\; 0 < p \leq 1 \tag{43.3}$$

$$\text{(b)} \quad |||\log(A + \mathbf{1}) - \log(B + \mathbf{1})||| \leq |||\log(|A - B| + \mathbf{1})||| \tag{43.4}$$

B. Simon, *Loewner's Theorem on Monotone Matrix Functions*, Grundlehren der mathematischen Wissenschaften 354, https://doi.org/10.1007/978-3-030-22422-6_43

Since $x \mapsto x^p$, $p > 1$, $x \mapsto e^x - 1$ and $x \mapsto x^p \log(x+1)$ are all compositional inverses of functions in $\mathcal{M}_\infty(0, \infty)$ (see Example 10.13), we have that

Corollary 43.4 *For any $A, B \geq 0$ and any unitarily invariant norm $|||\cdot|||$, we have that*

$$\text{(a)} \quad \left|\left|\left| e^{|A-B|} - \mathbf{1} \right|\right|\right| \leq \left|\left|\left| e^A - e^B \right|\right|\right| \tag{43.5}$$

$$\text{(b)} \quad |||\,|A-B|^p\,||| \leq |||\,|A^p - B^p|\,||| \text{ for } 1 \leq p \tag{43.6}$$

$$\text{(c)} \quad |||\,|A-B|^p \log(|A-B| + \mathbf{1})\,||| \leq |||A^p \log(A + \mathbf{1}) - B^p \log(B + \mathbf{1})||| \tag{43.7}$$

We'll derive Theorem 43.2 from Theorem 43.1 by applying (43.1) with A, B replaced by $g(A), g(B)$ and then using Corollary 42.7. We begin the proof of Theorem 43.1 with three lemmas. The first is the special case of the theorem when $f(x) = x/(x+1)$ and $0 \leq B \leq A$:

Lemma 43.5 *Let $0 \leq B, C$ be two self-adjoint operators. Then, for each k*

$$\left|\left|\left|(B+C)(B+C+\mathbf{1})^{-1} - B(B+\mathbf{1})^{-1}\right|\right|\right|^{(k)} \leq \left|\left|\left|C(C+\mathbf{1})^{-1}\right|\right|\right|^{(k)} \tag{43.8}$$

Proof Since

$$X(X+\mathbf{1})^{-1} = \mathbf{1} - (X+\mathbf{1})^{-1} \tag{43.9}$$

it suffices to prove that

$$\left|\left|\left|(B+\mathbf{1})^{-1} - (B+C+\mathbf{1})^{-1}\right|\right|\right|^{(k)} \leq \left|\left|\left|\mathbf{1} - (C+\mathbf{1})^{-1}\right|\right|\right|^{(k)} \tag{43.10}$$

We begin by noting that

$$(B+\mathbf{1})^{-1} - (B+C+\mathbf{1})^{-1} = \\ (B+\mathbf{1})^{-1/2}\{\mathbf{1} - (\mathbf{1} + (B+\mathbf{1})^{-1/2}C(B+\mathbf{1})^{-1/2})^{-1}\}(B+\mathbf{1})^{-1/2} \tag{43.11}$$

Since $\|(B+\mathbf{1})^{-1/2}\| \leq 1$, we can use (42.2) to see that

$$\text{LHS of (43.10)} \leq \left|\left|\left|\mathbf{1} - (\mathbf{1} + (B+\mathbf{1})^{-1/2}C(B+\mathbf{1})^{-1/2})^{-1}\right|\right|\right|^{(k)} \tag{43.12}$$

$$= \left|\left|\left|\mathbf{1} - (\mathbf{1} + C^{1/2}(B+\mathbf{1})^{-1}C^{1/2})^{-1}\right|\right|\right|^{(k)} \tag{43.13}$$

since XY and YX have the same non-zero eigenvalues including multiplicity (see, for example, [329, Problem 2.2.5]).

Moreover, $C^{1/2}(B+\mathbf{1})^{-1}C^{1/2} \leq C$, so, we have that

$$0 \leq \mathbf{1} - \mathbf{1} - C^{1/2}(B+\mathbf{1})^{-1}C^{1/2} \leq \mathbf{1} - (C+\mathbf{1})^{-1} \tag{43.14}$$

so, RHS of (43.13)≤RHS of (43.10) proving (43.10). □

Lemma 43.6 *If* $X_1, X_2, Y_1, Y_2 \geq 0$*,* $X_1X_2 = Y_1Y_2 = 0$*, and for* $j = 1, 2$ *and all* k*, we have that* $\left|\left|\left|X_j\right|\right|\right|^{(k)} \leq \left|\left|\left|Y_j\right|\right|\right|^{(k)}$*, then*

$$|||X_1 + X_2|||^{(k)} \leq |||Y_1 + Y_2|||^{(k)} \tag{43.15}$$

Proof $X_1X_2 = 0$ implies that $\ker(X_1)^{\perp}$ is orthogonal to $\ker(X_2)^{\perp}$ so the singular values of $X_1 + X_2$ is the union of the singular values of X_1 and X_2 reordered in decreasing order.

Fix $k \in \{1, 2, \dots\}$. Then there are k_1 and k_2 with $k_1 + k_2 = k$ so that

$$|||X_1 + X_2|||^{(k)} = |||X_1|||^{(k_1)} + |||X_2|||^{(k_2)} \tag{43.16}$$

$$\leq |||Y_1|||^{(k_1)} + |||Y_2|||^{(k_2)} \tag{43.17}$$

$$\leq |||Y_1 + Y_2|||^{(k)} \tag{43.18}$$

since the right side of (43.21) is the sum of k singular values (not necessarily the k smallest) of Y_1+Y_2. □

Lemma 43.7 *If* X *and* Y *are self-adjoint with* $X \leq Y$ *and if* $C^+ = \frac{1}{2}(|C| + C)$*, then for all* k

$$\left|\left|\left|X^+\right|\right|\right|^{(k)} \leq \left|\left|\left|Y^+\right|\right|\right|^{(k)} \tag{43.19}$$

Proof If $\lambda_1 \leq \cdots \leq \lambda_n$ are the eigenvalues of X and $\kappa_1 \leq \cdots \leq \kappa_n$ are the eigenvalues of Y, then $X \leq Y \Rightarrow \forall_j\, \lambda_j \leq \kappa_j$. Since $x \mapsto x^+$ is scalar monotone, $\lambda_j^+ \leq \kappa_j^+$ which implies that $\mu_j(X^+) \leq \mu_j(Y^+)$ for all j from which (43.19) is immediate. □

Proof of Theorem 43.1 Suppose first that $A \geq B \geq 0$ so that $C \equiv A - B \geq 0$. We can write f as (3.68):

$$f(x) = a + bx + \int_0^\infty \frac{x\lambda}{x+\lambda}\, d\mu(\lambda) \tag{43.20}$$

where $a \equiv f(0) \geq 0$ and $b \geq 0$.

Since f is monotone and positive, for any $X \geq 0$, we have that

$$\mu_j(f(X)) = f(\mu_j(X)) \tag{43.21}$$

Since $C = |A - B| = A - B \geq 0$, we have that

$$|||f(|A-B|)|||^{(k)} = \sum_{j=1}^{k} \mu_j(f(C)) = \sum_{j=1}^{k} f(\mu_j(C)) \tag{43.22}$$

$$= ak + b\sum_{j=1}^{k} \mu_j(C) + \int_0^\infty \sum_{j=1}^{k} \frac{\mu_j(C)t}{\mu_j(C)+t}\, d\mu(t) \tag{43.23}$$

$$\geq b|||C|||^{(k)} + \int_0^\infty t\left|\left|\left| C(C+t\mathbf{1})^{-1}\right|\right|\right|^{(k)} d\mu(t) \tag{43.24}$$

where to get (3.27), we used $a > 0$, and that since $x \mapsto x(x+1)^{-1}$ is positive and monotone, we have that $\mu_j(C(C+\mathbf{1})^{-1}) = \mu_j(C)/(\mu_j(C)+1)$.

By (43.20) and the triangle inequality for the norm $|||\cdot|||^{(k)}$, we see that

$$|||f(B+C) - f(B|||)^{(k)} \leq b|||C|||^{(k)} +$$
$$\int_0^\infty \left|\left|\left| (B+C)(B+C+t\mathbf{1})^{-1} - B(B+t\mathbf{1})^{-1}\right|\right|\right|^{(k)} d\mu(t) \tag{43.25}$$

so, by (43.24), to prove (43.1) for this case, we need only show that for all t

$$\left|\left|\left| (B+C)(B+C+t\mathbf{1})^{-1} - B(B+t\mathbf{1})^{-1}\right|\right|\right|^{(k)} \leq \left|\left|\left| C(C+t\mathbf{1})^{-1}\right|\right|\right|^{(k)} \tag{43.26}$$

For $t = 1$, this is (43.8). In that formula, if we replace B and C by $t^{-1}B$ and $t^{-1}C$ and multiply by t, we get (43.26). This proves (43.1) when $B \leq A$, or, by symmetry, when $A \leq B$.

Now consider the general case of $A, B \geq 0$. Clearly, $0 \leq A \leq B + (A-B)^+$ so, since f is monotone,

$$f(A) - f(B) \leq f(B + (A-B)^+) - f(B) \tag{43.27}$$

so by Lemma 43.7,

$$\begin{aligned} |||(f(A) - f(B))^+|||^{(k)} &\leq |||f(B + (A-B)^+) - f(B)|||^{(k)} \\ &\leq |||f((A-B)^+)|||^{(k)} \end{aligned} \tag{43.28}$$

by the special case of (43.1) we proved above when $A \leq B$.

Interchanging A and B, we get that

$$|||(f(B)-f(A))^+|||^{(k)} \leq |||f((B-A)^+)|||^{(k)} \tag{43.29}$$

Let $X_1 = (f(A) - f(B))^+, X_2 = (f(B)-f(A))^+, Y_1 = f((A-B)^+)$, $Y_2 = f((B-A)^+)$. Then $X_1X_2 = Y_1Y_2 = 0$ so we can apply Lemma 43.6. Noting that

$$X_1 + X_2 = |f(A) - f(B)| \tag{43.30}$$

$$Y_1 + Y_2 = f(|A-B|) \tag{43.31}$$

(43.15) is (43.1) for this general case. □

Proof of Theorem 43.2 Fix k and $A, B \geq 0$. Let f be the compositional inverse of g. Applying (43.1) with A replaced by $g(A)$ and B by $g(B)$ and $|||\cdot||| = |||\cdot|||^{(k)}$, we see that

$$\sum_{j=1}^{k} \mu_j(A-B) \leq \sum_{j=1}^{k} \mu_j(f(|g(A)-g(B)|) \tag{43.32}$$

$$\leq \sum_{j=1}^{k} f(\mu_j(g(A)-g(B))) \tag{43.33}$$

where we used (43.21) for scalar monotone functions to get (43.33). Since k is arbitrary, in terms of majorization defined by (42.9), we have that

$$\mu.(A-B) \prec f(\mu.(g(A)-g(B))) \tag{43.34}$$

Since $f \in \mathcal{M}_\infty(0,\infty)$, f is C^∞ with $f' > 0$ and $f'' < 0$ (see Corollary 14.5 or Theorem 10.2), so g is also C^∞. Differentiating $g(f(x)) = x$ twice, using the chain and Leibniz rules, we find that

$$g'(f(x))f'(x) = 1 \Rightarrow g''(f(x))f'(x)^2 + g'(f(x))f''(x) = 0 \tag{43.35}$$

so $g > 0$ and, since f is scalar concave, g is scalar convex.

Thus, we can apply Corollary 42.7 to (43.34) to get

$$g(\mu.(A-B)) \prec \mu.(g(A)-g(B)) \tag{43.36}$$

Using (43.21) again, we get

$$\mu.(g(|A - B|)) \prec \mu.(g(A) - g(B)) \tag{43.37}$$

which is (43.2). □

Notes and Historical Remarks The results in this chapter are due to Ando [16]. We follow his proofs fairly closely. As he remarks, he was motivated by a theorem (proven using results of Powers–Størmer [276] and Kittaneh [183]) of Bhatia [31] that $\left|\left|\left|(A - B)^{2^k}\right|\right|\right| \leq \left|\left|\left|A^{2^k} - B^{2^k}\right|\right|\right|$ and the question raised by Bhatia whether this extends to $|||\,|A - B|^p\,||| \leq |||A^p - B^p|||$ for all $p \geq 1$ (see Corollary 43.4).

Appendix A
Boutet de Monvel's Note

In 1989, Louis Boutet de Monvel wrote a small note in French—for both this and the English follow-up, I have corrected typos. I also note that elsewhere in this book, inner products are linear in the second factor but Boutet de Monvel uses the convention of linearity in the first factor. Here is the 1989 French paper.

Fonctions croissantes hilbertiennes

L. Boutet de Monvel, Analyse Algébrique
IMJ, Université Pierre et Marie Curie
4, place Jussieu F-75252 Paris Cedex 05

Soit f une fonction borélienne positive sur $\mathbb{R}^+$. Alors $f(a)$ est bien défini pour tout opérateur hermitien positif a sur un espace de Hilbert $\mathcal{H}$. On dit que f est H-croissante si $a \leq b$ implique $f(a) \leq f(b)$. Alors f est croissante (mais la réciproque est fausse, par exemple la fonction $f(x) = x^2$ n'est pas H-croissante). Dans les lemmes suivants on suppose f H-croissante.

Lemme 1 *Soit* $a = \begin{pmatrix} a_{11} & a_{12} \\ a_{21} & a_{22} \end{pmatrix}$ *une matrice par blocs: on a* $f(a)_{11} \leq f((1+\epsilon)a_{11})$ *pour* $\epsilon > 0$.

En effet on a $\begin{pmatrix} \epsilon a_{11} & -a_{12} \\ -a_{21} & \frac{1}{\epsilon} a_{22} \end{pmatrix} \geq 0$, d'où $a \leq b = \begin{pmatrix} (1+\epsilon)a_{11} & 0 \\ 0 & (1+\frac{1}{\epsilon})a_{22} \end{pmatrix}$ et $f(a) \leq f(b)$ implique $f(a)_{11} \leq f(b)_{11} = f((1+\epsilon)a_{11})$.

Pour une matrice 2×2: $u^* x u$ avec $u = \begin{pmatrix} \cos(\theta) & -\sin(\theta) \\ \sin(\theta) & \cos(\theta) \end{pmatrix}$,

B. Simon, *Loewner's Theorem on Monotone Matrix Functions*, Grundlehren der mathematischen Wissenschaften 354, https://doi.org/10.1007/978-3-030-22422-6

$x = \begin{pmatrix} x_0 & 0 \\ 0 & x_1 \end{pmatrix}$, $t = \sin^2\theta$ on obtient

$$(1-t)f(x_0) + tf(x_1) \leq f(y) \text{ si } y > (1-t)x_0 + tx_1$$

Corollaire 2 *f est concave.*
En particulier f est continue, et on peut remplacer ϵ par 0 dans le lemme 1.

Corollaire 3 *Si u est une application linéaire de norme ≤ 1 on a $u^* f(a)u \leq f(u^*au)$.*

En effet u^*au est le premier coefficient de U^*aU si $U = \begin{pmatrix} u \\ v \end{pmatrix}$ est une isométrie ($v = \sqrt{1 - u^*u}$).

Corollaire 4 *Si $u^*u \leq 1$ et $u^*au \leq a$ on a aussi $u^* f(a)u \leq f(a)$. Autrement dit f définit un foncteur d'interpolation hilbertien (si $A, B : \mathcal{H} \to \mathcal{H}^*$ sont des normes hilbertiennes, la norme interpolée est $F(A, B) = Af(A^{-1}B)$).*

Remarque 5 Si f est H-croissante, définie sur $\mathbb{R}$ tout entier, elle est linéaire (car f et $-f(-x)$ sont concaves).

Remarque 6 Dans tout ce qui précède on a supposé l'argument (opérateur a ou réel x) > 0; on peut toujours prolonger $f(x)$ à l'origine par n'importe quel nombre tel que $0 \leq f(0) \leq f(+0)$: ça reste H-croissant sur $[0, \infty[$.

La condition d'interpolation ci-dessus ($u^*u \leq 1$ et $u^*au \leq a$ impliquent $u^* f(a)u \leq f(a)$) a une formulation infinitésimale équivalente:

Lemme 7 *f est hilbertienne croissante ssi* $\text{Re}\, h \geq 0$ *et* $\text{Re}\, Ah \geq 0 \Rightarrow \text{Re}\, f(a)h \geq 0$.

Si f est dérivable la matrice de la dérivée de $a \to f(a)$ est $\frac{\text{ad} f(a)}{\text{ad}\, a}$, de matrice "par blocs" $\frac{f(x)-f(y)}{x-y}$ dans la décomposition spectrale de A (remplacer par $f'(x)$ pour $x = y$). Alors f est H-croissante ssi cette matrice est positive.

Si f, g sont H-croissantes il en est de même de $\lambda f + \mu g$ (λ, μ positifs) et de $\frac{fg}{f+g} = (\frac{1}{f} + \frac{1}{g})^{-1}$ (parce que $1/x$ est H-décroissante). En particulier les fonctions $1, x, \frac{x}{x+a}$ sont H-croissantes ($a > 0$); elles sont extrémales car pour ces fonctions la matrice $\frac{f(x)-f(y)}{x-y}$ est de rang ≤ 1.

Théorème 8 *L'ensemble des fonctions H-croissantes est un cône convexe à base compacte. Les génératrices extrémales sont les demi-droites engendrées par 1, x et les $\frac{x}{x+a}$, $a \geq 0$. En particulier toute fonction H-croissante f (continue pour $x = 0$) est de la forme $f(x) = \int_0^1 \frac{x}{(1-a)x+a} \, d\mu(a)$ où μ est une mesure positive; elle se prolonge holomorphiquement à $\mathbb{C} \setminus \mathbb{R}_-$.*

Preuve: il faut montrer que si μ est une mesure (ponctuelle, à support compact, ...) telle que $\int \frac{x}{x+a} d\mu \geq 0$, alors $\int f d\mu \geq 0$ si f est H-croissante. Il suffit de le montrer lorsque f est C^∞ et $f = O(\frac{x}{1+x})$ parce que dans le cas général f est somme d'une constante et d'une limite de telles fonctions.

La condition sur μ est $g = \mu * \frac{1}{1+x} \geq 0$ (le produit de convolution multiplicatif), de façon équivalente, après transformation de Fourier–Mellin: $\hat{\mu}(s) = \int x^s \, d\mu(x) = \sin(\pi s) F$ où $F = \frac{1}{\pi} \hat{g}$ est une fonction de type positif.

On peut reconstituer μ à partir de g (transformée de Mellin inverse ou formule de Cauchy):

$$\mu = \frac{dx}{x} \frac{1}{2i\pi} [g(e^{-i\pi} x) - g(e^{i\pi} x)]$$

Si μ est une mesure ponctuelle, $g = \mu * \frac{1}{1+x}$ est une fraction rationnelle

$$g(x) = \frac{P(x)}{\prod (x + a_k)}$$

où $P(x)$ est un polynôme positif pour $x \geq 0$ ($a_k > 0$). Un tel polynôme est combinaison à coefficients positifs de polynômes de la forme

$$x^j Q(x)^2$$

comme on voit par récurrence sur le degré de P: les facteurs irréductibles réels sont de la forme $(x+\alpha)$ ou $((x+\alpha)^2+\beta^2)$; chaque facteur du premier type ($(x+\alpha)$, $\alpha \geq 0$) ou $((x + \alpha)^2 + \beta^2)$ se décompose en somme; au bout de la récurrence on reste avec un polynôme de la forme $x^j Q^2 P'$ où P' est positif pour $x \geq 0$ et ses racines sont réelles positives, donc P' est un carré.

On est ainsi ramené au cas $g = \frac{x^j Q(x)^2}{\prod (x+a_k)}$. Posons alors

$$v = \sqrt{g} = \frac{x^{j/2} Q(x)}{\prod \sqrt{x + a_k}}$$

de sorte que v est toujours holomorphe dans le plan coupé $\mathbb{C} \setminus \mathbb{R}_-$.

Soient $\mathcal{H} = L^2(\mathbb{R}_+, \frac{dx}{x})$ et a est l'opérateur de multiplication par x. Parmi les opérateurs tels que $\operatorname{Re} h, \operatorname{Re} ah \geq 0$, distinguons l'opérateur de convolution H défini par

$$\widehat{Hu} = \tan(\pi s) \hat{u}$$

(H est positif comme voulu parce que $\operatorname{Re} \tan \pi s \geq 0$ pour $0 \leq \operatorname{Re} s \leq 1$). De façon heuristique, posons

$$u = \frac{1}{2} [v(e^{-i\pi} x) + v(e^{i\pi} x)] \qquad (\hat{u} = \cos(\pi s) \hat{v})$$

de sorte qu'on a

$$Hu = \frac{1}{2i}[v(e^{-i\pi}x) - v(e^{i\pi}x)] \qquad (\widehat{Hu} = \sin(\pi s)\hat{v})$$

On a alors $u.Hu = \mu$ (u et Hu sont réels), donc $\int d\mu f = \langle f(a)Hu|u\rangle \geq 0$.

La démonstration ci-dessus est abusive parce que u ainsi construit n'est pas de carré intégrable (et Hu n'est pas borné); pour rendre tout ceci correct il convient de tronquer et de régulariser; ceci ne présente pas de difficulté - le seul point délicat de la preuve était l'éxtraction de la racine carrée v:

Introduisons la contraction H_α de H: $\widehat{H_\alpha u} = \tan(\alpha s)\,\hat{u}\,(\alpha < \pi)$ et les "troncatures"

$$v_\epsilon = e^{-\epsilon(\log x)^2}v, \qquad g_\epsilon = e^{-2\epsilon(\log x)^2}g$$

Alors $u_{\alpha,\epsilon} = \frac{1}{2}[v_\epsilon(e^{-i\alpha}x) + v_\epsilon(e^{i\alpha}x)]$ est de carré intégrable (à décroissance rapide). On a $H_\alpha u_{\alpha,\epsilon} = \frac{1}{2i}[v_\epsilon(e^{-i\alpha}x) - v_\epsilon(e^{i\alpha}x)]$ de sorte qu'avec

$$\mu_{\alpha,\epsilon} = \frac{dx}{x}\frac{1}{2i\pi}[g_\epsilon(e^{-i\alpha}x) - g_\epsilon(e^{i\alpha}x)] = H_\alpha u_\alpha.u_\alpha$$

on a bien $\int d\mu_{\alpha,\epsilon} \geq 0$.

Pour $\epsilon \to 0$, $\mu_{\alpha,\epsilon}$ tend vers $\mu_\alpha = \frac{dx}{x}\frac{1}{2i\pi}[g(e^{-i\alpha}x) - g(e^{i\alpha}x)]$ et on obtient $\int d\mu_\alpha\, f \geq 0$: l'intégrale passe à la limite, cela ne pose pas de problème parce que $f = \mathrm{O}\left(\frac{x}{1+x}\right)$ et $\mu_{\alpha,\epsilon} = \mathrm{O}(\frac{dx}{x}\frac{1}{1+x})$ uniformément pour $\alpha < \pi$ fixé.

Le deuxième passage à la limite ne pose pas plus de problèmes: μ_α reste convenablement uniformément majorée à l'infini et près de 0; et à distance finie elle converge vers μ en tout cas au sens des distributions (en fait vaguement au sens des mesures: la masse sur tout compact reste bornée).

Par transformation homographique, on voit que les fonctions hilbertiennes croissantes (mod. les constantes) sur un intervalle I de $\mathbb{R}$ forment un cône convexe de base compacte, dont les points (génératrices) extrémaux sont les fonctions homographiques croissantes. Il est immédiat que si f est dérivable, c'est une fonction homographique ssi la matrice $\left(\frac{f(x)-f(y)}{x-y}\right)$ est de rang ≤ 1.

Y a–t–il une démonstration directe qu'une telle matrice $\left(\frac{f(x)-f(y)}{x-y}\right)$ positive de rang ≥ 2 est décomposable, ce qui redémontrerait le théorème de façon plus simple?

This completes the 1989 note. When he sent it to me in 2005, Boutet de Monvel added an "end remark" which read: "hilbertienne croissante" is an abominable terminology. "n-monotone" ($1 \leq n \leq \infty$) is better. The note only concerns positive monotone functions, which are related to interpolation. But it is obvious that any monotone function is a limit of (positive monotone + constant). For these (on the positive half line) the extremal functions (rays) are the constants and the functions $\frac{x-1}{(1-a)x+a}$, $\quad 0 \leq a \leq 1$.

After some email exchange, he sent a note in English which he described as "a revised and shortened version with more details". It is not quite self-contained referring to the earlier note on some points and to an early draft of this book on one point ("your monograph"). Here is this English note:

Interpolation Functions

L. Boutet de Monvel,
IMJ, Université Pierre et Marie Curie,
boutet@math.jussieu.fr

1. Ingredients

- Matrix-monotone functions on an interval I (your monograph).
- Interpolation functions. If E is a Hilbert space, A a positive Hermitian operator, we note $\|x\|_A^2 = \langle Ax|x\rangle$ and $\|u\|_A = \sup_{\|x\|_A \leq 1} \|ux\|_A$ the induced norm on $L(E)$. A positive function f on $\mathbb{R}_+$ is an interpolation function if $\|u\| \leq 1$ and $\|u\|_A \leq 1$ imply $\|u\|_{f(A)} \leq 1$. Equivalently

$$u^*u \leq 1 \text{ and } u^*Au \leq A \Rightarrow u^*f(A)u \leq f(A)$$

It is convenient to extend f to the closed half-line $x \geq 0$ by putting $f(0) = 0$ (another possibility is $f(0) = \lim_{x\to+0} f(x)$—which does not commute with limits).

If f and g are n-monotone or interpolation functions, then so are $f + g$, $f \circ g$, and $\frac{fg}{f+g} = (f^{-1} + g^{-1})^{-1}$ if $f, g \geq 0$. In particular an increasing homographic function $\frac{ax+b}{cx+d}$ with no pole in I is monotone; a homographic change of variable reduces the study of monotone functions on an interval $I \neq \mathbb{R}$ to the study on $\mathbb{R}_+$ or $(-1, 1)$.

Proposition 1 *Interpolation functions are the same as positive* ∞*-monotone functions on* $\mathbb{R}_+$.

Proof Let f be an interpolation function and $A \leq B$. Choosing

$$u = \begin{pmatrix} 0 & 1 \\ B^{-\frac{1}{2}}A^{\frac{1}{2}} & 0 \end{pmatrix} \text{ on } E \oplus E \text{ equipped with } \begin{pmatrix} A & 0 \\ 0 & B \end{pmatrix}$$

we get $f(A) \leq f(B)$ and $B^{-1}f(B) \leq A^{-1}f(A)$ so f is monotone, so is $xf(x^{-1})$.

Conversely if f is monotone and $\|u\| \leq 1$, we have $u^*f(A)u \leq f(u^*Au)$ (cf. first note), which implies that f is an interpolation function. □

The interpolation condition has an infinitesimal version:

Lemma 2 *f is an interpolation function iff for any positive A,* $\operatorname{Re} h \geq 0$ *and* $\operatorname{Re} Ah \geq 0$ *imply* $\operatorname{Re} f(A)h \geq 0$.

Indeed for any $A > 0$ the condition Re $Ah \geq 0$ is equivalent to $\|\frac{1-h}{1+h}\|_A \leq 1$ (Cayley transform).

If f is an analytic function (e.g., a polynomial), the derivative of the map $A \mapsto f(A)$ is $\frac{\operatorname{ad} f(A)}{\operatorname{ad} A}$ whose block matrix in the spectral decomposition of A is $(\frac{f(x)-f(y)}{x-y})$ (replace by $f'(x)$ for $x = y$). This is still true if f is smooth enough (e.g., if the Fourier transform of f' is integrable). Then f is monotone iff this matrix is Hermitian positive.

2. Extremal Points

Theorem 3 *Interpolation functions form a convex cone with compact basis whose extremal rays are the half-lines generated by* 1, x *and the* $\frac{x}{x+a}$ $(a > 0)$. *In particular, any interpolation function is of the form* $f(x) = \int_0^1 \frac{x\,d\mu(\alpha)}{(1-\alpha)x+\alpha}$, μ *a positive measure on* $[0, 1]$.

The fact that interpolation functions or n-monotone functions mod. constants (for $n \geq 2$) form a convex cone with compact basis is immediate (basis: interpolation functions such that $f(1) = 1$ (this implies $\frac{x}{x+1} \leq f \leq x$), or n-monotone functions such that $f(1) = 0, f'(1) = 1, n \geq 2$ (this implies $\frac{x-1}{x} \leq f \leq x$)). Increasing homographic functions are extremal, since for those the matrix $\left(\frac{f(x)-f(y)}{x-y}\right)$ is of rank ≤ 1.

The measure μ is unique (the maps $\alpha \mapsto \frac{x}{(1-\alpha)x+\alpha}$ span a dense subspace of $C([0, 1])$). f extends as a holomorphic function on $\mathbb{C} \setminus \mathbb{R}_-$ such that Im $f(x + iy)$ has the sign of y.

Proof of Theorem 3 We will use the Hilbert spaces $\mathcal{H}_0 = L^2(\mathbb{R}_+, \frac{dx}{x})$, with A the multiplication by x, $\mathcal{H}_1 = L^2(\mathbb{R}_+, dx)$. Mellin transformation $(\hat{f}(s) = \int x^s f(x) \frac{dx}{x})$ identifies $\mathcal{H}_0$, resp. $\mathcal{H}_1$ to $L^2(i\mathbb{R})$, resp. $L^2(\frac{1}{2} + i\mathbb{R})$, and $\mathcal{H}_0 \cap \mathcal{H}_1$ with the space of holomorphic functions on the band $0 \leq \operatorname{Re} s \leq \frac{1}{2}$ which are square integrable on the boundary ($\operatorname{Re} s = 0$ or 1).

Let H be the convolution operator such that $\widehat{Hf} = \tan(\pi s)\hat{f}$. We have $\operatorname{Re} \tan(\pi s) \geq 0$ on the band $0 \leq \operatorname{Re} s \leq \frac{1}{2}$, so Re $H \geq 0$, Re $AH \geq 0$ (in fact the closed convex cone of holomorphic functions such that Re $\phi \geq 0$ for $0 \leq \operatorname{Re} s \leq \frac{1}{2}$ is spanned by the $\tan(\pi s + i\sigma)^{\pm 1}$).

Thus for any interpolation function f we have

$$\operatorname{Re} \int_0^\infty f \bar{u} H u \frac{dx}{x} \geq 0 \quad \text{for any} \quad u \in \mathcal{H}_0 \cap \mathcal{H}_1 \tag{A.1}$$

(The fact that H is not bounded on $\mathcal{H}_1$ does not matter for the preceding statement, since one can take limits in an obvious manner (e.g., replacing $\tan(\pi s)$ by $\tan(1 - \epsilon)\pi s)$).

Suppose first $f = O(\frac{x}{1+x})$. Then $\hat{f}$ is holomorphic in the band $-1 \leq \operatorname{Re} s \leq 0$, and (A.1) can be rewritten (the integral being taken over any vertical line in the band $0 < \operatorname{Re} s < 1$):

$$\mathrm{Re}\int \hat{f}(-s)\bar{u}\widehat{Hu}(s) \geq 0 \tag{A.2}$$

We choose $\hat{u} = v \cos \pi s$ for a suitable holomorphic function $v(s)$, so that $\widehat{Hu} = v \sin \pi s$. Then the Mellin transform of $\bar{u}Hu$ is $\frac{1}{4i}(e^{i\pi s}v^* + e^{-i\pi s}v^*) * (e^{i\pi s}v - e^{-i\pi s}v)$, and its real part is $\frac{1}{2}\sin \pi s(v * v^*)$ (with $v^*(s) = \bar{v}(\bar{s})$, $v * v^*(s) = \int v(s-t)\bar{v}(\bar{t})$, integral over some vertical line).

The condition on v is that $v \cos \pi s$ (so also $v \sin \pi s$) is a Mellin transform from $\mathcal{H}_0 \cap \mathcal{H}_1$. There are sufficiently many such functions, e.g. $v = e^{s^2}P(s)$ with P a polynomial, so that we can conclude that $\hat{f}(-s)\sin \pi s$ is of positive type (on any vertical line)—equivalently: for any family (z_j) of points in the band, the matrix $(f(z_j + \bar{z}_k))$ is Hermitian positive.

It follows that $\hat{f}(-s)\sin \pi s$ is the Mellin transform of a positive measure ËIJṽ ($\int x^s\, d$ËIJṽ $< \infty$ if $0 < \mathrm{Re}\, s < 1$); symmetrically, $\hat{f}(s)(-\sin \pi s)$ is the Mellin transform of a measure ν ($\int x^s\, d\nu < \infty$ if $-1 < \mathrm{Re}\, s < 0$).

The Mellin transform of $\frac{x}{x+1}$ is $-\frac{\pi}{\sin \pi s}$ (for $-1 < \mathrm{Re}\, s < 0$), so we get finally

$$f = \frac{1}{\pi}\nu * \frac{x}{1+x} = \frac{1}{\pi}\int d\nu(a)\frac{x}{x+a} \tag{A.3}$$

This can be rewritten $f(x) = \int_0^1 \frac{x\, d\mu(a)}{(1-\alpha)x+\alpha}$ $(a = \frac{\alpha}{1-\alpha}, \mu = \frac{1}{\pi}(1-\alpha)\nu)$.

For a general interpolation function f we have $f = \lim f_n$ with $f_n = O(\frac{x}{1+x})$ (e.g., $f_n = \frac{f g_n}{f+g_n}$ with $g_n = \frac{nx}{x+1}(g_n \to \infty)$). The corresponding measures μ_n are bounded ($\int_0^1 d\mu_n = f_n(1) \to f(1)$), so we can extract a subsequence weakly converging to a limit $\mu \geq 0$: in the limit, we get $f(x) = \int_0^1 \frac{x\, d\mu(\alpha)}{(1-\alpha)x+\alpha}$ for $x > 0$.

(The $f_\alpha = \frac{x}{(1-\alpha)x+\alpha}$ $(f_0 = 1)$ form a compact set of interpolation functions and are extremal, so they are the only extremal points.) □

Appendix B
Pictures

As we explained in the Preface, this book is a love poem to the subject of Loewner's theorem and like many mathematical love poems, we want to include pictures of some of the main players:

(1) Charles Loewner (1893–1968) founded the subject of this book with his 1934 paper written while he was a professor at Charles University of Prague where he'd been a student of Pick. He spent the last years of his career at Stanford. The picture, taken in 1963, was placed in the public domain by his daughter.
(2) Tsuyoshi Ando (1932–) is a Japanese mathematician with numerous contributions to the analytic side of matrix theory including many of the applications of Loewner theory. He was a student of Nakano and spent most of his career in the Division of Applied Mathematics in the Research Institute of Applied Electricity of Hokkaido University. The picture is used by permission of Prof. Ando and the RIAE of Hokkaido University.
(3) Seymour Sherman (1917–1977) was the senior author of the Bendat–Sherman paper written while he was a professor at the University of Southern California. He spent the end of his career at the University of Indiana. The picture is from the Oberwolfach Institute collection used under a creative commons license.
(4) William Donoghue (1921–2002) was a functional analyst who wrote his own book on Loewner's theorem and many papers on the subject. He got his PhD under the supervision of Eberlein at Wisconsin and spent the end of career at UC Irvine where he wrote the aforementioned book. He was a high school classmate of my thesis advisor, Arthur Wightman. The picture, from his retirement party, was supplied to me by Svetlana Jitomirskya and is used with her permission.
(5) Eugene Wigner (1902–1995) was a Nobel prize winning physicist who was clearly the principal coauthor of the Wigner–von Neumann paper on Loewner's theorem. He was born in Hungary. From 1930 onwards, Wigner was at Princeton where he was eventually a professor of mathematics and physics. The picture has been placed in the public domain by the Nobel Foundation.

B. Simon, *Loewner's Theorem on Monotone Matrix Functions*, Grundlehren der mathematischen Wissenschaften 354, https://doi.org/10.1007/978-3-030-22422-6

(6) Gert Pedersen (1940–2004) was a Danish mathematician. He wrote several seminal joint works with his student Frank Hansen which are connected to Loewner theory. He spent most of his career at the University of Copenhagen. The photograph was taken by Grethe Lissner in 2000 in connection with Pedersen's sixtieth birthday. It is owned by a group of his friends who have given permission for its use and supplied to me by the editors of Math. Scand., who used it for a memorial issue.

(7) Louis Boutet de Monvel (1941–2014) was a distinguished French mathematician whose unpublished manuscript is included as an Appendix to this book. He was a student of Laurent Schwartz, a member of Bourbaki, and spent most of his career at Pierre and Marie Curie University, aka Paris VI. The picture was provided by his wife Anne, herself a talented mathematical physicist and analyst. The picture is used with her permission.

(8) Otte Heinävaara (1994–) is a young Finnish mathematician who did the work presented in this book while an undergraduate and master's student at the University of Helsinki. He should be beginning graduate school in the United States at about the time this book is published. The picture is supplied to me by Otte and used with his permission.

Charles Loewner

Tsuyoshi Ando

Seymour Sherman

William Donoghue

Eugene Wigner

Gert Pedersen

Louis Boutet de Monvel

Otte Heinävaara

Appendix C
Symbol List

$\langle \cdot, \cdot \rangle$	Hilbert space inner product, linear in second factor and antilinear in the first; see pg. 3
$\langle \cdot, \cdot \rangle_L$	Loewner matrix induced inner product; see (27.1), pg. 283
$\vert\vert\vert \cdot \vert\vert\vert$	General unitary invariant norm; see (42.1), pg. 399
$\vert\vert\vert \cdot \vert\vert\vert^{(k)}$	Ky Fan norm; see (42.6); pg. 400
$\vert\vert\vert \cdot \vert\vert\vert_{\mathfrak{s}}$	Operator norm associated to a symmetric sequence norm; see (42.27), pg. 406
$A \leq B$	Order for matrices; see (1.1) and (1.3), pg. 3
$a \prec b$,	b majorizes a; see (42.9), pg. 402
$[x_1, \ldots, x_\ell; f]$	Divided differences; see (5.40), pg. 52
$f \wedge g$	Multiplicative convolution; see (34.2), pg. 346
$f_y^{(1)}(x)$	Shorthand for $[x, y; f]$; see (9.3), pg. 96
$f^{[N,M]}$	Padé approximant; see (21.2), pg. 221
$\mathcal{H}_1 \otimes \mathcal{H}_2$	Tensor product of Hilbert spaces; see [329, Section 1.3]
$A : B$	Parallel sum of two matrices, i.e. $(A^{-1} + B^{-1})^{-1}$; see (36.2), pg. 374
$A!B$	Harmonic mean of two matrices, i.e. $2(A^{-1} + B^{-1})^{-1}$; see (36.2), pg. 374
$A\#B$	Geometric mean of two matrices, i.e. $A^{1/2}\left[A^{-1/2}BA^{-1/2}\right]^{1/2}A^{1/2}$; see (36.9), pg. 376
$A\nabla B$	Arithmetic mean of two matrices, i.e. $\frac{1}{2}(A + B)$; see (36.14), pg. 378
$A\#_\alpha B$	Power mean of two matrices, $A^{1/2}(A^{-1/2}BA^{-1/2})^\alpha A^{1/2}$; see pg 383
$A\sigma B$	Generic symbol for general operator mean of two matrices; see (36.15), pg. 378

B. Simon, *Loewner's Theorem on Monotone Matrix Functions*, Grundlehren der mathematischen Wissenschaften 354, https://doi.org/10.1007/978-3-030-22422-6

$X \odot Y$	Schur (or Hadamard) product given by $(X \odot Y)_{mk} = X_{mk}Y_{mk}$; see (5.4), pg. 44
M/Y	Schur complement of a matrix Y in a block matrix M; see (13.6), pg. 136
$f \mapsto f_n^\#$	Fourier series coefficients of $f \in L^2(\partial\mathbb{D})$; see pg. 190
$f \mapsto f^\#$	Function transform used in the study of matrix convexity; see (10.20), pg. 112
$N \mapsto N^\#$	Matrix transform used in the Krein variational principle; see (13.13), pg. 138
$f \mapsto f^\flat$	Function transform used in the study of matrix convexity; see (10.20), pg. 112
$f \mapsto f^*$	Function transform used in the study of matrix convexity; see (10.20), pg. 112
$f \mapsto \hat{f}$	Fourier transform of $f \in L^2(\mathbb{R}, dx)$; see (19.19), pg. 194
$f \mapsto \tilde{f}$	Mellin transform of $f \in L^2(\mathbb{R}_+, \frac{dx}{x})$; see (34.6), pg. 347
$A \mapsto f(A)$	Functional calculus for self-adjoint matrices; (1.5) or (1.7), pg 4
$A \mapsto A_P$	Restriction of a matrix to $\operatorname{Ran} P$; see (10.5), pg. 109
$A \mapsto \lvert A\rvert$	Absolute value of a matrix; see [329, Sections 2.4 and 3.5]
$\sigma \mapsto \sigma'$	Transpose of an operator mean; see pg. 383
$A_n(x_1, \dots, x_n; f)$	Multipoint Loewner matrix; see (5.72), pg. 59
$b(z, w)$	Blaschke factor; see (19.10), pg. 192
$B(z, \{z_j\}_{j=1}^\infty)$	Blaschke product; see (19.12), pg. 192
$B_n(x; f)$	Dobsch matrix; see (5.38), pg. 51
$B_m(x)$	Bernstein polynomials; see (5.8), pg. 45
$B(\mu)$	Barycenter of a measure μ; see (28.2), pg. 290
$\mathcal{B}_n$	Functions on $(0, \infty)$ whose $n \times n$ Loewner matrices are all conditionally negative definite; see pg. 124
$\mathcal{B}_\infty$	Functions on $(0, \infty)$ whose Loewner matrices are all conditionally negative definite, i.e. $\cap_n \mathcal{B}_n$; see pg. 124
$C(t, X)$	Heinävaara matrix, which turns out to be a change of basis; see (6.3), pg. 75
$\mathbb{C}$	The complex numbers
$\mathbb{C}_+$	The upper half plane = $\{z \in \mathbb{C} \mid \operatorname{Im}(z) > 0\}$
$\widehat{\mathbb{C}}$	Riemann sphere; see [326, Sections 7.1, 7.2 and 10.1]
$\mathcal{C}_n(a, b)$	Functions concave on $n \times n$ matrices with eigenvalues in (a, b); see (9.2), pg. 95

$\mathcal{C}_\infty(a,b)$	Matrix concave functions on (a,b), i.e. $\cap_n \mathcal{C}_n(a,b)$; see (9.2), pg. 95
Δ_{n-1}	$n-1$ dimensional standard simplex $=\{\mathbf{s}\in\mathbb{R}^n_+ \mid \sum_{j=1}^n s_j = 1\}$; see pg. 53
dE_λ	Spectral resolution; see [329, Chapter 6]
$d_I(A)$	Principal determinants; see pg. 49
$\mathbb{D}$	The open unit disk = $\{z\in\mathbb{C} \mid \lvert z\rvert < 1\}$
$\mathbb{E}_{m,x}$	Binomial distribution expectation; see (5.9), pg. 45
$\mathcal{E}(S)$	The set of extreme points of a convex set, S; see pg. 291
$\mathcal{F}$	The set of Sparr functions, see (29.5), pg. 301
$\gamma(T)$	The convex hull of the support of a distribution, T; see pg 210
$\mathcal{G}_n$	Group of permutations and sign flips on $\mathbb{R}^n$; see pg 403
h_n	Heine determinant of a set of moments; see (21.16), pg. 224
H_f	Hankel operator with symbol f; see pg. 195
$H(\{a_n\}_{n=0}^\infty)$	Hankel matrix; see pg. 195
$H_n(x;f)$	Hansen–Tomiyama matrix; (5.71), pg. 59
$H^p(\mathbb{D})$	Hardy spaces on $\mathbb{D}$, especially for $p=2,1,\infty$; see pgs. 190–193
$H^p(\mathbb{C}_+)$	Hardy spaces on $\mathbb{C}_+$, especially for $p=2,\infty$; see pgs. 193–195
$\mathcal{H}(-1,1)$	Functions analytic and real on $(-1,1)$ with Herglotz continuations to $\mathbb{C}_+$; see pg. 25
$\iota(f)$	$\inf \operatorname{supp}(f)$; see pg. 207
I_ν	Interval of convergence for the Mellin transform of ν; see pg. 347
$I_X(t)$	Heinävaara kernel (which turns out to be a Peano kernel); see (6.3), pg. 75
$I_\nu(z)$	Singular inner function; see (19.15), pg. 193
$\mathcal{I}_n$	Interpolation functions of order n; see pg. 154
$\mathcal{I}_\infty$	Interpolation functions of infinite order, i.e. $\cap_n \mathcal{I}_n$; see pg. 154
$J_{n,F}$	Truncated Jacobi matrix; see (21.35), pg. 228
$J_{x_0,X}(t)$	Heinävaara kernel for convexity (which turns out to be a Peano kernel); see (6.49), pg. 83
$K_n(x_0;x_1,\dots,x_n;f)$	Kraus matrix; see (5.69), pg. 58
$K_n^{(f)}(t_1,\dots,t_n)$	Kwong matrix; see pg. 134
$K(e^{i\theta},z)$	Complex Poisson kernel, $K(w,z)=(w+z)/(w-z)$; see (3.2), pg. 17
$K_2(t,\varphi)$	Peetre quadratic functional, see (15.11), pg. 156
$K_1(t,\varphi)$	Peetre K-functional, pg. 159

K_Θ	Where Θ is an inner function, orthogonal complement of ΘH^2 in H^2; see pg. 198
$K_{n-1}(x, x)$	Christoffel–Darboux kernel; see (21.40), pg. 229
$L_n(x_1, \ldots, x_n)$	Loewner matrix, written as $L_n^{(f)}$ when function needs to be made explicit; see (5.1), pg. 43
$L_n^{(e)}(\lambda_1, \ldots, \lambda_n; \mu_1, \ldots, \mu_n)$	Extended Loewner matrix; see (5.100), pg. 66
$\mathcal{L}(\mathcal{H})$	The space of bounded operators on a Hilbert space $\mathcal{H}$; see [325, Section 3.3]
$\mu_j(A)$	Singular values of a matrix; see [320]
$m(z)$	Stieltjes transform of a measure, aka m-function; see (21.5), pg. 222
M_{jk}	Pick matrix for Pick's problem for Herglotz functions; see (16.10), pg. 166
$M_{jk}^{(C)}$	Pick matrix for Pick's problem for Carathéodory functions; see (16.17), pg. 168
$M_{jk}^{(S)}$	Pick matrix for Pick's problem for Schur functions; see (16.19), pg. 168
$\mathcal{M}_n(a, b)$	Functions monotone on $n \times n$ matrices with eigenvalues in (a, b); see (1.9), pg. 5
$\mathcal{M}_\infty(a, b)$	Matrix monotone functions on (a, b), i.e. $\cap_n \mathcal{M}_n(a, b)$; see (1.9), pg. 5
$\widetilde{\mathcal{M}}_\infty(-1, 1)$	Functions C^∞ in a neighborhood of $[-1, 1]$ whose Loewner matrices are all strictly positive; see pg. 215
$\mathcal{M}_\infty(\Omega)$	Matrix monotone functions on a general open set Ω; see pg. 359
$\mathcal{M}_\infty^{(r)}(\Omega)$	Matrix monotone functions in restricted sense on a general open set Ω; see pg. 359
$N_j(x, \{x_k\}_{k=1}^j)$	Newton polynomials; see (22.10), pg. 237
$P_r(\theta, \varphi)$	Poisson kernel for $\mathbb{D}$, $P_r(\theta, \varphi) = (1 - r^2)/(1 + r^2 - 2r\cos(\theta - \varphi))$; see (3.9), pg. 18
$P(x; x_1, \ldots, x_n; f)$	Lagrange interpolation polynomial to f; see (5.63), pg. 57
$\mathcal{P}_n$	Real polynomials of degree at most $n + 1$; see pg. 90
$\mathcal{P}$	Strictly positive functions in $\mathcal{M}_\infty(0, \infty)$; see (28.4), pg. 292
$\mathcal{P}_1$	Those $f \in \mathcal{P}$ with $f(1) = 1$; see (28.5), pg. 292
$\mathbb{Q}$	The rational numbers
$\mathbb{R}$	The real numbers
$\sigma(X, Y)$	Y-weak topology on X; see [325, Section 5.7]
$s_j(z)$	Szegő functions; see (19.45), pg. 202
$S(p)$	Classical entropy; see (38.2), pg. 385
$S(\rho)$	Quantum entropy; see (38.5), pg. 386

$S(A\|B)$	Relative entropy; see (38.19), pg. 389
S_ν	Fundamental strip for the Mellin transform of ν; see pg. 347
$\mathfrak{s}_{\|\|\|\cdot\|\|\|}(a_1, \dots, a_n, 0, \dots)$	Sequence norm generated by a unitarily invariant norm; see (42.3), pg. 399
$\mathcal{S}_\infty(a, b)$	Strongly operator convex functions on (a, b), i.e. which obey (13.2); see pg. 135
T_f	Toeplitz operator with symbol f; see pg. 195
T_f^V	Toeplitz operator with symbol f for a subspace $V \subset H^2(\mathbb{D})$; see pg. 195
T_f^Θ	Where Θ is an inner function, Toeplitz operator with symbol f for a subspace $V = \Theta H^2(\mathbb{D})$; see pg. 198
$T(\{a_n\}_{n=0}^\infty)$	Toeplitz matrix; see pg. 195
$T_{z_0}(z)$	Basic Möbius maps of $\mathbb{D}$; see (16.28), pg. 169
$\mathrm{Tr}(\cdot)$	Trace of a trace class operator; see [329, Section 3.6]
$\mathrm{Tr}^{J\setminus K}$	Partial trace; see (38.10), pg. 387
V	Volterra operator; see (19.57), pg. 207
$V_n(x_1, \dots, x_n)$	Vandermonde determinant; see (22.1), pg. 235
$W(z)$	Wronskian of two solutions of a multipoint rational interpolation problems; see (23.26), pg. 255
$\mathbb{Z}$	The integers

Bibliography

1. V. Adamjan, D. Arov, M.G. Krein, *Infinite Hankel matrices and generalized Carathéodory–Fejer and Riesz problems*, Functional Analysis and its Applications, **2(1)** (1968), 1–18, 1968, Russian original: Funkcional. Anal. i Prilozen. **2** (1968), 1–19.
2. J. Agler and J. E. McCarthy, *Pick Interpolation and Hilbert Function Spaces*, Graduate Studies in Math., **44**, American Mathematical Society, Providence, RI, 2002.
3. J. Agler, J. E. McCarthy and N. J. Young, *Operator monotone functions and Löwner functions of several variables*, Ann. of Math. **176** (2012), 1783–1826; Corrigendum: Ann. of Math. **180** (2014), 403–405.
4. S. Agmon, *Sur une problème de translations*, C. R. Acad. Sci. Paris 229 **11** (1949) 540–542.
5. N. I. Akhiezer, *The Classical Moment Problem and Some Related Questions in Analysis*, Hafner, New York, 1965; Russian original, 1961.
6. E. Y. Ameur, *Interpolation of Hilbert spaces*. Thesis (Ph.D.) Uppsala Universitet, Sweden, 2002. Online at www.maths.lth.se/matematiklu/personal/yacin/avh.pdf
7. E. Y. Ameur, *The Calderón problem for Hilbert couples*, Ark. Mat. **41** (2003), 203–231.
8. E. Y. Ameur, *A new proof of Donoghue's interpolation theorem*, J. Funct. Spaces Appl. **2** (2004),253–265.
9. E. Y. Ameur, *Interpolation Between Hilbert Spaces*, to appear in *Analysis of Operators on Function Spaces, The Serguei Shimorin Memorial Volume*, edited by A. Aleman, H. Hedenmalm, D. Khavinson and M. Putinar, Birkhäuser, 2019.
10. E. Y. Ameur, S. Kaijser, and S. Silvestrov, *Interpolation classes and matrix monotone functions*, J. Operator Theory **57** (2007), 409–427.
11. W. N. Anderson and R. J. Duffin, *Series and parallel addition of matrices*, J. Math. Anal. Appl., **26** (1969) 576–594.
12. W. N. Anderson and G. E. Trapp, *Matrix operations induced by electrical network connections–a survey*, Constructive Approaches to Mathematical Models (1979), pp. 53–73, Academic Press.
13. T. Ando, *On a pair of commutative contractions*, Acta Sci. Math. (Szeged) **24** (1963), 88–90.
14. T. Ando, *Topics on operator inequalities*, Lecture Notes, Sapporo, 1978.
15. T. Ando, *Concavity of certain maps on positive definite matrices and applications to Hadamard products*, Linear Algebra Appl. **26** (1979), 203–241.
16. T. Ando, *Comparison of norms* $|||f(A) - f(B)|||$ *and* $|||f(|A - B|)|||$, Math. Z. **197** (1988), 403–409.
17. Araki, H. *Relative Entropy of States of von Neumann Algebras*, Publ. RIMS **11** (1976), 809–833.

B. Simon, *Loewner's Theorem on Monotone Matrix Functions*, Grundlehren der mathematischen Wissenschaften 354, https://doi.org/10.1007/978-3-030-22422-6

18. D. H. Armitage, *The Riesz-Herglotz representation for positive harmonic functions via Choquet's theorem*, Potential Theory—ICPT 94 (Kouty, 1994), pp. 229–232, de Gruyter, Berlin, 1996.
19. N. Aronszajn, *On a problem of Weyl in the theory of singular Sturm–Liouville equations*, Amer. J. Math. **79** (1957), 597–610.
20. J. S. Aujla, *A simple proof of Lieb concavity theorem*, J. Math. Phys. **52** (2011), 043505.
21. T. Banachiewicz, *Zur Berechnung der Determinanten, wie auch der Inversen, und zur darauf basierten Auflösung der Systeme linearer Gleichungen*, Acta Astronomica, Series C, **3** (1937), 41–67.
22. J. Bendat and S. Sherman, *Monotone and convex operator functions*, Trans. A.M.S. **79** (1955), 58–71.
23. C. Bennett and R. Sharpley, *Interpolation of Operators*, Academic Press, Inc., Boston, MA, 1988.
24. J. Bergh and J. Löfström, *Interpolation spaces. An introduction*, Springer-Verlag, Berlin-New York, 1976.
25. S. N. Bernstein, *Démonstration du théorème de Weierstrass fondée sur le calcul des probabilités*, Comm. Soc. Math. Kharkow **13** (1912/1913), 1–2.
26. S. N. Bernstein, *Leçons sur les propriétes extrémales et la meilleure approximation des fonctions analytiques d'une variable réelle*, pp. 196–197, Paris, 1926.
27. S. N. Bernstein, *Sur les fonctions absolument monotones*, Acta Math. **52** (1928), 1–66.
28. L. Bers, *The action of the universal modular group on certain boundary points*, Contributions to Algebra (collection of papers dedicated to Ellis Kolchin), pp. 25–36, Academic Press, New York, 1977.
29. J. Bertrand, P. Bertrand, and J. Ovarlez, *The Mellin Transform*, chapter in *The Transforms and Applications Handbook*: Second Edition; Ed. Alexander D. Poularikas CRC Press LLC, Boca Raton, 2000.
30. A. Beurling, *On two problems concerning linear transformations in Hilbert space*, Acta Math. **81** (1948) 239–255.
31. R. Bhatia, *Some inequalities for norm ideals*, Comm. Math. Phys. **111** (1987), 33–39.
32. R. Bhatia, *First and second order perturbation bounds for the operator absolute value*, Linear Algebra Appl. **208/209** (1994), 367–376.
33. R. Bhatia, *Matrix Analysis*, Graduate Texts in Mathematics, 169, Springer-Verlag, New York, 1997.
34. R. Bhatia, *Positive definite matrices*, Princeton Series in Applied Mathematics. Princeton University Press, Princeton, NJ, 2007.
35. R. Bhatia, S. Friedland and T. Jain, *Inertia of Loewner matrices*, Indiana Univ. Math. J. **65** (2016), 1251–1261.
36. R. Bhatia and J. A. Holbrook, *Fréchet derivatives of the power function*, Indiana Univ. Math. J. **49** (2000), 1155–1173.
37. R. Bhatia, and T. Sano, *Loewner matrices and operator convexity*, Math. Ann. **344** (2009), 703–716.
38. R. Bhatia, and T. Sano, *Positivity and conditional positivity of Loewner matrices*, Positivity **14** (2010), 421–430.
39. M. Sh. Birman and M. Solomyak, *Double Stieltjes operator integrals, I–III*, Probl. Math. Phys. **1** (1966), 33–67; **2** (1967), 26–60; **6** (1973), 27–53; English translations of I and II: Topics in Mathematical Physics, **1** (1967), 25–54; **2** (1968), 19–46.
40. M. Sh. Birman and M. Solomyak, *Remarks on the spectral shift function*, Zap. Nauchn. Sem. Leningrad. Otdel. Mat. Inst. Steklov **27** (1972), 33–46; English translation in J. Soviet Math. **3** (1975), 408–419.
41. M. Sh. Birman and M. Solomyak, *Double operator integrals in a Hilbert space*, Integral Equations Operator Theory **47** (2003), 131–168.
42. R. P. Boas, *Functions with positive derivatives*, Duke Math. J. **24** (1941), 163–172.

43. L. Boltzmann *Lectures on gas theory*, Translated by Stephen G. Brush University of California Press, Berkeley-Los Angeles, Calif. 1964 German original: *Vorlesungen über Gastheorie*, J. A. Barth, Leipzig, Part I: 1896; Part II: 1898.
44. A. Böttcher and B. Silbermann, *Analysis of Toeplitz operators*, Second edition. Springer-Verlag, Berlin, 2006. First edition, 1990.
45. L. Boutet de Monvel, unpublished note, 1989; included as an appendix to this book.
46. C. Brezinski, *History of Continued Fractions and Padé Approximants*, Springer Series in Computational Mathematics, 12, Springer-Verlag, Berlin, 1991.
47. M. S. Brodski, *On a problem of I.M. Gel'fand*, Uspekhi Mat. Nauk **12** (1957) 129–132.
48. L. G. Brown, *Semicontinuity and multipliers of C^*–algebras*, Canad. J. Math. **40** (1988), 865–988.
49. L. G. Brown, *A treatment of strongly operator convex functions that does not require any knowledge of operator algebras*, Ann. Funct. Anal. **9** (2018), 41–55.
50. L. G. Brown and M. Uchiyama, *Some Results on Strongly Operator Convex Functions and Operator Monotone Functions*, Linear Alg. Appl. **553** (2018) 238–251.
51. F. Burstall, F. Pedit, and U. Pinkall, *Schwarzian derivatives and flows of surfaces*, Differential Geometry and Integrable Systems (Tokyo, 2000), pp. 39–61, Contemp. Math., **308**, American Mathematical Society, Providence, RI, 2002.
52. J. W. Calkin, *Two–sided ideals and congruences in the ring of bounded operators in Hilbert space*, Ann. Math **42** (1941), 839–873.
53. C. Carathéodory, *Über den Variabilitätsbereich der Koeffizienten von Potenzreihen die gegebene Werte nicht annehmen*, Math. Ann. **64** (1907), 95–115.
54. C. Carathéodory, *Über den Variabilitätsbereich der Fourier'schen Konstanten von positiven harmonischen Funktionen*, Rend. Circ. Mat. Palermo **32** (1911), 193–217.
55. C. Carathéodory and L. Fejér, *Über den Zusammenhang der Extremen von harmonischen Funktionen mit ihren Koeffizienten und über den Picard-Landau'schen Satz*, Rend. Circ. Mat. Palermo **32** (1911), 218–239.
56. E. A. Carlen and E. H. Lieb, *Some trace inequalities for exponential and logarithmic functions*, Bull. Math. Sci, to appear, arXiv:1709.05450.
57. N. N. Chan and M. K. Kwong, *Hermitian matrix inequalities and a conjecture*, Amer. Math. Monthly, **92** (1985), 533–541.
58. J. D. Chandler, *Extensions of monotone operator functions*, Proc. A.M.S. **54** (1976), 221–224.
59. M. Choi, *A Schwarz inequality for positive linear maps on C^*–algebras*, Illinois J. Math. **18** (1974), 565–574.
60. G. Choquet, *Theory of capacities*, Ann. Inst. Fourier **5** (1955), 131–295.
61. G. Choquet, *Lectures on Analysis, Volume I: Integration and Topological Vector Spaces; Volume II: Representation Theory; Volume III: Infinite Dimensional Measures and Problem Solutions*, W.A. Benjamin, New York-Amsterdam, 1969.
62. M. Chuaqui, P. Duren, and B. Osgood, *Curvature properties of planar harmonic mappings*, Comput. Methods Funct. Theory **4** (2004), 127–142.
63. A. Connes, *Une classification des facteurs de type III*, Ann. Sci. École Norm. Sup. **6** (1973), 133–252.
64. H. Cordes, *Spectral Theory of Linear Differential Operators and Comparison Algebras*, Cambridge Univ. Press, London, 1987.
65. M. M. Crum, *On the resultant of two functions*, Quarterly J. Math. Oxford Ser. (2) **12** (1941), 108–111.
66. H. B. Curry and I. J. Schoenberg, *On Pólya frequency functions. IV. The fundamental spline functions and their limits*, J. Analyse Math. **17** (1966), 71–107.
67. Ju. L. Daleckiǐ *Integration and differentiation of functions of hermitian operators depending on a parameter*, Amer. Math. Soc. Transl. **16** (1960) 396–400, Russian original: Uspehi Mat. Nauk **12** (1957), 182–186
68. Ju. L. Daleckiǐ and S. G. Krein, *Formulas of differentiation according to a parameter of functions of Hermitian operators*, Dokl. Akad. Nauk SSSR **76** (1951), 13–16.

69. Ju. L. Daleckiĭ and S. G. Krein, *Integration and differentiation of functions of Hermitian operators and applications to the theory of perturbations*, Voroneđ. Gos. Univ. Trudy Sem. Funkcional. Anal. **1956** (1956), 81–105.
70. C. Davis, *A Schwarz inequality for convex operator functions*, Proc. A.M.S. **8** (1957), 42–44.
71. C. Davis, *Notions generalizing convexity for functions defined on spaces of matrices*, Proc. Sympos. Pure Math., **VII** (1963), 187–201, Amer. Math. Soc., Providence, R.I.
72. C. de Boor, *Divided Differences*, Surveys in Approximation Theory, **1** (2005), 46–69.
73. C. de Boor, *A practical guide to splines*, Revised edition, Springer-Verlag, New York, 2001; original: 1978.
74. P. A. Deift, *Applications of a commutation formula*, Duke Math. J. **45** (1978), 267–310.
75. A. de Morgan, *The Differential and Integral Calculus*, Baldwin and Cradock, London, 1842.
76. O. Dobsch, *Matrixfunktionen beschränkter Schwankung*, Math. Z. **43** (1937), 353–388.
77. W. F. Donoghue, *The lattice of invariant subspaces of a completely continuous quasinilpotent transformations*, Pacific J. Math. **7** (1957) 1031–1035.
78. W. F. Donoghue, *On the perturbation of spectra*, Comm. Pure Appl. Math. **18** (1965), 550–579.
79. W. F. Donoghue, *The interpolation of quadratic norms*, Acta Math. **118** (1967), 252–270.
80. W. F. Donoghue, *The interpolation of Pick functions*, Rocky Mountain J. Math. **4** (1974), 169–173.
81. W. F. Donoghue, *Monotone Matrix Functions and Analytic Continuation*, Springer, New York-Heidelberg-Berlin, 1974.
82. W. F. Donoghue, *Monotone operator functions on arbitrary sets*, Proc. A.M.S. **78** (1980), 93–96.
83. W. F. Donoghue, *Another extension of Loewner's theorem*, J. Math. Anal. Appl. **110** (1985), 323–326.
84. R. Doss, *An elementary proof of Titchmarsh's convolution theorem*, Proc. A.M.S. **103** (1988), 181–184.
85. J. Dufresnoy, *Sur le produit de composition de deux fonctions*, C. R. Acad. Sci. Paris **225** (1947), 857–859.
86. P. L. Duren, *Theory of H^p Spaces*, Academic Press, New York/London, 1970.
87. A. Ebadian, I. Nikoufar and M. Eshaghi Gordji, *Perspectives of matrix convex functions*, Proc. Natl. Acad. Sci. USA **108** (2011), 7313–7314.
88. E. Effros, *A matrix convexity approach to some celebrated quantum inequalities*, Proc. Natl. Acad. Sci. USA **106** (2009), 1006–1008.
89. E. Effros and F. Hansen, *Non-commutative perspectives*, Ann. Funct. Anal. **5** (2014), 74–79.
90. H. Epstein, *Remarks on two theorems of E. Lieb*, Comm. Math. Phys. **31** (1973), 317–325.
91. N. Eraser, *Newton's Interpolation Formulas*, J. Institute Actuaries **51** (1919), 211–232.
92. K. Fan, *On a Theorem of Weyl Concerning Eigenvalues of Linear Transformations. I*, Proc. Natl. Acad. Sciences (U.S.) **35** (1949), 652–655.
93. K. Fan, *On a Theorem of Weyl Concerning Eigenvalues of Linear Transformations. II*, Proc. Natl. Acad. Sciences (U.S.) **36** (1950), 31–35.
94. K. Fan, *Maximum Properties and Inequalities for the Eigenvalues of Completely Continuous Operators*, Proc. Natl. Acad. Sciences (U.S.) **37** (1951),760–766.
95. P. Fatou, *Séries trigonométriques et séries de Taylor*, Acta Math. **30** (1906), 335–400.
96. L. Fejér, *Über trigonometrische Polynome*, J. Reine Angew. Math. **146** (1916), 53–82.
97. H. Feshbach, *A unified theory of nuclear reactions, I, II*, Ann. Phys. **5** (1958) 357–390, **19** (1962) 287–313.
98. E. Fischer, *Über das Carathéodory'sche Problem, Potenzreihen mit positivem reellen Teil betreffend*, Rend. Circ. Mat. Palermo **32** (1911), 240–256.
99. C. FitzGerald *On analytic continuation to a schlicht function*, Proc. Am. Math. Soc. **18** (1967), 788–792.
100. M. S. Floater and T. Lyche, *Two chain rules for divided differences and Faa di Bruno's formula*, Math. Comp. **76** (2007), 867–877.

101. C. Foiaş, A. Frazho, *The commutant lifting approach to interpolation problems*, Operator Theory: Advances and Applications, **44**. Birkhäuser Verlag, Basel, 1990.
102. C. Foiaş and J.-L. Lions, *Sur certains théorèmes d'interpolation*, Acta Sci. Math. Szeged **22** (1961), 269–282.
103. G. Frobenius, *Über die Entwicklung analytischer Functionen in Reihen die nach gegebenen Functionen fortschreiten*, J. reine angew. Math. **73** (1871), 1–30.
104. J. Fujii and M. Fujii, *A norm inequality for operator monotone functions*, Math. Japon. **35** (1990), 249–252.
105. J. Fujii, M. Fujii, T. Furuta, and R. Nakamoto, *Norm Inequalities Equivalent to Heinz Inequality*, Proc. A.M.S, **118** (1993), 827–830.
106. M. Fujii, *Furuta's inequality and its mean theoretic approach*, J. Operator theory, **23** (1990), 67–72.
107. M. Fujii and T. Furuta, *Löwner-Heinz, Cordes and Heinz-Kato inequalities*, Math. Japon. **38** (1993), 73–78.
108. M. Fujii, T. Furuta and E. Kamei, *Operator functions associated with Furuta's inequality*, Linear Algebra Appl., **179** (1993), 161–169.
109. M. Fujii and E. Kamei, *Furuta's inequality and a generalization of Ando's Theorem*, Proc. A.M.S., **115** (1992), 409–413.
110. M. Fujii, J. Jiang, and E. Kamei, *Characterization of chaotic order and its application to Furuta inequality*. Proc. A.M.S. **125** (1997), 3655–3658.
111. M. Fujii and R. Nakamoto, *Extensions of Heinz-Kato-Furuta inequality*, Proc. A.M.S. **128** (2000), 223–228.
112. M. Fujii, *Furuta inequality and its related topics*, Ann. Funct. Anal. **1** (2010), 28–45.
113. T. Furuta, *$A \geq B \geq 0$ assures $(B^r A^p B^r)^{1/q} \geq B^{(p+2r)/q}$ for $r \geq 0$; $p \geq 0$; $q \geq 1$ with $(1+2r)q \geq p+2r$*, Proc. A.M.S. **101** (1987), 85–88.
114. T. Furuta, *Elementary proof of an order preserving inequality*, Proc. Japan Acad., **65** (1989), 126.
115. T. Furuta, *Norm inequalities equivalent to Löwner-Heinz theorem*, Rev. Math. Phys. **1** (1989), 135–137.
116. T. Furuta, *Invitation to Linear Operators – From matrices to bounded linear operators on a Hilbert space*, Taylor and Francis, London-New York, 2001
117. T. Furuta, *Concrete examples of operator monotone functions obtained by an elementary method without appealing to Löwner integral representation*, Linear Algebra Appl. **429** (2008), 972–980.
118. T. Furuta, and J. Hakeda, *Applications of norm inequalities equivalent to Löwner-Heinz theorem*, Nihonkai Math. J. **1** (1990), 11–17.
119. J. B. Garnett, *Bounded Analytic Functions*, Pure and Applied Mathematics, **96**, Academic Press, New York-London, 1981; Revised first edition. Graduate Texts in Mathematics, **236**. Springer, New York, 2007.
120. I. M. Gel'fand, *A problem*, Uspekhi Mat. Nauk **5** (1938) 233.
121. A. Genocchi, *Relation entre la différence et la dérivée d'un même ordre quelconque*, Archiv Math. Phys. **49** (1869), 342–345.
122. A. Genocchi, *Intorno alle funzioni interpolari*, Atti della Reale Accademia delle Scienze di Torino **13** (1878), 716–729.
123. A. Genocchi, *Sur la formule sommatoire de Maclaurin et les fonctions interpolaires*, C. R. Acad. Sci. Paris **86** (1878), 466–469.
124. Ya. L. Geronimus, *On polynomials orthogonal on the circle, on trigonometric moment problem, and on allied Carathéodory and Schur functions*, Mat. Sb. **15** (1944), 99–130.
125. J. W Gibbs, *Elementary principles in statistical mechanics: developed with especial reference to the rational foundation of thermodynamics*. Dover publications, Inc., New York 1960. Originally published by Scribner's Sons, New York, 1902.
126. I. Gohberg, ed., *I. Schur Methods in Operator Theory and Signal Processing*, Birkhäuser, Basel, 1986.

127. I. C. Goh'berg and M. G. Krein, *Introduction to the theory of linear nonselfadjoint operators*, Translations of Mathematical Monographs, American Mathematical Society, Providence, RI, 1969.
128. R. C. Gunning, *Special coordinate coverings of Riemann surfaces*, Math. Ann. **170** (1967), 67–86.
129. P. R. Halmos, *Shifts on Hilbert spaces*, J. Reine Angew. Math. **208** (1961), 102–112.
130. H. Hamburger, *Über eine Erweiterung des Stieltjesschen Momentproblems*, Math. Ann. **81** (1920), 235–319; **82** (1921), 120–164, 168–187.
131. F. Hansen, *Operator monotone functions of several variables*, Math. Ineq. Appl, **6** (2003), 1–17.
132. F. Hansen, *Extensions of Lieb's Concavity Theorem*, J. Stat. Phys., **124** (2006), 87–101.
133. F. Hansen, *The fast track to Löwner's theorem*, Linear Algebra Appl. **438** (2013), 4557–4571.
134. F.Hansen, G. Ji and J. Tomiyama, *Gaps between classes of matrix monotone functions*, Bull. London Math. Soc., **36** (2004), 53–58.
135. F. Hansen, M. S. Moslehian and H. Najafi, *Operator maps of Jensen-type*, Positivity **22** (2018), 1255–1263.
136. F. Hansen and G. K. Pedersen, *Jensen's inequality for operators and Löwner's theorem*, Math. Ann. **258** (1982), 229–241.
137. F. Hansen and G. K. Pedersen, *Jensen's operator inequality*, Bull. London Math. Soc. **35** (2003), 553–564.
138. F. Hansen and J. Tomiyama, *Differential analysis of matrix convex functions*, Linear Algebra Appl. **420** (2007), 102–116.
139. F. Hansen and J. Tomiyama, *Differential analysis of matrix convex functions, II,* J. Inequal. Pure Appl. Math. **10** (2009), Article 32, 5 pp.
140. G. H. Hardy, J. E. Littlewood, and G. Pólya , *Some simple inequalities satisfied by convex functions*, Messenger of Math. **58** (1929), 145–152.
141. G. H. Hardy, J. E. Littlewood, and G. Pólya , *Inequalities*, second edition, Cambridge University Press, Cambridge, 1952; original 1934.
142. H. Hasegawa, *α–divergence of the noncommutative information geometry*, Rep. Math. Phys. **33** (1993), 87–93.
143. F. Hausdorff, *Eine Ausdehnung des Parsevalschen Satzes über Fourierreihen*, Math. Z. **16** (1923), 163–169.
144. E. V. Haynsworth, *Determination of the inertia of a partitioned Hermitian matrix*, Linear Algebra Appl. **1** (1968), 73–81.
145. E. V. Haynsworth, *Reduction of a matrix using properties of the Schur complement*, Linear Algebra Appl. **3** (1970) 23–29.
146. O. Heinävaara, *Local characterizations for the matrix monotonicity and convexity of fixed order*, Proc. A.M.S. **146** (2018), 3791–3799.
147. O. Heinävaara, *Matrix monotone functions*, University of Helsinkii Thesis, unpublished; available at https://helda.helsinki.fi/handle/10138/273502.
148. O. Heinävaara, *Characterizing matrix monotonicity of fixed order on general sets*, Paper in preparation.
149. O. Heinävaara, private communication,
150. E. Heine, *Handbuch der Kugelfunctionen, Theorie und Anwendungen, Vols. 1, 2*, G. Reimer, Berlin, 1878, 1881.
151. E. Heinz, *Beiträge zur Störungstheorie der Spektralzerlegung*, Math. Ann. **123** (1951), 415–438.
152. H. Helson, *Harmonic analysis*, Addison-Wesley, Reading, Mass., 1983; Wadsworth, Pacific Grove, Ca., 1991.
153. G. Herglotz, *Über Potenzreihen mit positivem, reellem Teil im Einheitskreis*, Ber. Ver. Ges. wiss. Leipzig **63** (1911), 501–511.
154. C. Hermite, *Sur la formule d'interpolation de Lagrange*, J. Reine Ang. Math. **84** (1878), 70–79.

155. F. Hiai, *Matrix analysis: matrix monotone functions, matrix means, and majorization*, Interdiscip. Inform. Sci. **16** (2010), 139–248.
156. F. Hiai, H. Kosaki, D. Petz and M. B. Ruskai, *Families of completely positive maps associated with monotone metrics*, Linear Algebra Appl. **439** (2013), 1749–1791.
157. F. Hiai and D. Petz, *Introduction to Matrix Analysis and Applications*, Springer, 2014.
158. F. Hiai, and T. Sano, *Loewner matrices of matrix convex and monotone functions*, J. Math. Soc. Japan **64** (2012), 343–364.
159. C. Hidaka, and T. Sano, *Conditional negativity of anti-Loewner matrices*, Linear Multilinear Algebra **60** (2012), 1265–1270.
160. D. Hilbert, *Grundzüge einer allgemeinen Theorie der linearen Integralgleichungen, I–VI*, Gött. Nachr. (1904), 49–91; (1904), 213–259; (1905), 307–338; (1906), 157–227; (1906), 439–480; (1910), 355–419.
161. A. C. Hindmarsh, *Pick's conditions and analyticity*, Pacific J. Math. **27** (1968), 527–531.
162. J. B. Hiriart-Urruty and C. Lemaréchal, *Convex Analysis and Minimization Algorithms, I and II*, Springer Verlag, Berlin, Germany, 1993; an abridged one volume textbook version of the books was published by Springer in 2001 as *Fundamentals of Convex Analysis*.
163. D. T. Hoa and O. E. Tikhonov, *On the theory of operator-monotone and operator-convex functions*, Izv. Vyssh. Uchebn. Zaved. Mat. (2010), 9–14; translation: Russian Math. (Iz. VUZ) **54** (2010), 7–11.
164. F. Holland, *The extreme points of a class of functions with positive real part*, Math. Ann. **202** (1973), 85–87.
165. E. Hopf, *Über die Zusammenhänge zwischen gewissen höheren Differenzenquotienten reeller Funktionen einer reellen Variablen und deren Differenzierbarkeitseigenschaften*, Universität Berlin dissertation, 1926.
166. R. A. Horn, *The theory of infinitely divisible matrices and kernels*, Trans. A.M.S. **136** (1969) 269–286.
167. R. A. Horn, *Schlicht mappings and infinitely divisible kernels*, Pacific J. Math. **38** (1971), 423–430.
168. R. A. Horn, *The Hadamard product*, in Matrix theory and Applications, Proc. Sympos. Appl. Math. **40**, pp. 87–169, American Mathematical Society, Providence, RI, 1990.
169. R. A. Horn and C. R. Johnson, *Matrix Analysis*, Cambridge University Press, Cambridge, 1990; corrected reprint of the 1985 original.
170. A. Jenčovà and M. B. Ruskai, *A unified treatment of convexity of relative entropy and related trace functions, with conditions for equality*, Rev. Math. Phys. **22** (2010), 1099–1121.
171. J. L. Jensen, *Sur les fonctions convexes et les inégalités entre les valeurs moyennes*, Acta Math. **30** (1906), 175–193.
172. J. L. Jensen, *Bidrag til Kædebrøkenes teori*. Festskrift til H. G. Zuhten, 1909.
173. G. K. Kalisch, *A functional-analytic proof of Titchmarsh's convolution theorem*, J. Math. Anal. Appl. **5** (1962), 176–186.
174. E. Kamei, *A satellite to Furuta's inequality*, Math. Japon., **33** (1988), 883–886.
175. T. Kato *On the convergence of the perturbation method*, J. Fac. Sci. Univ. Tokyo **6** (1951), 145–226.
176. T. Kato, *Notes on some inequalities for linear operators*, Math. Ann. **125** (1952), 208–212.
177. T. Kato, *A generalization of the Heinz inequality*, Proc. Japan. Acad. Ser. A. Math. Sci. **6**, 305–308 (1961).
178. T. Kato, *Perturbation Theory for Linear Operators*, 2nd ed., Grundlehren der Mathematischen Wissenschaften, Band 132, Springer, Berlin-New York, 1976.
179. Y. Katznelson, *An Introduction to Harmonic Analysis*, John Wiley, New York, 1968; Second edition, Dover, 1976; Third edition, Cambridge University Press, 2004.
180. J. L. Kelley, *Note on a theorem of Krein and Milman*, J. Osaka Inst. Sci. Tech. Part I, **3** (1951), 1–2.
181. J. L. Kelley, *General Topology*, Graduate Texts in Math., **27**, Springer-Verlag, New York-Berlin, 1975; Reprint of the 1955 edition [Van Nostrand, Toronto].
182. J. Kiefer, *Optimum experimental designs*, J. Roy. Statist. Soc. **B21** (1959), 272–319.

183. F. Kittaneh, *Inequalities for the Schatten p–norm, IV*, Comm. Math. Phys. **106** (1986), 581–585.
184. F. Kittaneh, *Norm inequalities for fractional powers of positive operators*, Lett. Math. Phys. **27** (1993), 279–285.
185. P. Koosis, *Introduction to H^p spaces*, 2nd ed., Cambridge Tracts in Mathematics, **115**, Cambridge Univ. Press, Cambridge, 1998.
186. A. Korányi, *On a theorem of Löwner and its connections with resolvents of selfadjoint transformations*, Acta Sci. Math. Szeged **17** (1956), 63–70.
187. A. Korányi, *On some classes of analytic functions of several variables*, Trans. A.M.S. **101** (1961), 520–554.
188. O. Kozlovski, *Structural stability in 1D dynamics*, Topics in Dynamics and Ergodic Theory, pp. 107–124, London Math. Soc. Lecture Note Ser., **310**, Cambridge University Press, Cambridge, 2003.
189. F. Kraus, *Über konvexe Matrixfunktionen*, Math. Z., **41** (1936)18–42.
190. M. G. Krein, *The theory of self-adjoint extensions of semi-bounded Hermitian transformations and its applications. I*, Rec. Math. [Mat. Sbornik] N.S. **20** (1947), 431-495.
191. M. G. Krein, *Perturbation determinants and a formula for the traces of unitary and selfadjoint operators*, Soviet Math. Dokl. **3** (1962), 707–710; Russian original in Dokl. Akad. Nauk SSSR **144** (1962), 68–271.
192. M. Krein and D. Milman, *On extreme points of regularly convex sets*, Studia Math. **9** (1940), 133–138.
193. F. Kubo and T. Ando, *Means of positive linear operators*, Math. Ann., **246** (1980), 205–224.
194. M. K. Kwong, *Some results on matrix monotone functions*, Linear Algebra Appl. **118** (1989), 129–153.
195. J. L. Lagrange, *Leçons élémentaires sur les mathématiques données à l'École Normale en 1795.* Séances des Écoles normales an III (1794/95). Reprinted in Journal de l'École Polytechnique **2**, 173-278.
196. O. Lanford, and D. W. Robinson, *Mean entropy of states in quantum statistical mechanics*, J. Math. Phys. **9** (1968), 1120–1125.
197. P. Lax, *Translation invariant spaces*, Acta Math. **101** (1959), 163–178.
198. P. Lax, *Linear algebra and its applications*, Wiley-Interscience, Hoboken, NJ, 2007; first edition, 1997.
199. O. Lehto, *Univalent Functions and Teichmüller Spaces*, Graduate Texts in Mathematics, **109**. Springer-Verlag, New York, 1987.
200. E. H. Lieb, *Convex trace functions and the Wigner-Yanase-Dyson conjecture*, Adv. Math. **11** (1973), 267–288.
201. E. H. Lieb, *Some convexity and subadditivity properties of entropy*, Bull. A.M.S. **81** (1975), 1–13.
202. E. H. Lieb and M. B. Ruskai, *A fundamental property of quantum mechanical entropy*, Phys. Rev. Lett. **30** (1973), 434–436.
203. E. H. Lieb and M. B. Ruskai, *Proof of the strong subadditivity of quantum mechanical entropy*, J. Math. Phys. 14 (1973), 1938–1941.
204. G. Lindblad, *Expectations and entropy inequalities*, Comm. Math. Phys. **39** (1974), 111–119.
205. J. Lindenstrauss, G. H. Olsen, and Y. Sternfeld, *The Poulsen simplex*, Ann. Inst. Fourier Grenoble **28** (1978), 91–114.
206. J. L. Lions, *Supports de produits de compositions*, C. R. Acad. Sci., **232** (1951), 1530–1532.
207. J. L. Lions, *Supports de produits de compositions. I*, C. R. Acad. Sci., **232** (1951), 1622–1624.
208. J. L. Lions, *Supports dans la transformations de Laplace. II*, J. Anal. Math., **2** (1953), 369–380.
209. K. Löwner, *Über monotone Matrixfunktionen*, Math. Z. **38** (1934), 177–216.
210. C. Loewner, *Semigroups of conformal mappings*, Seminar on analytic functions, vol. 1, Princeton: Institute for Advanced Study 1957.
211. C. Loewner, *On semigroups in analysis and geometry*, Bull. A.M.S. **70** (1964) 1–15.

212. C. Loewner, *On Schlicht-Monotonic Functions of Higher Order*, J. Math. Anal. Appl. **14** (1966), 320–325.
213. C. Loewner, *Collected Papers*, edited and with an introduction by L. Bers, Contemporary Mathematicians, Birkhäuser, Boston, MA, 1988.
214. P. Maréchal, *On a functional operation generating convex functions. I. Duality; II. Algebraic Properties*, J. Optim. Theory Appl. **126** (2005), 175–189; 357–366.
215. A. S. Markus, *Eigenvalues and singular values of the sum and product of linear operators*, Dokl. Akad. Nauk SSSR **146** (1962),34–36; English translation: Soviet Math. Dokl **3** (1962), 1238–1241.
216. A. W. Marshall and I. Olkin, *Inequalities: Theory of Majorization and Its Applications*, Academic Press, New York–London, 1979.
217. D. E. Marshall, *An elementary proof of the Pick-Nevanlinna interpolation theorem*, Michigan Math. J. **21** (1975), 219–223.
218. R. Mathias, *Concavity of monotone matrix functions of finite order*, Linear and Multilinear Algebra, **27** (1991), 129–138.
219. H. Mellin *Über die fundamentelle Wichtigkeit des Satzes von Cauchy für die Theorie der Gamma und hypergeometrischen Funktionen* Acta Soc. Sci. Fennica, **21** (1896), 1–115.
220. H. Mellin *Über den Zusammenhang zwischen linearen Differential und Differenzengleichungen*, Acta Math., **25** (1902), 139–164.
221. J. G. Mikusinski, *On generalized exponential functions*, Studia Math. **13** (1951), 48–50.
222. J. G. Mikusinski, *A new proof of Titchmarsh's convolution theorem*, Studia Math. **13** (1952), 56–58.
223. J. G. Mikusinski, *Operational calculus*, Pergamon Press, 1959.
224. J. G. Mikusinski and C. Ryll-Nardzewski, *A theorem on bounded moments*, Studia Math. **13** (1952), 51–55.
225. B. S. Mitjagin, Normed Ideals of intermediate type, Izv. Akak. Nauk SSSR Ser. Mat. **28** (1964), 819–832; English translation: Amer. Math. Soc. Transl. 63 (1967), 180–194.
226. M. S. Moslehian and H. Najafi, *Around operator monotone functions*, Integral Equations Operator Theory **71** (2011), 575–582.
227. M. Moslehian, H. Najafi and M. Uchiyama, *A normal family of operator monotone functions*, Hokkaido Math. J. **42** (2013), 417–423.
228. M. Nagisa, *A note on rational operator monotone functions*, Sci. Math. Jpn. **78** (2015), 207–214.
229. M. Nagisa and S. Wada, *Operator monotonicity of some functions*, Linear Algebra Appl. **486** (2015), 389–408.
230. H. Najafi, *More on operator monotone and operator convex functions of several variables*, Linear Alg. Appl. **532** (2017), 127–139.
231. Y. Nakamura, *Classes of operator monotone functions and Stieltjes functions*, Oper. Theory Adv. Appl., **41** (1989), 395–404.
232. H. Narnhofer and W. Thirring, *From Relative Entropy to Entropy*, Fizika **17** (1985), 258–262.
233. S. Nayak, *Monotone matrix functions of successive orders*, Proc. A.M.S. **132** (2004), 33–35.
234. Z. Nehari, *The Schwarzian derivative and Schlicht functions*, Bull. A.M.S. **55** (1949), 545–551.
235. Z. Nehari, *Conformal Mapping*, McGraw-Hill, New York-Toronto-London, 1952; reprinted by Dover, New York, 1975.
236. Z. Nehari, *On bounded bilinear forms*, Ann. Math. **65**, (1957), 153–162.
237. E. Nelson, *A quartic interaction in two dimensions*, 1966 Mathematical Theory of Elementary Particles (Dedham, MA, 1965), pp. 69–73, M.I.T. Press, Cambridge, MA.
238. J. von Neumann, *Mathematical foundations of quantum mechanics*, Translated from the German and with a preface by Robert T. Beyer. Princeton University Press, Princeton, NJ, 1996. German original: *Mathematische Grundlagen der Quantenmechanik*, Springer, Berlin, 1932.

239. J. von Neumann, *Some matrix inequalities and metrization of matrix spaces*, Tomsk. Univ. Rev. **1** (1937), 286–300.
240. R. Nevanlinna, *Über beschränkte Funktionen, die in gegebenen Punkten vorgeschrieben Werte annehmen*, Ann. Acad. Sci. Fenn. Ser. A **13** (1919), no. 1.
241. R. Nevanlinna, *Asymptotische Entwickelungen beschränkter Functionen und das Stieltjessche Momentenproblem*, Ann. Acad. Sci. Fenn. (A) **18** (1922).
242. I. Newton, *Methodus differentialis*, Jones, London, 1711.
243. N. K. Nikolski, *Treatise on the shift operator. Spectral function theory*. Translated from the Russian by Jaak Peetre. Springer-Verlag, Berlin, 1986. Russian original, 1980.
244. N. K. Nikolski, *Operators, functions, and systems: an easy reading. Vol. 1. Hardy, Hankel, and Toeplitz*. Translated from the French by Andreas Hartmann. American Mathematical Society, Providence, RI, 2002.
245. N. K. Nikolski, *Operators, functions, and systems: an easy reading. Vol. 2. Model operators and systems*. Translated from the French by Andreas Hartmann. American Mathematical Society, Providence, RI, 2002.
246. F. Oberhettinger, *Tables of Mellin transforms*. Springer-Verlag, New York-Heidelberg, 1974.
247. T. Ogasawara, *A theorem on operator algebras*, J. Sci. Hiroshima Univ. **18** (1955), 307–309.
248. M. Ohya and D. Petz, *Quantum Entropy and Its Use* Springer, 1993; second edition, 2004.
249. H. Oliveira and J. Sousa Ramos, *On iterated positive Schwarzian derivative maps*, Dynamical Systems and Functional Equations (Murcia, 2000), Internat. J. Bifur. Chaos Appl. Sci. Engrg. **13** (2003), 1673–1681.
250. H. Osaka, Gap Problems of Matrix Monotone Functions and Matrix Convex Functions and Their Applications, Unpublished Lecture Notes at http://www.ritsumei.ac.jp/se/~osaka/gaps(02142015)(final)(2).pdf.
251. H. Osaka, S. Silvestrov and J. Tomiyama, *Monotone operator functions on C^*–algebras*, Internat. J. Math. **16** (2005), 181–196.
252. H. Osaka, S. Silvestrov and J. Tomiyama, *Monotone operator functions, gaps and power moment problem*, Math. Scand. **100** (2007), 161–183.
253. H. Osaka and J. Tomiyama, *Double piling structure of matrix monotone functions and of matrix convex functions*, Linear Alg. Appl. **431** (2009) 1825–1832.
254. H. Osaka and J. Tomiyama, *Double piling structure of matrix monotone functions and of matrix convex functions, II*, Linear Alg. Appl. **437** (2012) 735–748.
255. B. Osgood, *Old and new on the Schwarzian derivative*, Quasiconformal Mappings and Analysis (Ann Arbor, MI, 1995), pp. 275–308, Springer, New York, 1998.
256. M. Pálfia, *Löwner's Theorem in several variables*, Preprint, arXiv:1405.5076.
257. K. R. Parthasarathy and K. Schmidt, *Positive definite kernels, continuous tensor products and central limit theorems of probability theory*, Springer-Verlag, Berlin, 1972.
258. J. Partington, *Interpolation, Identification, and Sampling*, Oxford University Press, New York, 1997.
259. J. E. Pascoe and R. Tully-Doyle, *Free pick functions: representations, asymptotic behavior and matrix monotonicity in several noncommuting variables*, J. Funct. Anal. **273** (2017), 283–328.
260. G. Peano, *Resto helle formule di quadratura espresso con un integrale definito*, Atti della Reale Accademia dei Lincei, Rendiconti **22** (1913), 562–569.
261. G. Peano, *Residuo in formulas de quadratura*, Mathesis **4** (1914), 5–10.
262. G. K. Pedersen, *Some operator monotone functions*, Proc. A.M.S. **36** (1972), 3090–310.
263. H. L. Pedersen and M. Uchiyama, *Inverses of operator convex functions*, in *Ordered structures and applications, Papers from Positivity VII*, ed. M. de Jeu, B. de Pagter, O. van Gaans and M. Veraar, pp 363–370, Birkhäuser/Springer, 2016.
264. J. Peetre, *Nouvelles propriétés d'espaces d'interpolation*, C. R. Acad. Sci. Paris **256** (1963) 1424–1426.
265. J. Peetre, *A theory of interpolation of normed spaces*, Notas de Matemática, No. **39** (1968), Instituto de Matemática Pura e Aplicada, Rio de Janeiro.
266. V. Peller, *Hankel operators and their applications*, Springer-Verlag, New York, 2003.

267. O. Perron, *Die Lehre von den Kettenbrüchen*, Chelsea Publishing Co., New York, 1950; originally published 1913 by B.G. Teubner, Leipzig.
268. P. A. Perry *Scattering Theory by the Enss Method*, Harwood Academic Publishing, 1983.
269. S. S. Petrova, A. D. Solov'ev, *The calculus of finite differences*, in *Mathematics of the 19th century*, edited by A. N. Kolmogorov and A. P. Yushkevich, 261–310, Birkhäuser, Basel, 1998.
270. D. Petz, *Quasi-entropies for finite quantum systems*, Rep. Math Phys. **21** (1986), 57–65.
271. D. Petz, *Quantum Information Theory and Quantum Statistics*, Springer, 2008.
272. R. R. Phelps, *Integral representations for elements of convex sets*, Studies in Functional Analysis, pp. 115–157, MAA Stud. Math., **21**, Math. Assoc. America, Washington, DC, 1980.
273. G. Pick, *Über die Beschränkungen analytischer Funktionen, welche durch vorgegebene Funktionswerte bewirkt werden*, Math. Ann. **77** (1916), 7–23.
274. T. Popoviciu, *Sur quelques propriétés des fonctions d'une ou de deux variables réelles*, dissertation presented to the Faculté des Sciences de Paris (1933), published by Institutul de Arte Grafice "Ardealul" (Cluj, Romania) .
275. E. T. Poulsen, *A simplex with dense extreme boundary*, Ann. Inst. Fourier Grenoble **11** (1961), 83–87.
276. R. T. Powers and E. Størmer, *Free states of the canonical anticommutation relations*, Commun, Math. Phys. **16** (1970), 1–33.
277. S. Puntanen and G. Styan, *Historical Introduction: Issai Schur and the Early Development of the Schur Complement*, [370], pp 1–16.
278. W. Pusz and S. L. Woronowicz, *Functional calculus for sesquilinear forms and the purification map*, Rep. Math. Phys., **8** (1975), 159–170.
279. M. Reed and B. Simon, *Methods of Modern Mathematical Physics, I: Functional Analysis*, Academic Press, New York, 1972.
280. M. Reed and B. Simon, *Methods of Modern Mathematical Physics, II: Fourier Analysis, Self–Adjointness*, Academic Press, New York, 1972.
281. M. Reed and B. Simon, *Methods of Modern Mathematical Physics, IV: Analysis of Operators*, Academic Press, New York, 1978.
282. F. Rellich, *Störungstheorie der Spektralzerlegung. I–V*, Math. Ann. **113** (1937), 600–619, 677–685; **116** (1939), 555–70; **117** (1940) 356–382; **118** (1942), 462–484.
283. F. Riesz, *Sur certains systèmes singuliers d'équations integrales*, Ann. École Norm. Sup. (3) **28** (1911), 33–62.
284. F. Riesz, *Über ein Problem des Herrn Carathéodory*, J. Reine Angew. Math. **146** (1916), 83–87.
285. D. W. Robinson and D. Ruelle, *Mean entropy of states in classical statistical mechanics*, Comm. Math. Phys. **5** (1967) 288–300.
286. M. Rosenblum and J. Rovnyak, *Two theorems on finite Hilbert transforms*, J. Math. Anal. Appl. **48** (1974), 708–720.
287. M. Rosenblum and J. Rovnyak, *Restrictions of analytic functions. I*, Proc. A.M.S. **48** (1975), 113–119.
288. M. Rosenblum and J. Rovnyak, *An operator-theoretic approach to theorems of the Pick-Nevanlinna and Loewner types. I*, Integral Equations Operator Theory **3** (1980), 408–436.
289. M. Rosenblum and J. Rovnyak, *Hardy classes and operator theory*, Oxford Univ. Press, Oxford, 1985; Dover, Mineola, 1997.
290. W. Rudin, *Real and Complex Analysis*, 3rd edition, McGraw-Hill, New York, 1987.
291. M. B. Ruskai, *Inequalities for quantum entropy: a review with conditions for equality*, J. Math. Phys. **43** (2002),4358–4375; erratum: **46** (2005) 019901.
292. L. A. Sakhnovich, *On the reduction of Volterra operators to the simplest form and on inverse problems*, Izv. Akad. Nauk SSSR **21** (1957), 235–262.
293. D. Sarason *A remark on the Volterra operator*, J. Math. Anal. Appl. **12** (1965), 244–246.
294. D. Sarason, *Generalized interpolation in H^∞*, Trans. A.M.S. **127** (1967), 179–203.

295. D. Sarason, *Moment problems and operators in Hilbert space*, Moments in mathematics (San Antonio, Tex., 1987), pp. 54–70, Proc. Sympos. Appl. Math., **37** (1987), Amer. Math. Soc., Providence, RI.
296. D. Sarason, *New Hilbert spaces from old*, Paul Halmos, pp. 195–204, Springer, New York, 1991.
297. D. Sarason, *Commutant lifting*, A glimpse at Hilbert space operators, pp. 351–357, Oper. Theory Adv. Appl., **207** (2010), Birkhäuser Verlag, Basel.
298. A. Sard, *Integral representations of remainders*, Duke Math. J. **15** (1948), 333–345.
299. T. Sasaki and M. Yoshida, *Schwarzian derivatives and uniformization*, The Kowalevski Property (Leeds, 2000), pp. 271–286, CRM Proc. Lecture Notes, **32**, American Mathematical Society, Providence, RI, 2002.
300. R. Schatten, *The cross–space of linear transformations*, Ann. Math. **47** (1946), 73–84.
301. R. Schatten, *A Theory of Cross-Spaces*, Annals of Mathematics Studies, Princeton University Press, Princeton, NJ, 1950.
302. R. Schatten, Norm Ideals of Completely Continuous Operators, Springer–Verlag, Berlin , 1960.
303. R. Schatten and J. von Neumann, *The cross–space of linear transformations, II*, Ann. Math. **47** (1946), 608–630.
304. R. Schatten and J. von Neumann, *The cross–space of linear transformations, III*, Ann. Math. **49** (1948), 557–582.
305. R. L. Schilling, R. Song and Z. Vondraček, *Bernstein functions. Theory and applications*, de Gruyter Studies in Mathematics **37**, Walter de Gruyter & Co., 2010.
306. E. Schmidt, *Zur Theorie der linearen und nichtlinearen Integralgleichungen. I. Teil: Entwicklung willkürlicher Funktionen nach Systemen vorgeschriebener; II. Teil:Auflösung der allgemeinen linearen Integralgleichung*, Math. Ann. **63** (1907), 433–476; **64** (1907), 161–174.
307. I. Schur, *Bemerkungen zur Theorie der beschränkten Bilinearformen*, J. Reine Angew. Math. **140** (1911), 1–29.
308. I. Schur, *Über Potenzreihen, die im Innern des Einheitskreises beschränkt sind, I*, J. Reine Angew. Math. **147** (1917), 205–232. English translation in "I. Schur Methods in Operator Theory and Signal Processing" (edited by I. Gohberg), pp. 31–59, Operator Theory: Advances and Applications, 18, Birkhäuser, Basel, 1986.
309. I. Schur, *Über Potenzreihen, die im Innern des Einheitskreises beschränkt sind, II*, J. Reine Angew. Math. **148** (1918), 122–145. English translation in "I. Schur Methods in Operator Theory and Signal Processing" (edited by I. Gohberg), pp. 66–88, Operator Theory: Advances and Applications, 18, Birkhäuser, Basel, 1986.
310. H. A. Schwarz, *Über einige Abbildungsaufgaben*, J. Reine Angew. Math. **70** (1869), 65–83.
311. H. A. Schwarz, *Démonstration élémentaire d'une propriété fondamentale des fonctions interpolaires,* Atti Accad. Sci. Torino **17** (1881–1882), 740–742.
312. I. E. Segal, *A non-commutative extension of abstract integration*, Ann. of Math. **57** (1953), 401–457.
313. C. E. Shannon, *A Mathematical Theory of Communication*, The Bell System Technical Journal, **27** (1948), 379–423, 623–656.
314. H. S. Shapiro, *Operator theory and harmonic analysis*, Twentieth century harmonic analysis-a celebration (Il Ciocco, 2000), pp. 31-56, NATO Sci. Ser. II Math. Phys. Chem., **33** (2001), Kluwer Acad. Publ., Dordrecht.
315. M. Singh and H. L. Vasudeva, *Monotone matrix functions of two variables*, Linear Alg. Appl. **328** (2001), 131–152.
316. B. Simon *The bound state of weakly coupled Schrödinger operators in one and two dimensions*, Ann. Phys. **97** (1976), 279–288.
317. B. Simon *Analysis with weak trace ideals and the number of bound states of Schrödinger operators*, Trans. A.M.S. **224** (1976), 367–380.
318. B. Simon, *Lower semicontinuity of positive quadratic forms*, Proc. Roy. Soc. Edin. **29** (1977), 267–273.

319. B. Simon, *A canonical decomposition for quadratic forms with applications to monotone convergence theorems*, J. Funct. Anal. **28** (1978), 377–385.
320. B. Simon, *Trace Ideals and Their Applications*, London Mathematical Society Lecture Note Series, **35**, Cambridge Univ. Press, Cambridge-New York, 1979; second edition, American Mathematical Society, Providence, RI, 2005.
321. B. Simon, *Spectral averaging and the Krein spectral shift*, Proc. A.M.S. **126** (1998), 1409–1413.
322. B. Simon, *The classical moment problem as a self-adjoint finite difference operator*, Adv. in Math. **137** (1998), 82–203.
323. B. Simon, *Orthogonal Polynomials on the Unit Circle: Part 1: Classical Theory*, American Mathematical Society, Providence, RI, 2005.
324. B. Simon, *Convexity: An Analytic Viewpoint*, Cambridge Tracts in Mathematics, **187**, Cambridge University Press, Cambridge, 2011.
325. B. Simon *A Comprehensive Course in Analysis, Part 1: Real Analysis*, American Mathematical Society, Providence, RI, 2015.
326. B. Simon *A Comprehensive Course in Analysis, Part 2A: Basic Complex Analysis*, American Mathematical Society, Providence, RI, 2015.
327. B. Simon *A Comprehensive Course in Analysis, Part 2B: Advanced Complex Analysis*, American Mathematical Society, Providence, RI, 2015.
328. B. Simon *A Comprehensive Course in Analysis, Part 3: Harmonic Analysis*, American Mathematical Society, Providence, RI, 2015.
329. B. Simon *A Comprehensive Course in Analysis, Part 4: Operator Theory*, American Mathematical Society, Providence, RI, 2015.
330. B. Simon and R. Høegh-Krohn, *Hypercontractive semigroups and two dimensional self-coupled Bose fields*, J. Funct. Anal. **9** (1972), 121–180.
331. Ju. L. Šmul'jan, *Monotone operator functions on a set consisting of an interval and a point*, (Russian) Ukrain. Mat. Ž. **17** (1965), 130-136; English translation: Amer. Math. Soc. Transl. (2) **67** (1968), 25–32.
332. G. Sparr, *A new proof of Löwner's theorem on monotone matrix functions*, Math. Scand. **47** (1980), 266–274.
333. J. F. Steffensen, *Note on divided differences*, Danske Vid. Selsk. Math.-Fys. Medd. **17(3)** (1939), 1-12.
334. S. Steinerberger, *Refined Heinz-Kato-Löwner Inequalities*, J. Spec. Th. **9** (2019), 1–20.
335. T. Stieltjes, *Recherches sur les fractions continues*, Anns. Fac. Sci. Univ. Toulouse **8** (1895), J1–J122; **9** (1895), A5–A47.
336. D. Sutter, M. Berta and M. Tomamichel, *Multivariate Trace Inequalities*, Comm. Math. Phys. **352** (2017), 37–58.
337. V. E. S. Szabó, *A class of matrix monotone functions*, Linear Algebra Appl. **420** (2007) 79–85.
338. B. Sz.-Nagy, *Perturbations des transformations autoadjointes dans l'espace de Hilbert*, Commentarii Math.Helv. **19** (1947), 347–366.
339. B. Sz.-Nagy, *Sur les contractions de l'espace de Hilbert*, Acta Sci. Math. Szeged **15** (1953), 87–92.
340. B. Sz.-Nagy and C. Foiaş, *The Lifting Theorem for intertwining operators and some new applications*, Indiana Univ. Math. J. 20 (1971), 901–904.
341. B. Sz.-Nagy, C. Foiaş, H. Bercovici, and L., *Harmonic Analysis of Operators on Hilbert Space*, Second edition, Springer, New York, 2010.
342. B. Sz.-Nagy and A. Korányi, *Operatortheoretische Behandlung und Verallgemeinerung eines Problemkreises in der komplexen Funktionentheorie*, Acta Math. **100** (1958), 171–202.
343. E. C. Titchmarsh, *The zeros of certain integral functions*, Proc. London Math. Soc. **25** (1926), 283–302.
344. E. C. Titchmarsh, *The Theory of Functions*, Oxford University Press, 1968; second edition 1939.

345. O. Toeplitz, *Über die Fouriersche Entwickelung positiver Funktionen*, Rend. Circ. Mat. Palermo **32** (1911), 191–192.
346. H. Triebel, *Interpolation theory, function spaces, differential operators*. Second edition. Johann Ambrosius Barth, Heidelberg, 1995. (First edition 1978).
347. M. Uchiyama, *Powers and commutativity of selfadjoint operators*, Math. Ann. **300** (1994), 643–647.
348. M. Uchiyama, *Operator monotone functions which are defined implicitly and operator inequalities*, J. Funct. Anal. **175** (2000), 330–347.
349. M. Uchiyama, *Construction of operator monotone functions*, Rend. Circ. Mat. Palermo (2) Suppl. No. **73**, (2004), 137–141.
350. M. Uchiyama *Operator Monotone Functions, Positive Definite Kernels and Majorization*, Proc. A.M.S. **138** (2010), 3985–3996.
351. M. Uchiyama, *Operator monotone functions, Jacobi operators and orthogonal polynomials*, J. Math. Anal. Appl. **401** (2013), 501–509.
352. M. Uchiyama and M. Hasumi, *On some operator monotone functions*, Integral Equations Operator Theory **42** (2002), 243–251.
353. A. Ulhmann *Endlich-dimensionale Dichtematrizen, II*, Wiss. Z. Karl-Marx-Univ. Leipzig, Math.-Nat. R. **22** (1973) 139–177.
354. A. Ulhmann, *Relative Entropy and the Wigner–Yanase–Dyson–Lieb Concavity in an Interpolation Theory*, Comm. Math. Phys. **54** (1977) 21–32.
355. H. Umegaki, *Conditional expectation in an operator algebra. IV. Entropy and information*, Ködai Math. Sem. Rep. **14** (1962) 59–85.
356. E. B. Van Vleck, *On the Convergence of Continued Fractions with Complex Elements*, Trans. A.M.S., **2** (1901), 215–233.
357. A. Vershynina, E. A. Carlen and E .H. Lieb, *Strong Subadditivity of Quantum Entropy*, Scholarpedia, **8(4)**:30920 (2013) http://www.scholarpedia.org/article/Strong_Subadditivity_of_Quantum_Entropy
358. H. S. Wall, *Continued fractions and bounded analytic functions*, Bull. A.M.S. **50** (1944), 110–119.
359. H. S. Wall, *Analytic Theory of Continued Fractions*, Van Nostrand, New York, 1948; AMS Chelsea, Providence, RI, 2000.
360. E. P. Wigner, *On a class of analytic functions from the quantum theory of collisions*, Ann. of Math. **53** (1951), 36–67.
361. E. P. Wigner, *Simplified derivation of the properties of elementary transcendentals*, Amer. Math. Monthly **59** (1952), 669–683.
362. E. P. Wigner, *On the connection between the distribution of poles and residues for an R function and its invariant derivative*, Ann. of Math. **55** (1952), 7–18.
363. E. P. Wigner and J. von Neumann, *Significance of Loewner's theorem in the quantum theory of collisions*, Ann. of Math. (2) **59** (1954), 418–433.
364. E. P. Wigner, M. Yanase, *Information contents of distributions*, Proc. Natl. Acad. Sci. U.S.A. **49** (1963), 910–918.
365. E. P. Wigner, M. Yanase, *On the positive semidefinite nature of a certain matrix expression*, Canad. J. Math. **16** (1964), 397–406.
366. T. Yoshino, *Note on Heinz's inequality*, Proc. Japan Acad. Ser. A Math. Sci. **64** (1988), 325–326.
367. T. Yoshino, *A modified Heinz's inequality*, Linear Algebra Appl. **420** (2007), 686–699
368. K. Yosida, *Functional analysis*, Reprint of the sixth (1980) edition. Springer-Verlag, Berlin, 1995, First Edition, 1965.
369. D. Zagier, *Introduction to modular forms. From number theory to physics* 238–291 in (Les Houches, 1989), Springer, Berlin, 1992. (Appendix on Mellin Transforms available at people.mpim-bonn.mpg.de/zagier/files/tex/MellinTransform/fulltext.pdf)
370. F. Zhang, ed., *The Schur Complement and Its Applications*, Springer, 2005.

Name Index

B. Simon, *Loewner's Theorem on Monotone Matrix Functions*, Grundlehren der mathematischen Wissenschaften 354, https://doi.org/10.1007/978-3-030-22422-6

Subject Index

B. Simon, *Loewner's Theorem on Monotone Matrix Functions*, Grundlehren der mathematischen Wissenschaften 354, https://doi.org/10.1007/978-3-030-22422-6

GPSR Compliance

The European Union's (EU) General Product Safety Regulation (GPSR) is a set of rules that requires consumer products to be safe and our obligations to ensure this.

If you have any concerns about our products, you can contact us on ProductSafety@springernature.com

In case Publisher is established outside the EU, the EU authorized representative is:

Springer Nature Customer Service Center GmbH
Europaplatz 3
69115 Heidelberg, Germany

Batch number: 10372223

Printed by Printforce, the Netherlands